Chemical Manufacturing for the 21st Century 2013

Topical Conference at the 2013 AIChE Spring Meeting and 9th Global Congress on Process Safety

San Antonio, Texas, USA
28 April - 2 May 2013

ISBN: 978-1-62748-451-0

Printed from e-media with permission by:

Curran Associates, Inc.
57 Morehouse Lane
Red Hook, NY 12571

Some format issues inherent in the e-media version may also appear in this print version.

Printed by Curran Associates, Inc. (2013)

For permission requests, please contact AIChE
at the address below.

AIChE
3 Park Avenue
New York, NY 10016-5991

Phone: (203) 702-7660
Fax: (203) 775-5177

www.aiche.org

Additional copies of this publication are available from:

Curran Associates, Inc.
57 Morehouse Lane
Red Hook, NY 12571 USA
Phone: 845-758-0400
Fax: 845-758-2634
Email: curran@proceedings.com
Web: www.proceedings.com

TABLE OF CONTENTS

() Delivering Innovative Solutions for an Energy Future

Aamir Farid, Shell, Houston, TX; Shell, Houston, TX

Abstract to come

() Rebirth of Manufacturing in the U.S.: What are the Implications for Chemical Engineers?

Dan Romasko, Tesoro Corporation, San Antonio, TX

Technology advances have increased production of oil and natural gas from reserves that were previously unprofitable. This rapid increase in production is restructuring the energy industry and providing more cost advantaged feedstocks and fuel to the nation's manufacturing base, particularly the petrochemical industry. This improved cost structure leads to increased competitiveness in the global markets and supports growth in the domestic industry. Along with this growth comes the need for skilled engineers at a time when the demographic changes point toward a declining supply of engineering talent.

Business and Regulatory Aspects of Effective Near-Miss Management Implementation

Ulku G. Oktem, PhD
Near-Miss Management LLC
Philadelphia, PA USA (Tel: 267-603-2378; e-mail: oktem@nearmissmgmt.com)

Wharton Risk Management and Decision Processing Center
University of Pennsylvania, Philadelphia, PA 19104 (Email: oktem@wharton.upenn.edu)

Iclal Atay, PhD
Chief Bureau of Release Prevention, Division of Environmental Safety and Health
Department of Environmental Protection, New Jersey, USA (Email: Iclal.Atay@dep.state.nj.us)

Prepared for Presentation at
American Institute of Chemical Engineers
2013 Spring Meeting
9th Global Congress on Process Safety
San Antonio, Texas
April 28 – May 1, 2013

Business and Regulatory Aspects of Effective Near-Miss Management Implementation

Ulku Oktem, PhD
Co-Founder and President, Near-Miss Management LLC
Philadelphia, PA USA (Tel: 267-603-2378; e-mail: oktem@nearmissmgmt.com)

Adjunct Professor and Senior Research Fellow, Risk Management Center
Wharton Business School, University of Pennsylvania (Email: oktem@wharton.upenn.edu)

Iclal Atay, PhD
Chief Bureau of Release Prevention, Division of Environmental Safety and Health
Department of Environmental Protection, New Jersey, USA (Email: Iclal.Atay@dep.state.nj.us)

Abstract

Accidents typically have a history of warning signals that are leading indicators or precursors of major events causing business disruptions. Collectively these signals are called near-misses. The definition of a near-miss was debated until about a decade ago, but now the concept is much better understood by most corporate safety, health and environmental groups. Yet, its effective implementation still lags behind standard safety practices.

In this paper, principles of an effective near-miss management system are discussed and both regulatory and business aspects of near-miss management are addressed. The regulatory review targets various near-miss related regulations and the challenges faced in their implementation. The business aspect focuses on cost effective and results oriented methodologies to meet regulatory obligations. Examples are presented to facilitate implementation process.

1. Introduction

Every year chemical process industry experiences significant financial losses due to accidents and unplanned shutdowns. This number for chemical and petrochemical industry in US alone is greater than $10 billion per year excluding major accidents, such as 2010 BP Deepwater Horizon accident at Gulp of Mexico and 2005 Texas City refinery accident. A quick look at Chemical Safety Board's (www.csb.gov) website will be sufficient to be reminded of major accidents – between January 2010 and February 2013, there are 14 completed investigations whose reports are released and 7 ongoing investigations. All these accidents are tip of the pyramid and almost all cases have numerous near-misses associated with it.

The near-miss concept has long been understood in chemical industry but integration of near-miss management into the culture and day to day operations in a manner that makes everyone in the organization both responsible and accountable from near-miss management is still in its infancy in many organizations. This is one area where regulations are not very effective beyond encouraging management.

In this paper, we will first present the regulatory space mainly focusing on New Jersey. Next, we will talk about key components of integrating near-miss management into the business fabric and give some examples, and finally, we will bring up some ideas on connection (or disconnection) between the two systems.

2. Regulatory Landscape

Federal Regulations

Under the authority of section 112 (r) of the Clean Air Act, the United States Environmental Protection Agency (USEPA) promulgated the Chemical Accident Prevention (CAP) rules in 1998.

The CAP rules are built upon existing industry codes and standards and require facilities that use certain flammable and toxic substances at or above specified threshold quantities to develop a Risk Management Program (RMP). The RMP must include the following: hazard assessment including an evaluation of worst-case and alternative accidental releases that details the potential effects of an accidental release, an accident history of the last five years; registration information for the facility including identification of its processes, regulated substances handled, and amounts; prevention program that includes safety precautions and maintenance, monitoring, and employee training measures; and emergency response program that spells out emergency health care, employee training measures and procedures for informing the public and response agencies (e.g. the fire department) should an accident occur.

An important element of the prevention portion of the RMP is to have procedures for investigating incidents. This RMP requirement is promulgated at 40 CFR 68.81. This provision requires the owner or operator of a covered facility to investigate each incident which resulted in or could have reasonably resulted in a catastrophic release of a regulated substance. 40 CFR 68.81 does not have a definition for the term "incident."

New Jersey's Regulations

Following the 1984 catastrophic release of methyl isocyanate at Bhopal India, New Jersey enacted the Toxic Catastrophe Prevention Act (TCPA) in 1986 and the implementing regulations in 1988 (see N.J.A.C. 7:31). Following the promulgation of the CAP rules, New Jersey amended its TCPA rules and obtained delegation from USEPA. In lieu of the undefined term "incident," New Jersey proposed its rules using the term "Extraordinarily Hazardous Substance (EHS) accident" and provided the following definition: "EHS accident means an unplanned, unforeseen, or unintended incident, situation, condition, or set of circumstances which directly or indirectly results in an EHS release." The definition of EHS release must also be looked at to understand the meaning of EHS accident. The TCPA rules provide the following definition: "EHS release means a discharge or emission of an EHS from a piece of EHS equipment in which it is contained, excluding discharges or emissions occurring pursuant to and in compliance with the conditions of any State permit or regulation". In summary, as proposed New jersey rules would have required investigation of incidents which resulted in an unpermitted release of a regulated hazardous substance regardless how small the quantity of the release. To delegate the implementation of the CAP rules, USEPA required New Jersey to also require investigation of the incidents that did not result in any release and the incidents that resulted in a permitted but unplanned release. In response, New Jersey adopted amendments to the TCPA rules to require investigation of "EHS accidents" and "Potential catastrophic events." The rules provided the following definition: "Potential catastrophic event" means an incident that could have reasonably resulted in a catastrophic release of an EHS."

"Near-Miss" defined as "Potential Catastrophic Event"

New Jersey clarified in the adoption documents of the TCPA rules published in the July 20, 1998 and March 16, 2009 New Jersey Registers that the term potential catastrophic event is equivalent to the term "near miss" by using both terms interchangeably (see 30 N.J.R. 2733 Response to comment 53; and 30 N.J.R. 2740 Summary of Agency-Initiated Changes, items 2., 4., and 6.) New Jersey further clarified that the term near miss has the same meaning as the term near miss used in the publications by the Center for Chemical Process Safety of the American Institute of Chemical Engineers (AIChE-CCPS) (see 41 N.J.R. 1218 and 12 19, response to comments 65 and 66).

The AIChE-CCPS provides definitions and examples of process safety near misses in two of its publications: the CCPS Book, Guidelines for Investigating Chemical Process Incidents, Second Edition, March 2003, page 61; and Process Safety Leading and Lagging Metrics – You Don't Improve What You Don't Measure, January 2011, page 36. Some of the examples include the following:

- Excursions of process parameters beyond pre-established critical control limits.
- Activation of layers of protection such as relief valves, interlocks, rupture discs, blowdown systems, halon systems, vapor release alarms, and fixed vapor spray systems.
- Activation of emergency shutdowns.

USEPA found the approach of New Jersey outlined above to be at least equivalent to CAP rules requiring investigation of incidents.

Summary of New Jersey EHS Reports

Regulated facilities located in New Jersey are required to submit annual reports to the Department of Environmental Protection. Prior to 2010 some facilities were submitting triennial reports. The reports include a description of significant changes to the management system, a process hazard analysis report if one was completed in the past year, an inherently safer technology report if one was completed in the past year, the compliance audit report, and a summary of EHS accidents and potential catastrophic events. The summary of incidents includes the name and quantity of EHS released if any, the date and time of the incident, the equipment involved, and the basic contributory causes.

Table 1 below summarizes the incidents reported to the New Jersey Department of Environmental Protection between 2002 and 2011. Each year less than fifty percent of the facilities reported that they had one or more incidents. Most of the incidents reported involved the release of an EHS. The number of near misses reported average less than 1 per facility. This may not be an accurate representation of the near misses that actually occur. The cause for not reporting near misses could be a misunderstanding of the reporting requirements or a company culture that does not encourage reporting of near misses. It may also be a concern with perception because the term used for near misses is "potential catastrophic events." A list of potential catastrophic events in an annual report may be misleading to people who are not directly involved in the process.

Table 1. EHS Incidents Reported in TCPA Annual/Triennial Reports

Calendar Year	Total # of Incidents Reported	# of Near Misses (no release)	# of Facilities that Reported	Ave. # of Incidents per Facility	Max. # of Incidents Reported by a Facility	# of Facilities that Reported an Incident	Max. Quantity Released (lbs.) *See Note 1
2002	169	13	90	1.9	30	34	42,717
2003	209	12	79	2.6	58	34	8,900
2004	199	5	87	2.3	53	24	75,000
2005	267	25	76	3.5	82	24	51,666
2006	298	37	72	4.1	93	28	63,704
2007	327	39	80	4.1	97	28	7,219
2008	244	18	67	3.6	67	25	122,000
2009	209	12	68	3.1	54	24	7,600
2010	220	20	77	2.9	68	21	54,800
2011	228	9	79	2.9	89	22	17,955

*Note 1: Each year, the maximum quantity of an EHS released was a Flammable Substance or Flammable Substance Mixture.

3. Business Approach to Near-Miss Management

Fundamentals of Near-Miss Assessment

Near-misses have three fundamental characteristics: (a) frequency, (b) actual damage, and (c) maximum potential damage of related worst-case scenarios that could result from specific categories of near-misses. All of these should be recorded, monitored, and used to track progress in safety, operability, and reliability of any manufacturing process. These three fundamental characteristics form the base of leading indicators. From a process control perspective, the use of the process performance indicators as a basis for prediction and control has been a central research area since the work of Pontryagin and Kalman in the 1950's and 1960's, leading to the revolution in real-time control.[1]

"Near-Miss Management" concept is based on identification of early stress signals and taking the proper action to eliminate the stressor. There are numerous examples of major accidents in manufacturing industry whose earlier signals were ignored. After listing several major accidents in chemical industry Michael Dolan in his key note speech at 2012 AIChE Spring Meeting reiterated the important fact: "Each and every one was preventable with good engineering practice and attention to detail. And beneath each of these is a pyramid of near misses in many plants that could have been equally disastrous".

There are numerous examples of major accidents in manufacturing industry whose earlier signals were ignored. In this section, a methodology is presented for integration of near-miss management into the culture and day-to-day operations in a manner that underlines the critical connections between near-misses (both personal and process related) and behavior (both employees and management). This was also emphasized by Michael Dolan in his same speech as "the need to make the safety and security of our operations even more central to our profession than we already do". This methodology, which resulted from observations of successful near-miss management in various industries and research on the principles of risk management, is based on three principles, all of which are equally essential for successful reduction in accidents and significant improvements in operational reliability:

a. Organic integration of near-miss management into the operational fabric

Effective utilization of near-misses requires its components to be organically integrated into the operational fabric. That is, every employee and manager has an active role and participates fully in the near-miss management process (such as the eight-step process described in the next section).

b. Categorization of near-misses

To be able to track and monitor in a manner that helps with decision making, the overall risk has to be divided into meaningful, track-able categories, such as technology, facility, personnel

c. Tracking and monitoring of near-misses in each category

[1] For a concise summary of the history of optimal control theory and its applications to business problems, see Sethi and Thomson (2000).

As mentioned above, the tracking of near-misses in each of the categories should be done based on their three characteristics:

a) *Frequency* refers to the number of events per unit period.

b) *Actual damage* is the observed damage and applies only to those cases where an incident has occurred.

c) *Potential damage* is associated with each and every near-miss regardless of the presence of actual damage.

In a recent article, J.B. Titus poses the question: "Which factor has the highest priority for assessing hazard risk levels: possibility, severity, or frequency?" We think to measure and monitor the overall performance, if it will be helpful – not required, one can assign different weights to various categories and calculate their weighted sums. But, it's most important that the underlying details of the summed value be transparent, that is, directly addressable, since they point to the key problem areas.

The above categorization, combined with the track-able characterization of near-misses, would make a powerful tool to manage near-misses so as to reduce accidents. The most accurate near-miss information is available at the basic worker level. Operators, who are responsible for day-to-day operations of a process, are the best people who know the disturbances, causal factors of these disturbances, and the difficulties in overcoming such disruptions.

Eight-Step Near-Miss Process Management

An eight-step near-miss management process, which was developed as part of the Wharton Risk Center study (Phimister et al., 2003; Oktem U., 2003; Muermann and Oktem, 2002), incorporates all the key factors necessary to address near-misses in a facility to reduce accidents and catastrophes. Below is brief explanation of each step:

1. Identification

Identification is the first step of the process where an individual recognizes an incident or a condition as a "near-miss". To execute this step successfully each facility should have a clear definition of a near-miss which is communicated to each employee effectively, at all times.

2. Disclosure (Reporting)

Once a near-miss is identified it must be disclosed, preferably in a written form. This can be done either by the person who identified the near-miss or by a supervisor to whom a near-miss is reported verbally. Having a clear, simple, and possibly multi-channel, procedure for reporting would encourage this process and would increase the probability of reporting most near-miss observations.

3. Prioritization

Once an incident is reported, it needs to be prioritized. This very critical step determines the path to be followed in the subsequent steps; the level of attention that will be given to the incident, the depth of analysis that will be performed in finding causes, the resources that will be dedicated to finding and implementing solutions, and the extent to which the information about this incident will be disseminated. It is important for each corporation to set their own criteria for prioritization, based on the above mentioned factors: (a) frequency, (b) actual damage, and (c) maximum potential damage of related worst-case scenarios.

4. Distribution

Based on the priority and the nature of a near-miss, the information is distributed to the people who would be analyzing the causes of the incidents. The distribution mechanism for various priority levels must be determined during system development by the management team.

5. Identification of Causes (Causal Analysis)

This step includes identification of both direct- and root-causes of a near-miss. The implementation of this step can be as simple as the reporter inputting his/her ideas to identify the causes. On the other hand, for the highest priority near-misses, a committee may be formed to undergo an extensive root-cause analysis.

6. Solution Identification

The most important feature of this step is looking for a solution for each identified cause. Sometimes, several causes can be corrected with a single solution. In other cases, there may not be a feasible, effective solution, and hence, a less than ideal corrective action may be taken. All solution decisions, even "no solution for the time being", should be noted. In the end, each cause must have been addressed.

7. Dissemination

Once solutions are identified, they should be communicated to the people who will execute these solutions, assuming they have not participated in the solution identification process. This step also includes an important intermediate function, which, if overlooked, can stall the system; i.e., obtaining permission from the manager with resources to implement the solutions. Another function of the dissemination step is to inform all possible interested parties of the particular near-miss. This may extend well beyond the site or corporate functions and may include customers, contractors, suppliers, etc.

8. Resolution (Tracking)

Once solutions are identified and implementers are informed, it is important to track all suggested changes to ensure they are properly executed. Also, when all of the changes are completed, to encourage future participation, the reporters of near-misses should be informed of the results from their identifications of near-misses.

How Effective Near-Miss Management in Reducing Accidents:

A well managed near-miss system is expected to help identify problems at the early stages and eliminate causal factors, reducing the probability of having major accidents/releases. Below are three examples which demonstrate that at facility level paying attention to safety issues does reduce accidents.

Example 1: Norsk Hydro

Figure 1 shows the results of a study published by Jones et al. (1999) on Norsk Hydro data. This is an offshore drilling and on-shore refining company in Norway. They started to implement a near-miss program in 1985 at their off-shore drilling facility where they employed 3500 people. As we can see from the data on this chart, between 1985 and 1997, their lost time injuries decreased by tenfold. During the same period, their Near-Miss reporting rate went from zero to 0.5 near-miss per person per year.

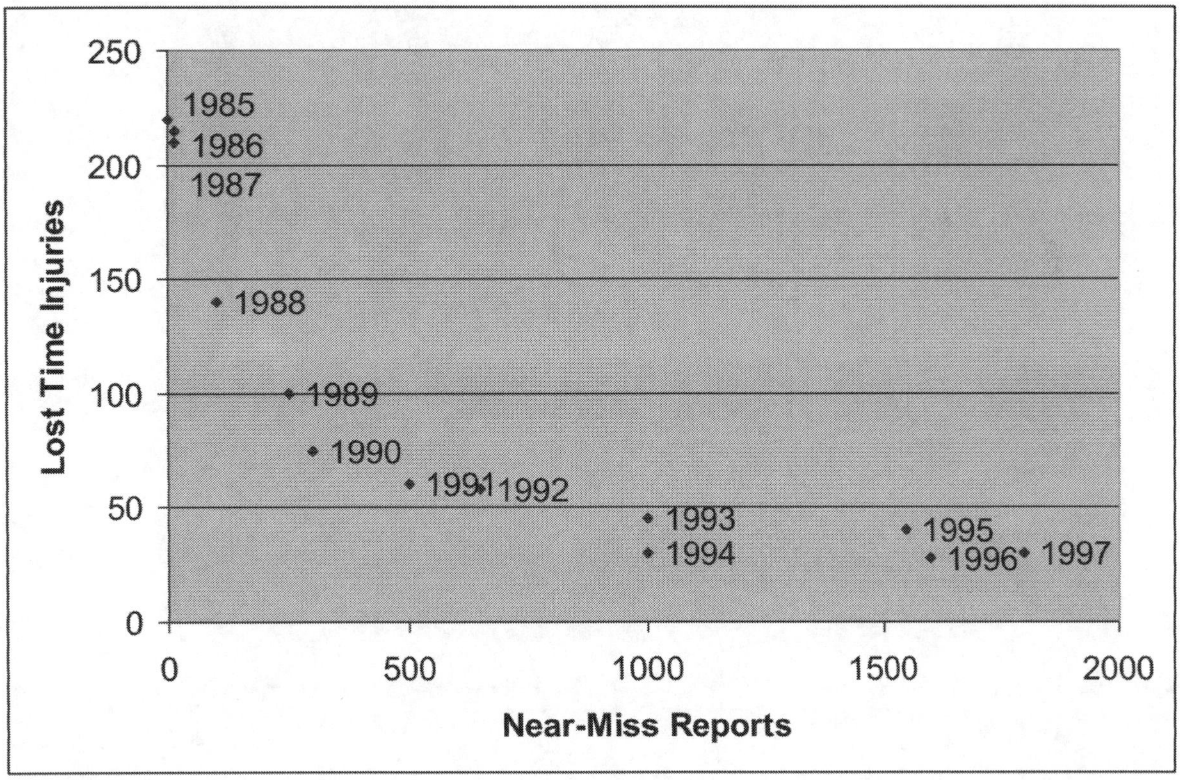

Figure 1: Norsk Hydro data (Jones et al., 1999)

Example 2: Dow Chemical

A recent study (Nesmith et al., 2012) reported by Dow Chemical show that a pioneering site has the same trend as Norsk Hydro, that is, as the number of reported process safety near-misses increased the number of incidents decreased – confirming that a well-managed near-miss program can help reduce the number of accidents. In this case, a strong and personal leadership support for PSNM reporting, including creation of a network for sharing experiences, accelerated implementation of the program.

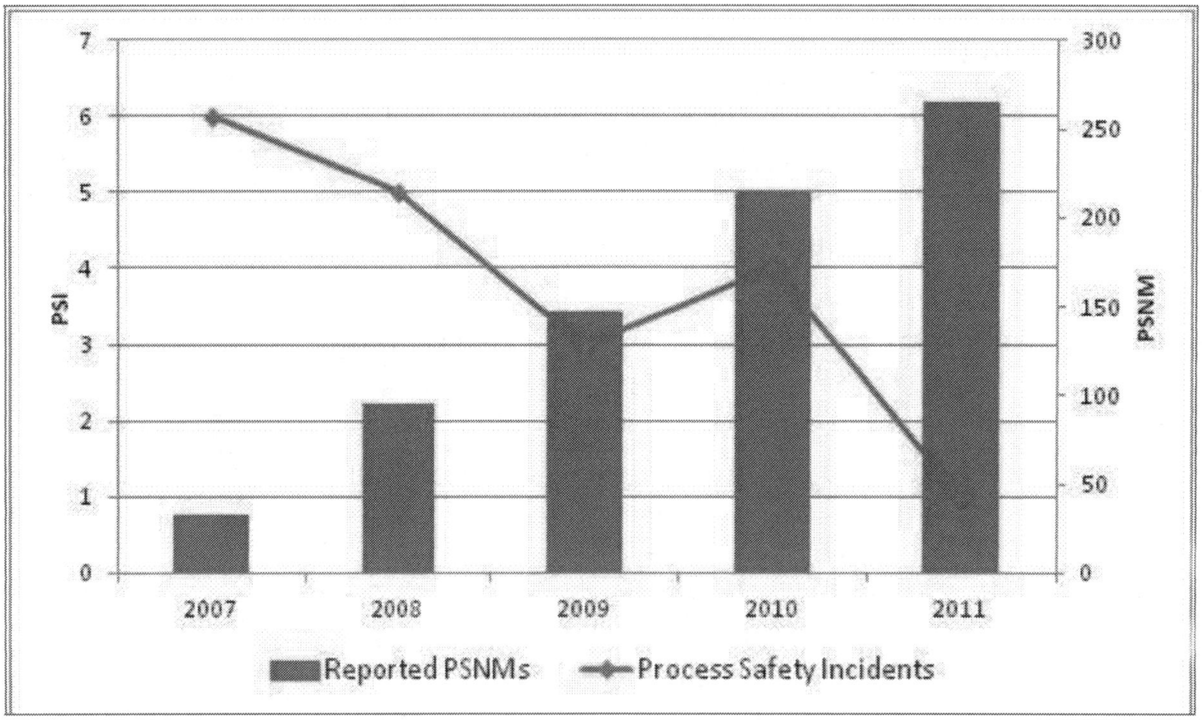

Figure 2: One of Dow Chemical's North American Sites (Nesmith et al., 2012)

Example 3: DuPont

In this example for DuPont data, even though there is no specific number of near-miss reporting presented, however, if one assumes that the increase in the audit scores has some relevance to paying attention to process safety issues, bringing them to forward and addressing them such as the near-miss program again, a similar trend on reduction of incidents is observed.

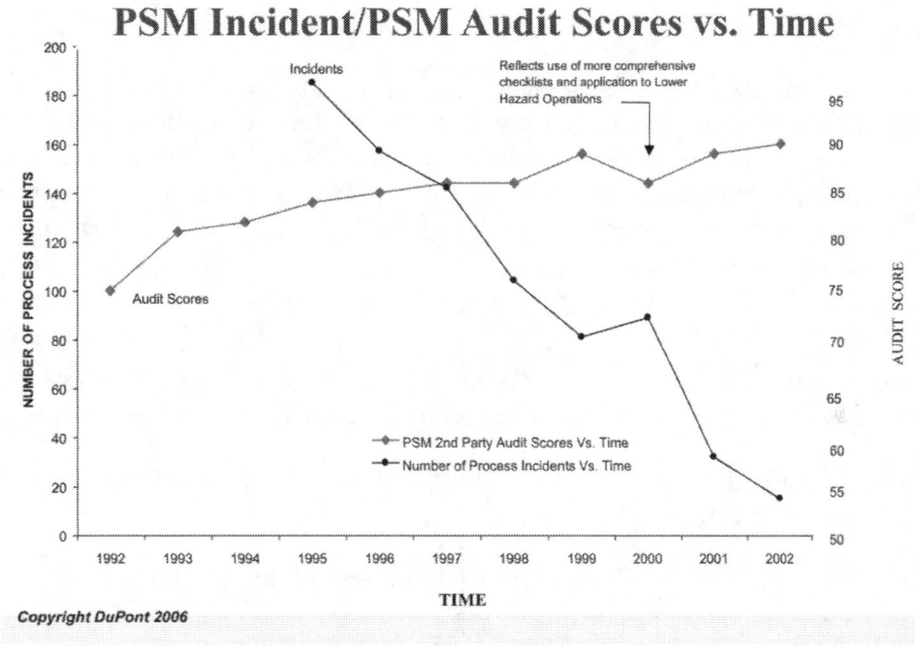

Figure 3: DuPont data (Kleindorfer, 2006)

4. Can Facilities' Success be Repeated at the State Level?

Looking at New Jersey DEP's data , it is difficult to draw conclusions. There are many variables, and data that is reported is very limited.

New Jersey has facilities in the following sectors: chemical manufacturing, refineries, food processing, power generation, water and waste water, and miscellaneous other types. Refineries are the most complex facilities with many regulated processes. Some chemical manufacturing facilities are also very complex with multiple regulated processes. Warehouses are the least complex regulated entities.

Every year more than 50 % of the regulated facilities report that they had no near misses or accidents. The reasons for not reporting any accidents or near misses may vary from facility to facility. They may not be doing a good job at identifying and addressing incidents and near misses, or they may be doing a good job but electing not to report for concerns of public perception or other reasons. Hence, there is not enough information about incidents to draw any conclusions.

Of the facilities that report any accidents and near misses, the largest number of accidents and near misses are reported by complex multi process chemical manufacturing facilities and refineries. This is expected.

What incidents are considered significant?

In this case the largest releases are reported by the facilities that also reported the largest number accidents/near misses. Hence, a straight forward extrapolation of results obtained by individual facilities cannot be confirmed at the state level. To encourage facilities run a more rigorous near-miss management program, identifying and reporting higher number of near-misses, state may evaluate the reasons why large percentage of facilities are not reporting near misses and take actions to address those reasons.

Table 2. Quantity of the extraordinarily hazardous chemical releases reported to New Jersey DEP from 2007 through 2011 (There is no reported significant damage to equipment or facilities.)

Quantity of Extraordinarily Hazardous Substance (Chemical) Releases (Q in lbs.)	**Number of Facilities Reported Releases**
100,000 < Q	1 (from a refinery)
10,000 < Q < 100,000	5 (all from refineries)
1,000<Q <10,000	25 (21 from refineries)
100 < Q < 1,000	60 (50 from refineries)

5. Conclusions

Often, near-miss management has played an ex post forensic role in risk management rather than an alerting one Above examples show that an effective near-miss management program, which identifies leading indicators and precursors of hazardous conditions can help improve safety and environmental releases significantly. This requires strong commitment and integration of the concept into the culture and day to day operations in a manner that underlines the critical connections between near-misses and behavior.

While this paper provides some roadmap for an effective near-miss management at the facility level it points out that regulatory landscape appears to not provide a very conducive environment to help flourish the concept. A solution to this issue may be achieved by forming a public private partnership group to facilitate companies to report near-misses and get support in improving their safety and environmental releases.
The Center for Chemical Process Safety (CCPS) developed a membership based Process Safety Incident Database (PSID). The purpose of the CCPS PSID is to pool process safety incident experience among participating companies so they can learn from the experiences of others without suffering the consequences of failures, while minimizing corporate liability. PSID includes process safety incidents with a potentially important lesson to be learned from incidents that did or could have resulted in a fire, explosion, fatality, multiple injuries, significant release of hazardous materials, or any other unique process safety incident (including near-misses). Currently a fraction of the CCPS member entities are members of the PSID data base. The PSID database currently has information about more than 700 incidents.

6. References

Jones S., Kirchsteiger C., Bjerke, W. (1999), "The importance of near miss reporting to further improve safety performance." *J. Loss PreVention Process Ind., 12* (1), 59.

Kleindorfer P.R. (2006), "Workplace Safety at Alcoa Case Discussion", Presentation at ARKEMA Executive Program, INSEAD.

Muermann A., U. G. Oktem, (2002). "The Near-Miss Management of Operational Risk", *Journal of Risk Finance*, Fall, 25-36.

Nesmith G., Keating J. T., and Zacharias L. A. (2012). "Process Safety Near Misses in Dow Chemical: A Proactive Program to Improve Process Safety Performance," 2012 AIChE Spring Meeting, April 1-4.

Oktem, U. G. (2003). "Near-miss: a tool for integrated safety, health, environmental and security management". 37th Annual AIChE Loss Prevention Symposium. March 30–April 3.

Phimister J. R., U. G. Oktem, P. Kleindorfer, and H. Kunrether (2003). "Near-Miss Incident Management in the Chemical Process Industry", *Risk Analysis*, **23** (3), 445-459.

Sethi, S. P. and G. L. Thompson (2000). *Optimal Control Theory: Applications to Management Science and Economics*, Second Edition, *Kluwer Academic Publishers*, Boston.

Titus J.B., "Risk level assessment priority: Possibility, severity, or frequency?," *Control Engineering*, January 2013, p25.

9TH GLOBAL CONGRESS ON PROCESS SAFETY

Process Near misses Identification, Investigation and Challenges

Pawan Mittal / Parminder Sawhney
Reliance Industries Limited
Jamnagar
pawan.mittal@ril.com/parminder.sawhney@ril.com

Prepared for Presentation at
American Institute of Chemical Engineers
2013 Spring Meeting
9th Global Congress on Process Safety
San Antonio, Texas
April 28 – May 1, 2013

UNPUBLISHED

Process Near misses Identification, Investigation and Challenges

Pawan Mittal / Parminder Sawhney
Reliance Industries Limited
Jamnagar
pawan.mittal@ril.com / **parminder.sawhney@ril.com**

Keywords: Process Safety near miss, reporting, investigations

Abstract

The benefits of reporting and investigating Near misses have long been established in the safety management systems, as they result in prevention of actual incidents from taking place. At Reliance in line with this belief, reporting and investigation of Process safety Near miss is given top priority by the management.

The paper presents the systems and benefits of process safety near miss reporting and investigation at Reliance. Few of the key elements of it are: i) IT tool for reporting of Near miss, ii) categorization of near miss based on the risk rating, iii) investigation based on risk rating using the same procedure which is being used for Process Safety Incidents, iv) review of investigations results by the top management, v) reporting of near miss one of the key parameters in team KRAs (Key Result Appraisal).

Also the paper discusses few near misses which have helped in identifying opportunities for strengthening our Process Safety management systems. The incidents range from challenges to Equipment integrity, errors in operating procedures, faults in instrumented systems etc.

Note: Do not add page numbers. Do not refer to page numbers when referencing different portions of the paper

1. Introduction

Process Safety near miss incidents are the most important indicators of safety performance. Process safety incidents are actual encounter of the consequences whereas Process Safety near miss incidents are considered “weak signals”, some of which contain the signature of more serious and adverse effects. Therefore, addressing Process Safety near miss incidents in a timely and disciplined manner reduces the frequency and consequences of larger problems. Hence, reporting and analyzing them, especially those with high potential, is a pre requisite for maintaining and sustaining strong Safety Management Systems within any organization.

2. Definition:

2.1 Process Safety Incidents:

The Process Safety Incidents are those incidents that represent failures in a line of defense in the overall HSE program but *with* consequence.

2.2 Process Safety Nearmiss:

The Process Safety Nearmiss are those incidents that represent failures in a line of defense in the overall HSE program but *without* consequence.

2.3 Examples of some Process Safety Nearmiss:

A safety interlock that discovered during a routine interlock check (say during a turnaround) was not function properly at the correct set point. During the normal operations, it would have caused a Process Safety Incident.

A PSM critical valve (either a relief valve or an emergency shut-off valve) is found to be defective during a routine maintenance inspection; non-functioning of that valve during normal operations could lead to serious consequences.

The interlock set point for a safety interlock is found in the field to not be as is documented in the control copy of the unit's Process Technology. It could have triggered a Process upset and if interlock set point is at a higher set point than the actual, it could cause a Process Safety Incident also.

A conservation vent or an overflow line on a tank is found to be plugged and would not work properly if the tank experienced a change in pressure or level. Failure of pass flow leads to low flow on tripping of circulating pump. It would cause over pressurization/rupture of tank.

Any time a safety instrumented system malfunction and operates when demand is not placed on the system (i.e. nuisance failure) resulting into process upset.

Failure to open of a rupture disk, a pressure control valve to flare or atmospheric release, or a pressure safety valve when the system conditions reach or exceed the prescribed trigger point could lead to over pressurization of equipment and can result in explosion/fire.

System pressure exceeded the Integrity Operating Window[1] (IOW) during normal operations can not only cause immediate incident, it can lead to undesirable effect on equipment and process (corrosion, stress etc.)

3. Nearmiss Reporting & Investigation

1. **Identifying of Process Safety Nearmiss & Reporting:** Once the nearmiss is identified, it should immediately be reported / logged. There should be well defined system /procedure in place in order to report near miss incidents. At Reliance, an online portal system is there wherein any employee can report a near miss and subsequently, the near miss is validated and approved by the Plant-HOD (Head of Department) and the HSEF (Health, Safety, Environment & Fire) head. Even if a near miss is not approved and is purely an unsafe condition, necessary actions are initiated for corrective/preventive actions.

2. **Formation of Investigation team: -** Once a Process Safety near miss incident, is identified, a committee shall be constituted to investigate the Process Safety near miss incident. This team should be a cross functional team with a `Root Cause Analysis facilitator who is trained on the subject. The team consists of a chairman, coordinator and two-three members based on criticality of the near miss.

3. **Corrective /Preventive action:-**Take immediate corrective action, (if cause seems apparent) even prior to the formation of formal team for investigation and without waiting for their recommendation.
   For Citation, during loop checking it was found that the PSV installed in the loop was of lower rated than the design. Same had been replaced immediately with the required design rated capacity. To avoid reoccurrence, investigation team was formed to find out the root cause.

[1] As per API RP 584

4. **Communication of the Incident: -** Discuss the Process Safety near miss incident in daily meetings at plant level, interplant level and site level so that they are well communicated across the site to Operations & Maintenance group.

5. **Verify the gaps:-**All Operation & Maintenance leads to verify, whether similar problem / gap exists in their area through sharing of Process Nearmiss from other areas. At Reliance, we have following system to achieve this-
   a. Discussion on process safety near miss at different levels such as plant level meeting, interplant level meeting and site level meeting, so that concerned persons of the plant and other plants are aware of the near miss and review the system in their area also.
   b. After investigation, LFI (Learning from Incident) is being circulated to all plant managers and maintenance leads for review of recommendation applicable to their system and follow up. For this we have well developed LRTS system (Learning from incidents and Recommendations Tracking System) – The identified actions for the process safety near miss are entered in Corrective Action Tracking (CAT) system for ensuring closure of the recommendations.

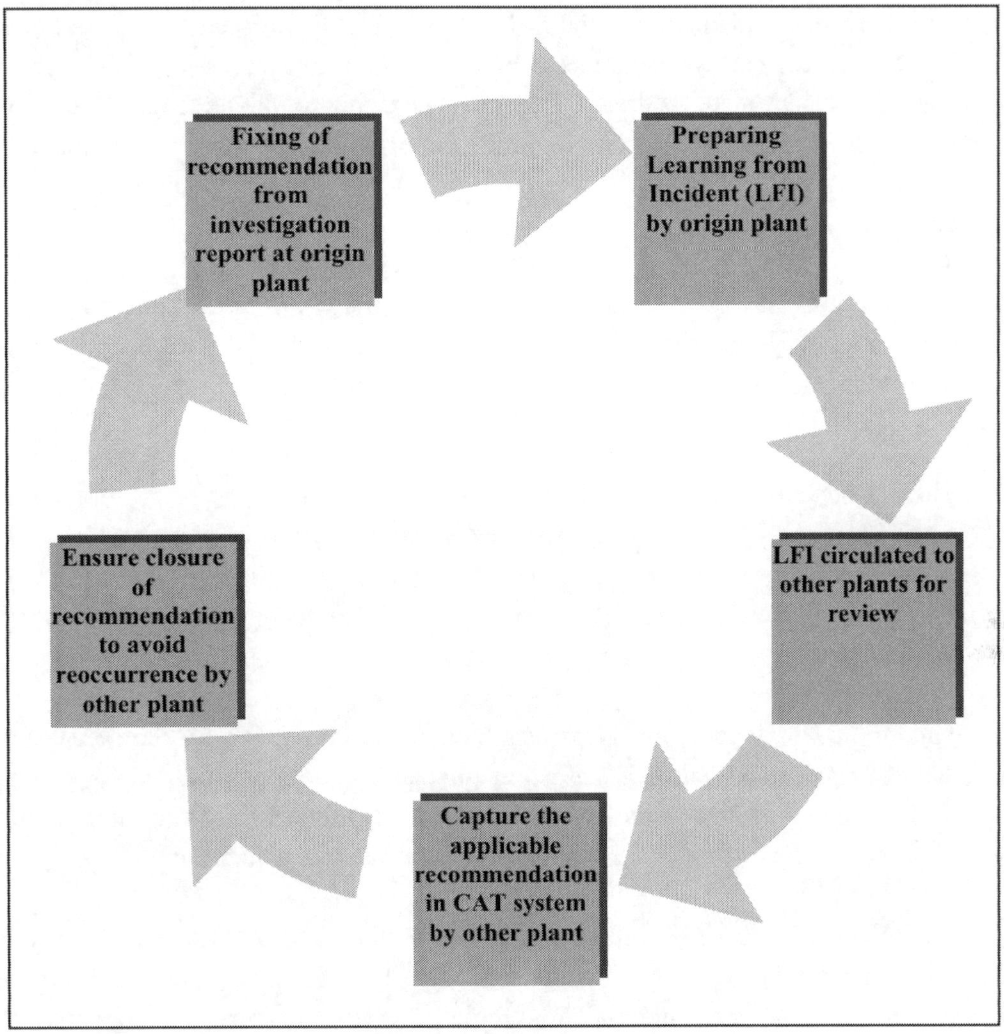

6. **Feedback about the actions: -** Concerned Operation & Maintenance leads should confirm to the site management that either such gap doesn't exist has been rectified or is being identified in their respective areas.

7. **Inputs to Investigation team:** - It is to ensure that the corrective actions taken are shared with the investigation team. This will facilitate investigation team to fix the key factors and do a thorough and fair investigation. At Reliance, we have well designed and defined web based reporting portal (iRIL- Incident Reporting, Investigation and Learning). The corrective actions taken are captured while reporting near miss in web based portal. Same will be visible to investigation team while they initiate the investigation process.

8. **Summary of Root causes & Corrective action:-** Maintain summary of the corrective action taken and identified root causes for future reference & analysis.

At Reliance, the identified failed PSM[2] and OD[3] (Operational Discipline) element for the process safety near miss is being captured and categorized in web based portal. The summary is being fetched at six monthly bases and analyzed and report is generated. Based on this report action plan is developed for implementation and strengthening the process safety management and operational discipline.

9. **Preparation of LFI (Learning from Incident):-**It is ensure that timely preparation of Learning from incident (LFI) and publish the same for sharing in other plants for identifying and correcting similar issues.

Various benchmarking methods are followed at Reliance for Nearmiss Reporting - like Nearmiss Reported/Employee, Maximum number of Nearmiss reported in an area etc. to promote Nearmiss reporting and infuse a healthy reporting culture. The selected Nearmiss reporters are also awarded to recognize their effort and timely action.

The employees who report Nearmiss have to be convinced by the actual actions that their reported Nearmiss are being tracked and actions initiated for closure to prevent re-occurrence. It not only motivates employees for continuous reporting but it also brings a positive Safety Culture within an organization.

At Reliance, Process Nearmiss are investigated (based on criteria) like any other incident, root cause analysis is done to identify root causes (physical, human & system) and actions taken. The whole idea is to prevent re-occurrence as Nearmiss could easily convert into an incident. The difference between a near miss and an accident often is a fraction of a second or an inch.

[2] PSM elements: Process Technology, Process Hazard Analysis, Operating Procedure and Safe Practices, Management of Change, Quality Assurance, Pre Startup Safety review, Mechanical Integrity, Training and Performance, Contractor Safety performance, Incident investigation, Emergency Planning and Response, Auditing

[3] OD element: Leadership by Example, Common Shared Values, Strong Team Work, Pride in the Organization, Sufficient and Capable Resources, Employee Involvement, Active lines of Communication, Up to date documentation, Practice Consistent with Procedures, Absence of Short cuts, Excellent Housekeeping

4. Typical Business Flow to identify the Process Safety Nearmiss incidents & incidents

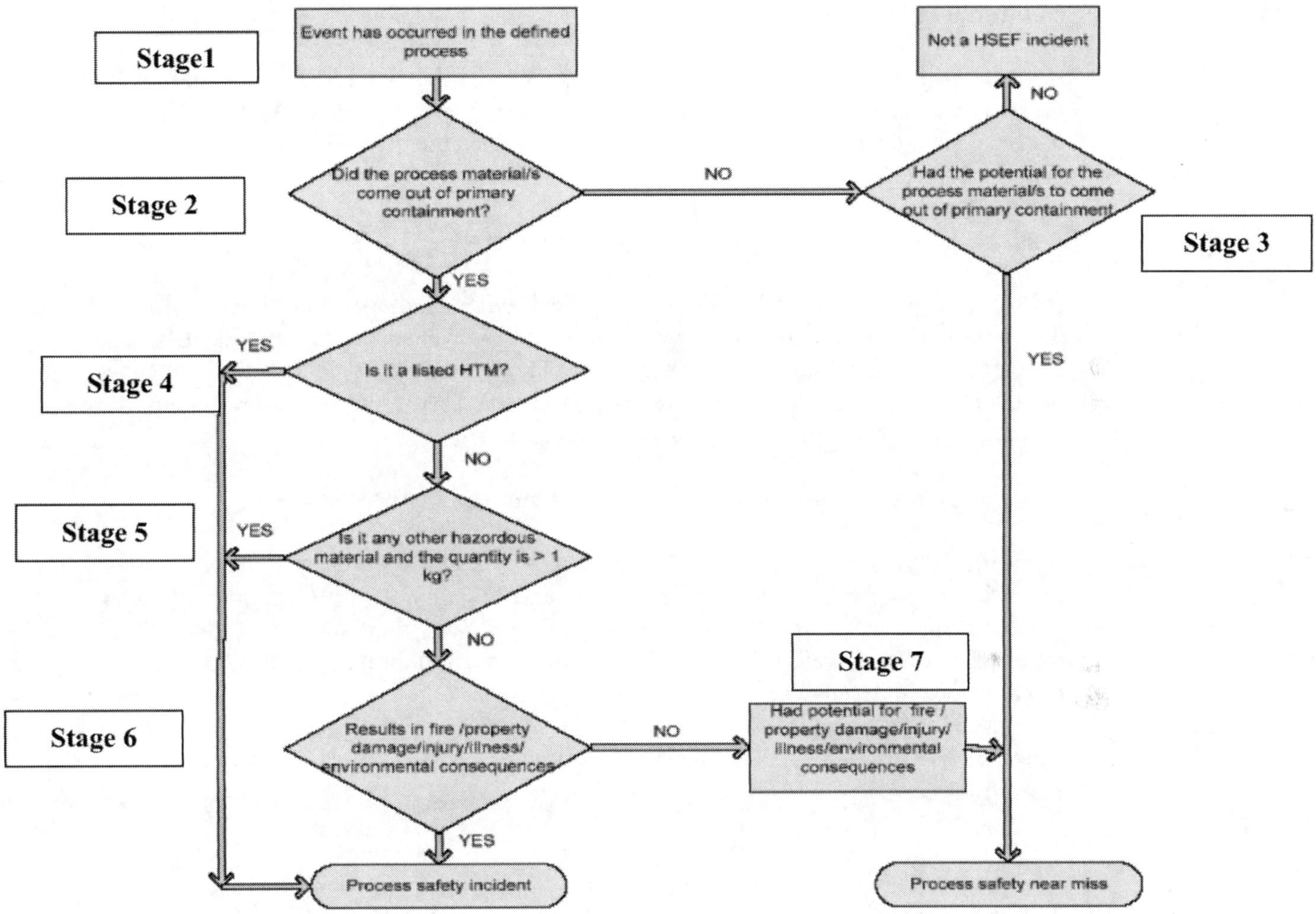

Stage-1: Event has occurred in the defined process

The incident or event that has occurred must be in the defined process to qualify as a process incident or near miss. For example an electrical tube light fell from height in the warehouse is not considered as process event whereas over pressurization of process vessel due to malfunction of instrument is considered as process event.

Stage-2: Did the process material/s come out of primary containment

The intention of this question is to check whether the material involved in the event had come out from primary containment or not.

Stage-3: Had the potential for the process material/s to come out of primary containment

If the decision of stage-2 is "No", then check whether the event had a potential for the process material to come out of primary containment. If it is no then it should not be a considered as a HSEF incident and if it is yes then it will be considered as Process Safety Nearmiss.

Stage-4: Is it a listed HTM (Highly Toxic Material)

If the decision of stage-2 is "Yes", then check whether the material involved in the event is listed as highly toxic material or not. At Reliance, we have identified 14 chemicals as "Highly Toxic Material"[4]. If the decision of stage-4 is "Yes" then the event will be reported as Process Safety Incident irrespective of the quantity of material that came out from primary containment.

Stage-5: Is it any other hazardous material and the quantity is > 1 kg.

If the decision of stage-4 is "No", then check whether the released material is listed in any other hazardous material list and whether the release quantity is greater than 1 kg. If the decision of stage-5 is "Yes" then the event shall be reported as Process Safety Incident. At Reliance, each plant has identified the list of hazardous materials in the Process Technology package.

Stage-6: Result in the fire/ property damage/ injury/ illness/ environment consequences

If the decision of stage-5 is "No", then check whether the consequence of event is fire/ property damage/ injury/ illness/ environmental impact. If it is yes then the event will be reported as Process Safety Incident.

Stage-7: Had potential for fire/ property damage/ injury/ illness/ environment consequences

If the decision of stage-6 is "No" then check whether the event had a potential for fire/ property damage/ injury/ illness/ environmental impact. If it is yes then it will be considered as Process Safety Nearmiss.

---

[4] Highly Toxic Materials: Ammonia (anhydrous), Chlorine, Dimethyl amine, Di Methyl Di Sulphide (DMDS), Ethylene Oxide, HF (anhydrous), HCl Anhydrous, Hydrogen Cyanide, Hydrogen Sulphide (anhydrous liquid), Methyl Mercaptan, Monomethylamine, Sulphurdioxide (anhydrous), Titanium Tetra Chloride, Trimethylamine

5. Challenges

5.1 Reporting Recriminations

Employee may be reluctant to report near misses due to potential recriminations such as –

5.1.1 Peer pressure
5.1.2 Investigation Style
5.1.3 Direct Disciplinary Action
5.1.4 Unintended Disciplinary Action

These issues are addressed by management in the following ways-

- Rights of near miss reporting to all employees.
- Demonstrate leadership by example by reporting near miss and involvement in investigation.
- Discussion of each and every process safety near miss at appropriate management level.
- Take immediate corrective actions and inform to employee/ contractor who had reported the near miss.
- Conduct training program for effective investigation.
- Not to write name of any person in the investigation report.
- Drill down to system factors in the investigation and not limit the investigation to physical or human factor only.

5.2 Lack of Incentives: Those employees and contractors should be awarded suitably who reports nearmiss incidents specifically high potential work place or process safety neramiss incidents. An organisation should have a structured programme / scheme to recognize such employees / contractors. Actually, these types of schemes facilitate in higher reporting and later this translates into a cultural change wherein people start reporting even without any awards / incentives. There are various ways to give awards or incentives to employees or contractors who reported the process safety near miss.

- Appreciation to employees/ contractors at appropriate level such as plant level meeting, interplant level meeting and site level meeting.
- Reporting of near miss should be considered during year end performance.
- Display of name & photo of employee/ contractor who reported the best process safety near miss.
- Incentives to employee/ contractor

5.3 Individual Confusion: There are occasions, when employees and contractor personnel may be having some confusion in identifying incidents and nearmiss incidents; hence there should be clearly defined guidelines for better understanding the concept of near miss and incidents. Awareness campaign and awareness session should be arranged especially for new entrants. In fact, employees should be motivated to observe and report any unsafe act/condition even if it is not deemed fit as a `Nearmiss'.

Number of activities can be carried out during awareness campaign on identification and reporting of process safety near miss such as Special training programs on guidelines of identification and reporting of process safety near miss; Special tool box talk by senior management to motivate the employees; Display of posters and banners; Screening of films; Quiz contest; Appreciation or awards distribution to employees who reported process safety near miss; etc.

5.4 Lack of Management Commitment: Commitment of the management plays a vital role in the process of reporting near miss and incidents as well. Therefore, leaders from all disciplines should demonstrate their commitment for strengthening the reporting process. Motivation and encouragement of employees /contractors by the leaders of the company certainly plays a vital role in enhanced nearmiss reporting. If necessary corrective actions are not implemented in a timely manner, it may create a sense of demotivation in the employees and they feel that no action is being taken on their reported observations. It may actually hamper the overall Nearmiss reporting. This signifies the role of monitoring and tracking actions generated from the investigation of Nearmiss.

Repeat Nearmiss is not a very healthy sign and it somewhere indicates that the System is not robust enough for sustenance and needs to improve.

Management commitment on near miss reporting is demonstrated by the followings-

- Rewarding the employees/ contractors for near miss reporting.
- Showing sense of commitment in fixing corrective and preventive actions.
- Provide necessary resources for addressing the corrective and preventive actions appropriately.
- Widely communicate and discuss the learning's from other plant in appropriate forums.
- Ensure framing of appropriate action plan based on analysis report of near miss reported to prevent reoccurrence.

6. Maximizing the Near Miss Reporting

Near miss reporting provides valuable data for improving the process safety management systems at a facility. The following processes can maximize the benefits from a process safety near miss program.

1. Use process safety lagging indicator, process safety near miss, and management system leading indicators to build a process safety performance pyramid.

2. When evaluating process safety near misses, consider the potential adverse impacts. The level of response to a near miss (i.e. investigation, analysis, and follow-up) should be determined using the potential as well as the actual consequences of the event.

3. Tie the near miss data to the deficient management system in order to drive system improvements from near misses as well as from actual incidents.

4. Place value upon reporting near misses. Consider reward / recognition for reporting near misses as well as rewards for bottom line performance.

7. Conclusion

Article proposes a near miss management framework that adds operational and strategic value to corporate environmental, health and safety practice. To improve process safety near miss management system one must focus on the details of the process by which near miss are identified, reported and analyzed. Definitely there are some challenges such as peer pressure, disciplinary actions, lack of incentives, lack of knowledge, etc. but management commitment towards reporting and investigation of near miss, leadership by example, follow up of investigation and recommendations, incentive scheme, appreciation to employee/ contractor at appropriate level are some of the ways to override these challenges.

"Learning from near misses is much cheaper than learning from accidents."

Near Misses – Private or Public Concern?

Jay Brabson, P.E.
Accidental Release Prevention Program – State of Delaware
715 Grantham Lane
New Castle, DE 19720
Jay.Brabson@state.de.us

Prepared for Presentation at
American Institute of Chemical Engineers
2013 Spring Meeting
9th Global Congress on Process Safety
San Antonio, Texas
April 28 – May 1, 2013

UNPUBLISHED

Near Misses – Private or Public Concern?

Jay Brabson, P.E.
Accidental Release Prevention Program – State of Delaware
715 Grantham Lane
New Castle, DE 19720
Jay.Brabson@state.de.us

Keywords: Near Miss, RMP, Incidents, Regulation, Regulator, PSM

Abstract

Near miss incidents include those releases and/or fires that do not result in significant employee injury or costs. In many cases there are federal reporting requirements that include reports to the National Response Center (NRC). Loss of containment releases are often near misses for catastrophic incidents and provide evidence of process safety program weaknesses. Facilities have an interest in conducting incident investigations to find root causes, lessons learned, and make safety improvements. Small sites often do not have the expertise to conduct good investigations while some large sites may tend to blame operator error rather than admit management system errors or process design issues [9].

This paper explains 2002 amendments of the Delaware requirement to report releases. The amendment adds the reporting of flammable gas releases and expands the existing written follow up report requirements to explain the 'facts and circumstances leading to the release and the measures proposed to prevent the future releases and remedy any deficiencies in the prevention program.' Having this 'public' examination of near miss investigations puts pressure on both industry and our regulatory program to seek root causes and make practical improvements in site process safety programs – without adding any substantial regulatory burden to industry.

1. Introduction

There is a positive story concerning improvements to industry process safety management (PSM) programs and adoption of better safety cultures at large chemical companies, small manufacturing sites and even oil refineries. Yes, there are still near miss accidental and environmental releases occurring and the Chemical Safety Board (CSB) has their hands full investigating continuing catastrophic accidents. As the CSB can attest, many of these releases occur at small sites, often not even involving extremely hazardous substances (EHS) [1]. And unfortunately, there are still many releases occurring at refineries and other 'PSM covered' sites. But these releases should not hide the fact that better process safety programs and safety management cultures have improved process safety at most of the facilities that are regulated under the Occupational Safety and Health Administration (OSHA) PSM program and the Risk Management

Program (RMP) under the Environmental Protection Agency (EPA). Some facilities, especially large corporations, are adopting process safety program elements even when there is no regulatory requirement – it just makes good business sense. But, with accidental releases still occurring and attracting media attention or environmental concerns, there continues to be discussion of potential new regulatory initiatives or new regulations.

In the real world of resource constraints, there are not enough regulatory resources and expertise to adequately cover the universe of stakeholders subject to regulatory requirements now, let alone the many sites covered under only 'general duty' requirements due to low quantity or no EHS onsite. This story is about working smarter as regulators to deploy regulatory resources into process safety issues that need attention. How does this relate to near misses?

Most accidental releases reported under EPA federal rule 40 CFR Part 355 – *Emergency Planning and Notification* are near misses involving no injuries and no significant equipment damage. There is also a recognition that if catastrophic releases are at the top of the pyramid, serious near misses include minor injuries as well as minor equipment damage[2][3]. There are numerous papers and studies explaining the importance of doing good incident investigations on accidents and near misses as a means for improving process safety programs. This paper explores the concept of applying regulatory resources to accidental releases, most of which are near misses, in order to improve the overall process safety of all manufacturing facilities including those not directly covered by process safety regulations.

2. Finding an appropriate regulatory 'tool'

The EPA's federal rule 40 CFR Part 355 is not a useful tool for tracking and preventing accidental releases [4] due to its emphasis on trigger reporting and emergency notification. Reporting chemical substances released above a certain quantity may be of some value to various community right-to-know entities, but this information alone, reported to the NRC without further investigation has limited utility as a means to determine a facility's ongoing process safety performance. We are fortunate in Delaware to have a state regulation entitled *Report of Discharge of a Pollutant or an Air Contaminant*, promulgated in 1991, that mirrors and expands upon these federal reporting regulations [5]. In 2002 the regulation was amended to update reporting quantities. At that time we made several additional changes that make this Delaware regulation a useful tool for investigating release incidents by our Accidental Release Prevention program. Amendments included:

- Adding RMP flammables and Delaware explosive substances (reporting quantity of 1000 pound release within 15 minutes to an hour).
- Adding a general duty to provide a written follow-up report, regardless of the substance or quantity as requested if there is concern for public health or safety or environmental welfare.
- Adding, beyond the 40 CFR Part 355 written follow-up report requirements, two report requirements; 'facts and circumstances leading to the release and the

measures proposed to prevent the future releases and remedy any deficiencies in the prevention program'.

While this may sound like a lot, these amendments only build on federal requirements already in place. Assuming facilities conduct incident investigations after accidental releases anyway, Delaware requirements do not add to the existing federal regulatory burden. We knew from experience that accidental releases occur at manufacturing sites using a variety of hazardous substances and we simply wanted to ensure that good quality incident investigations with practical recommendations for improvement were being conducted.

This idea is not anything new under the sun [6]. The Center for Chemical Process Safety (CCPS) of the American Institute of Chemical Engineers (AICHE) has a fee based private and confidential project; the Process Safety Incident Database (PSID) for 'sharing incident information among members with a focus on lessons learned from the incident'[7]. While many industry leaders recognize the value of these incident investigations, there are also many manufacturers that are not members of CCPS [8]. This paper suggests there is a place for regulators representing interests of workers, the public and the environment, to review a subset of incident investigations for all facilities, after 'reportable' releases of hazardous substances.

3. Results

These written follow-up reports submitted after reportable incidents and accidental releases allow the ARP program (and potentially air and water environmental programs) to get a better idea of the quantity released, what happened, and the incident investigation recommendations. This includes facilities both large and small that might have a release involving any of over 1100 hazardous substances. With this Delaware regulatory reporting regulation in place for over 10 years we have found a reduction in releases, improvements to process equipment and design and improvements to process safety programs. Admittedly, improvement in safety programs can be hard to measure and antidotal based on inspection reports, compliance audits, and recommendations. Measurable reductions in reportable releases include reductions to 4 or 5 per year in 2003-2005 down to 2 per year by 2010-2012. (This ignores oil refinery flaring reports which can number over 100 per year.) Would this have been the result even if there were no regulatory hazardous substance release reporting program? Maybe, as noted earlier, many facilities see the value in improving their process safety programs and safety culture. However, common sense suggests that having facility incident investigation reports and recommendations open to review by regulators is a strong motivator for facilities to conduct quality incident investigations and allocate dollars for equipment improvements in the face of tight facility budgets. Anecdotally, we know from experience that just having a regulator show up to review an incident and the accompanying investigation recommendations catches upper management's attention. And, where appropriate, we have added to a facility's recommendations providing technical recommendations for further improvements to process safety programs, equipment and management control systems. We have even had some cases where facilities, particularly smaller sites, use this written follow-up report as sent to regulators to increase leverage for having recommendations implemented and resources deployed by corporate executive management [9].

It has been my experience as a process safety regulator in a small state that many of the covered facilities we inspect have well documented process safety programs. During full RMP inspections it can take days of reviewing documentation to find even minor recommendations for improvements. Near miss incidents involving accidental releases of EHS are opportunities for both the facility and our program to uncover weaknesses and holes in process safety programs. We have also helped smaller sites find best process and equipment practices particularly when they have no real engineering staff. As regulators we see a variety of processes and can share non-confidential ‘best practices’ we observe in our inspections. We have found this reporting tool to be a win-win for both industry and the regulators representing the public and environment.

4. Observations

Clearly, complex manufacturing processes increase the probability of accidental releases, and there is evidence that chemical companies and petro-chemical companies with more complex processes lead to a greater risk of accidental releases. Upon reviewing the EPA nationwide database for risk management plan submissions on their Chemical Data Exchange (CDX) webpage, simpler manufacturing processes involving chlorine (most often used for water treatment) and ammonia (most often used in refrigeration or fertilizers) have approximately seven percent of facilities with reportable accidents. They make up 70% of the total RMP submissions and not surprisingly, the total report of accidents is also about 7 percent for all facilities. The chart below shows that the much smaller universe of chemical and petro-chemical facilities report higher accident rates (20 and 38 percent respectively when I checked in 2012) [10].

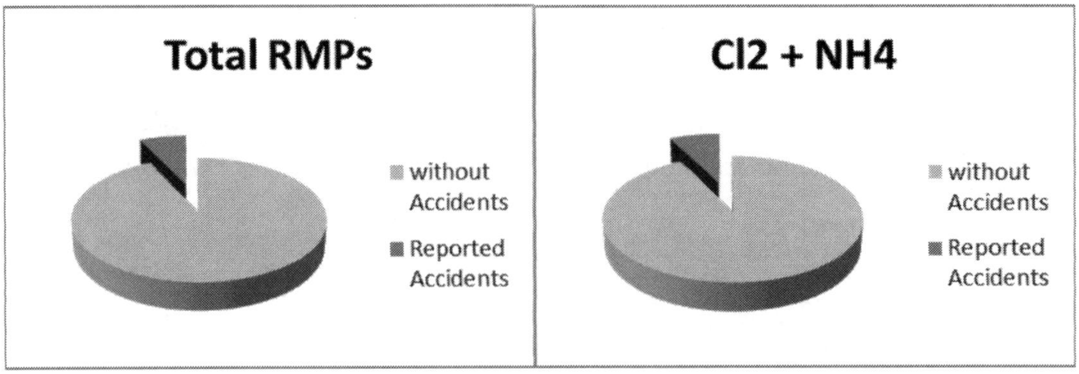

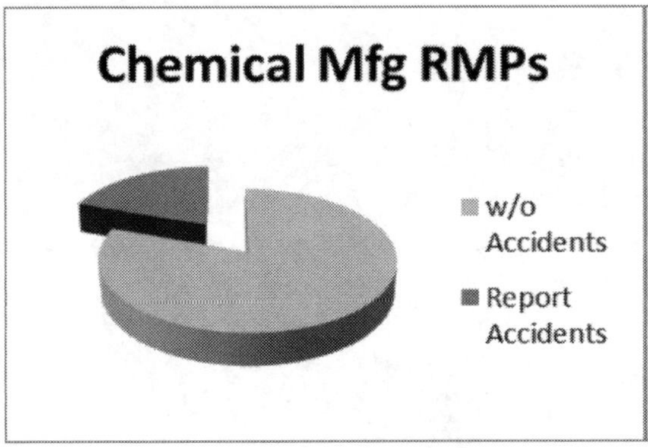

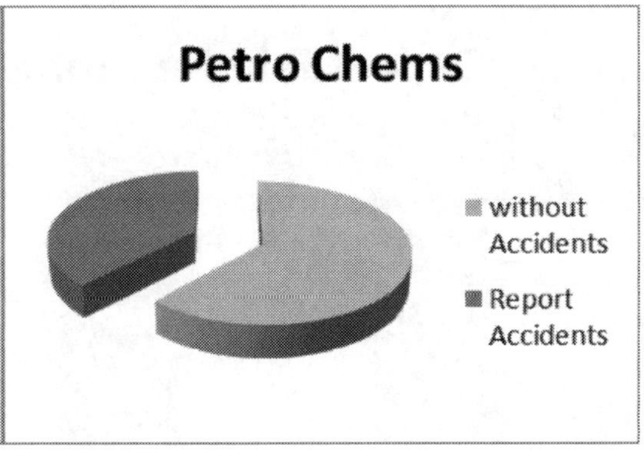

Also, as evidenced by past CSB accident data, many small (and some larger) sites may not recognize certain risk and safety issues when dealing with substances not categorized as EHS including dust, reactive substances, and flammables.

And finally, we find that no matter how good the process safety program, outdated process designs and older equipment can contribute to accidental releases. At this time in 2013, most refineries have improved process safety programs. But while equipment has been replaced and updated, much of the piping and associated supports and some equipment and process designs are over 30 years old with associated 'lean' maintenance crews. Refineries may have good pressure vessel and pipe testing programs. They may relentlessly look for root causes after incidents and provide good recommendations. But like an old house or an old car, there is always something ready to fail due to age (wear and tear), corrosion or vibration.

In making this point in the past, people asked me for concrete examples. I want to share an interesting example of a robust process safety program at a Delaware facility that was still having accidental releases/near misses. It involved a recent ammonia refrigeration inspection. A review of all Delaware ammonia releases reported to the NRC from 2005 through 2012 indicated a total of 10 'reportable' ammonia releases. Of those 10, six came from just one particular site. There are 15 separate sites in Delaware using ammonia, and 13 of those sites have ammonia refrigeration. Six out of 10 releases from one site would certainly not be expected. This particular ammonia refrigeration facility has two different sites in Delaware. We generated a facility 'report card' through a review of the Delaware required incident reports database which indicated that of those 6 releases, 5 came from just one of the four separate ammonia systems at this one site. Not surprisingly, these five releases came from the oldest of the four refrigeration systems.

(See graph below)

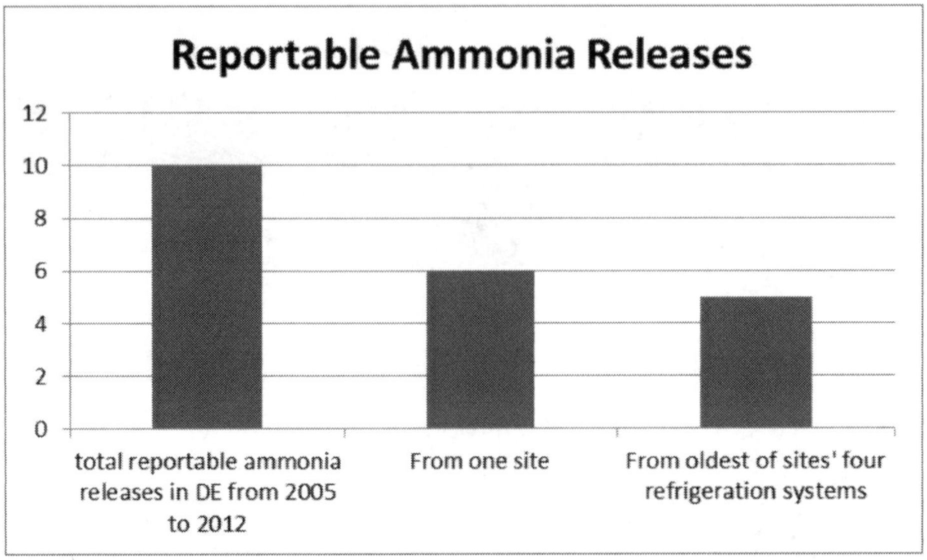

During our inspection we found an organized process safety program led by competent facility and corporate PSM experts with good management support. They appeared to have a good mechanical integrity program with well trained and knowledgeable refrigeration engineers. We gave them this 'report card' on near miss accidental releases to review and recommended their PSM team 'discuss releases and incidents associated with this system and explore options for improvements'. After review, the PSM committee and corporate managers budgeted over 600k for process safety improvements. This case is further evidence that older process equipment can lead to accidental releases despite good process safety programs. This is an interesting local case study that could have national applicability – the idea that there is outdated equipment and process designs out there that could potentially be preemptively targeted for inspection based on a history of accidental releases reported to the NRC.

The above example represents the core message of this paper; promoting industry process safety improvements through the use of near miss reportable accidental releases to measure a site's performance and the use of incident investigations to improve process safety programs. Continuous process safety improvement is not only a public and environmental mandate; it is a sound corporate investment. Delaware's ARP program works with industry to reduce accidental releases in the spirit of partnership, performance, and efficiency.

Federal regulators continue to be focused on regulatory compliance and enforcement with limited interest in partnerships and limited technical or practical processing knowledge. Their inspection programs are governed by high profile accidents, 'emphasis' programs, and worst case scenarios as reported in RMP submissions.

Therefore, based on our experience, state and local governments have an important role to play in this process safety improvement story. Being 'closer to the action' so to speak, local government can play an important role in developing constructive safety partnerships with industry. However, states embarking on such a path need to be sure to hire regulators with sufficient industry experience to genuinely make government-industry partnerships work successfully. The CSB would appear to agree that local jurisdictions have a better track record. In their August 28, 2008 investigation report on the Bayer CropScience, LP release in Institute, West Virginia they make an eloquent two page recommendation (recommendation 8.3) to the Kanawha-Charleston Health Department to establish a 'Hazardous Chemical Release Prevention Program' [11].

5. Conclusion

There are still accidental releases occurring, both many near misses and occasional catastrophic accidents. There is room for continuous improvement and near miss incident investigations are a powerful tool for process safety improvement whether used by regulators or facilities. Based on the EPA risk management plan accident data presented earlier, some releases and accidents appear to be due to process complexity; an issue not directly addressed as part of a typical Process Hazard Analysis (PHA). This is an issue that might yield additional insight in further study. Why does risk analysis on all the parts of processes not result in an overall elimination of releases? Are there subtle interactions or intersecting variables in complex processes that are not adequately covered in a typical risk analysis? Are mechanical integrity program staffing and resource allocations legitimate questions for PHA? Or, are there just black swan accidents that defy risk analysis [12]?

Facilities with older equipment and outdated process designs also have more releases. There is evidence this is the case in spite of high-quality PSM programs. Facilities need to step back and take an honest look at where their near miss accidents are occurring over time. Sometimes improving the process safety program isn't enough – engineer's need to point out the need for capital improvements based on real performance data.

Finally, in a world of limited resources and technical knowledge, regulatory programs representing the public should be identifying ways to utilize resources that encourage sustainable process safety improvements. Statistically, there will always be those facilities at lowest end of the bell curve that need some level of enforcement to encourage compliance with regulations. But the good news is that most facilities have taken process safety seriously and developed quality PSM and RMP programs. Regulators should wake up to the fact there is a bell curve of facility performance and consider using near miss accidental releases to measure process safety program performance. Reviewing near miss incident investigations for reportable releases is much less technically and resource oriented than full scale multi-day, week or month inspections. Taking into account the limited manufacturing experience of many process safety regulators, this represents a better use of resources and technical abilities.

6. References

[1] John W. Herber, "Smaller Companies Struggle with Process Safety", Process Safety Progress (2012), 346-349

[2] James R. Phimister, Ulku Oktem, Paul Rl Kleindorfer, and Howard Kunreuther, "Near-Miss Management Systems in the Chemical Process Industry," Risk Management and Decision Processes Center Operations and Information Management the Wharton School of Management, University of Pennsylvania, 2003

[3] Heinrich, H.W., "Industrial Accident Prevention", New York, McGraw-Hill, 1931.

[4] Environmental Protection Agency (EPA), 40 CFR Parts 300 and 355 extremely hazardous substances list and threshold planning quantities; emergency planning and release notification requirements, first published Federal Register 52:77, pp. 133378-13410, 22 April, 1987

[5] State of Delaware, Administrative Code 1203 Effective date April 1, 1991, Revised December 11, 2002 and July 11, 2004

[6] Riwik U., Kuwait National Petroleum Co., "Risk-Based Approach to near miss," Hydrocarbon Processing October 2002.

[7] CCPS Process Safety Incident Data Base (PSID) Project http://www3.aiche.org/PSID/PDF/PSID.pdf (accessed December 2012)

[8] "Process Safety Performance Indicators for the Refinery and Petrochemical Industries," ANSI/API Recommended Practice 754 First Edition, April 2010

[9] Jay Brabson, P.E., "Regulatory Initiative for Improving Industry Process Safety Performance," Process Safety Progress (2010), 305-307

[10] EPA Central Data Exchange https://cdx.epa.gov/ (Accessed January 2012)

[11] U.S. Chemical Safety and Hazardous Investigation Board, Report No. 2008-080WV, 'Pesticide Chemical Runaway Reaction Pressure Vessel Explosion', January 2008

[12] John W. Murphy and Jim Conner, "Beware of the Black Swan: The Limitations of Risk Analysis for Predicting Extreme Impact of Rare Process Safety Incidents," Process Safety Progress (2012), 330-333.

9TH GLOBAL CONGRESS ON PROCESS SAFETY

Are You Prepared for the Coming Economic Recovery?

John W. Herber
Process Hazard Management Services, LLC.
11 Moonlight Bay
Stillwater, MN 55082
john@phms.us.com

Prepared for Presentation at
American Institute of Chemical Engineers
2013 Spring Meeting
9th Global Congress on Process Safety
San Antonio, Texas
April 28 – May 1, 2013

UNPUBLISHED

Are You Prepared for the Coming Economic Recovery?

John W. Herber
Process Hazard Management Services, LLC.
11 Moonlight Bay
Stillwater, MN 55082
john@phms.us.com

Keywords: Economic recovery, Restarting equipment, PSSR, Training, Preparation

Abstract

The world economy has been depressed for several years. During that time, changes have occurred; these include reductions in funding, headcount reduction, reduced production, equipment temporarily out of service, and reduced equipment integrity inspections to name a few. Some of these changes may have been reviewed, but many especially those involving human resources may not have been thoroughly analyzed or were only reviewed to reassign duties until a replacement can be named. Restarting mothballed equipment and a return to "full" production will make these gaps more obvious.

Gaps may be the availability of qualified workers, training resources for new employees as production returns to pre-recession level. If these gaps are not recognized, the economic upswing will challenge the existing resources & systems and potentially lead to unsafe or even catastrophic situations. Also, there may be shortages of raw materials if suppliers are not prepared to meet your production levels.

This paper serves to alert the petrochemical industry to the impending problems that may be encountered during the economic recovery, to encourage planning to address resource gaps, and to review existing PS systems to verify their ability to handle the influx of changes that will occur.

1. Introduction

"It is worth discussing radical changes, not in the expectation that they will be adopted promptly, but for two other reasons: One is to construct an ideal goal so that incremental changes can be judged by whether they move the institutional structure toward or away from that ideal. The other reason is very different. It is so that if a crisis requiring or facilitating radical change does arise, alternatives will be available which have been carefully developed and fully explored." Milton Friedman, U.S. economist. [1]

This describes the situation the U.S. chemical manufacturing industry is in today. It is teetering on the edge of a recovery, although when it will tip into more consistent improvement is unknown. This paper will explore the various management and economic factors prevalent in the U.S. economy and how they impact key decisions that affect process safety.

1.1 history

Over my 35 + year chemical engineering career, the U.S. and many other economies have experienced various downturns. They may have affected only a small sector of the economy or a single region of a nation's economy. I was the beneficiary of the upturn following a 2+ year economic recession in the mid-1970s. When I was hired, the plant was recovering from staff reductions across many areas – production, maintenance, and support functions. Many remaining employees had been placed in positions that were not their primary skill area. At that time, as during the present "recession", 'a job was a job'. However as the economy recovered the plant began to return to pre-downturn operations, they needed to hire and train many new workers. In addition, as other more suitable positions opened up in other areas, people moved onto new job responsibilities. By the end of my first year, 25-35% of the operators had less than 1 year of experience.

Today, the general economy and the petrochemical industry in general are in a similar position. When they are pushed for production, they must pull in more resources. This may mean hiring temporary workers or utilizing overtime to stretch the existing resources. Companies are reluctant to hire permanent resources due to the uncertain future which may result in yet another layoff. The "on again/off again" nature of a recovery can also place suppliers of material and equipment in that same operating mode.

1.2 economic situation

It is at best difficult to project future economic trends. The current economic recession has baffled economists since its beginning in 2007, so this article will not attempt to predict when the current economy will improve. It may be doubtful that things will even return to what was 'normal' prior to the recession. This paper cannot define what the second decade of this millennium holds. However, with the number of changes in economic and management practices it is not difficult to foresee problems and pitfalls that may present themselves as the economy improves.

A recent article in Chemical Engineering Progress [2] predicted growth for the U.S. economy and the chemical industry in general. Those predictions are summarized in Figure 1 below. Another consistent prediction from economists is that this recovery will be slow and long in duration. On the surface this may seem to be the best situation, a steady 3-5% growth rate. It is relatively easy to forecast single-digit growth and to adapt business plans production schedules and capital forecasts to meet this pace. However, what if the economists are wrong? What if the most aggressive rate of improvement is what really happens? Further consider that compounding a 3 to 5% growth rate over the course of five years becomes a 16-28% percent increase. Somewhere in that period, resources and capacity are going to be overtaxed.

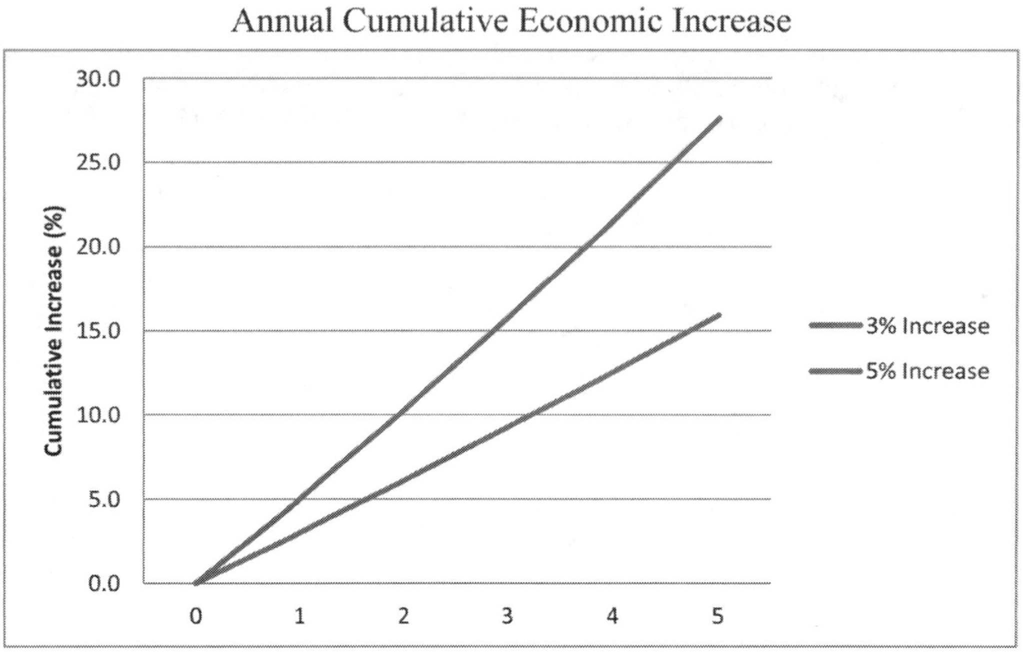

Figure 1.

The chemical industry is capital intensive requiring highly specialized equipment. There is no inventory of such equipment sitting on a show room floor waiting to be installed in any petrochemical plant. Lead time for equipment, particularly large specialized equipment can be over two years. This means that the chemical industry must be looking ahead at least four or five years to implement what growth they anticipate today. For large continuous operations such as refineries single-digit gains in productivity may be achieved by planned efficiency or de-bottlenecking efforts. For smaller specialized or batch operations, this may mean installing another reactor system very similar to those that exist today. These two scenarios are easy to predict, estimate and have approved by management. However when the smaller incremental improvements no longer meet the projected demand, a major capital project and even perhaps an entirely new site may be required. Projects of this magnitude require many years to design, acquire land and permits, install, check-out and eventually operate. That is why it is necessary to prepare for economic recovery before existing equipment and resources are overtaxed.

This paper is intended to spark thinking and discussion around the topic of the possibility that the economy may improve more quickly. If the common theory that the U.S. stock market leads the general economy by six to nine months, the time for planning was the last quarter!

"It is generally true that, the more preparations one has for an event, the more inconveniently fast the event will occur." (Monica Fairview, Darcy Cousins, 2010) [3]

What if the economic recovery comes later than planned? Consider yourself lucky! A good plan for restarting process can always be improved and some pre-start-up activities can be performed well in advance of start-up. A good start-up plan is developed by a cross-functional team that knows the process and equipment well. The start-up plan requires procedures written with sufficient detail to allow operators who have not performed those tasks in many years to complete them safely. Test methods and results need to be documented in case there is a delay or change of assignment before start-up.

2. General Approach

The difficulty in presenting this topic is that the operation of any manufacturing plant is like a spider web. Any perturbation in one area affects most others to some degree. For instance, a change to operating schedules not only impacts operations, but also the entire supply chain, maintenance, and other staff functions. Other factors at work in today's economy are its global nature, "just-in-time" inventories, and a higher reliance on contractors and outsourced workers to name a few.

Also, to understand how manage this situation, we need to understand what has happened over the last two to five years that has made this issue much different from other economic recessions. To best dissect this, the paper will sort the issues by personnel, equipment, and materials factors. As noted above, there may not be clear delineation among these areas, but this is one way to cover the important points.

3. Personnel Issues

Many of the observations made here derive from personal experience, contact with petrochemical companies, and general articles in newspapers and magazines.
For at least 10 years, the U.S. economy and perhaps others have been using a headcount mentality and have made generalizations about the "correct" number of people a given operation requires. Unfortunately, this may be based on not science, but on loosely related ratios of the number of people required to some production metric. It may be salespeople per sales dollar or the ratio of engineers to the capital budget. This very short-sighted thought process ignores many key risk based process safety principles such as Process Safety Knowledge, Process Management Systems, Management of Changes and Process Safety Competency. Consider the very ends of operation: When production is zero, i.e., during a turnaround or shutdown, are all employees laid off? Actually the maximum number of people working a given area probably occurs immediately before operations actually begin. This principle also does not account for the need for highly specialized skills in many areas of the chemical industry plus a significant amount of experience.

Compounding this management trend is yet another aspect of the U.S. economy. Over the past five years, many companies have encouraged more experienced employees to consider retirement and have provided incentives for early retirement. This strategy would not have significant impacts if the company were entirely exiting a particular business and would no longer need that experience or expertise in the foreseeable future. Many companies have mothballed equipment or reduced production rates and therefore reassigned the people from particular production area. Some companies do not apply management of change to personnel issues. Some may review differences in production shift schedules, but loss of technical expertise across several years may not have been evaluated for its impact and probably not considered a threat to future production.

Looking forward, demographics indicate that the "graying of America" is a sobering reality. Many companies are stocked with people ready to retire as soon as their investments seem sufficiently secure. Even if the workforce's average age is under 55, a major turnover of personnel may occur as workers migrate to new positions opened by other career opportunities within or outside their current company.

The unemployment rate in the U.S. has ranged between 8 and 10 percent over the last two years, and stood at 7.8% at the end of 2012 [4] dropping about one percentage point per year. The current shortage of highly skilled workers will worsen. When an influx of new workers is required, a company may struggle to find sufficient workers who meet their basic selection criteria for open positions. One strategy to address this problem is to use contract resources, outsource the work to a different region, in the country or offshore. If the economy in general is improving, there will be competition for skilled laborers and the ability of contractors to supply skilled labor will also decline. Additionally, transferring operations overseas would require even more planning and near-flawless implementation to meet a sudden increase in demand.

3.1 personnel change backlog

Over the past three to five years, job mobility has been limited by the number of openings in companies. As a result, when recovery begins that there may be a number of employees who are looking for new opportunities either inside their company or at a different company. This is yet another risk factor when looking at staffing in an organization. Expansion may create new positions and employees may be promoted. This generates more demand for entry-level positions as experienced workers move to new or different positions. There may be also a loss of experienced employees who take positions in other companies. These factors alone are difficult to predict but are obvious threats to steady operation.

3.2 recommended personnel change actions

1. Review the seniority and job posting policies to ensure they are robust enough to handle the expected shuffle of job openings.
2. Renew/revise the policies for promotions, job bidding, etc. in the changing employment environment. This may require discussions with labor representatives.
3. Communicate the policies to personnel at the first sign of increased job change activity so that all understand the system's function.
4. Utilize Management of Change when modifying staffing level, either up or down and when assigning new personnel to a process area.
5. Conduct an open discussion with technical staff to spotlight their internal career options before they look elsewhere to meet their career goals.
6. Acknowledge the value of their expertise and loyalty shown during earlier economic difficulties.

3.3 recruiting new personnel

3.3.1 operations & craft skilled employees

In a perfect world, there would be a ready supply of highly trained and skilled people, just waiting to be hired. If there were this pool of candidates, how can you ensure they are skilled, qualified, or willing to work for you? The local labor representative may manage the seniority pool and also provide classes to keep those in the pool current on their training and certifications. However, in many markets, there is a shortage of skilled workers, or the only available people lack acceptable work histories.

3.3.2 technical employees

All comments noted above for operations and craft-skilled personnel, also apply to technical employees. They fall into two categories – experienced and new grads; each has its pluses and minuses. Experienced technical staff come with a practiced set of skills and proven work habits. They arrive with installed hazard awareness and risk acceptance that may be different from your company's. On the other hand, those skills may not be state-of-the-art. More experienced workers may be more set in their ways and less adaptable to your systems and methods.

New grads are the flip-side of the experienced engineer with less hazard/risk awareness, but more adaptable and probably have more current technical skills especially in computerized areas.

3.3.3 recommended personnel recruiting actions:

1. Consult with the local labor representatives to determine what sources of labor exist and compare this to foreseeable labor needs.
2. Look beyond the local labor pool; if possible to find other workers who can fill hourly or technical resource needs. These resources may require a different orientation or training plan.
3. Use existing employees to recruit acquaintances who may be underemployed or looking for new opportunities.
4. Do not assume that new grads or experienced workers understand your processes and hazards. They must all receive training that enables them to perform safely even if it means some redundant training.

3.4 personnel training

New employees will require basic training. However, if the minimum worker proficiency cannot be found, new employees may require far more training than in the past. Secondly, training resources in a company may have included some of those people who retired early or transferred to other areas. That jeopardizes the ability to provide adequate training to a significant number of new employees.

3.4.1 recommended training actions:

1. Inventory existing training requirements to ensure all the required training materials are still available, current, and include new training requirements that were developed. This is also an excellent time to cull out any training that has exceeded its effectiveness.
2. Once the necessary training 'courses' have been defined, assess if sufficient resources exist inside or outside the organization to provide the courses on the required schedule.

3. Using the gaps identified in 1 & 2 above, develop a more complete training strategy that utilizes internal and external resources to meet the training needs.
4. Use Chemical Safety Board (CSB) videos [5] and other incident-based examples to increase employees' sense of vulnerability of serious events.
5. Set a long-term goal of establishing an e-learning library. Develop new training materials for this format in mind.

4. Equipment

Of the areas to analyze in preparation for economic recovery, the equipment issue may be the easiest. Equipment, unlike people has a nameplate capacity and can be factually evaluated for its fitness for duty. When needed, a preparation plan can be developed based on past experience and a PreStart-up Safety Review (PSSR).

There are a number of industry practices that can be employed to inspect and prepare equipment. However, they are only effective if they are performed well and necessary corrective actions are taken. Some inspections are based on runtime, but others, such as corrosion under insulation, have been at work even while equipment has been idle. Despite pressures to restart equipment, these inspections must all be performed to avoid a catastrophic event or merely a nuisance shutdown.

Other key areas for review are the instruments and PLC logic. All field instruments need to be tested to verify they still function as expected. Similarly, the control logic should not have changed over time, but how do you know? A vice president that I worked with had an excellent quotation for this, "In God we trust. All others bring data." [6] All control logic needs a functional review that may include revisiting the initial check-out and start-up plans. Likewise, mechanical equipment also needs to be checked out. Even basic attributes such as proper rotational direction are important. Since it is very difficult to determine what is critical, the most practical approach may be to check-out everything in detail. Remember that shipping containers may also be delinquent in their required inspections. Containers, from intermediate bulk containers (IBCs) through tank trailers to railcars, need regular inspection to protect products in transit. These inspections may not have occurred while the shipping containers were inactive.

4.1 recommended equipment actions

1. Develop a check-out and restart plan with the assistance of workers experienced in that area, even if it means moving some people back to old positions temporarily.
2. Equipment that has been inactive for months or years doesn't magically repair itself. Review repairs and inspections that may have been deferred and address them as part of the re-start/start-up plan.
3. Cost-cutting measure over previous years may have led to "cannibalization" of equipment, spare parts or components. Include this potential in the inspection procedures.
4. Revisit PHAs as part of the re-start efforts and use hazard identification techniques to evaluate restart activities.

5. Never assume an erroneous reading is 'just a bad instrument'. Restarting equipment takes the process into territory it has not experienced recently or ever. Reasonable estimates of process parameters must be made to provide the operators with ranges that can be expected until steady-state is attained. Start-up procedures need to provide remedial action for when the expected range is exceeded.
6. Renew process hazard training for equipment that has been inactive for longer than a few months.
7. Include mitigative safeguards and equipment in the restart plan since there is a higher risk of leaks, spills or off-spec material.

6. Materials

During the years since full scale production, materials in storage have aged, and former sources of supply may have disappeared. As a rookie process engineer, I was tasked with starting up a batch fine chemicals reactor. The planned check-out of the equipment went as expected; some minor repairs were needed. The first batch produced an unexpectedly high yield. The second batch was more normal even though all the raw materials came from the same sources. What was probably the fourth batch had a drastically low yield. What happened?? The raw material tank had stratified over the three years the system was inactive, so the first batch had a higher concentration of reactant and the fourth was deficient. This was proven by lab analysis after the fact. Wouldn't it have been smarter to sample the tank before the first batch and recognize the need to mix the material even though mixing had never been required in the past?

If tanks or pipelines have been used for other material over a shutdown period, residual contaminants or cleaning materials may still be present. Flushing and testing of the lines may seem like overkill, but in reality it is cheap insurance against quality issues or a major process upset.

Sources of supply that provided key raw materials may have gone out of business or altered their range of products. That key material once readily available may require the supplier to restart a process and mothballed equipment. For certain materials, the manufacturing location may have changed, and subtle differences may exist even though the containers are still labeled the same way with the same company logo.

When new sources of raw materials are required, they need particular evaluation. One of the best evaluation methods is to progress from lab scale through pilot scale evaluations before making a full scale material change. Leaks, spills and off-spec material may cause new and/or different waste streams. Like other suppliers, waste handling contractors will need to be ready and capable of handling the influx of wastes.

1. Do not assume that materials which have been in inventory maintain the same potency as when initially purchased; test and confirm. Lower potency may not correlate to lower reactivity; inhibitors and stabilizers also lose effectiveness over time.
2. Verify past suppliers are still viable and are prepared to meet your renewed demand.
3. Where necessary, use the company's practices to search for new suppliers and to verify the quality of their products. This may take several iterations to successfully complete.
4. Any material changes should receive a thorough Management of Change (MOC) review.

5. Be prepared to handle new and different waste streams during cleaning, checkout and start-up. This includes collection, storage, and disposal.
6. When inventorying raw materials, double check that spill response chemicals and equipment are also at proper levels.

6. Conclusions

1. "Better to prepare and prevent than repair & repent" – Ezra Taft Benson U.S. Secretary of Agriculture
2. "Before anything else, preparation is the key to success". Alexander Graham Bell U.S. (Scottish-born) inventor (1847 - 1922)
3. The ability to react quickly and correctly to the future depends on the degree of anticipation and preparation that can be made in advance.
4. To be prepared requires a reasonably accurate vision of the future and a plan on how your business will fit there.
5. After years of being denied funding, expect more, positive responses to funding proposals. Therefore, be careful what you ask for; you just might get it.
6. Well-thought-out proposals will be those that get approved and implemented.
7. Your suppliers may also struggle with the same restart and resource issues. Use extra diligence to avoid problems that been noted in.

7. Acknowledgements

The author would like to acknowledge the reviews by Ms. Kathleen Clemmer, whose editing and constructive comments definitely improved this paper.

8. References

1. Milton Friedman, Capitalism & Freedom 2nd ed. (1982) pg ix
2. Chemical Engineering Progress January 2013 pg 26. " Shale Gas Brightens Outlook for U.S. Chemicals" T. Kevin Swift & Martha Gilchrist Moore
3. Monica Fairview, Darcy Cousins, (2010)
4. Bureau of Labor Statistics (BLS) website. (http://www.bls.gov/bls/unemployment.htm) last accessed 3/3/2013
5. Chemical Safety Board website (http://www.csb.gov/videoroom/detail.aspx?) last accessed 3/3/2013
6. Chuck Oesterlein, Vic President, 3M Company (retired) (unpublished)

Additional references not cited:

Guidelines for Risk Based Process Safety, CCPS (2007) Wiley

Recognizing Catastrophic Incident Warning Signs in the Process Industry, CCPS (2012) Wiley

Guidelines for Hazard Evaluation Procedures, 3rd ed. CCPS (2008) Wiley

Guidelines for Mechanical Integrity Systems, CCPS (2006) Wiley

Guidelines for Performing Effective Pre-Startup Safety Reviews, CCPS (2007) Wiley

9. Author's Biography

John is a Staff Consultant for the Center for Chemical Process Safety (CCPS) and principal consultant for Process Hazard Management Services LLC (PHMS).
John concluded a 33 year career with 3M Company in 2009 that included engineering, production operations and corporate safety. In Corporate Safety Services, John developed programs including PSM metrics, Process Hazard Management policy and implementation guidance. John's diverse experience has shaped his personal passion for process safety.

John has made presentations at 2 CCPS global congresses, participated on writing committees for several CCPS Guidance books and chairs the Process Safety Beacon and Process Safety Messages committees, and is a Process Safety Boot Camp instructor. He was named a CCPS Fellow in 2009.

John has a BS in Chemical Engineering from Purdue University. He resides in Stillwater Minnesota and is married with 2 married children and 2 grandchildren.

How to Create a Sustainable Safety Culture
Moving from Pound and Blame to Partnering

Richard N Knowles, Ph.D.
The Safety Sage
Richard N. Knowles and Associates, Inc.
6083 Bahia Del Mar Circle
Unit 564
Saint Petersburg, FL 33715
RNKnowles@aol.com
www.SafetyExcellenceForBusiness.com

Prepared for Presentation at
American Institute of Chemical Engineers
2013 Spring Meeting
9th Global Congress on Process Safety
San Antonio, Texas
April 28 – May 1, 2013

UNPUBLISHED

How to Create a Safety Culture
Moving from Pound and Blame to Partnering

Richard N. Knowles, Ph. D.
The Safety Sage
Richard N. Knowles and Associates, Inc.
6083 Bahia Del Mar Circle
Unit 564
Saint Petersburg, FL 33715
RNKnowles@aol.com

Keywords: Safety in the Workplace, Employee Engagement, Leadership, Safety Culture

Abstract

Safety cultures do not spontaneously occur. They are created and nourished by honest, authentic operating managers and leaders. These people must insist on high standards and foster trust and interdependence within their organization. This requires open, freely flowing information that people at all levels in the organization can access, analyze and use to keep their workplace, the hazardous materials and energies under control. This work is focused and relentless, requiring discipline on the part of the whole organization, including the top level managers, front line supervisors, engineers and hands-on employees. This paper presents an operating manager's experience on how he evolved his "pound and blame" managerial culture into a "partnering" culture and successfully improved the organization's safety performance on both the plant's personal safety and process safety metrics. This effort contributes to the well-known "business case" argument for a robust process safety culture, as measured by the significant increase in the plant's operating profit and corresponding process safety-related improvements.

1. Introduction

Personal and process safety are critical factors in safely operating a chemical plant and must be conducted at the highest levels of competence and integrity. As a plant manager I had total responsibility and accountability for the operation of our facility; this was an awesome experience. Any major mistake could hurt people on the plant, the neighbors and cause a major financial impact on the company.

Total safety performance, and process safety in particular, are core technologies for those operating a chemical plant using hazardous materials. Sloppy operations with injuries, leaks and explosions are totally unacceptable. We have no right to be careless, hurting our people, the neighbors and the environment.

In managing a plant where highly hazardous materials, like phosgene, anhydrous ammonia, chlorine, oleum, hydrogen cyanide, were used, I placed high value on personal and process safety management. I also respected the engineers who did the analyses, the managers who had to make tough decisions, the operating people and mechanics who worked safely and did the operations, inspections, maintenance and all the follow-up. It takes a focused operating team to thoroughly address process hazards and to ensure the safety, health and welfare of everyone inside and outside the plant.

I had to help to create a culture of operating discipline, openness, honesty, respect, integrity, listening, sharing, supporting each other and doing the things we agreed needed to be done. Excellence in personal safety flows into and supports excellence in process safety management and visa-versa; it is truly a team effort so that we can operate with the highest levels of safety and process integrity and standards.

This is an effort that also reaches out into the communities surrounding our plants. Our neighbors need to have the confidence that we are operating in a way that will protect them from the potential problems we could face. We have to lower both injury rates, and the probabilities of an incident or release to the very lowest levels. This is an issue of our freedom to operate and avoid excessive regulations. These are issues of survival for our industry.

The Challenge:

As an example of the success of shifting the culture to excellence in both total safety and in process safety management, I want to share the story of our work in the Kanawha Valley in West Virginia. When I was assigned as the Plant Manager of the DuPont Belle Plant in April, 1987, the initial challenge was to restore safety and operating discipline and performance.

When the safety gets right, everything gets right. The results that the Belle Plant people achieved were extraordinary. Total Recordable Injury Rates dropped by over 95%, emissions dropped by 88%, productivity rose by 45% and earnings rose 300%. In the first 2-3 years we achieved significant reductions in injuries and incidents. Then we had to move beyond the initial successes putting systems and processes into place over the next several years to sustain this work

2. Our Story

Safety Leadership is the Key to Sustainable Safety Excellence

Everything that happens in an organization happens through people. We depend on people doing everything within their capabilities to do their best. We need to have open, honest, knowledgeable, competent, people of high integrity doing the safety and process hazards work and in carrying out the various analyses, installations, inspections and all the routine follow up work like relief valve inspections, pressure testing and thickness measurements as scheduled. We can have the most sophisticated analyses possible, but if people don't take the initiative to follow through, then all is for nothing. Cutting corners, ignoring the use of PPE, covering up, not doing the inspections, keeping inadequate records, not following up, bullying and lying are unacceptable.

My mantra was "I don't have a right to make my living where it is okay for you to get hurt. Now we have to get on top of the safety stuff and build a strong business."

Figure 1 illustrates these ideas of how the technology and the people are both critical to build a sustainable safety culture. Most organizations put most of their emphasis on the technological side of things doing excellent analyses and studies. They develop recommendations for the best practices in managing the processes and in minimizing the risks. This is often called "the hard side" and is critical to do well.

A few organizations combine this work with the people side. All the technological work and recommendations must flow through people. The engineers and designers, those doing the installations, the supervisors, the operators, the mechanics and the safety people who must train and monitor things all need to be involved and committed to things being done right. The processes put into place to do the proper follow-up inspections and checks must be rigorous all the way through the design, installation and onward for the life of the equipment. The best technical work will be wasted if these things are not done. The people side is often called the "soft side". I've heard it said "the hard side is the easy side and the soft side is the hard side".

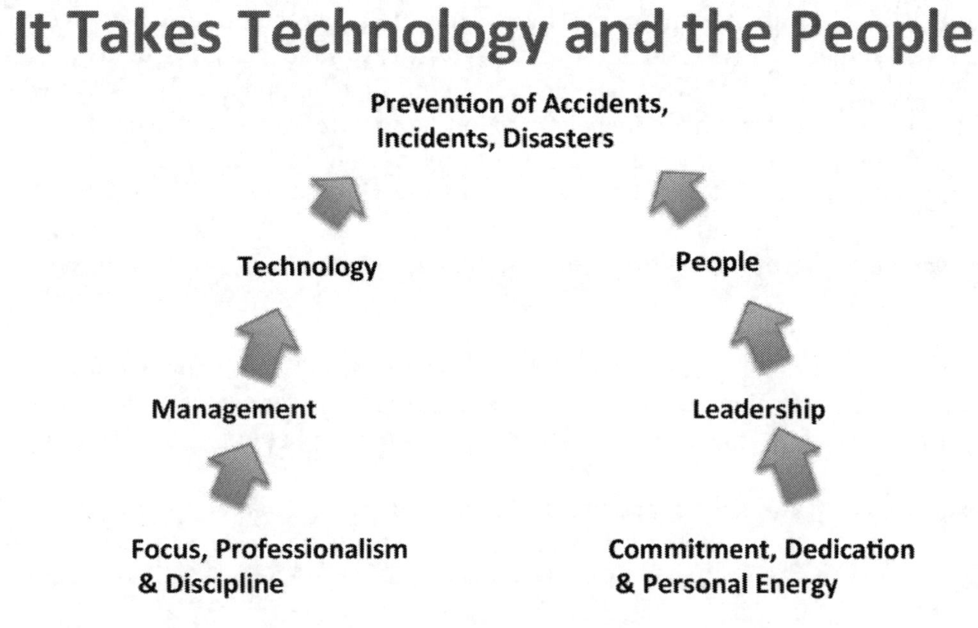

Figure 1

The shift in the safety culture is focused, relentless work, requiring discipline on the part of the whole organization.

One of the most important tasks for the operating manager is to create a culture where:

- information is freely flowing so people can find and use it in their work and analyses,
- there is a climate of trust and interdependence,
- people can and do tell the truth as they see it,
- we listen to each other seeking understanding,
- people, at any level, including first line supervisors and hands-on employees, can take appropriate leadership action to fix a problem that they encounter,
- People act on the information so that the plant can be the safest it can be

Many people know critical information but it is often not shared or discussed. This can lead to disasters. Incidents like the Challenger Disaster, where the engineers knew the low temperatures could cause the O-Rings to fail but their managers rejected or ignored their cautions, are unacceptable. Similar cultural conditions were present in the Deep Water Horizon disaster.

When I was first assigned as manager of the DuPont plant in Belle, WV in 1987 there were about 1,300 people working there and the plant's injury performance was very poor. This is quite apparent in Figure 2. The standards were terrible, things were being covered up, people would not tell the truth about incidents, various groups did not get along with other groups, management didn't get along with the hourly people and visa versa, everyone blamed everyone, else, the bullies blocked improvements etc. I was expected to get the performance much better before we killed someone. I went on the warpath and raised a lot of fuss. I pounded, verbally, on people and pushed hard. I talked with as many as I could find, but most tried to stay away from me. I had lots of arguments with the managers reporting directly to me. I chastised some people publically, which was poor on my part. I spent hours and hours each day in the plant and many evenings and nights as well. I walked everywhere. I wore glove so I could pick up trash. When people refused to work more safety or were bullying their work mates, they went into the discipline system. We terminated those who just didn't get the message…working safety was important!

In a climate like this everyone was under a lot of pressure, Healthy communications were non-existent. I was not happy about the plant or myself, and neither were most of the others. But our injury performance, safety standards and housekeeping got a lot better.

In two years the injury rate had dropped to 0.74 and leveled out. This injury rate was close to the DuPont rate at the time and that was pretty good. But we could not sustain this without a lot of effort and pressure. We were all sick of the fighting.

As things stabilized we began to talk together about how to shift our culture to one where the performance could become sustainable and better. In 1991 we shifted to teams across the entire plant. You can see in Figure 2 that the injury rate improved, dropping to 0.26. We learned to talk together more effectively, we learned together more effectively and the sustainability in our performance showed up as the years rolled by. The depth of our total safety effort significantly improved so process safety was better, uptime improved, earnings increased. Figure 2 shows that this level of performance between about 0.3 and 0.0 sustained for many years after I was reassigned in 1995. My role shifted from being a tough taskmaster, driving and pounding all the time to become a cheerleader. We were becoming partners. This was a lot more fun.

In all this work we shared information abundantly, trust and interdependence built, we talked a lot and listened even more. Good ideas emerged from all over the place about how we could improve customer service, improve our environmental performance, shorten the turn-around time between product campaigns from 3-4 weeks to just 3-4 days through better planning and cooperation. Resistance to change almost disappeared as people came up with good ideas and implemented them. In my first few years at the plant we struggled with change making only 3-4 major changes. As we all learned together creating the new ideas, resistance to change dropped to where we were making 2-3 changes a month and I was slowing them down so that we would not out-run our technology. Management of change is critical in process safety management.

Figure 2

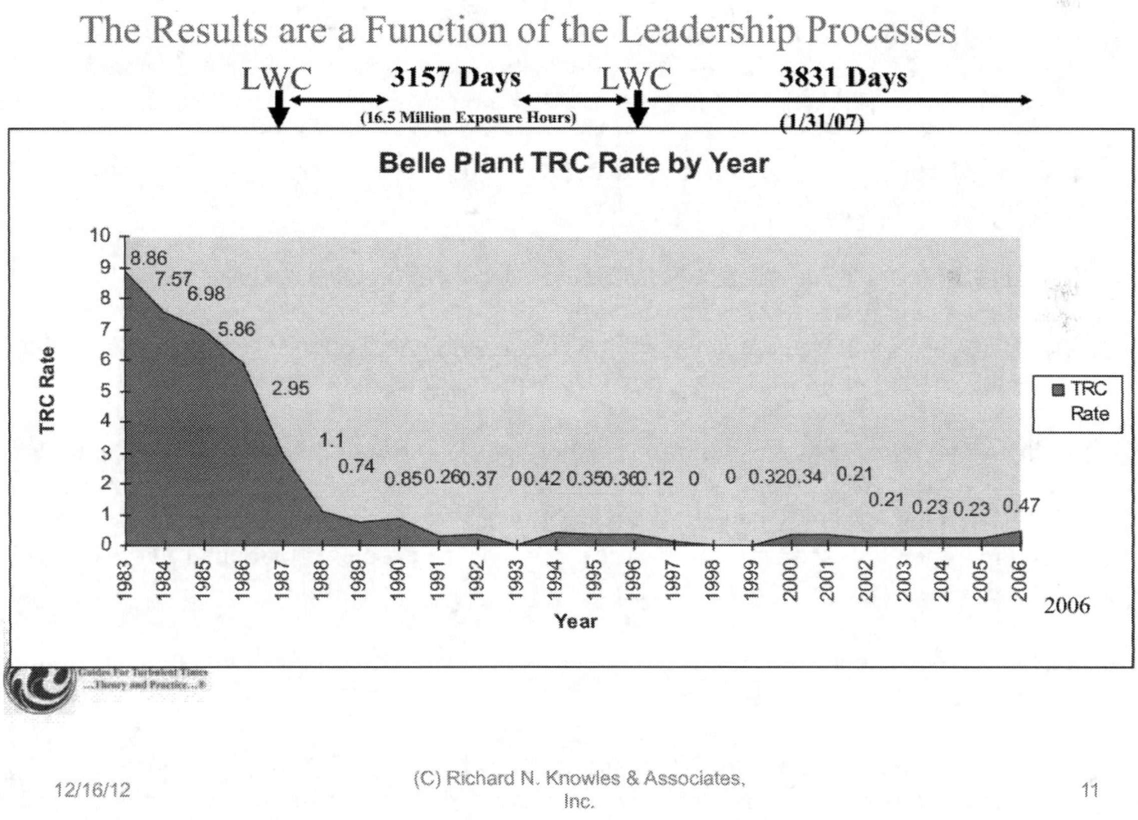

Note: LWC means lost workday case and TRC means total OSHA recordable injury rate.

In reflecting on all this work to improve safety and our over-all performance, and in the consulting I've done around the world since retiring 17 years ago, I have been able to develop a simple, 4-step process that summarizes our work at Belle and works quickly in every organization in which we've worked. In working this way the injury rates drop quickly, the communications open up, people develop more energy and are more creative. They experience a cultural change like we did at Belle. This shift does not require any capital investment.

This cultural shift begins with a 4-step Safety Leadership Process™.

1. The top leaders have to talk together, openly and honestly, about what they really want to have happen and develop the commitment to do it. They need to be very clear and coherent about what they are really going to do to move towards sustainable safety excellence. Are they really going to live and do the critical work? The Process Enneagram© tool is used to hold the focused, open conversations. This tool is the only tool that I know of that helps people solve complex problems, make the social connections they need to get the work done and releases the emotional energy and commitment to do the work quickly…all at the same time. My book, The Leadership Dance, provides a lot of information on the creation and development of this tool [1].

   This first workshop takes about a day to develop their Safety Strategic Plan. During the course of this first step of the work, the fierce conversations are held and trust and interdependence build. The top leaders develop their Principles and Standards, and ground rules as to how they will work together and with the others in the organization. They agree to share information and to hold each other accountable to live up to their Principles and Standards. They agree to "walk the talk".

2. Having co-created a Safety Strategic Plan, the leadership goes into their workplaces to talk about the plan and share their thinking, listen, and to look for improvements in the Plan with the operating personnel. This work is visible and open for all to see, and with everyone sharing and listening together trust and interdependence build.

3. Each day involves listening, talking together, and exploring new and better ways to do the safety work more effectively. They learn to conduct simple, effective Safe Acts Audits, actually seeing what people do and learning together and acting on what they are learning to prevent incidents and injuries. Most of the problems that emerge are related to systems and communications problems.

4. Then they help create and support the environment where the people can make changes that they have discovered together.

The Safety Leadership Process includes listening, talking, sharing, and co-creating the future together. This is done over and over as the organization and its environment evolve. This Safety Leadership Process is developed more fully in my new book, "Partnering for Safety and Business Excellence" [2]. In this process we partner with the people to achieve excellent, sustainable performance.

Once the leaders are clear and unified in their thinking about safety, they must step out of their offices, step into their workplaces and step up to the challenges and ensure that problems are promptly fixed. During the course of this work, as everyone is learning, the organization becomes "leaderful" - that is, when anyone sees a problem, they step up to take the initiative to get it fixed.

An example of the sort of environment we were able to create occurred one morning when I was walking around the plant. A methylamines operator came up to me to tell me he'd shut down the amines plant, on his own initiative, because he'd found a small, weeping type of leak at the base of a relief valve on top of the 150-foot distillation column. He wanted to prevent a more serious leak so he took the initiative to shut-down to fix it. I felt that this was a great contribution when an operator can take this sort of initiative and prevent a problem.

Just look at the way our safety performance changed and improved over 23 years. Our safety leadership process before 1987 was one of talking about safety but doing little to make it happen. When someone did get hurt, there was a lot of blame and shame for the operators or mechanics. When I came into the plant (1987-1990), our process was a typical, hard, top-down process that I applied to everyone. We drove things hard, argued a lot and reduced the injury rates to about 0.7, but could not sustain this level of performance. As soon as the pressure was released, things grew worse. While we were at this level of compliance, the system was full of holes and people kept getting hurt. When we learned to partner with the operating personnel in 1991, look at how well they performed and sustained this excellent level of performance for years after I was reassigned in 1995.

As we re-established safety and operating discipline our process safety improved as well. Sharing information, building trust and interdependence, and helping everyone see the importance of their work and how it all fit together spread to other aspects of our work. The inspections were done as scheduled, and things got repaired in a timely way. More and more people at the operating level got involved as the story of the amines operator illustrated. The number of releases went down and the severity was significantly reduced. The engineers were doing the analyses with the appropriate technology, thoroughly and with the highest integrity. When they would review their work with me, I would often ask, "Is your analysis strong enough that you would be okay if your family lived at our fence line?" We had families living next to our fence line and we were responsible for their safety. If the engineers hesitated in their answer, they went back to do the analyses again.

Having learned to shift the safety culture to one of sustainability and excellence the positive impacts showed up everywhere. When the safety gets right, everything gets right. The results that the Belle Plant people achieved were extraordinary. Total Recordable Injury Rates dropped by over 95%, emissions dropped by 88%, productivity rose by 45% and earnings rose 300%.

3. Conclusion

The top operating managers have to create the desired culture, spend time in the plant to know the people, see what is happening, pick up any clues and be sure that people are doing their best. They have to listen openly to each other working with the highest level of respect and integrity.

Sometimes tough judgments have to be made about just what has to happen to improve the safety. There may be disagreements between the engineers, operators, mechanics, supervisors and operating managers about the level of hazard and what needs to be done. Since we can never get away from this tension, everyone can live in it providing we listen for understanding with integrity, openly discuss everything and struggle to come to the best agreements we can make. We need to co-create our future. Safety cultures do not spontaneously occur. They must be created and nourished by operating managers. This includes partnering with all the people with high operating discipline, and fostering openness, the free flow of information, and working to build trust and interdependence within their organization.

4. References

[1] Knowles, Richard N., *The Leadership Dance, Pathways to Extraordinary Organizational Effectiveness.* Niagara Falls, N.Y. The Center for Self-Organizing Leadership. (2002)

[2] Knowles, Richard N., *Partnering for Safety and Business Excellence.* St. Petersburg, FL. The Center for Self-Organizing Leadership. (2013)

9TH GLOBAL CONGRESS ON PROCESS SAFETY

Process Safety: At the Heart of Operations

Chung Duc Nguyen
Gulf of Suez Petroleum Company (GUPCO)
Palestine Street, Maadi, 11511, P.O. Box 2400, Egypt

Prepared for Presentation at
American Institute of Chemical Engineers
2013 Spring Meeting
9th Global Congress on Process Safety
San Antonio, Texas
April 28 – May 1, 2013

UNPUBLISHED

Process Safety: At the Heart of Operations

Chung Duc Nguyen
Gulf of Suez Petroleum Company (GUPCO),
Palestine Street, Maadi, 11511, P.O. Box 2400, Egypt

Keywords: Process Safety, KPI, Operations, Management, Implementation

Abstract

Process safety excellence can only be achieved when it is placed at the heart of operations. The majority of oil and gas companies have well-defined process safety frameworks and process safety resources at the corporate level. However to make the program work at a site, additional factors are required. This paper will address aspects that are critical to a successful site process safety program, focusing on the key success factors, aspects of process safety which need the most attention and resources from operating managers, and methods to effectively communicate the most important aspects of process safety to the site team

1. Introduction

Site implementation will determine success of any process safety program. All frameworks, processes and procedures set at the corporate level will be implemented at a site by front line people. Having processes in place and disciplined operations are pre-requisite conditions for success. Additional factors are also required to make process safety truly owned by operations and driven by operations

2. Key success factors for a well-functioning process safety program

2.1 Operations discipline

Operations discipline is probably the most important factor for a successful process safety program. First and foremost, a just culture should be established with compliance being the expected behavior. The site team should be very well aware of the boundary between accepted and non-accepted operational behavior. Secondly, there should be a sense of ownership and pride for disciplined operations amongst the site team. Operations discipline has to become a personal thing for everyone at the site, something that people want to do, not something that they have to do. A good practice is to have the site team get involved in the development of operating procedures, assurance processes, and key performance indicators. Site leaders need to instill the belief within their team that they can achieve "world class operations" or "best in class process safety performance". They could also have the site team agree upon and sign a "How we operate" document which clearly defines expectations for safety and operations behaviors, and display it in the form of big posters in the control room, mess hall and public locations on-site.

2.2 Site leader is a passionate process safety role model

This could be a real challenge for companies that have never experienced significant losses caused by process safety incidents or those that are very much production-oriented. Passion for process safety has to come from a deep understanding about process safety, about the fact that safety is a long and continuous journey that never ends, and from genuine concerns for the safety of frontline people. This is reflected in a strong commitment for disciplined operation, process safety performance, and in giving process safety an equal place in the operations agenda with production priorities. Site leaders need to go to great lengths to understand site operations and not take things at face value. A passionate site leader will always ask questions to probe for potential issues, i.e. challenge control and mitigation measures for risk assessment, and question the reason for overdue critical maintenance or deferral requests.

One of the quickest ways to destroy commitment from the workforce for a process safety program is seeing leaders do not practice what they preach. Site leaders should never let production prevail if there is a conflict between production and safety. For example, if a site leader is willing to carry out a closure test on a critical pipeline isolation valve even though the test may lead to production loss if the valve fails to open after the test, the leader clearly demonstrates the priority of safety. With a strong role model, the site leader's people are much more likely to feel empowered to make critical decisions such as shutting the plant down in case of an emergency. If a leader constantly looks for opportunities to improve process safety performance, the whole organization will react in the same way. The whole site team looks to the site leader as the representative of the company, therefore it is important that site leaders walk the talk and set a role model for process safety

2.3 Process safety is part of the performance contract and reward program

If expectations are set for process safety but do not tie in with some form of annual performance contract and reward program, the workforce will have fewer incentives to improve process safety performance. Policies are set at the corporate level, however having a program that ties individual process safety contribution with performance evaluation will support the site leader in driving process safety performance. Site leaders need to ensure contribution to process safety is a significant item in the performance contract for site personnel. Department managers need to commit to carrying out a certain number of process safety audits, providing training to ensure competency of the workforce, and improving process safety key performance indicators (KPIs). Offshore installation managers (OIMs) and superintendents need ensure that all critical registers are up to date, procedures are complied with, and process safety performance is communicated to the crew. These are just a few examples of process safety priorities in performance contracts.

2.4 Process safety capability of the organization

A process safety adviser role should be established to support the site leader to deliver process safety programs such as providing process safety training for the site team, setting up reporting guidelines, determining KPIs and providing support for specific process safety issues.
The whole site team also needs to have a certain level of understanding about process safety and about their role and responsibility for safe and reliable operations. For example, an OIM should

be able to look at the process safety scorecard and explain the risks on his platform and areas that need focus of his crew. A shift supervisor knows that an effective shift handover should be carried out in both spoken and written formats and the handover should include critical process safety information. An operator should understand that operating and work procedures need to be complied with and that any anomalies found during the shift should be reported. All of this can be done through training and by reinforcing disciplined operations.

2.5 Process safety assurance processes and reporting tools

2.5.1 Processes and reporting tools

Key process safety assurance processes should be established and made available to site personnel, such as management of change, prestart-up assurance, deviation from normal operations, incident investigation, risk assessment, inspection and maintenance procedures. The site leader could request a gap analysis against the company's process safety framework or a well-established management system (OSHA Process Safety Management, for example) to identify if any important processes are missing and develop a plan to close the gap. Reporting tools should also be available and accessible for the frontline people. Common process safety reporting tools are incident and near miss reporting, process upset reporting, and anomalies reporting. The site leader needs to ensure that reports are followed up, actioned, and information is then fed back to the organization either in the form of lessons learned or follow up actions.

2.5.2 Process safety Key Performance Indicators (KPIs)

A fit for purpose set of KPIs provides a snapshot of site process safety performance and helps site leaders to identify areas that require intervention. Common mistakes when setting up KPIs are: defining too many KPIs such that they are not really "key" anymore, adopting KPIs from other sources that are not representative for site operations, or choosing KPIs without considering the organization's reporting capability. KPIs for a refinery may be different from KPIs for a modern platform and different from those for aging facilities. A process safety adviser who really understands processes, facilities, and the site team's capability would be able to help the site leader to develop a set of KPIs that are a true reflection of site operations.

2.5.3 Communication

Communication is very often the missing link in the process safety loop. Decisions and interventions are made at the site management level but very little information may be passed on to the front line workers who interact with the plant every day and play an important role in keeping the plant in a safe operating condition. A communication tool that transfers the most critical information to front line people should be in place. For example, top risks at site, long-term overrides, high risk wells, overdue safety actions, and temporary changes should be clearly communicated. When developing such a communication tool, one has to bear in mind that the information needs to be presented in a way that it is understandable for front line personnel, that and is concise and visual. An example of such a process safety communication tool is described in Section 5.

3. Process safety - important aspects for site leader

As mentioned earlier, a thorough understanding about all element of the process safety management framework is necessary for site leaders. Below are aspects of process safety that seem to be systematic issues at operating sites and need further attention.

Operating and work procedures: Site leaders need to ensure all procedures are up to date, accessible and followed. This is easier said than done. In the real world of operations, people often take shortcuts or violate procedures with the very best intention to get things done. This results in "procedures not up to date" or "procedure not followed" being common immediate causes in major incident investigation reports. It is a good practice to get operations teams involved in the development of operating procedures to ensure that the procedures are fit for purpose and increase ownership of procedures for operating personnel.

Hot work and breaking containment: Sources of ignition and fuel are two main elements of a recipe for disaster. Hot work and breaking containment jobs should have rigorous attention when it comes to writing procedures, assigning crews and performing work at the site. Compliance should be a strict requirement. A good practice is to identify critical hold points in the procedure that need to be signed off before moving on to the next steps. Site leaders should request compliance audits to determine if procedures are being followed.

Risk assessment: Two common pitfalls associated with risk assessment that the site leader should be aware of are: risk normalization and confirmation bias. Confirmation bias means the purpose of the risk assessment is to justify continued operations instead of to evaluate the risk. This leads to a number of issues: a poor quality of risk assessment, lower risk ranking of potential hazards, and inadequate control and mitigation measures. The site leader, therefore, needs to set a clear expectation for the risk assessment team and challenge the risk assessment before signing off.
Risk normalization tends to happen in organizations that have to manage risk on a more frequent basis, such as aging facilities. The level of risk that the team is willing to take increases gradually and accepting greater risks becomes the norm. The site leader needs to set a clear limit for the acceptable risk level, i.e. only a certain number of deviations from normal operations on a plant are allowed, any additional deviation will lead to plant shutdown. Examples of deviation from normal operations are running the plant outside of the operating envelope or with known integrity issues (for example pipeline thickness) and impaired critical safety systems. Another example is to set a time limit for continued operations with deviation, and shut the plant down when the limit expires.

Pre-startup safety review: Start-up of a plant creates significant risks compared to normal operations. Site leaders need to ensure that a start-up assurance process will be utilized to support safe and reliable start-up of plant and equipment. A set of fit for purpose prestart-up checklists for different start up scenarios may be required.

Shift handover: Shift handover needs to be carried out in a structured way, in both verbal and written formats. Site leaders need to ensure critical process safety information is recorded in the handover note and discussed, for example, impaired safety instrumentation or equipment, and operational status for the process. Audits of the shift handover process could be carried out to ensure compliance.

Incident investigation and learning: Learning is always the difficult part of the incident investigation process. Looking back at major process safety incidents, it is not uncommon to find the same root causes for multiple incidents i.e. overriding critical instrumentation, not following procedures, incompetent personnel, and inadequate engineering. Site leaders need to make sure that lessons from incidents that have happened on site and lessons from external sources that are relevant to the site are really internalized, not just by increasing awareness during safety stand downs, but through specific actions such as reviewing engineering design, updating procedures, and reinforcing operational discipline.

Management of change (MoC): One of the biggest challenges for process safety in operations is to manage changes. MoC has been covered in numerous process safety discussions; however, it is worth mentioning here again with a few key points that site leaders should consider. Firstly, a comprehensive MoC procedure that provides guidance for different types of changes should be in place. Secondly, the site leader needs to ensure all stakeholders in the MoC process receive adequate MoC training and are competent. Finally, ensure the process is followed, including: adequate engineering review and risk evaluation, personnel are trained on changes that occur, and procedures and engineering documents are updated.

4. How to know if your process safety program is working

4.1. KPIs trending in the right direction

The easiest way to see if a process safety program is working is to look at trends: improvement in leading indicators and reduction in lagging indicators. If there is no improvement or the trend is going in the reverse direction, site leaders need to dig deeper to find out what is happening and intervene if required, by: providing additional resources to clear safety critical maintenance backlogs, speeding up technician assessment programs to improve competency KPIs, and carrying out more emergency drills.

4.2 Level of engagement of the organization

This is all about a positive attitude towards process safety and a commitment for continuous improvement. This soft part of process safety is hard to measure, however there are different ways that a leader can read the level of engagement of the organization. For example, the practice of reporting all incidents, near misses, and process upsets in a timely manner becomes the expected behavior. Management of change meetings have the right level of attendance from the relevant departments. The workforce shows a strong compliance with work procedures. Department managers see monthly process safety meetings as a forum to discuss issues and an opportunity to drive process safety performance.

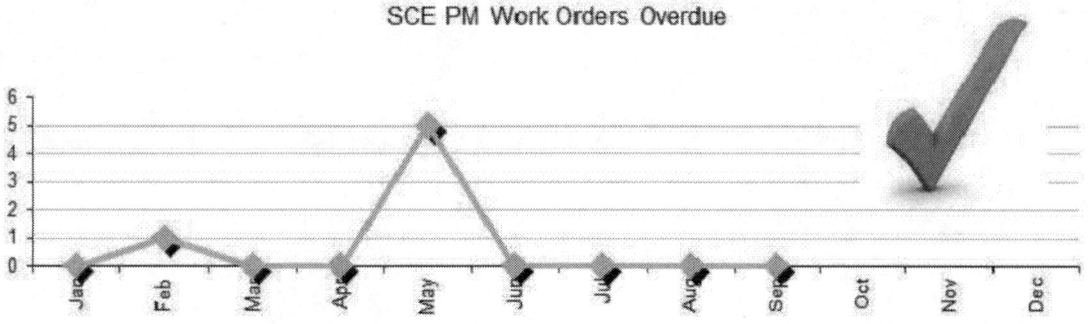

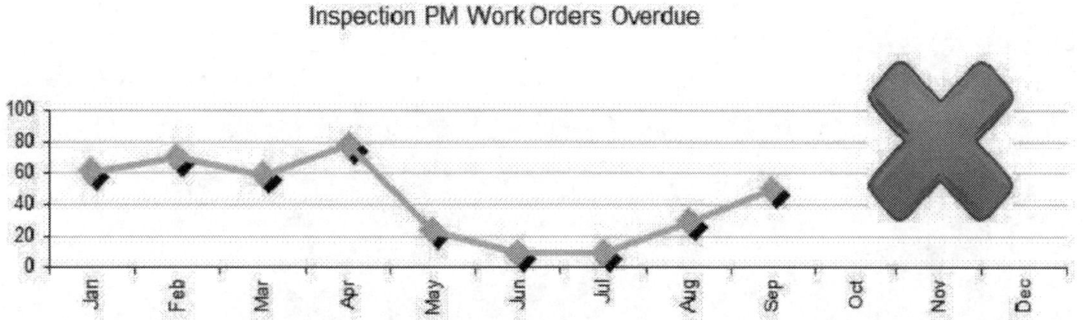

Figure 4.1.1 Examples of KPI trending for safety critical equipment and inspection preventive maintenance (PM)

4.3 Process safety audits

Process safety audits could be carried out to evaluate effectiveness of the process safety program at the site. A three year site audit against the corporate process safety management system is the practice in a number of organizations. The quality of the audit team is important to ensure that the audit will reveal any hidden issues and provide practical resolution to address them. At the site level, a site leader can request compliance audits on certain elements of the process safety framework on a more frequent basis. For example, compliance with the permit to work system, compliance with the management of change process, work procedures, and risk assessment can all be audited if deficiencies are suspected or to determine the general health of these management systems. A team led by the process safety adviser with participants from other disciplines and supporting functions should be able to deliver this.

5. Case study: Process safety scorecard in GUPCO

Process safety "at the heart of operations" means it is really owned by operations and driven by operations. In GUPCO's operations, process safety scorecards have been effectively used to facilitate the process (see example Figure 5.1). The site process safety adviser set up various sections with leaders, managers and site team to go through the process safety framework, layers of protection concept, and discuss key performance indicators for each department. This top-down and bottom-up enrolment of staff ensures clarity regarding KPIs and ownership by

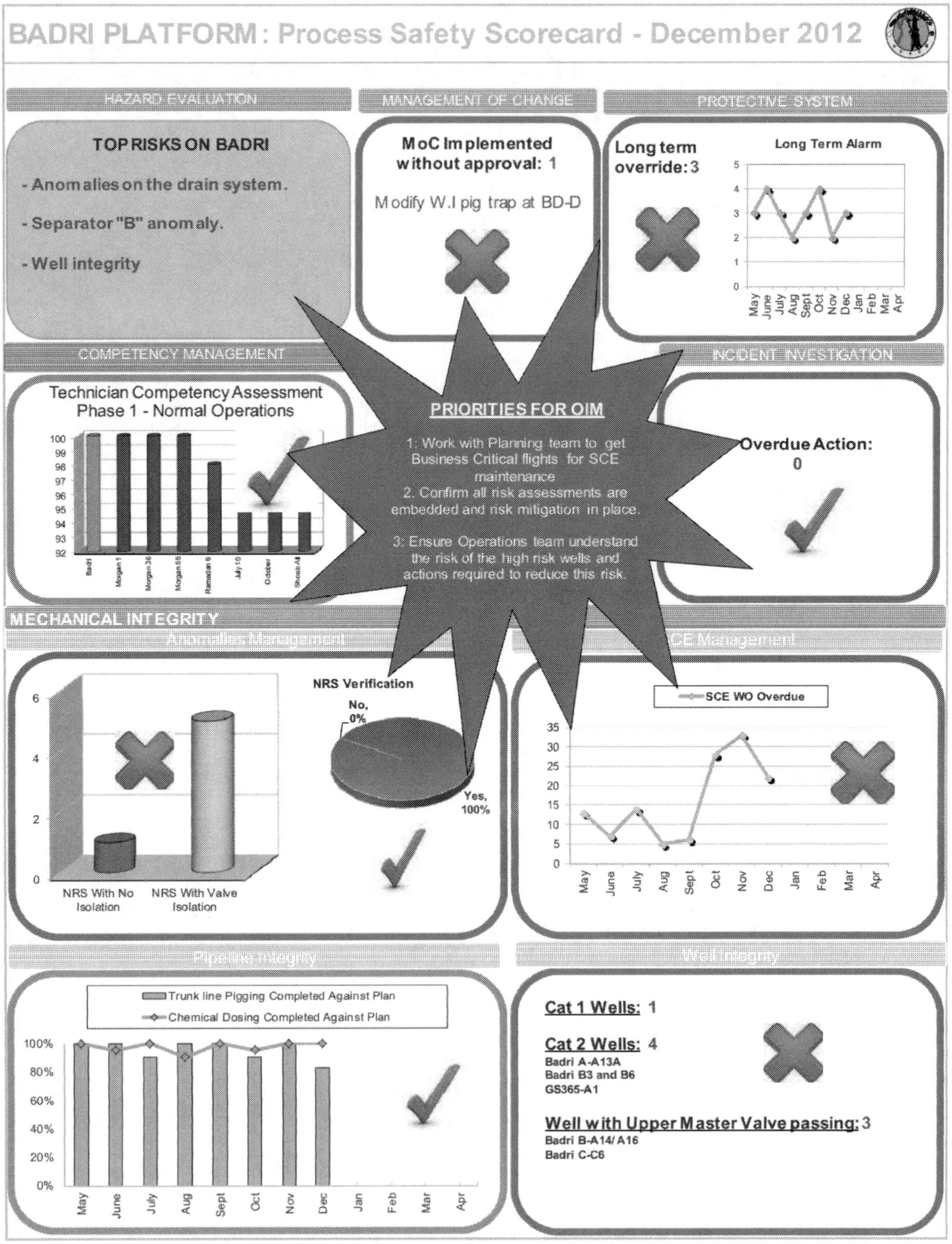

Figure 5.1 Example of Operations monthly process safety performance scorecard

functional departments who play a key role in "driving" these KPIs. A central database is created for functional departments to update their monthly key performance indicators. Operations managers then work with offshore installation managers (OIMs) to develop process safety performance scorecards using information from this central database. Elements and sub-elements of the process safety management framework are color-coded in the scorecard to highlight process safety issues and areas that need intervention and priorities. Monthly process safety meetings are established for OIMs and supervisors to communicate the scorecard to the team and for the crew to provide feedback for issues identified. The site leader and functional department managers also participate in the meetings. The scorecard is posted in the control room at each operations location, serving as a communication tool for the OIM to the crew, and as a reminder of risks and priorities. It is also used as a tool to assist site management for process safety audits.

6. Conclusion

Process safety is always at the heart of safe and reliable operations. The way to process safety excellence is a long journey with a disciplined operation being the pre-requisite condition. In addition to that, process safety assurance processes, leader commitments, a reward program that connects with process safety performance, and site process safety capability are required to ensure a successful implementation of process safety programs at site.

7. References

[1] Hopkins, Andrew, (2009) *Failure to Learn: The BP Texas City Refinery Disaster, Page 44.* Sydney Australia: CCH Australia Limited

[2] Reason, James, (2008) *Managing the risks of organizational accidents, Page 195.* England: Ashgate Publishing Limited

(27a) Life Cycle Assessment and Life Cycle Cost of the Pyrolysis of Switchgrass to Use As an Energy Source in a Conventional Power Plant

Nuttapol Lerkkasemsan, Chemical Engineering, Virginia Polytechnic Institute and State University, Blacksburg, VA

Biomass is expected to become an important energy source. This paper considers both life cycle assessment and life cycle cost of the pyrolysis of switchgrass to use as an energy source in a conventional power plant. Specifically the greenhouse gas (GHG) emission for power generation was investigated through life cycle assessment. The process consisted of cultivation, harvesting, transportation, storage, pyrolysis, transportation and power generation. Here pyrolysis oil is converted to electric power through co-combustion in conventional fossil fuel power plants. The conventional power plants which are considered in this work are diesel engine power plant, natural gas turbine power plant, coal-fired steam-cycle power plant and oil-fired steam-cycle power plant. Several scenarios were conducted to determine the effect of selected design variables on the production of pyrolysis oil and type of conventional power plants. The set of design variables consisted of land fraction, land shape, the distance needed to transport switchgrass to the power plant, the distance needed to transport pyrolysis oil to where, and the pyrolysis plant capacity.

The results show that with an existing average agriculture land fraction in the United State which is 44.1 percent of all the United State land area, the cost of electricity from pyrolysis of switchgrass is lowest at $1.2 per gallon with a capacity of 5000 ton of switchgrass per day of pyrolysis plant. However, power generation from switchgrass required huge land especially for an existing average agriculture land fraction which is 0.441. In order to provide enough switchgrass for a pyrolysis plant capacity of 5000 ton per day, a large land area with a size of 1706.11 square miles is required; note that the land area is bigger than the state of Rhode Island. Using natural gas turbine power plant for electricity generation, the price of electricity can go as low as 7.70 cent/kwh; contrast this with the average electric price of 10 cent/kwhin the United States. Based on 5000 tons of switchgrass per day, a natural gas turbine power plant can produce 5011.86 million kilowatthours, which is enough to supply the state of Hawaii. The main advantage in using a pyrolysis plant is the negative GHG emission from the process. This conclusion came from Life cycle assessment analysis of power generation using pyrolysis oil, which showed that it was environmentally friendly and would therefore reduce the GHG emission to the atmosphere. If and when carbon credit becomes significant, pyrolysis oil could be even more competitive.

(27b) Natural Gas as a Transportation Energy Resource: Life Cycle Considerations for Personal Mobility Options

Christian M. Lastoskie, Civil & Environmental Engineering, University of Michigan, Ann Arbor, MI

Presently the U.S. is experiencing a significant boom in domestic natural gas production due to the application of hydraulic fracturing and horizontal drilling to Marcellus and other formations with an abundance of shale gas. The combination of decade-low natural gas prices and increased restrictions on coal utilization for electric power has led to a displacement of coal-fired generation with electricity from natural gas.

With the prospect of the United States undergoing a transition to being a net energy exporter within the next decade, it is timely to consider scenarios wherein a displacement of petroleum as the predominant domestic transportation energy resource occurs, with steadily increasing utilization of cheap abundant natural gas as the source fuel for personal mobility. In this presentation, we consider three options for natural gas-derived personal mobility: compressed natural gas automobiles, battery electric vehicles, and hydrogen fuel cell cars. The life cycle environmental impacts are compared for the operation of these three types of vehicles, including the fuel processing infrastructure required to deliver the respective fuel for each vehicle. We consider the impact of potential future constraints on carbon emissions on the environmental footprint of each mode of natural gas-powered personal mobility, on a per mile driven basis.

(27c) Environmental Controls in Coal-Based Electrical Energy Production

Noah Meeks, Research & Environmental Affairs, Southern Company Services, Inc., Birmingham, AL

Over the past half-century, the coal-fired electric power plant has evolved from a simple machine to a complex chemical process plant. While the heart of the industry remains the same: turning turbines with coal-generated steam, the treatment of coal combustion emissions has grown considerably and more processes are under development. There is a significant role for chemical engineers to play in meeting these current and future needs. This tutorial will cover the environmental impacts and pollution control unit operations of coal-fired electric power plants, both current and future. The four main unit ops are selective catalytic reduction of NOx, electrostatic precipitation, fabric filtration ("baghouse"), and flue gas desulfurization ("scrubber"). These incorporate a number of chemical engineering principles including separation processes and reactor design. In addition, new compliance drivers are pushing additional pollution controls for mercury, which is typically done as an adsorption process. Finally, concerns over global climate change have resulted in extensive research into carbon dioxide capture and sequestration using absorption processes. Also water and wastewater impacts from the power plants will be covered.

From Research to Practice: A Story of Mathematically Predictive Actionable Process Safety Leading Indicator Index

AICHE - 2013
Abstract ID: 294440
Submitted to: ENVIRONMENTAL DIVISION

Author Contact Details

Speaker: Joe Stough, **VP Innovation & Technology**
Company: IHS
Email: joseph.stough@ihs.com
Phone: +1 713 499 4288

Author Bio:
Joe Stough is the visionary and strategist behind the IHS Operational Excellence Forum methodology. With a professional background in large-scale database software design a graduate degree in Applied Statistics from the University California, Stough founded Syntex Management Systems and led the company in providing solutions that successfully combine the disciplines of enterprise-level business software design and advanced statistical analysis to dramatically impact Enterprise Risk on a global basis. Syntex is now a part of IHS, Inc., where on an annual basis, Stough is continuing to refine his findings in the context of the Operational Excellence Forum.

ABSTRACT:

For the past several years, a group of global Energy companies, including nearly all of the super majors, has been working on a research initiative with a common mission to find an actionable leading indicator index to drive Safety and eventually overall Operational Excellence performance.

This group has been collaborating to provide a large multi-year data set, which contains millions of data records from events such as incidents, investigations, near misses, audits, observations, assessments, and many other routine field-level activities. Since 2008, a rigorous statistical analysis process has been iteratively applied to this data to identify the mix of leading metrics which most effectively predict Safety performance outcomes.

This paper will review the leading indicator research findings drawn from analyzing this large Energy industry data set. It will reveal the key components of an actionable leading index which has been found to be uniquely strong in organizations who continue to sustain top-level Safety lagging performance.

From the analytical research and careful review with this consortium of industry QHSE executives and Subject Matter Experts, the index that was found to best predict the Safety outcome performance of an organization included measurements of (a) proactive event reporting (Reporting Culture), (b) the discipline and consistency in execution of QHSE business processes, (c) the timeliness in closure of important corrective / preventive actions, and finally (d) the responsiveness of supervisors and line managers.

The paper will provide an overview of the analytical process that produced these results and will outline alternatives for applying these insights on leadership performance dashboards. Although the leading indicator research and analytics activities continue annually to strengthen the effectiveness of the leading indices, companies are applying the index to provide actionable leading metrics on executive scorecards. We will outline a sample process for implementing practical, "actionable" leading metrics in each key area to drive line manager / leadership behaviour and ultimately improve QHSE performance outcomes.

Overview and Background: Researching and Solving the Problem

An organization attaining true Operational Excellence has combined best-of-class business and financial performance with zero loss of production up-time, asset integrity, the environment, and safety. However, the frequency and severity of operational losses are significantly above zero for most corporations. Contributing to this is the lack of a standardized measurement of operational loss to enable performance comparisons and benchmarking.

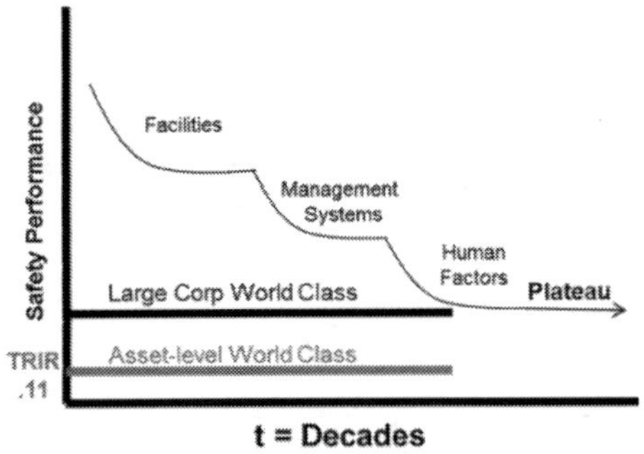

Although it is difficult to standardize the measurements of production up-time, asset integrity, and environmental outcomes across varying types of assets, the one area that does enable a relatively standard view of performance is **safety**. And since many industry executives hold the belief that the safest operating assets are also the best run operations, our research views the Operational Excellence problem through the lens of safety performance.

Figure 1: The Problem

Over the past few decades, the industry as a whole has seen safety incident frequency rates trend toward zero. But two key problems remain:

1. Even for the world-class performers, this trend has reached a plateau; and
2. The rate of high-severity injuries and fatalities has NOT followed the trend toward zero.

With these issues in mind, the focus is not only to drive Total Recordable Injury Rates (TRIR) to zero but more importantly to drive **severity-Weighted** Total Recordable Injury Rates (WTRIR) to zero. Whereas this goal of zero is perceived by most to be lofty at the aggregate or corporate-level, our research data set revealed operating assets (as depicted in Figure 1) which have sustained performance very close to zero.

In addition, the research data set revealed that the top 10% of assets as measured by WTRIR had the following interesting characteristics in comparison to the bottom 10%:

- WTRIR was roughly two orders of magnitude better
- Rate of near misses reported was 78% higher
- Incidents were reported 3 times more promptly
- Action items were authorized by leaders over twice as fast
- Rate of completed action items was 79% higher

Since these assets are sustaining near zero losses in the same risky environment with a similar level of safety exposures, this indicates the possibility of a much higher degree of Operational Excellence than is routinely achieved in industry.

A Path to Solving the Problem

With the top-performers raising the bar, a key challenge is to discover what the leaders, the workforce, and the overall organizations of these assets are doing differently. And the answer needs to come in a form that is measurable and actionable so that it can be repeated and applied to lower performing assets to continuously drive toward zero loss.

Specifically, the question that needs to be answered is: "What are the actionable, measurable factors which yield the near zero operational loss performance that is being achieved by these top performers?"

Using a severity-weighted safety incident frequency rate as the standardized outcome measure of operational losses, the research studies assets with performance varying from near zero to orders of magnitude worse to understand what creates high performing assets. The inputs in this research include data from workforce involvement in proactive activities such as audits, inspections, management system assessments, and observations as well as reactive events such as incidents, near misses, investigations and corrective actions.

By applying a rigorous statistical analysis process, the research reveals correlations between multiple leading factors and severity-weighted safety outcome performance. The focus is to discover those leading metrics which are not only predictive of outcome performance, but also are the most practical for use by top leadership as actionable, routinely measurable management controls. The process is continuously repeated with the aim of creating the optimum set of leading metrics which can be combined into a single, predictive multi-variable leading index.

The intended use of this index is two-fold:

1. To provide operating executives with actionable management controls for stewarding the process of continuously improving Operational Excellence performance; and
2. To deliver benchmarking of statistically proven leading indicators which reveal performance gaps and enable measurable goals and annual performance improvement plans.

Additional Obstacle: Management Addiction to Lagging Indicators

A primary obstacle to realizing the benefits of the above approach is top-level management's historical addiction to lagging indicators. Key Performance Indicators (KPI's) are management metrics used by business leaders to measure operational performance and to compel leadership to make performance improving operational changes. Lagging KPI's for QHSE are calculated from the occurrences of medium to high consequence loss incidents and provide a historical view of outcome performance. An often-referenced metaphor infers that using lagging indicators to drive performance is like "driving by looking in the rear view mirror". But in our research, it is extremely rare at the top-executive level to see anything other than lagging metrics of Safety performance used to judge performance.

Figure 2: Lagging KPI's from loss events

Per this depiction of the "Heinrich triangle", reporting data from incident events of the highest severity is routinely governed by a corporate standard (and often regulated) - thus rendering lagging data widely applicable and measurable across the enterprise.

In most companies, managers still use TRIR or a similar lagging metric as the primary KPI for judging an operation's Safety performance. One reason for this dominance is the practicality of having a standard in producing a normalized performance metric which can deliver an "apples to apples" comparison of these lagging KPI's across the enterprise (as described above). In addition to being routinely measurable, lagging measures are believed by management to be more quantifiable, tangible, objective, and generally more reliable than any leading KPI's proposed by the business.

Criteria of Leading KPI's as Management Controls

For business leaders to overcome this addiction and enable a leading KPI to truly rival lagging

KPI's as the preferred metric on a corporate-wide scale, the leading KPI must meet the following criteria to become a management control:

Actionable

To drive leadership behaviour, the leading KPI must offer the operational leader with a practical solution which can be acted upon to affect operational changes.

Objective

Surveys and opinion based assessments are too subjective. The best leading KPI's involve objective calcutions drawn from detailed activity or measurable events.

Normalized

TRIR enables and "apples to apples" comparison. Similarly, a leading KPI must be normalized to enable management comparisons across operating units.

Routinely measurable

The most effective and accurate leading KPI's are calculated from measuring leadership and worker involvement in routine operational activity on a relatively frequent (i.e. monthly) basis – as opposed to annual opinion-based surveys.

Believable and predictive

To attain the full commitment of field-level leaders, the leaders must believe that outcome performance can be affected with proper changes to performance in the leading KPI. Data and math can be used to assist leaders in developing this belief.

Throughout the aforementioned research and statistical process, the above criteria were applied to assure the resulting index of leading KPI's would likely be a cure. The rest of this paper will describe the research and business practices designed to overcome management's lagging indicator addiction.

Research Hypothesis

Per the premise described above, businesses are involved in the research to find the keys to preventing operational losses. These companies believe that an operational loss is a function of an organization's inability, either via failures in systems or behaviors, to effectively mitigate the risks involved in the loss. Therefore, the following is the research hypothesis:

"Operating units that are the best at reducing risks will sustain the best lagging Safety performance."

Risk Reduction Cycle (RRC) Database

An important characteristic of the research initiative is the availability of data pertaining to an organization's efforts, business processes to identify and **reduce the risks** to operational losses. The companies involved in the research have implemented a common enterprise-wide database application to consolidate data from business practices that fit the "risk reduction cycle" pattern. These companies are able to draw data from the database to observe the execution of these routine operational activities and are minimizing the effort of gathering leading data along the way. The following risk reduction process pattern of work practice activities applies to many QHSE business processes found in routine operations.

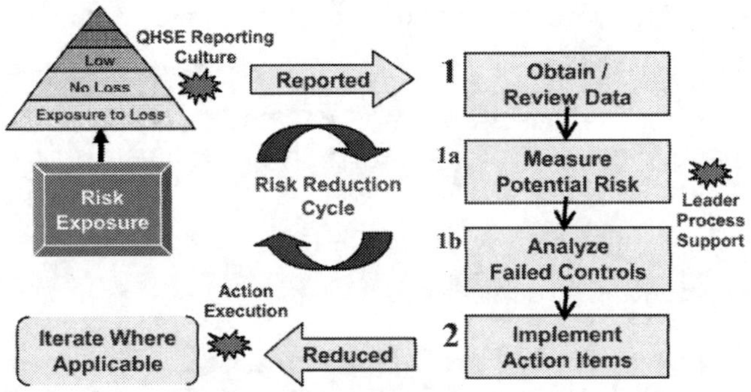

Figure 3: Risk Reduction Cycle pattern

As depicted, there are 2 key elements which are common to all risk reduction activities: (1) obtaining awareness via "reported" events and (2) implementing corrections to reduce the risk – i.e. "reporting and fixing" reported items.

Through researching operating facilities in the Energy industry, hundreds of event types have been discovered as risk reduction events – and thus contribute to the data set. The list ranges from multiple different types of incidents, near misses, and investigations to the many different types of inspections, observations, audits, etc...

The Core Elements of the Risk Reduction Cycle

The following describes how each element of the risk reduction cycle generally maps to the key steps in work practices resulting from both reactive (incident-based) events and proactive (assessment-based) events.

1. Obtain / review data (event reporting)
If you don't know about the risk, you can't reduce it. This element maps to the initial incident and assessment-based process steps whereby events are reported and people are assigned to contribute expertise / input and to review / approve the report.

1a. Measure potential risk (risk score)
For incident events, this is an optional / advanced practice of classifying "how bad it could have been" by using a risk matrix. Risk assessments, PHA's, and various other proactive processes include this as a key process step where the risk level is formally scored.

1b. Identify failed controls (root cause analysis / investigations)
For incident events, this element maps to the investigation process. Since many companies only investigate high-severity or high-risk events, this step is optional as well. An advanced practice is to apply an informal investigation to classify root causes and map those causes to Management System elements for ALL incident events – including near misses. For assessment-based processes, this element often represents the core purpose of the event – i.e. to identify the areas where risk control needs improvement.

2. Implement / repair controls (action item management)
For incident events and proactive events alike, this element maps to the most important step in all risk reduction processes – the execution of tasks to repair broken controls or implement new ones to ultimately reduce exposure to risk.

The research involves defining and calculating metrics across each of the above aspects of risk reduction process execution. Per the hypothesis, the intent is to identify the appropriate mix of metrics which both effectively measure an organization's risk reduction capability and are predictive of safety performance.

Researching a World-wide Data Set of Risk Reduction Processes

The companies involved in the research project provided their data on a scale that spans thousands of operating facilities in over 100 countries. Engaged in this initiative are world-wide operations from multiple Energy companies which have world-class QHSE performance... as well as leading companies in Oil Field Services, Chemicals, Manufacturing, and other industries where managing operational risks is a material concern.

Although most of the companies referenced in this study deploy unique company-specific programs for managing QHSE performance, they share one thing in common. Each of them is applying a common technology solution to collect and analyze data from a myriad of risk reduction process events such as incident investigations, near miss reports, management system audits, risk assessments, assurance reviews, behavioral observations, field-level inspection programs, hazard analysis, and many other processes. As depicted in the below diagram, the intended scope of this database is to be the organization's collection of all events that result in risk reduction actions.

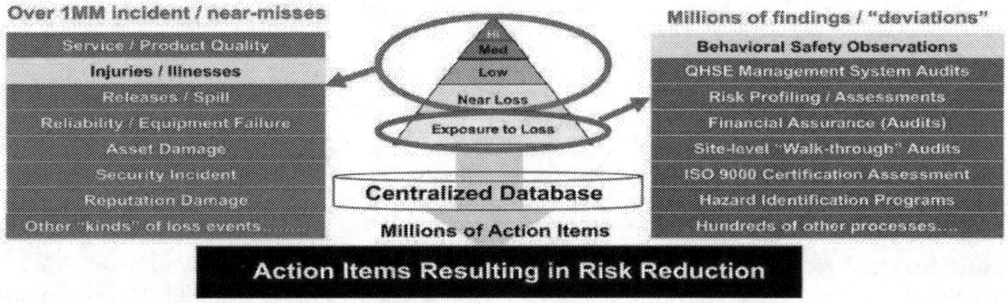

Figure 4: Risk reduction cycle events and actions database

On the surface, these organizations have simply been using a common "mechanism" to manage their own unique set of QHSE risk reduction processes and ultimately analyze the resulting data. However, at a deeper level, these companies are not only collecting data resulting from the event occurrences (e.g. incident reports, injury details, spill quantities, near miss types, root causes, audit results, assessment scores, inspection findings, etc.) but also the "work practice behaviors" reflecting the organization's tendencies in executing the processes behind such events (e.g. mean-times between completion of "critical" process steps, rate of leadership involvement, ratio of near-misses reported to high-consequence incidents, etc.). Applying statistical analysis to this vast data set has revealed the measurable organizational tendencies which are unique to top performers. The ultimate goal is to convince top business leaders to use these types of leading measurements.

From Research to Practice

Going forward in this paper, we will unveil some of the math-grounded process-based metrics which were discovered via the research initiative. Then, we review the critical role that each of these measurable areas play in the mission to proactively drive Safety performance. We'll end by showing the best approaches for applying these leading indicators on Key Performance Indicator (KPI) dashboards through an iterative measurement process to shift leadership focus, continuously improve QHSE performance, and attain Operational Excellence.

The Research Initiative: A Statistically Significant Data Set

Over the last several years, a collective data set of millions of records of data covering over a million direct and contractor employees of customer companies has been accumulated. As conveyed in the below diagram, the risk reduction cycle database manages both lagging and leading data. This has afforded the opportunity to study the relationships between lagging and leading data elements – with the purpose of finding statistically proven leading indicators.

The Opportunity to Correlate

The data set includes lagging metrics derived from high-consequence incidents as well as multiple sources of potential leading metrics including (a) leading events such as near-misses, low-consequence incidents, assessments, observations, action items, etc. and (b) the organizational tendencies in managing the business processes associated to such events.

NOTE: Although as depicted in Figure 4 low-consequence incidents and near misses are part of the incident event database, the organizational behaviours associated to reporting and managing such "non-mandatory" events are categorized in Figure 5 as leading areas of measurement.

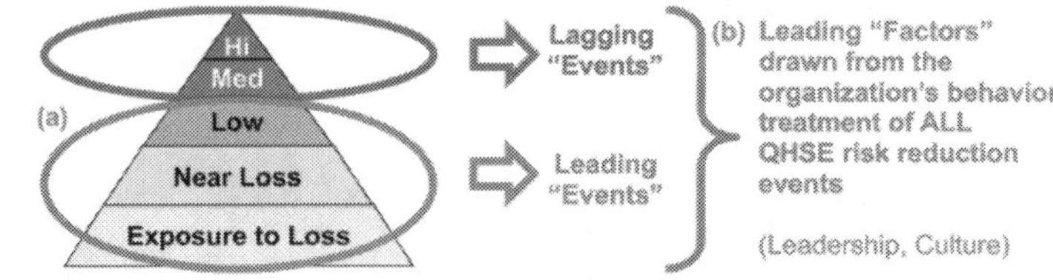

Figure 5: Sources of Data

With lagging and leading measurements coexisting in the same centralized database, a unique opportunity exists to practically, rigorously study correlations and identify math-proven leading indicators.

(a) A Database of Risk Reduction Events and Actions

As previously described, a risk reduction event is any event that starts with the (lagging or leading) discovery of a risk exposure and ends with the assignment and completion of action items to strengthen protections and ultimately reduce risk exposure. Examples of "lagging" events contained in the data set include near misses, incidents, and associated causal investigations from safety, security, environmental, quality, reliability (maintenance), and other types of loss events. Also included are "leading" events such as audits, inspections, behavioural observations, management system assessments, and over 400 other types of assessment-based events. Collectively millions of incidents, assessments, and action items have been accumulated over the past several years – establishing a rich, statistically significant data set for the research initiative.

(b) Data to Measure "how organizations manage risk reduction events"

The goal of these business process software tools is not only to collect the detailed data from risk reduction events but also more importantly to manage the full process lifecycle from initiation (reporting) of events through each key step in processing the event all the way to completion of the final corrective action. The research applies statistical analysis to this vast set of data to study the organizational tendencies in how the various parts of the organization are managing their roles in the risk reduction process.

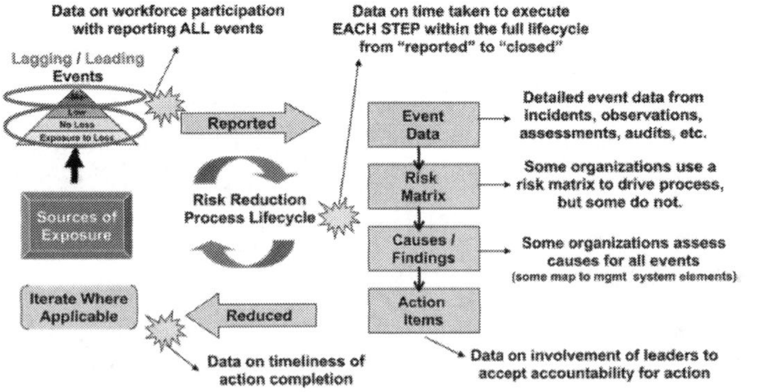

Figure 6: The Process-based Data

The diagram to the right depicts the many organizational touch points throughout the full "risk reduction process lifecycle" from which the metrics are being drawn. Through collaboration with Subject Matter Experts (SME's) from participating companies, over 200 different metrics have been defined in the research thus far – ranging from measures of reporting culture to various types of leadership and process-based metrics to measures of action item performance.

Research Premise: Find Opportunities for Improvement

The premise of the research initiative is to study relationships between the ordinary lagging metrics and the multitude of leading metrics as described above. The objective is to reveal measurable differences in the organizational factors which convey how organizations treat risk reduction events. These differences are reflected by substantial variation in metric scores when calculated across a group of organizations. Although the vast data set (described above) includes many different types of data elements for study, it is the leading metrics which expose such differences that represent the truly meaningful opportunities for improvement.

Getting to the Meaningful Data

The result of an enterprise-wide implementation of such a risk reduction process database is a practical means of collecting and analyzing data spawned from the event processes within the scope of the project. After establishing such a platform, some company executives have asked the following question:

"How do we get the meaningful data to enable our managers to lead performance improvement?"

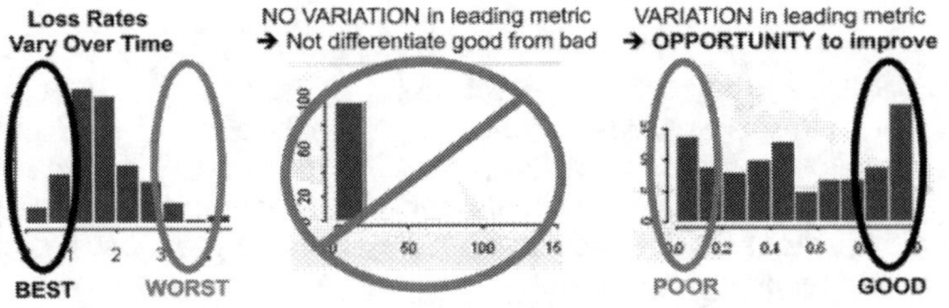

Figure 7: Meaningful Data

The best leading metrics are the ones for which the top performers (in terms of loss rates) are doing well and the worst performers are doing poorly. Because this variation exists in loss performance, a meaningful leading metric must show variation (e.g. example metric to far right) to represent an opportunity to improve.

Answering the Question: "What are the BEST performers doing differently?"

As conveyed in the above diagram, the histogram to the far right shows a metric with substantial variation – i.e. a material number of organizations have scores in each major segment of the performance spectrum. In this example metric (near misses as % of total incident events), organizations with a score near zero show up on the left side of the histogram and those with a score near 100 show up on the right side – i.e. some good performers and some bad.

For the best leading metrics, the majority of those organizations which show up on the "good" side of the metric's histogram are also present on the list of the BEST performers in terms of loss rates. These metrics represent the unique measurable characteristics of top performers and are the target of the research initiative.

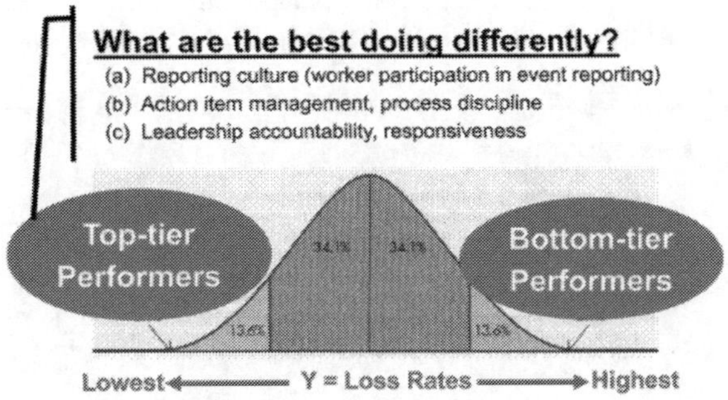

Figure 8: Finding the unique characteristics of the top performers

As conveyed in the diagram to the left, the research and benchmarking initiative has revealed a list of factors that are collectively predictive of performance and meet all of the previously listed criteria.

These factors were revealed after many repeated cycles of statistical analysis and review by SME's. The final list of the most meaningful metrics was then calculated in quartiles. From there indices were created to establish a leading performance measurement for each of the three factors. Finally the indices were combined into an overarching index to provide a single score which was found to be most predictive of WTRIR.

The Ultimate Leading Indicator: A Composite Safety Performance Index

The composite index is made up of the three dimensions: ***Reporting Culture Composite Index***, ***Action Item Composite Index*** and ***Leadership / Process Composite Index***. Each dimension is calculated from multiple metrics to form a unique measurement that results from the "tension" between the combined metrics.

Index from "tension" between metrics

For example, by combining a measure of the volume of action creation with a measure of discipline in on-time completion, there is tension between volume and disciplined execution – i.e. it is easier to complete a very short list of actions on-time than it is to carry such disciplined execution across a much longer list of action items.

Reporting Culture Composite Index = "culture of reporting and fixing"

This index results from the tension between the volume of "voluntary" event reporting and the organization's tendency to include actions as follow-up to reported events. In order to perform well in this index, an organization must not only sustain a substantial rate of near miss reporting (normalized by work hours) but also must follow-up reported events with actions. An organization must be strong in BOTH areas to be a good performer in the Reporting Culture Index.

Action Item Composite Index = "rate of action with timely execution"

This index results from the tension between the volume of actions generated and the organization's tendency to complete actions in a timely manner per the planned due date. In order to perform well in this index, an organization must BOTH sustain a relatively high rate of action AND continue to complete actions in a timely manner.

Leadership / Process Composite Index = "responsive, disciplined leadership"

This index results from the combination of supervisor / leadership involvement in two key steps in the risk reduction process lifecycle – i.e. the inital response to a reported event and the initial authorization of actions derived from events. To be a top performer in this area, an organization must have low lag times in BOTH of these areas.

The Importance of Reporting Culture

A common view of reporting culture is that it is an organization's total event reporting volume. Per the above index components, it is evident that event reporting "volume" has a critical affect on both of the first 2 above indices. Event reporting volume directly affects the Reporting Culture Index and indirectly affects the Action Item Index in that total action volume is minimized if reported events are minimized. Thus two-thirds of the total safety performance index discovered through the research project is affected by the volume of reported events – i.e. Reporting Culture.

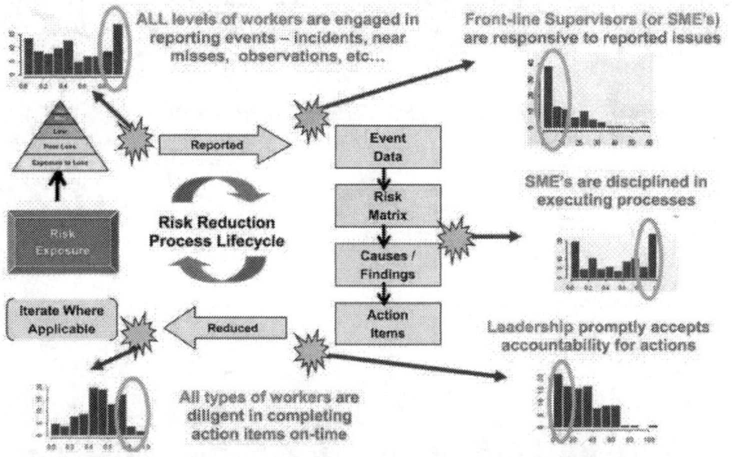

Figure 9: The unique tendencies of top performing organizations

Reading the histograms (to the right) in a clockwise manner not only portrays a list of math-grounded leading metrics but also tells a practical and believable story about organizational tendencies.

By studying the statistical relationships between the many leading metrics and WTRIR, it was revealed that organizations which perform well in each of these areas were much more likely to be top-level QHSE performers. Since these types of leading metrics are contained in the same database tool, they are applied as KPI's via the same tool to continuously improve the factors that lead to better performance outcomes.

Applying the Research Findings in Practice

Some companies have corporate databases which are justified on the ability "to produce better pareto charts of event data". But if these tools could directly DRIVE better QHSE performance, the true return on investing in these programs would be achieved. Many companies are preparing analytical reports for which some do include "meaningful" leading metrics. But are they getting the commitment of operations leaders to use those reporting tools to directly monitor and improve (i.e. measure and act upon) the math-proven factors that result in better performance? Leaderhip commitment to implement such leading KPI's as management controls is critical to drive improvements in QHSE outcome performance. Helping to get this commitment is a key purpose of the analytical research and the overall KPI measurement strategy.

Figure 10: Leadership as "Users" to continuously monitor and improve

To improve, leaders must be the KEY USERS of the continuous improvement solution.

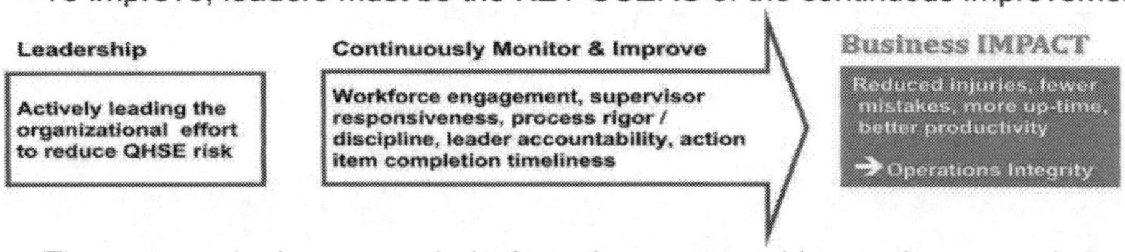

The research has revealed that the measurable math-proven keys to improvement are "organizational" in nature – i.e. workforce engagement, leadership accountability, etc.. Active management sponsorship is required to truly lead the required improvements in these areas. So, to achieve continuous improvement, it is imperative to get the data and math to help obtain management buy-in as well as the iterative benchmarking measurement tools to sustain their focus year over year.

Managed KPI Scorecard			
	Recordable Rate (Lagging)	Safety Triangle Ratio (as %)	% High Actions On Time
Div 1	2.50	37%	67%
Div 2	3.10	22%	62%
Div 3	1.70	83%	95%
Div 4	1.30	75%	77%
Div 5	0.95	93%	75%

Figure 11: Measure leading KPI's to DRIVE performance

What "leaders" measure and act upon gets improved. To obtain such active response from operations leadership, the KPI's must be carefully selected. Only those metrics which resonate at all levels of business management are included on the "managed KPI" scorecard. An example of such broadly applicable metrics is provided to the right. The managed metrics are typically calculated on a 12 month rolling basis and cascaded through multiple levels of the organization to assure improvements are realized enterprise-wide. But these only represent a sub-set of the leading indicators which have potential to affect performance outcomes.

Executive Analysis ONLY			
Predictive Index	Report Culture	Action Execution	Leader Response
4.79	2.10	24%	11%
5.25	1.40	17%	41%
17.10	7.50	54%	57%
7.10	4.10	72%	3%
22.0	18.30	46%	21%
5.20	6.10	64%	12%
10.24	6.58	45%	24%
4.10	1.50	93%	8%
9.70	6.30	71%	31%

Figure 12: Preparing top leadership to evolve the "managed" KPI's & continuously improve

To evolve management's focus year over year, another required component entails an additional set of metrics included only for analysis by members of the top-level executive steering committee. This example (to the right) displays a sample list of the math-grounded metrics discovered through the research. Although many organizations may not quite be ready to hold managers accountable for such a list of metrics, these additional benchmark KPI's may reveal valuable trends and help to prepare the top leadership to consider such metrics for inclusion as "managed" KPI's in the subsequent measurement cycle – establishing a continuous improvement process.

Iterating with Benchmarks to Continuously Improve

The traffic light metrics are derived from benchmark calculations. Companies may choose from various types of benchmark calculations during design of the KPI reports. Internal benchmarks provide comparisons to other organizational units' performance within the same company. And external benchmarks provide comparisons to the aggregate scores per other companies using the same database tools and involved in the collective benchmarking efforts. Additionally, a discrete scale may be used to establish the traffic light scoring effect.

Division/Site	TRIR	WTRIR	Preventive Index	Reporting Culture	Action Execution	Leader Response
Org Unit 1	0.60	0.60	1	1	1	3
Org Unit 2	0.44	0.44	1	2	1	2
Org Unit 3	0.47	0.47	2	2	2	4
Org Unit 4	0.76	0.76	2	4	2	2
Org Unit 5	0.00	0.00	1	1	1	2
Org Unit 6	0.41	0.41	2	2	3	2
Org Unit 7	0.78	0.78	2	2	3	2
Org Unit 8	0.42	0.42	3	4	2	2
Org Unit 9	0.23	0.23	1	1	1	2
Org Unit 10	0.91	0.91	3	2	3	4
Org Unit 11	0.51	0.51	2	1	2	2
Org Unit 12	1.80	1.80	3	1	2	2
Org Unit 13	0.55	1.64	3	3	3	2
Org Unit 14	1.42	1.42	3	1	2	4
Org Unit 15	1.81	1.81	4	4	3	3
Org Unit 16	0.63	2.31	3	4	2	3
Org Unit 17	1.00	4.98	4	2	4	4

Figure 13: Example annual benchmark report

Each year the research members receive a benchmarking report at the operating unit level. The purpose of the report is to identify performance gaps and improvement opportunities in the areas measured by the index of leading measurements.

By updating these statistics periodically, a continuous improvement platform is established to leverage the knowledge gained across all member companies. As more companies get involved in the benchmarking program and more business leaders are engaged in improving the math-proven leading indicators, the statistical inferences are continually sharpened and the overall community of collaborating companies benefits with improved QHSE performance results.

Summation

Through applying structured statistical processes and subject matter expert interpretations to a large statistically significant data set, it has been discovered that top-performing organizations (in terms of Safety outcomes) are the ones which have their leadership and front-line workers engaged in a collective effort to reduce operational risks.

This discovery may not be a surprise but rather a validation of existing beliefs. However by deriving these facts from a system which manages daily operational activities (e.g. incidents, near misses, investigations, observations, inspections, action items, etc.), a new opportunity is afforded operations managers to continuously measure and thus improve these organizational factors and ultimately to gain control of Operational Excellence and Safety performance outcomes.

In conclusion, the research discoveries regarding the leadership and cultural dynamics of top-performing organizations are certainly not new to the Energy industry. However the ability to institute management controls which enable all levels of leadership to measurably improve these factors on a routine, daily basis is likely NEW to most companies. But for those companies who are measuring and improving broadly applicable QHSE process-based KPI's and are engaged in the collective benchmarking programs to continuously improve, this ability has become a key part of their ongoing strategy to achieve and sustain world-class QHSE performance.

(34a) Process Development for Savannah River Site High Level Radioactive Waste

Dan P. Lambert[1], Michael E Stone[1], David C. Koopman[2] and Bradley R. Pickenheim[1], (1)Savannah River National Laboratory, Aiken, SC, (2)Process Tech Programs, Savannah River National Laboratory, Aiken, SC

The Defense Waste Processing Facility (DWPF) treats legacy nuclear waste generated at the Savannah River Site (SRS) during production of plutonium and tritium required during the Cold War. Approximately 35 million gallons of nuclear waste is first treated via a complex sequence of controlled chemical reactions and then vitrified into a borosilicate glass form and poured into stainless steel canisters. Converting the nuclear waste into borosilicate glass canisters is a safe, effective way to reduce the volume of the waste and stabilize the radionuclides.

Radioactive processing in DWPF began in 1996.

To date 3.7 million gallons of waste have been processed and 2,562 two-ton canisters of glass (10 million pounds) have been produced.

However, higher facility throughput is needed to shorten the processing life of DWPF.

The current processing flowsheet is unable to handle this higher throughput.

In addition, there are safety (hydrogen generation) and processing issues (pH and anion depletion due to catalytic formate decomposition) that are impacting production.

Savannah River National Laboratory was asked to investigate potential flowsheet improvements.

Flowsheet development was completed in four stages, (1) identification of alternative reductants and acids in simple screening tests, (2) extensive testing of two best alternative flowsheets, (3) selection of the best alternative using a systems engineering evaluation, and (4) lab-scale testing to determine optimal processing conditions and to evaluate downstream processing impacts.

Bench-scale flowsheet testing was performed to develop the new nitric-glycolic acid flowsheet as an alternative to the existing nitric-formic acid flowsheet currently used by the DWPF. Testing established that the new flowsheet can effectively reduce mercury and remove it by steam stripping with minimal catalytic hydrogen generation. All other processing objectives were met, including greatly reducing the Slurry Mix Evaporator (SME) product yield stress and increasing throughput compared to the baseline nitric/formic flowsheet.

The vast majority of the experiments performed to date have been completed using nonradioactive simulants.

A successful demonstration of the flowsheet with radioactive waste was completed in August 2011.

The paper will describe the process flowsheet development and the remaining work required prior to implementing the flowsheet in DWPF.

(34b) NIST Thermodata Engine: Tool for Chemical Product Design

Vladimir Diky[1], Robert D. Chirico[1], Chris Muzny[1], Andrei Kazakov[1], Ilmutdin Abdulagatov[1], Joseph W. Magee[1], Kenneth Kroenlein[1], Jeong Won Kang[2], Rafiqul Gani[3] and Michael Frenkel[1], (1)Applied Chemicals and Materials Division, National Institute of Standards and Technology, Boulder, CO, (2)Department of Chemical and Biological Engineering, Korea University, Seoul, South Korea, (3)CAPEC, Department of Chemical and Biochemical Engineering, Technical University of Denmark, Kgs. Lyngby, Denmark

ThermoData Engine (TDE, NIST Standard Reference Database 103) is the first product that implements the concept of Dynamic Data Evaluation in the fields of thermophysics and thermochemistry. That includes maintaining of the comprehensive and up-to-date database of experimentally measured property data and expert system for data analysis and generation of recommended property values along with uncertainties on demand. Different areas of chemical engineering, including product design, benefit from the services provided by TDE.

Chemical product design can be based on experimental data or property prediction methods based on chemical structure. TDE provides unprecedented collection of evaluated experimental data and uncertainty estimation that distinguishes it from any other product design tool. TDE product design functions include pure-compound design and solvent selection. Pure-compound design is search of a pure compound possessing required properties at specified conditions. A dynamically grown collection, now about 30,000 compounds, is available for search. User can restrict selection by experimental data or allow property prediction for the compounds with no experimental data. Properties supported are normal melting and boiling temperature, heat capacity, density viscosity, vapor pressure, etc.

Solvent Design function serves three tasks: (1) selection of best solvent for a solid solute, (2) search for a selective solvent for separation of a solid binary mixture, and (3) selection of best solvent for extraction. Solvents are selected from a pool containing more than 27,000 compounds. Selection is made by best efficiency (solubility or selectivity, depending on the task) and matching other requirements requested by the user. Efficiency criteria are evaluated based on experimental data for binary mixtures or predictive models (original and NIST-KT UNIFAC). Underlying experimental data can be accessed and inspected in order to prove the reliability of the models used in the selection.

Examples will be shown and benefits of ThermoData Engine for product design demonstrated.

(34c) Acetylation of Wood - New Tricks for 100 Year Old Technology

David C. Attride, Eastman Chemical Company, Kingsport, TN

The acetylation of cellulose was first practiced in the early 1900s. Application of the reaction is more complex in a solid, heterogeneous natural composite such as wood. The acetylation of southern yellow pine (SYP) imparts improved properties that are of significance in the building & construction area. This paper will provide background on how the process enhances and alters certain properties.

THREE KEYS TO SUCCESSFUL NEW PRODUCT and PROCESS DEVELOPMENT

Douglas H. Lenz, PhD

SABIC
SABIC Innovative Plastics
1 Noryl Avenue
Selkirk, NY 12158

Prepared for Presentation at the Spring 2013 AIChE Meeting, San Antonio, Texas, April 28- May 2

Unpublished

ABSTRACT

Managing the development and implementation of new technology or a new process from lab through start-up is an exciting challenge that is certainly the goal of many process engineers. However, very few engineers and chemists actually have the opportunity to manage and successfully implement new technologies that start out as an idea all the way through to manufacturing a profitable product. More common is the challenge to manage the re-development of an existing process or technology with the goal of making a product or process more profitable, safer, more environmentally compliant, etc. but the same approach can be successful.

Three key techniques to the successful management of process development will be discussed, using examples from a coated product / process R&D development program to illustrate the techniques. First, intellectual downstreaming was used, which helped to envision the ideal product and process to produce it. A cross-functional team comprising engineers, operators, management and consultants from the plant, divisional, and corporate levels accomplished this step. Second, a systematic program was managed by focusing initially on how to make the "ideal product" from both a fundamental first principles basis and ideal plant basis compared to the new customer requirements or the existing plant basis. Managing this "gap analysis" then lead to either the development of the R&D plan with decision points or "gates" or the incorporation of the correct and sufficient capital improvements necessary for the existing process to produce the desired material. Third, managing the superb execution of the all resources working in parallel or consecutively was required to minimize the process development or re-development time-frame while pursuing ideas and options that can eventually lead to revolutionize the process. These diverse resources to be managed included teams of chemists and external consultants at the lab-scale, teams of engineers & co-op students at the pilot and plant-scale, state-of-the-art surface science / powder technology techniques & analytical methods, and sophisticated computer models.

These techniques for the development of new and existing technology will be discussed in a case study. This case study describes R&D New Product & Process Development for a novel coated product that could consistently meet key customer quality criteria for a new application. Meeting constantly evolving market projections and new product specifications required equipment for this coating process that was larger than anything previously built and operated. Success was achieved through the utilization of the intellectual downstreaming, a systematic program using gap analysis and gates, and superb execution.

INTRODUCTION

Managing the development and implementation of new product or a new technology or both from lab through start-up is an exciting challenge that is certainly the goal of R&D. However, very few engineers and chemists actually have the opportunity to manage and successfully implement new technologies that start out as an idea all the way through to manufacturing a profitable product. Three key techniques to the successful management of new product design and process development will be discussed. The first key concept is intellectual downstreaming which helps to envision the ideal product and process to produce it. A cross-functional team is best for accomplishing this step. The second concept is seamless engineering which incorporates a systematic program with decision points or gates for managing how to make the "ideal product" from both a fundamental first principles basis and ideal plant basis compared to the new customer requirements or the existing plant basis. An equally important aspect of seamless engineering is managing the superb execution of the all resources working in parallel or consecutively in order to minimize the product design and process development time-frame. The final key consideration is the use of the combination of fundamental science, statistical tools and methods and sophisticated computer simulations and models throughout the design and development process.

Several[i ii iii] process innovation / process development articles have described many of the above elements and their benefits. These techniques for the development of new products and technology will be illustrated in a case study. This case study[iv] describes R&D new product & process development for a novel coated product that could consistently meet key customer quality criteria for a new application. An additional challenge to the project was the constantly evolving market projections and new product specifications that required equipment for this coating process that was larger than anything previously built and operated. Success was achieved through the application of intellectual downstreaming, seamless engineering, and statistical & modeling tools to supplement the science.

DISCUSSION: THREE KEYS FOR A SUCCESSFUL PRODUCT DESIGN\PROCESS DEVELOPMENT PROGRAM

While many factors are important and indeed necessary for the successful development of a new product or process, from lab to commercialization, this paper will highlight three important concepts that may not always be a part of an R&D program. These 3 factors are: 1) Intellectual Downstreaming; 2) Seamless Engineering; and 3) the 3 S's – Science, Statistics & Simulations.

The first key concept can be thought of as INTELLECTUAL DOWNSTREAMING. The critical objective of this R&D factor is to define the ideal product or process that will be the measurable end result of the R&D program and commercialization effort. It is an element of Steven Covey's 7 Habits of Highly Effective People – "Begin with the End in Mind"[v]. Ultimately, the group or the R&D effort should be avoiding starting with the approach that if you don't know where you are going, any route will get you there! Therefore, intellectual downstreaming is like reading a map where you see where you want to go, know where you are starting from and can quickly find the easiest or best route to get there. For a corporate R&D program, it will help to envision the ideal product and process to produce it.

An important tool to get the most value from the intellectual downstreaming process is a FOCUSED TECHNOLOGY REVIEW. These are technical discussions, organized and managed similar to the systematic "HAZOP" style safety reviews, which focus on the state of the art related to the defined ideal product or process. To be most successful, these focused technology sessions include as broad and diverse a group of participants as possible. Ideally, this cross-functional team comprises engineers, operators, management and consultants from the plant, divisional, and corporate levels accomplished this step. A corporate or high level director, who will not be part of the working team, should facilitate this activity so that the resulting set of ideas and action plans is as objective as possible and independent of any one person's or groups bias.

FMC frequently conducted focused technology reviews facilitated by the corporate technology director. This meeting format assures that the plans and actions for technology improvement and/or development are understood and supported by corporate management. It is always important to get management support at the highest levels possible as early as possible for a successful project or program. These technology reviews bring together plant, divisional, and corporate technical personnel and often specific outside experts or consultants for broad, thorough, often state-of-the art technical discussions. Often, each area of the process' technology is reviewed in detail and as part of the entire system using an approach similar to a HAZOP (Hazard and Operability Review) session.

The outcome will therefore be a consensus action plan. An excellent tool known well by project managers to both communicate this plan and help the group manage the plan is a GANTT CHART[vi]. A Gantt chart is a compilation of all the project task timeline including the critical path or paths. In this way, each project member understands the importance of their individual task and how it impacts the other members and the project goals and objectives. So it can be a very valuable tool for this product improvement and process optimization program by helping communicate project progress and/or barriers and resource needs. An example is shown below with the critical paths highlighted in red.

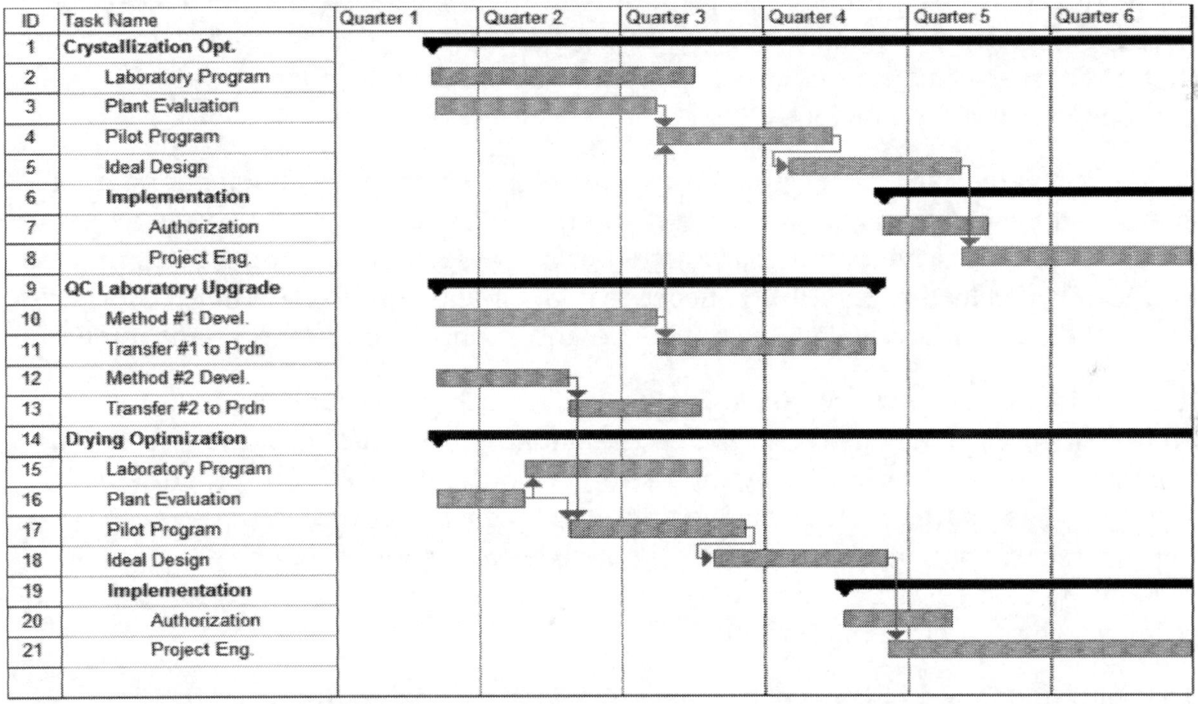

The second key concept is often referred to as SEAMLESS ENGINEERING. The "seamless engineering" concept was not developed at FMC, but with hindsight tends to be a part of many successful technical projects. Seamless engineering is more than just concurrent engineering or a group of individual projects. Seamless engineering can be thought of as a systematic program managed by focusing initially on how to make the "ideal product" from both a fundamental first principles basis and ideal plant basis compared to the new customer requirements or the existing plant basis. Managing this "gap analysis" then leads to either the development of the R&D plan with decision points or "gates" or the incorporation of the correct and sufficient capital improvements necessary for the existing process to produce the desired material. As demonstrated by the Gantt chart example above, it will be a combination of multiple, parallel or partial-parallel tasks like concurrent engineering.

The critical difference is that seamless engineering is an integrated program versus a series of individual projects handed over to the next team upon completion. The cross-functional, diverse team(s) that participated in the intellectual downstreaming now participate in making the defined "End" or Ideal Product/Process a commercial reality. To ensure success, this engineering effort must have a champion & sponsor leading and coordinating the multiple concurrent tasks. The program champion is involved in all efforts which facilitated communication of discoveries, ideas, problem resolution, project timing. The tools of first principle science, statistical analyses & methods and modeling or simulations (the 3 S's) are an integral part of a successful seamless engineering implementation. All these resources work in parallel or consecutively to minimize the program time frame. A final element of successful execution is superb communication fostered by frequent progress review meetings that often need to be no more than 1 hour on a biweekly or monthly basis or perhaps longer if on a bimonthly basis.

One of the values and benefits of a superbly executed seamless engineering effort is reduced development & implementation time. This result occurs because of the excellent communication and coordination from Production advising R&D during lab program which lead to the R&D concepts rapidly implemented by Production w/ minimal disruptions and all with confidence and support from upper management.

A second benefit is often obtained from reduced costs for recommended capital improvements. Since the technical fix accounting for "Whole" process or solution was developed and embraced by the team during the intellectual downstreaming, these designs are most often the best and most cost effective capital improvements, not necessarily requiring all new capital investment and certainly minimizing costly overdesigned solutions. In essence, it incorporates the tried and true concept of Right First Time (RFT) !!

The third and final key concept to be discussed is the incorporation of the 3 S's of SCIENCE & STATISTICS & SIMULATIONS or MODELING throughout the entire seamless engineering process. Many if not all R&D projects incorporate first principle science as part of the program. But to translate that science successfully and rapidly from the test tube to the customers application via a commercial process should also incorporate the tools of statistics and simulation or modeling.

What kind of statistics tools? Today many of these tools and approaches are found in the Six Sigma programs recognized and used by many corporations. These include designed experiments (DOEs) and statistical analyses on the lab, pilot, and plant levels. At every level, the baseline behavior of the process must be studied and documented to understand the measurement and process variability so the meaningful interpretations are obtained and statistically significant results can be found.

Often due to time or resource constraints, the fundamental science must come from plant level studies or the technology must be scaled directly from lab to plant scale. Now the cost of the experiment is high and the number and types of experiments must be very selective and well planned so as to not disrupt production or affect product quality. It is these situations where the 3 S's ensure a successful and useful result. Plant-scale DOEs are particularly effective as they can quickly determine the significant variables and/or interactions for that specific system from the resultant Scree plots, and estimate the effect of the variables on the process all while minimizing disruption to production.

Again, due to time and resource constraints, a full understanding or scale-up of the technology may not be achievable even with science and statistics. This situation is addressed with process simulation and/or product modeling. Examples of these can be heat & energy balance models, regression or multivariate analyses, fault-tree analysis, and dynamic behavior models.

Why model? Once a good model or simulation is developed and validated with actual data (ideally from multiple scales such as lab, pilot and plant data), then these simulations can improve process understanding, allow for an assessment of wider range of process conditions and atypical &/or new conditions without disrupting production. The end result is a more complete understanding and solution with reduced engineering time and costs!

UNUSUAL PRODUCT DESIGN & DEVELOPMENT CASE STUDY: CREATING A STABLE SOLID OXIDIZER USING AN AQUEOUS COATING AT HIGH TEMPERATURE IN THE "WORLD'S LARGEST" COATER

Extensive laboratory research determined that only a coated peroxygen product (CPP) could consistently meet key customer quality criteria for a new application. The preferred coating was applied as an aqueous solution onto a fluidized bed of this solid reactive oxidizer at high temperatures – a very difficult challenge! However, the commercial scale-up would be too capital intensive for the initially demonstrated Wurster, or bottom, coating technology (Figure 1). Top-coating technology was the method of choice for a commercial scale plant (Figure 2).

FIGURE #1: Wurster Technology

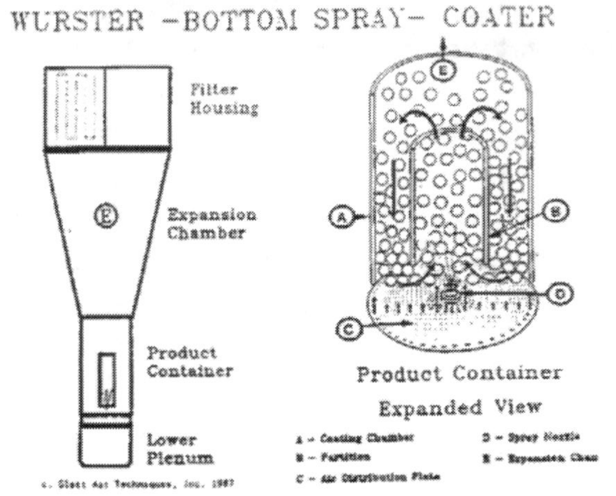

FIGURE #2: Top-Coater Technology

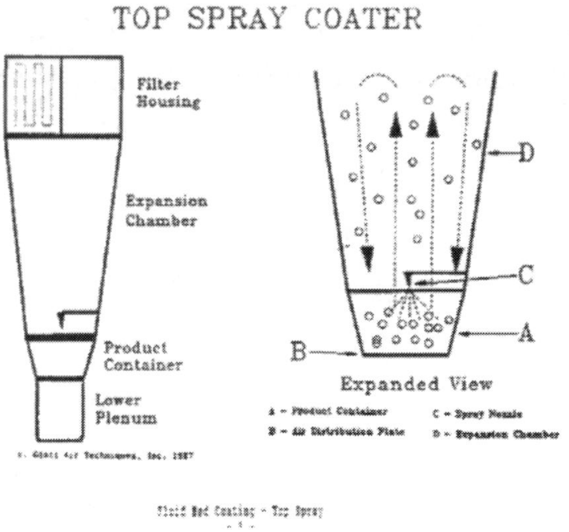

This new top-coater technology required a rapid demonstration and scale-up program for determining its viability as an alternative for coating the reactive material to keep the process / product development program on track. Success was achieved due to several factors including the 3 key factors described above: defining the ideal product; a cross-functional, responsive team took charge of the project up-front; the project had a champion; piloting, scale-up, lab work, and modeling were conducted very often in parallel both within FMC and at several equipment manufacturer's sites; fundamental science and statistical tools were utilized throughout the program; identified problems & issues were communicated early and resolved by the team rapidly.

Early in the program, the team created the ideal product definition for this application of a solid oxidizer (bleach) in a detergent box that could survive long periods at warm, humid conditions (a laundry room). These criteria included measurable specifications for 1) Shelf Life / Stability, 2) Active Oxygen, 3) Bulk Density and 4) Coating Level. Knowing this target upfront was important as the measurable specifications often changed throughout the project due to changing legal and customer requirements.

The cross-functional team comprised R&D chemists, R&D process engineers, project engineers, plant engineers, marketing & sales representatives, and legal representatives from within FMC. But as importantly for success, external consulting and support was provided by tollers and equipment manufacturers as well.

The program had a champion. The champion was responsible for 1) planning & coordinating all project activity; 2) ensuring resource availability; 3) coordinating the data sharing & analyses; 4) communicating program progress as well as issues and 5) understanding all aspects of the technology (very important!).

After an extensive program, top-coater development and scale-up was ultimately successful. Initial feasibility & scale-up trials were successfully conducted at a manufacturer's site. However, the initial semi-work scale-up trials were unsuccessful as an excessively agglomerated, unstable product was produced. The two main reasons for the set-back were investigated and identified. A designed experimental program was conducted to establish proper process conditions for successful scale-up and to understand the causes of the excessive agglomeration observed. A model was subsequently developed, based on a first principles understanding gained from the statistical designed experiments, to correlate key parameters to final bulk density (an agglomeration indicator). Extensive scale-up work was conducted with this technology for FMC's CPP product resulting in the successful installation and operation of semi-works plant. Whereas this semi-works size was considered commercial by the manufacturers, our commercial plant would require an innovative design & operation for an order-of-magnitude larger coater which had never been built at that time.

The top-coating technology development program was initiated after the successful scale-up at the Coating Place of the Wurster technology for coating CPP with FMC's preferred aqueous coating. If the Wurster technology had proven unfeasible, top-coating technology would not have even been considered.

Initial feasibility & scale-up trials were successfully conducted at Glatt Air Techniques, Inc. A basic three variable full factorial (8-run) designed experiment (DOE) was conducted at the lab scale which identified key parameters and conditions and potential problem areas. The real value of incorporating a statiscal-based tool early in the program, such as the DOE rather than conducting Edisonian trial & error testing is that one can quickly determine the significant variables and/or interactions, can estimate the effect of these variables on the process and reduce the development time & costs. Scree plots from this DOE are shown in Figures #3 and #4.

Figure #3: DOE Scree Plot for Coating Effectiveness

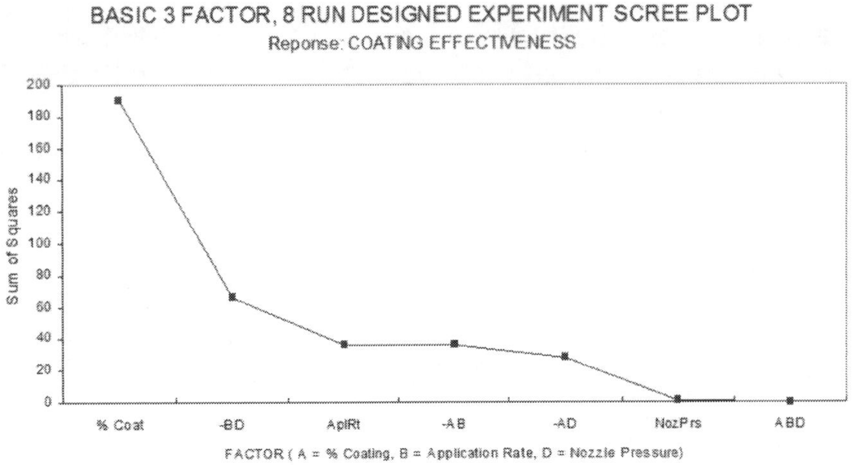

Figure #4: DOE Scree Plot for Coated Product Bulk Density

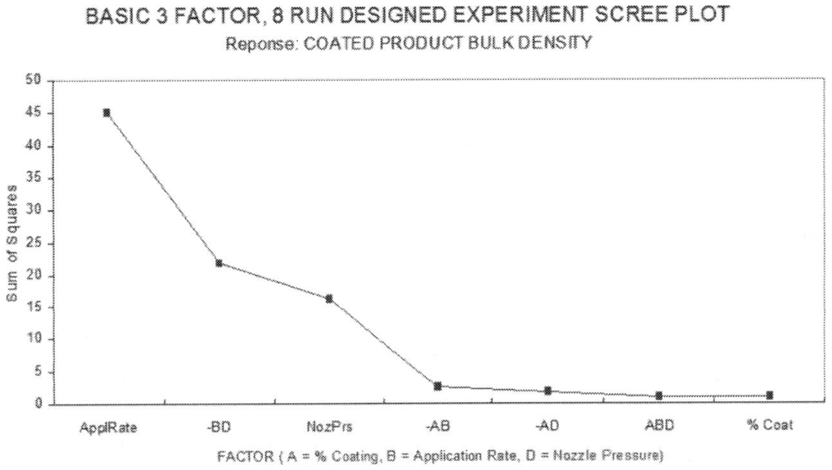

However, Murphy's Law often prevails at points throughout a development program but the use of the 3 KEYS can mitigate the consequences. The initial semi-work scale-up trials also conducted at Glatt were unsuccessful. An excessively agglomerated product was produced, with bulk densities well below the minimum target. The two main reasons for the set-back were; 1) using a nozzle design not suited for our product on this scale; 2) using semi-works scale top coater not properly designed for our product needs with respect to multiple nozzles, heat input, disengagement height, and filter design. The excessive agglomeration occurred because the coating zone was oversaturated due to the combination of process and design conditions. Two key causes of this oversaturation were the nozzle design and the unit design specifications. The likely mechanisms for oversaturation were either more overlapping of the zones than desired, or localized high humidity concentrated in the center of the coater only, or both. Heat input was limited due to design limitations of fluidizing air volume & temperature, disengagement height, and filter design encountered with our product. The coater design needs to take into consideration both the CPP particle size distribution and the coating solution properties. Attempts, listed above, to obtain adequate heat input resulted in filter pluggage with the semi-works unit, further limiting the amount of heat input.

Four major factors play a role in the dynamics of spray coating. These factors are 1) liquid/coating properties and characteristics; 2) substrate properties and characteristics; 3) bed humidity; 4) agglomeration versus coating dynamics.

The liquid/coating properties primarily affect the size and behavior of the droplet during the coating process. Due to properties of FMC's CPP substrate material, only aqueous solvents were considered. The physical properties of this coating solution as a function of temperature were measured and well understood which contributed to the success of this program.

Prediction of droplet size is well known[vii]. For a two-fluid atomizer a well-accepted equation[viii, ix, x] for predicting the Sauter Mean Diameter, D32, (in microns) is

$$D_{32} = \frac{1920}{U}\left(\frac{\sigma}{\rho_l}\right)^{0.5} + 597\left(\frac{\mu_l}{\sqrt{\sigma \rho_l}}\right)^{0.45}\left(1000\frac{L}{G}\frac{\rho_g}{\rho_l}\right)^{1.5} \qquad 1$$

where U = liquid to air relative velocity (ft/sec); L/G = liquid to gas mass ratio; s = liquid surface tension (dyne/cm); r = density, liquid - l or gas - g (g/cc); m = liquid viscosity (poise). Note that the equation is not dimensionally correct.

As can be seen, droplet size is a strong function of liquid velocity (spray rate & port diameter), atomizing air velocity (air pressure) and a weaker function of the temperature & physical properties of the liquid. For example, a 10 C increase in solution temperature decreases the mean diameter roughly 10% in the 40 C - 60 C range. The mean droplet size of the coating solution as a function of atomizing air pressure, liquid spray rate, and port diameter were all determined.

Particle size, friability, porosity, and solubility can all impact coating dynamics.

Particle size influences surface area. The smaller the particle, the more coating is required to achieve an acceptable film thickness, as determined by performance and customer specification. Larger particle sizes are preferred.

Particle friability can lead to particle attrition. The fine particles produced can accumulate in the layers of film, possibly decreasing its uniformity and effectiveness. The fines may also agglomerate with larger particles, requiring possibly higher percentage coatings to assure some minimum film thickness.

Porosity would only be a factor for solids with high surface porosity and large pore sizes. In this situation, it may be difficult to bridge the pores and additional coating would be required.

If the substrate is soluble in the spray solution, the chance of increased core penetration exists. One remedy for this is the concept of staged coating. Initial coating conditions can be low spray rates at possibly higher inlet air temperatures until the product is protected by a thin film. Then the spray rate can be increased and the temperature decreased to optimize coating quality without core penetration.

The humidity of the bed influences the evaporation rate of the coating solution both as a droplet and as a surface film. A balance is required, as discussed above. The bed humidity is a function of fluidization air volume, temperature, and dew point, plus coating solution concentration and spray rate.

The fluidized bed top spraying machine is designed to perform as either a top coater, an agglomerator / granulator, or a spray dryer. Therefore, it is critical to understand the mechanisms and conditions which cause or influence agglomeration and operate the top coater at conditions to minimize this effect.

Three major mechanisms are responsible for particle agglomeration or bed granulation to occur. Each mechanism involves material bridging and are the following:

1) Agglomeration by partial wetting/recrystallization;
2) Agglomeration by deposition of a film forming binder;
3) Agglomeration by layering

Agglomeration by wetting and recrystallization occurs at wet bed conditions of high humidity, high spray rate, low temperature. Highly porous and friable granules are produced.

Agglomeration by the deposition of a film forming binder is the most common mechanism. It is possible for solutions intended as coatings to become binders if the coating solution becomes viscous enough to inhibit proper spreading and remain tacky. Generally, granules produced are still porous but stronger, less friable.

Agglomeration by layering is the least common and controllable mechanism. It involves formation of a very wet agglomerate of fine particles engulfed by a larger droplet which become tightly packed upon drying. These small agglomerates coalesce and grow with a "snowballing" effect. Again, excessively wet bed conditions are required. Granules produced are typically very dense with minimal porosity.

Three key conditions are required to minimize agglomeration during the coating process. These conditions are small droplet size, low bed humidity, and short droplet travel distance from nozzle port to fluidized particle(s).

The concept of the coating zone is key to understanding the complex interactions involved during the coating process and to reliably scale-up the technology. The coating zone is the area in the immediate vicinity of the nozzle port where the liquid droplets mix with and contact the solid particles. Localized conditions within this zone influence the coating dynamics, not overall bed or process conditions. Once the proper combination of process conditions are determined for a single coating zone, then scale-up is accomplished primarily by multiplying the number of coating zones and scaling the conditions accordingly. The coating zone concept[xi] is shown schematically below.

Figure#5: Coating Zone Schematic

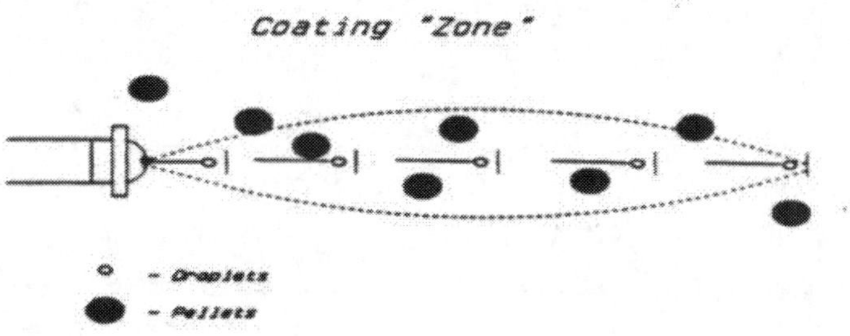

The coating solution spray or application rate is almost always dependent on the "saturation" of this zone and not the saturation of the fluidizing air. The same variables which control bed humidity influence the saturation of the coating zone. If the coating zone is saturated, then agglomeration will occur via one of the above mechanisms. This saturation condition can occur even if the fluidization air is not saturated and still has excess drying capacity. The remedy for this situation is to increase the number of zones by using multiple nozzles wands and/or multiple port nozzles. With this modification, the application rate can be significantly increased and/or the excess drying capacity can be utilized.

At this point, prior to further technology development, extensive technical and commercial discussions were held between FMC and the equipment manufacture personnel. Critical scale-up and design issues were discussed. These scale-up and design issues were:

- FLUIDIZATION AIR VOLUME/VELOCITY - More air is needed for fluidization and water removal.
- FLUIDIZATION AIR TEMPERATURE/HEAT INPUT - More heat is needed for drying. Increased air volume will allow for additional heat or greater flexibility in providing adequate heat.
- EXPANSION CHAMBER HEIGHT/TDH - More disengagement height is required to minimize fines loss and to minimize fines and/or coating solution droplets plugging the filter bags.
- FILTERS: MESH SIZE - Large, course filters are preferred (20-25 micron). Larger mesh sizes will decrease the pressure drop through the filter system allowing for higher air volumes. Larger mesh sizes will also minimize the filter plugging problem.
- NOZZLE DESIGN & CONFIGURATION - The key is to determine the conditions which provide for an effective coating zone at the nozzle, irrespective of application rate. Then if the application rate from one nozzle is less than required, the number of nozzles is increased until the required rate is obtained. A unit with multiple nozzle capabilities was required. Essentially, this was the approach learned in developing and scaling the Wurster technology. Once effective coating zone conditions were established on the lab and pilot scale, this same nozzle design was replicated to provide the desired application rates with the semi-works coater. It is important to recognize that technology scale-up set-backs can provide valuable learning experience and improved design data.

Incorporating the key concept of the 3 S's (Science, Statistics & Simulations) proved invaluable at this point in the program due to the complexity of the fundamental science and the shortened project timeline. Based on the first principles discussed above, both lab and pilot designed experiments were conducted. Besides the using the DOE methodology, another statistically-based method was used on the pilot scale know as self-directed optimization (SDO). The resulting data was further analyzed by simulation and statistical modeling to prepare as complete an understanding as possible prior to making a second semi-works trial which needed to succeed for the program to advance.

This pilot-scale designed experimental program was conducted to establish proper process conditions for successful scale-up relative to the coating zone concept and to understand the causes of the excessive agglomeration observed. The coating zone studies were designed as modified self-directed optimization (SDO) studies.

Effective coating zone conditions were determined using 8 runs by examining three fluidization air rates/velocities, two solution application rates, and two nozzle pressures. The application rate and nozzle pressure combinations yielded two ranges of droplet sizes. The air rate application rate combinations provided five drying capacities. The bed temperature and solution temperature remained constant. This set of conditions allowed for the replication of previous effective and poor coating zone found in the lab DOE while also exploring the limits and evaluating a preliminary bulk density prediction model. The SDO matrix of conditions is shown in Table 1, including run numbers and measured bulk densities.

TABLE #1: Modified Coating Zone SDO Matrix & Bulk Density Results

High Spray Rate (lb/hr/mm2)

Air Volume / Nozzle Pressure	Low (Bar)	High (Bar)	Drying Capacity
High (SCFM)	Run 4: 48 lb/ft3	Run 3: 56 lb/ft3	High (BTU/lb)
Middle (SCFM)	Run 2: 48 lb/ft3	Run 1: 52 lb/ft3	Middle (BTU/lb)
Low (SCFM)	Run 6: 44 lb/ft3	Run 5: 51 lb/ft3	Low (BTU/lb)
Droplet Size	Large (Microns)	Small (Microns)	

Low Spray Rate (lb/hr/mm2)

Air Volume / Nozzle Pressure	Low (Bar)	High (Bar)	Drying Capacity
High (SCFM)			High (BTU/lb)
Middle (SCFM)	Run 8: 50 lb/ft3	Run 7: 56 lb/ft3	Middle (BTU/lb)
Low (SCFM)			Low (BTU/lb)
Droplet Size	Large (Microns)	Small (Microns)	

As can be seen, higher bulk density (less agglomeration) is favored by higher nozzle pressure, lower spray application rate and higher air volumes/velocities. In terms of the coating zone, desired bulk densities are favored by smaller droplets combined with higher drying capacities. The data from all top-coating runs was also regressed and correlated into a bulk density model.

A model was subsequently developed to correlate key parameters to final bulk density (an agglomeration indicator). This bulk density model was developed for two reasons. First, to understand, control and minimize agglomeration by identifying the critical process parameters which lead to low bulk density. Second, to provide a predictive tool for evaluating coater designs and establishing future commercial process conditions.

The bulk density model was derived statistically. Several techniques were used. First, a database was constructed from all the top-coating trial process conditions and results. Using SYSTAT, Pearson correlations were calculated for bulk density with 28 variables and 22 potentially relevant interactions. The resulting correlation matrix primarily identified the variables & interactions to remove from further consideration. The matrix also provided a measure of the interdependence or intercorrelation of all the remaining variables. This information was useful in understanding possible compounding effects in the model. Multiple linear regression (MLG) was performed with the remaining variables and interactions. The goal was to derive as simple a model as possible which maximized the correlation (R, R2) and the F-Ratio (indicating minimized model residuals). Fifty (50) models were derived based on various combinations of the variables. In some cases, the regression was performed on large subsets of randomly chosen variables. In some cases, the regression was performed on smaller subsets of specific variables selected based on coating and agglomeration theory. Both forward and backward step-wise MLG's were done to validate the most promising models.

Ultimately, three models were generated in this manner. Each contains some subset of only eight variables and three of their interactions out of the original fifty. These key variables and interactions are:

1. Solution Application Rate per Port
2. Atomization Nozzle Pressure
3. Fluidization Air Volume
4. Fluidization Air Velocity
5. Fluidization Air Temperature
6. Total SPP charge
7. Solution Droplet Size
8. Heat Input
9. Zone Contacting Time(Interaction)
10. Droplet Evaporation Dynamic (Interaction)
11. Zone Drying Capacity (Interaction)

Model-by-model comparisons were made of the actual measured density versus the calculated density and of each model's residuals or fractional error. For the final models, the fractional errors are predominately less than 5%. The simplest and the best models are summarized below.

MODEL 1: R = 0.84 F = 45.3

BULK DENSITY = f (Air Flow, Mass Solid, Air Temperature, Droplet Size)

MODEL 25: R = 0.92 F = 87.6

BULK DENSITY = f (Air Flow , Mass & Volume Solid, Spray Rate, Nozzle Pressure, Drying Capacity)

Each model was tested for its predictive capabilities using a different top-coater trial bulk density results. Three models provided very good to excellent agreement with the Vector trial data, with Model #25 being the best. Each model predicted that the commercial designs can provide product with acceptable bulk density. The models assumed that nozzle positioning and height are properly designed.

This model was successfully used to determine scale-up conditions which provided a good coated product. With the information provided by the model, a successful semi-works scale-up trial was conducted using a different Glatt coater. This coater design was a closer match to our product needs with respect to multiple nozzles and additional heat input.

The coating process variables can be divided into six broad categories. Many of the variables have been identified above, but will be reiterated for completeness. The categories, including all relevant variables, are:

1) Coater Design - Wurster or Bottom Spray; Top Spray
2) Atomizing Air Conditions - Pressure; Volume
3) Nozzle Design - Port Size; Ports per Nozzle Head; # of Nozzle Wands; Spray Angle; Nozzle Position or Height (relative to the static bed)
4) Fluidizing Air Conditions - Volume; Superficial Velocity; Dew Point; Temperature
5) Coating Solution Conditions - Application/Spray Rate; Solids Concentration; Application Time; Temperature; Physical Properties (Viscosity, Surface Tension, Specific Gravity); Solvent
6) Solid Characteristics - Particle Size/Surface Area; Friability; Porosity; Solubility (relative to coating solution)

The critical variables for coating in general, either by Wurster or top coating, are atomizing air conditions, fluidization air dew point and temperature, and solution spray rate.

The atomizing air pressure and volume were selected so that the droplet size is small relative to the particle to be coated. The higher the pressure and volume, the smaller the droplets. However, above 3-4 bar (45-60 psi) pressure has a minimal effect on droplet size. Also, if the air velocity is too high at the nozzle tip, the shear forces may disrupt the fluidization pattern in the coating zone or damage friable particles.

The fluidization air temperature and dew point are critical for controlling the solvent evaporation rate and maintaining the desired product or bed temperature. These parameters in turn dictate the degree of saturation in the coating zone. The higher the air temperature, the less seasonal variation on drying capacity and the lower the bed humidity.

The spray rate influences both the droplet size and the bed humidity. The desire is to determine the maximum spray rate possible without saturating the coating zone. The trick is that at lower spray rates, the droplet size is smaller and bed humidity is lower. These are the conditions that favor coating over agglomeration with a top spray machine. For scale-up, the increase in spray rate is often proportional to the drying capacity.

Three additional key process variables for top coating are the fluidizing air volume, superficial velocity, and solution concentration.

The air volume/velocity control the fluidization pattern in the bed and the particle velocity through the coating zone as well as the excess drying capacity. It is important that the volume/velocity be high enough to provide smooth fluidization but not slugging. Slugging would disrupt the dynamics of the coating zone, effecting coating quality, and should be avoided. On scale-up, the bed depth should be maintained so as to prevent slugging. Expanding the bed diameter, while maintaining a minimum air velocity, is one way to accomplish scale-up.

The concentration of the coating solution needs to be optimized. Generally, lower viscosity and less water/solvent are preferred to minimize agglomeration. However, higher concentrations containing less water tend to have higher viscosities.

For design and scale-up of a coating technology system, the following factors need to be well understood:

A. BATCH SIZE - Typically the target is approximately 80% of the bowl capacity for final coated product. Process conditions are not scaled-up according to batch size.

A critical scale-up parameter is the nozzle height relative to the static bed. This distance should remain the same as in the lab/pilot units if the fluidizing velocity remains the same. If too high on start-up, excessive agglomeration or spray-drying can occur since the nozzle will be farther from the bed and spraying into a more disperse substrate pattern.

B. FLUIDIZATION PROPERTIES - Since bowl volume, bowl diameter, and bed depth increase on scale-up, the key is to maintain smooth fluidization and minimize slugging. If smooth fluidization is achieved in the lab/pilot units, scale-up is based on the corresponding air velocity (screen area) or air volume, whichever is lower to be conservative. Scale-up is not based on the increase in batch size. For example, if the scaled-up coater unit has 5x the bowl volume but 3x the bowl screen cross-sectional area, then the air volume is increased 3x to maintain the same face velocity, not 5x to maintain the same air/solids ratio.

Slugging is caused primarily by large air bubbles and is not usually seen in lab/pilot units due to the shallow bed depths and wall effects. Air bubble growth rate increases with increasing bed depth, particle size and density. In general, the SPP substrate properties were not in the range (diameter larger than 1 mm, density at or above 1 g/cc) which is known to cause fluidization problems for top-coating scale-up.

C. COATING ZONE: DROPLET SIZE - Droplet size is established from the spray rate, atomizing air pressure, and port size in the lab/pilot units. The droplet size should be maintained in the larger coater. This is done by simply using multiple nozzles wands with multiple heads while maintaining the same per port spray rate and atomizing pressure.

D. COATING ZONE: SATURATION - Agglomeration due to saturation of the coating zone often occurs well before the fluidization air is saturated. Again, the use of multiple nozzles can take advantage of the excess drying capacity. Key to start-up will be the positioning of the multiple nozzles so that minimum overlapping of the spray pattern occurs. Spray angle may need to be modified to achieve this minimum overlapping. Standard nozzles have a 60 degree spray pattern. However, special nozzles with 45 degree spray patterns exist.

E. SOLUTION SPRAY RATE - Spray rate is generally scaled proportionally to the increase in drying capacity. Drying capacity is determined by the fluidizing air volume and heat exchanger design. Since air volume scale-up is based on maintaining the same velocity, initial spray rate scale-up is based on the same factor. From the above example, the initial spray rate increase would be 3x. However, this factor may require modification if the coating zone is saturated. For example, additional nozzles and/or additional heat exchanger duty may be required to achieve the targeted or design application rates.

F. PRODUCT TEMPERATURE - The same product temperature should be maintained as used in the lab/pilot units. The inlet air temperature is adjusted or balanced to achieve this condition.

G. MASS EFFECTS - The larger batch sizes and bed depths associated with scale-up may lead to increased friability. The friability will be dependent on the strength or robustness of the substrate, and on the batch cycle time (dictated by the spray rate and coating level). Higher spray rates and lower coating levels would be beneficial for more friable material.

A phenomenon called "smushing" can also exist at larger scales. The mass of the bed may tend to compact weakly agglomerated particles, maintaining or increasing the bulk density. Since we want to minimize agglomeration, this effect may be beneficial for our product or may offset excessive agglomeration.

As an approach to reducing the scale-up risk and increasing the probability of commercial unit success for this technology, alternate top-coater vendors were evaluated to assure that the final commercial-size coater would be meet our design needs and be cost effective via competitive bidding. Vector Corporation and Fluid Air, Inc. were identified as potential vendors in addition to Glatt Air Technique. Both companies had different designs than Glatt and so scale-up trials were conducted by both at each manufacturing site. Vector's trials were successful whereas Fluid Air's trials were not for our particular product.

Commercial designs were obtained based on the results of the scale-up trials and additional engineering specifications. The final design was chosen based on its robustness and the manufacturer's experience in building industrial fluid-bed coater/granulators. Both manufacturers' designs addressed our capacity requirements based on current and potential optimized conditions. These potential optimized conditions included larger bed charges, higher solution concentrations, faster application rates, and/or lower coating levels either alone or in some combination.

CONCLUSIONS

1. Three Key Factors for successful product design and process development are intellectual downstreaming; seamless engineering and combined use of the 3 S's.
2. Good quality coated reactive peroxygen product was produced using top coating technology which had been successfully scaled-up from the lab to the semiworks.
3. Acceptable commercial scale top-coater designs for coating CPP were obtained from multiple fluid-bed coater manufacturers.
4. Successful process / product development of this innovative coating technology within the required project timeframe was due to the following key factors: defining the ideal product; a cross-functional, responsive team; a project champion; multiple parallel development efforts; systematic utilization of fundamental science, statistical tools and simulation & modelling; rapid resolution of process issues & problems.

---

[i] Basu, P.K., Mack, R.A, and Vinson, J.M., "Consider a New Approach to Pharmaceutical Process Development", Chem. Eng. Progress **95** (8), August 1999

[ii] Mukesh, D., "Succeed at Process Development", Chem. Eng.

Progress **95** (2), February 1999

[iii] Worstall, J.H., Adkison, W.L., and Durante, D.R., "Speed Your Process Development", Chem. Eng. Progress **94** (12), December 1998

[iv] Lenz, DH, Guiliano, BA, Terzian HD, Juelke, CV, and Hills,WA, "CREATING THE "WORLDS'S LARGEST COATER" FOR A NEW REACTIVE PEROXYGEN PRODUCT: A PROCESS INNOVATION CASE STUDY", Spring 2000 AIChE Meeting, Atlanta, GA, March 5-9

[v] Covey, S., "The 7 Habits of Highly Effective People"

[vi] Zeldman, M., "Keeping Technical Projects on Target", 3rd Ed., © 1998, EMZEE Associates, Inc.

[vii.] Perry's, 5th Edition, "Phase Dispersion", pages 18-58 to 18-67

[viii.] Nukiyama, S., and Tanasawa,.Y., "Experiments on the Atomization of Liquids in an Air Stream", TRANS. SOC. MECH. ENGRS. (Japan), 4, 5, 6, 1938-1940

[ix.] Laskowski, J.J., and Ranz, W.E., "Spray Quenching", AICHE J, 16,5, 802 (1970)

[x.] Wigg, L.D., "Drop-Size Prediction for Two-Fluid Atomizers", J. INST. FUEL, 37, 500 (1964)

[xi.] Ibid., Glatt Process Training Seminar

30, 60, 90 Day Plan for Operations Managers to Assess Process Safety and Prevent the Next Serious Incident

Brian D. Wilson, P.E.
Arkema, Inc
9502B Bayport Blvd
Pasadena, TX 77507
brian.wilson@arkema.com

Prepared for Presentation at
American Institute of Chemical Engineers
2013 Spring Meeting
9th Global Congress on Process Safety
San Antonio, Texas
April 28 – May 1, 2013

UNPUBLISHED

30, 60, 90 Day Plan for Operations Managers to Assess Process Safety and Prevent the Next Serious Incident

Brian D. Wilson
Arkema, Inc
Arkema, Inc
9502B Bayport Blvd
Pasadena, TX 77507
brian.wilson@arkema.com

Keywords: process safety, assessment, action plan, incident prevention, leadership, roadmap

Abstract

Operations managers today are faced with numerous challenges as they begin a new role and during their entire tenure in the role. Key job areas include: EH&S (Environmental, Health, and Safety) – especially process safety, operating cost, raw material budgets, logistics, reliable product supply, product quality, capital planning, and organizational stewardship. During the time the operations manager is coming up to speed in a new role, the organization is at higher risk as a key organization leader's focus can be diluted and distracted. Also, senior managers can be faced with similar "job overload" and lose focus on key priorities. It is absolutely critical that operations managers avoid these pitfalls and make process and personal safety a cornerstone of their day-to-day focus. This paper discusses why process safety must be a top priority for operations leaders and suggests key milestones over 30, 60, and 90 days to assess the site process safety culture and layers of protection. Practical tips are given on what and how to assess and what to do with gaps to prevent the next serious incident. These efforts might save a life, prevent harm to the community, or save the enterprise.

1. Operations managers – new leader challenges and no roadmap

Operations managers can have a very varied background. Some may have over 30 years of experience in multiple businesses, sites, regions, and technologies. For others, it can be their first assignment in a lead operations role. Regardless, this position is generally designed as a "linchpin" to integrate the process safety, EH&S, business, manufacturing, and supply chain to meet corporate goals. Even though this role is extremely critical and paramount to managing site and enterprise risk, training for the new role can be "ad hoc" and poorly defined. Many times the new operations manger must take the lead in "learning what they don't know," which can leave much to chance.

This paper outlines key process safety objectives and milestones as a new operations leader comes up to speed on the organization (people), technology, equipment, work processes, and management systems. This road map is intended to assist in prioritizing learning objectives and minimizing PSM (Process Safety Management) [1] oversight gaps during the leadership transition.

2. T-minus 60 days: due diligence of the new role – are you the right fit?

2.1 Due diligence

Preparation for the new operations role should not begin the day you start the new role. It should start as you begin your research for the new position. A due diligence review of a new role should include the familiar items: a site visit and tour, business review, discussion of key challenges of the new role, travel involved, location, compensation package, what success looks like, and your personal/family goals. The items below focus on the process safety challenges that should be investigated.

2.1.1 Preparing for the job interview

As you setup your site visit, you should compile as much information as possible on the health of the site's PSM systems and culture. The new position may be an area, site, region, or global role; however, many of the questions to answer will be the same.

Key questions to answer might include:

- How and who conducts your PHAs (Process Hazard Analysis) [2]? Are they current? Consider that the experience and skill level of PHA teams can vary greatly. If the initial PHAs were completed by a consulting group, they may or may not be available for discussion. Also, these reviews can often be "rubber stamped" by a fairly inexperienced team during the required 5-year renewals. Of course, the best case is to have a senior risk analyst professional with a strong background in the technology, past events, and RCIs (Root Cause Analysis) serve as the PHA facilitator and be a risk steward at the site or for the business.
- What is the status of the PHA action items? Who is driving closure?
- When was the last PSM audit [3] completed? Who did it, and what were the auditor's qualifications? What is the status of actions?
- What are the worst case scenarios in the RMP (Risk Management Plan) [4]? How many people on the site leadership team know what they are? Do the Emergency Response Team commanders know these scenarios?
- Is the mechanical integrity program functioning [5]? Are there established equipment inspection plans, inspection reports, notification of deficiencies, and high level management approvals for deferrals?
- How disciplined is the SWP (Safe Work Permit) process [6]? Does a good training, certification and audit process exist? What metrics are reported?
- Did the past leader and current staff understand and drive process safety improvements?
- What do the PSM leading and lagging metrics [7] look like at the site. What are the frequency and severity of process safety events? What type of leading indicators are being tracked?

It is best that you do not show up with a long list of questions and no answers. Site information can generally be found online, site contacts, calls to friends, using online RMP information, and through the OSHA (Occupational Safety and Health Administration) website. Therefore, do your homework and due diligence. This will make the interview process more productive and you will appear much more informed.

2.1.2 Understanding the needs of the position

It is critical that you perform an honest assessment of the new role, your knowledge and job skills, and your fit. Unfortunately, you will likely have a very incomplete picture of the PSM quality and health during your due diligence review. However, you should be able to define the process safety risk at the site and your process safety skills. A 3 x 3 matrix can help you identify the job risk vs. your skills (see Figure 1).

Figure 1. Site PSM hazards vs. your process safety/EH&S skills matrix

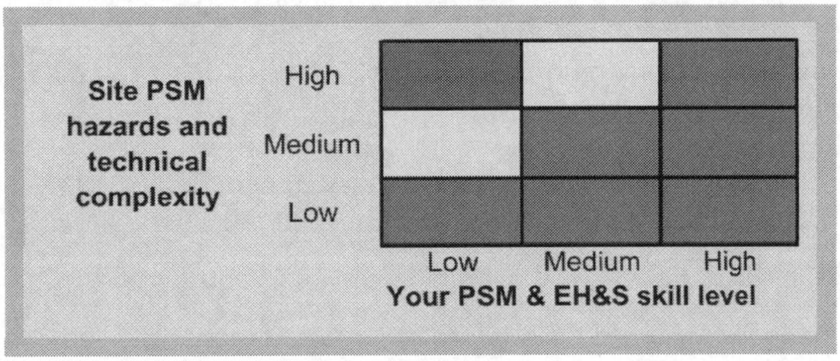

This table provides a warning (upper left quadrant) that a person with weak PSM/EH&S skills should not move to a role in a highly hazardous, complex plant or business. For some job sites, senior managers understand the process safety job risks are high and decline this high risk assignment. This sometimes results in a junior person being appointed to the role that may not fully understand the risk. Bottom line, go in with your eyes open, do your due diligence, and be willing to turn the job down if you are not a good fit for the organization's needs.

2.1.3 Managerial courage to ask the tough questions and set boundaries with the new boss

Following your due diligence review, you need to ask some direct questions of your potential new boss. Specifically, when have they had to make tough decisions between running a unit with high process safety risk vs. having to shutdown the unit? How did they deal with the business pressure to run the unit? What are the expectations of you to lead with process safety and EH&S as top priorities versus a "pounds first" mentality?

You need to be direct during the interview and communicate that you will not run an "unsafe" unit or plant. Some past examples where you have made tough decisions or took an unpopular stance to "do the right thing" are good to discuss. This may or may not help you get the job.

However, you should be willing to live with not getting a job working for a boss or business that does not support process safety.

3. T-minus 30 days: you've got the job – now what?

You asked the tough questions during the interview and you were offered the job. You feel based on the process safety risk analysis and your background, you are a good fit and you accept the job. Now what?

3.1. Wrap up the old job - quickly

It is critical to wrap up the old job as quickly as possible. The first step should be to document a summary of your key job tasks and DOA (Delegation of Authority) responsibilities. Generally, a table listing job task, frequency, and comments is a good place to start. Second, get the personnel MOC (Management of Change) process moving with your replacement on the old job. Begin bringing your replacement "up to speed" as quickly as possible.

3.2 Begin learning the new organization and processes, and prepare for the personnel MOC review (new leader review)

Learning the names of people in the organization is critical. Learning their names quickly shows them that they are important to you. It also allows you to begin the process of determining the organization's strengths and weaknesses.

If the organization has a defined "New Leader Review" process, ask to receive the study documents BEFORE you start the job. A new leader review typically contains items such as [8]:

- Organization overview and years of service in key roles
- Process overview
- Reactive chemistry hazards
- Reactive chemistry cardinal rules
- Past process safety incidents at the site and in the industry
- Status of key audits and action items (PSM, RMP, PHA)

3.3 Get clear vision of what success looks like

Have several discussions with the new leadership team on what they feel success would look like over the next year and then three years. This will help you begin formulating a roadmap to move the organization forward. This roadmap should include all key operations areas – but keep process safety and EH&S first on the list. Ensure the key elements of PSM are identified to allow you to make deliberate plans in each key area.

4. T+1-5 days: getting the first week right – everyone's watching

You have found a house in the new location, moved your family and home goods, and it is your first day at work. Do not underestimate what perception people take away based on your first few actions and words on the new job. If you say, "I'm here to return us to profitability", that will be their take away. The best leaders I have seen gather the leadership team together relatively quickly and give them an idea of the direction you will ask them to move in. The challenges of "low cost, reliable supply" are obviously a high priority for operating mangers. However, the leader makes it clear that process safety and EH&S performance are the absolute top priority. Personal examples of severe incidents the leader has seen can strengthen this call to champion process safety and general EH&S performance.

4.1 What you do speaks so loudly I can't hear what you are saying

Be careful of your actions – especially during the first few weeks. If all you do is mull over operating cost, the S&OP (Sales and Operations Planning) forecast, and business projections, folks will quickly feel your words on process safety and EH&S priority are hollow and not genuine. Therefore, spend time the first week doing field audits, speaking with process safety risk analysts, process safety and EH&S managers, SWP trainers, operators, and mechanics. Success in these conversations might give the impression that you are a bit of a zealot or fanatic when it comes to PSM and EH&S. There may even be a few jokes that begin with the new operations manager is the new process safety and EH&S manager. If that is what it takes to ensure the organization knows that safety is priority #1, then "so be it"!

4.2 Establishing a DOA timeline and communication

As you come on board, discuss with your new boss and who you are replacing the expectation of when you will receive the DOA to begin authorizing critical MOCs and process safety related escalations for approval. The organization needs to be clear that you should not be authorizing SWP escalations, deviations from normal work process, or SIS (Safety Instrumented System) deactivations until you are properly trained and properly understand the hazards.

4.3 Meetings with key leaders to get you up to speed

The following meetings should be set-up in the first 30 days to begin the learning process. These include:

Administrative leader in charge of on-boarding new employees: Review the new employee checklist – generally part of the on-boarding process that includes all administrative details and high level functional orientation.

EH&S manager. Questions should center on 1) health of the MOC process [9], 2) PSM concerns, 3) critical training you need, 4) key priorities over next 18 months, and 5) a review of the "cardinal rules" for safety, environmental operations, and reactive chemistry. These cardinal rules generally include:

> *Safety Cardinal Rules* – following MOC, SWP (especially the critical to life SWP standards), SIS deactivation, and procedure use policies

Critical to Life SWP standards generally cover LOTO (lock out/tag out), L&EO (line and equipment opening), confined space entry, hot work, and work at elevation

Environmental Cardinal Rules – following MOC policy, not bypassing critical environmental controls without proper authorization, ensuring all spills or releases are reported

Reactive Chemistry Cardinal Rules – following strict operating discipline to ensure chemicals are not inadvertently mixed, safe operating limits are not exceeded, or only proper materials of construction are used in highly hazardous service

Distributed Control System (DCS)/Safety Instrumented System (SIS) Control Engineer: Meet with the control engineer to determine if a clear SIS list has been created. Do testing procedures exist to test the systems, are the instruments used to test properly calibrated, how many cycles of testing have been completed over the past 5 years, and what have the findings been (as found/as left)? Do any test frequencies need to be adjusted based on the findings?

Ensure you understand how any SISs can be bypassed or deactivated. Who has the authority to bypass these devices, and what is the process used to ensure risks are managed? How are bypasses documented (when and how bypassed, when returned to service)?

Maintenance manager and the senior mechanical technicians (fixed, rotating, and Instrument and Electrical (I&E)): The goal of the meeting is to determine the health of the Mechanical Integrity (MI) system. Have PSM covered equipment been identified? Have equipment inspection plans been completed based on risks? Does a computerized maintenance management system (CMMS) exist to track the inspection preventive maintenance (PMs) tasks? Have all initial internal inspections been completed? Is there a mechanical integrity deficiency review process? How many deficiencies exist and who has approved continued operation of this equipment?

Senior foremen and technicians: These meetings are to ensure you fully understand the SWP procedures, especially for the critical to life SWPs (line and equipment opening, confined space entry, hot work, work at elevation, and lock-out/tag-out). You also need to understand the "escalation policy" for special hazards (when lines cannot be fully cleared, work in atmospheres above STEL (Short Term Exposure Limit)/IDLH (Immediately Dangerous to Life and Health), use of chemical hoses, etc.).

Training coordinator: Discuss how operators, maintenance personnel, and contractors are trained and tested for proficiency. It is critical that rigorous training and testing be completed.

Contractor manager [6]: Meet with the contractor manager and key contractor leaders to see what orientation and training contractors receive. Who do they go to with questions? Can they report hazards without being penalized? Does the contactor have a zero tolerance policy for major safety infractions?

5. T+ 5-30 days: understanding the hazards

5.1 Research the highly hazardous chemicals (HHCs) handled at the site

As you begin weeks 2-6, you need to understand the major hazards of the process. One potential place to start is to create a spreadsheet with all key chemicals, physical properties, chemical formula, and exposure limits (TWA – time weighted average, STEL, IDLH, ERPG – emergency response planning guideline). For senior managers, it is often good to add in a few familiar components for comparison to better ground yourself on the hazards of the new chemicals versus ones you have worked with in the past. This can also assist others in understanding how hazardous certain chemicals may be. For instance, the IDLH for hydrogen cyanide is 50 ppm. No one is surprised by the low number. However, the IDLH for more common materials such as acetic acid (50 ppm), ammonia (300 ppm), and for chlorine (10 ppm) are also quite low. The key is not to become complacent with "everyday" chemicals. They can be surprisingly toxic relative to other highly hazardous chemicals.

5.2 What do the RMP, air dispersion modeling, and quantitative risk analysis (QRA) identify as major hazards?

Now is the time to review past process hazard analyses (PHAs). This might include Hazard and Operability studies (HAZOPs), the RMP, air modeling results, facility siting studies, and any QRAs [10] and LOPAs (Layer of Protection Analysis) [11] that exist. It is unlikely you will be able to read through all of the documents. However, you should certainly read the management summaries and understand any gaps identified versus OSHA 1910.119 and your corporate standards.

5.2.1 Fatality versus frequency curves

For any completed QRAs, you should review the frequency vs. number of fatalities (FN) curves as compared to your corporate safety risk criteria [12]. Does your site meet this criteria? How do the FN curves fit with published industry guidelines? Who in your organization really understands how this risk compares to other process safety risk in the company.

5.2.2 ERPG hazard zone boundaries

Ensure you understand the key "worst case scenarios", evacuation distances, and how to alert people to evacuate. This should be part of the plant emergency response capabilities. If you have a corporate role, ensure you understand how corporate emergency response can support the corporation's and sites' emergency response plans [13 & 14].

5.3 Emergency response training and capabilities

Spend a day touring the facility focusing on site emergency response equipment. This includes field equipment (deluge systems, water monitors, foam systems, short stop systems, and SCBA (self contained breathing apparatus). It also includes what mobile fire fighting capability the site might have and the training of emergency response personnel. Training may include materials for dealing with fires, spills, and releases, as well as first aid and paramedic capabilities.

6. T+ 31-60 days: auditing the systems

Now that you have a better grasp of the key hazards, risks, work processes, and personnel at the site, it is time to begin auditing for compliance. The next section discusses what you can do to identify gaps and begin setting priorities for closure.

6.1 Key reviews: RMP, PSM audit, PHA summaries and action item status

Get ready to pull a few late nighters and weekends to review the PHA work. You need to review key summaries of PHAs, the RMP report, and the latest PSM audit. Understanding the key findings, gaps, and action items is critical to establishing a site PSM roadmap and setting goals for the next 12-18 months.

6.2 MOC discipline: MOC, PSSR (Pre-startup safety review), and RBS (required before start-up) documentation

Ensure you are invited and attend any MOCs that are occurring at your site. Once you understand the work process fully and the capabilities of the personnel, you may choose to delegate. The key is, don't delegate until you know who and what you are delegating.

We all understand an MOC review should be conducted for non-like-for-like changes and personnel reviews. However, how well-disciplined is this process at the site? Also, are the MOC leaders good at identifying the hazards and associated levels of risk for changes and how these risks will be mitigated? Do "first time" and infrequent activities receive a critical review? Are the right people involved? Are all actions documented?

For the PSSR, do the right people attend and can they identify any remaining hazards or risks? Are RBS actions clearly documented and does the management system check to ensure these are complete prior to start-up?

A strong MOC system is likely the single best defense for preventing a serious incident.

6.3 Safe work permit : how good are you on critical to life standards (line and equipment opening, lock-out/tag-out, confined space entry, hot work, work at elevation)?

Set up standing meetings with your SWP experts. This might be the SWP safety technician or a senior foreman or operator. Complete at least one audit per month on different SWPs (especially LOTO, CSE, hot work, and L&EO). In the author's experience, sites generally over estimate the accuracy and discipline used to issue SWP permits. At one site, the author found less than 70% of SWP permits are completed fully and accurately as measured by an independent safe work permit coach.

Several "watch out" areas include: (1) quality of the maintenance work package – how clear is the package on the scope of work and how it should be completed; (2) how clear are line-break points identified to prevent inadvertent opening of the wrong line; and (3) how well-trained is the maintenance team? Many times the maintenance team is oriented in a 4-8 hour session listing scores of chemicals and hazards. This can lead to poor retention of the chemical hazards involved and even the names of operating areas.

Field auditing will also help you see if the persons accepting safe work permits actually follow all of the instructions and restrictions on the permit (e.g. PPE (Personnel Protective Equipment) required, scope of work allowed, tools allowed, and when to shutdown work).

6.4 Procedure use policy and discipline

As boring as it may sound, use of procedures is absolutely critical to safe operations of a site. At least once per month, you should review if procedures that require each step to be signed-off are actually being properly authorized and if any step deviations are being approved at the right management level. Without oversight and proper discipline, the procedure use policy can quickly become "use when it is convenient". Don't let this happen at your facility!

6.5 Fatigue Management

Ensuring plant personnel are fit for duty is important and has been a challenge in industry. The most demanding area is ensuring personnel do not become overly fatigued by long work schedules. As the operations manager, you should set clear expectations on maximum allowable work hours per day and maximum days that can be worked co-currently [15, 16].

7. T+ 60-90 days: formal training sign-off by process safety, EH&S and management leadership and development of roadmap

In the 60–90 day period, your assessment and training should be completed. All formal sign-off should be completed at that time verifying that you are ready to properly manage the hazards and risks at the site.

You should also begin completing a 12-18 month roadmap to drive the key objectives at the site. Process safety and EH&S should be at the top of the list.

8. Evolution or revolution?

You now have the first 90 days behind you. They pass quickly. You should know the people, process, and systems. Along the way, you should have corrected any short term gaps you found in the key PSM systems. Now it's time to make some longer term decisions and implement the action plan. After firming up the action plan, you need to decide what should be completed in-house and where you might need outside help.

8.1 Evolution - driving continuous improvement

If you are fortunate enough to be working at a site that has a strong PSM system in place, that is great. Your focus and action plan then will be evolutionary. Specifically, how do you take this mature process safety management system and continue to improve on it? This is generally accomplished by continuing training, driving the right process safety metrics, and continuing to improve technology (especially inherently safer systems).

8.2 Revolution – when drastic change is needed

As you determine the health of the key systems (MOC, SWP, PHA, MI, SIS, process safety metrics), you may find serious gaps which cannot wait on a "continuous improvement" approach. A decision needs to be made on what can be completed "in house" and what may need to be outsourced. Where possible, staffing from within is generally best so you can retain the knowledge at the site. The next best choice is to get corporate resources to assist. If these are not available, you need to look outside. The key point is to get any critical gaps closed ASAP (as soon as possible) and be willing to make a strong case for the resources.

The chemical and petroleum industries have been able to find significant efficiencies via use of system automation, economies of scale, and energy integration. Continued efficiencies and leaner staffs are expected and can be implemented safely. However, these efficiencies must come after good PSM and reliability systems are in place. Ensure you do not blindly follow a "cost cutting" program. It may result in an extremely high price if the proper systems, organization, and technology are not in place.

8.3 Everyone needs a coach

Now that you know the organization, you should have a good feel for who would be a good mentor. This is a person to contact from time to time to discuss priorities, time schedules, and ideas. This can also be a person who can give you feedback on your performance and "what they are hearing" about your performance. Enlist the help of a mentor to help you on your journey to manufacturing and process safety excellence.

9. Conclusion: it ain't easy, but we have to get it right

The transition of an operations leader into a new role has many pitfalls. This can include organizational behavior missteps, poor emotional intelligence, bad financial decisions, and lack of supply to customers. However, no mistake or oversight can have the disastrous consequences of poor process safety management. Operational manager "on–boarding" should be structured and not left to chance. Ideally, the new organization has a detailed "on-boarding" program. However, the author's experience, in over a dozen operational roles, has seen most programs are lacking. A new operations leader will likely be well served by generating an on-boarding timeline with key milestones. This will help force the right prioritization based on the numerous demands on an operational leader's time. The clock is ticking and a new leader has to get it right.

10. Acknowledgements

Thanks to Karen Study, Phil Myers, and Don Rhea for their review and expert comments.

11. References

[1] *Process Safety of Highly Hazardous Chemicals (29 CFR 1910.119)*, U.S. Occupational Safety and Health Administration, May 1991

[2] *Hazard Evaluation Procedures*, New York: Center for Chemical Process Safety, American Institute of Chemical Engineers, P. 115, 2008

[3] *Guidelines for Auditing Process Safety Management Systems,* New York: Center for Chemical Process Safety, American Institute of Chemical Engineers, 1993

[4] *General Guidance on Risk Management Programs for Chemical Accident Prevention* (40 CFR Part 68), U.S. Environmental Protection Agency, 1994

[5] *Guidelines for Mechanical Integrity Systems,* New York: Center for Chemical Process Safety, American Institute of Chemical Engineers, 2006

[6] *Guidelines for Safe Process Operations and Maintenance,* New York: Center for Chemical Process Safety, American Institute of Chemical Engineers, P. 226, 239 1995

[7] *Guidelines for Process Safety Metrics,* New York: Center for Chemical Process Safety, American Institute of Chemical Engineers, 2010

[8] Study, Karen A., "Training Management of Change Approvers", American Institute of Chemical Engineers, 6th Global Congress on Process Safety, 2010

[9] *Guidelines for Management of Change for Process Safety,* New York: Center for Chemical Process Safety, American Institute of Chemical Engineers, 2008

[10] *Guidelines for Chemical Process Quantitative Risk Analysis,,* New York: Center for Chemical Process Safety, American Institute of Chemical Engineers, P. 395, 2000

[11] *Layer of Protection Analysis,* New York: American Institute of Chemical Engineers, 2001

[12] *Guidelines for Developing Quantitative Safety Risk Criteria,* New York: Center for Chemical Process Safety, American Institute of Chemical Engineers, 2009

[13] *Corporate Crisis Management*, New York: Center for Chemical Process Safety, American Institute of Chemical Engineers, 2005

[14] *Guidelines for Technical Planning for On-Site Emergencies*, New York: Center for Chemical Process Safety, American Institute of Chemical Engineers, 1995

[15] Baker, James, et al. *The Report of the BP U.S. Refineries Independent Safety Review Panel*, P. 87, 2007

[16] *Standard RP-755 - Fatigue Prevention Guidelines for the Refining and Petrochemical Industries*, American Petroleum Institute, Washington D.C., 2010.

9TH GLOBAL CONGRESS ON PROCESS SAFETY

Empowering Operating Managers with Decision Support System to Prevent Process Safety Incidents

Principal Author: Mr. Deepak Phanse

Centre for HSE Excellence
Reliance Industries Ltd.
Reliance Corporate IT Park,
Thane Belapur Road, Ghansoli,
Navi Mumbai, Maharashtra
India-410 007

deepak.phanse@ril.com

Other Authors: Ms. Prashanthi Bhupathi, Ms. Manasa Godavarthy, Mr.Rohan Vardekar

Centre for HSE Excellence
Reliance Industries Ltd.
Reliance Corporate IT Park,
Thane Belapur Road, Ghansoli,
Navi Mumbai, Maharashtra
India-410 007

prashanthi.bhupathi@ril.com, manasa.godavarthy@ril.com,
rohan.vardekar@ril.com

Prepared for Presentation at
American Institute of Chemical Engineers
2013 Spring Meeting
9th Global Congress on Process Safety
San Antonio, Texas
April 28 – May 1, 2013

UNPUBLISHED

Empowering Operating Managers with Decision Support System to Prevent Process Safety Incidents

Principal Author: Mr. Deepak Phanse
Centre for HSE Excellence
Reliance Industries Ltd.
Reliance Corporate IT Park,
Thane Belapur Road, Ghansoli,
Navi Mumbai, Maharashtra
India-410 007

deepak.phanse@ril.com

Other Authors: Ms. Prashanthi Bhupathi, Ms. Manasa Godavarthy, Mr. Rohan Vardekar
Centre for HSE Excellence
Reliance Industries Ltd.
Reliance Corporate IT Park,
Thane Belapur Road, Ghansoli,
Navi Mumbai, Maharashtra
India-410 007

prashanthi.bhupathi@ril.com, manasa.godavarthy@ril.com, rohan.vardekar@ril.com

Keywords: Operation engineer, incident investigation, near-miss, metrics, alarm rationalization, IT enabled, risk

Abstract

The operating manager can be considered as a "forward" or "finisher" as in a football (US soccer) team. He is an essential link between the strategic and tactical level of management and the shop floor personnel who are in close proximity to the process and its hazards.

Having a robust Safety Management System is essential for an organization. However, effective methods to get the operating managers to understand their role, the parts of the system that they need to deeply understand, the way to evaluate their plant's progress or lack of it, and their ability to get the operating team to do the right things is a crucial link in the delivery of good process safety.

This paper broadly outlines Reliance Industries Limited's (RIL's) endeavor to sensitize and facilitate the operating managers in their efforts to effectively manage process safety. Some of the programs initiated by RIL include:

- Establishing a key list of leading and lagging metrics to be monitored on a regular basis
- Bringing clarity on the need to achieve production targets while being within the stipulated safe limits
- Using a Job Cycle Check to improve compliance to operating procedures
- Developing and monitoring ways to share and implement learning from internal and external incidents.
- Using a Safety Observation program to influence safe behaviours at the workplace.
- Establishing non-negotiable actions such as completion of 2nd party audits, timely incident investigations, and adherence to Life Protection rules

This paper explains the usefulness of these efforts and the challenges faced.

1. Introduction

Reliance Industries Limited's (RIL) commitment to excellence in Health, Safety and Environment (HSE) performance is one of the Group's core values. The company is unwavering in the policy of "Safety of persons overrides all production targets", which drives all employees to work towards continual improvement in the safety management systems for the benefit of our people, communities and protection of property in which we operate. In bringing this vision to reality, the operating manager is a vital link between the strategic, the tactical levels of management and the shop floor personnel who are in close proximity to the process and its hazards.

The HSE vision drove the organization to define its strategies to achieve excellence through establishing standards and procedures in line with the global practices, involving all levels of employees with an accountability to ensure increased involvement, measuring performance at various stages, dedicating sufficient resources, increasing HSE competency, establishing a governance system and continual improvement tracked by various metrics. The initial stages of instituting the process safety program involved extensive emphasis on the development of various processes and procedures for the safe operations of the plants involving cross functional teams to establish a standardized program across all RIL operations.

To enable the operating manager to take decisions for safety enhancement while he/she is confronted with conflicting priorities, it is essential to sensitize them to the pertinent aspects of process safety.

The foundation of the RIL HSE Management Framework is "Health, Safety and Environment is the responsibility of the Line Organization". RIL believes that the only proven way to excel in Health, Safety and Environment management is for all members of the line organization to accept responsibility for their personal safety and for the safety of their team.

At Reliance Industries Limited, the Process Safety Management program stands as a stepping stone for the behavioral and the cultural shift towards a safer environment at the workplace. The involvement of the line management in the implementation of PSM is done by adopting the subcommittee approach at the sites. This also includes other initiatives like inculcating behavioral changes through safety observation programs. This has proven to be successful in the few years of our PSM journey.

The establishment of a robust process safety management program at Reliance follows a top-down approach. This is being demonstrated by the active involvement of all levels of employees, from the top management to the shop floor employees in all initiatives like the safety observation programs to the participation in the subcommittees.

2. Initiatives for enhancing employee involvement

Analyses of internal and external incidents over the years reveal that 'compliance to procedures' is one of the major contributing factors. Such incidents cannot be attributed to merely non-compliance by operating personnel, but the root causes could be systemic factors like unclear instructions through the laid down procedures.

These analyses prompted RIL to launch the 'Project Health Check' initiative. This involved identification of critical procedures across the sites by a team consisting of operating managers and shop floor personnel. These critical procedures were then studied for completeness and clarity of instruction.

A Job Cycle check program was introduced to validate the practice against procedures, thereby identifying a scope of improvement in the existing standard operating procedures. These checks prompt continual improvement in the procedures, making them safer and user-friendly. Job cycle check created a sense of ownership among the operating personnel about the procedures. This has now emerged as a tool for continual improvements of the procedures as a cyclic process. As a part of the Job Cycle Check, evaluation of the preparatory measures taken for the job and execution of the job as per the procedures is done, covering aspects like differences between procedures and practice, unsafe conditions and human factors like accessibility, clear identification of the equipment etc. This enables the operating personnel to capture major improvement opportunities and also enhance the compliance to procedures. Any revisions to the procedures based on the feedback from a job cycle check are made by a team of trained and competent operating personnel.

This time targeted initiative helped sensitize the operating managers to the importance of not only compliance to procedures but the importance of adequacy of procedures. While excellence in this area still remains a challenge in the journey, a step in the right direction was taken.

3. Establishing data systems to monitor and analyse leading and lagging metrics for improving process safety performance.

The performance of the Safety Management System is monitored by a set of leading and lagging indicators. RIL has set up monitoring, reporting and tracking systems for deviations and other performance parameters critical to process safety that are derived from global best practices advocated by CCPS and the local regulations. These leading and lagging metrics enable the operating management to measure, improve and sustain reliable execution of process safety management systems.

The data for some of the leading indicators used at RIL for performance assessment are from the safety observation process, near miss incidents, metrics that showcase progress in various PSM elements, with the lagging indicators being the incident and injury metrics.

The Process, Fire, Distribution, Environment and Work place incidents are reported and logged into a home-grown Information Technology (IT)-enabled system across all businesses. The cycle from the reporting of an incident, investigation, generation of recommendations, to action tracking is being done through this system. The output of the investigation generates a report stating various contributing factors (system, physical and human factors), the failed PSM elements and the operational discipline characteristics. This data is available for the operating manager, which enables him/her to identify the system deficiencies in the plant or area and take corrective actions to prevent recurrence of such incidents.

A formal process of learning from incidents is also enabled through this IT system where internal and external incidents are analyzed for learnings that may be applied at similar processes or plants and shared across the organization. Learnings are shared through various means, including displays at strategic locations, through e-communications, discussed at the beginning of meetings, etc. The recommendations arising from learnings from incidents are tracked through an IT-enabled system, where in many cases; the operating manager is the action party.

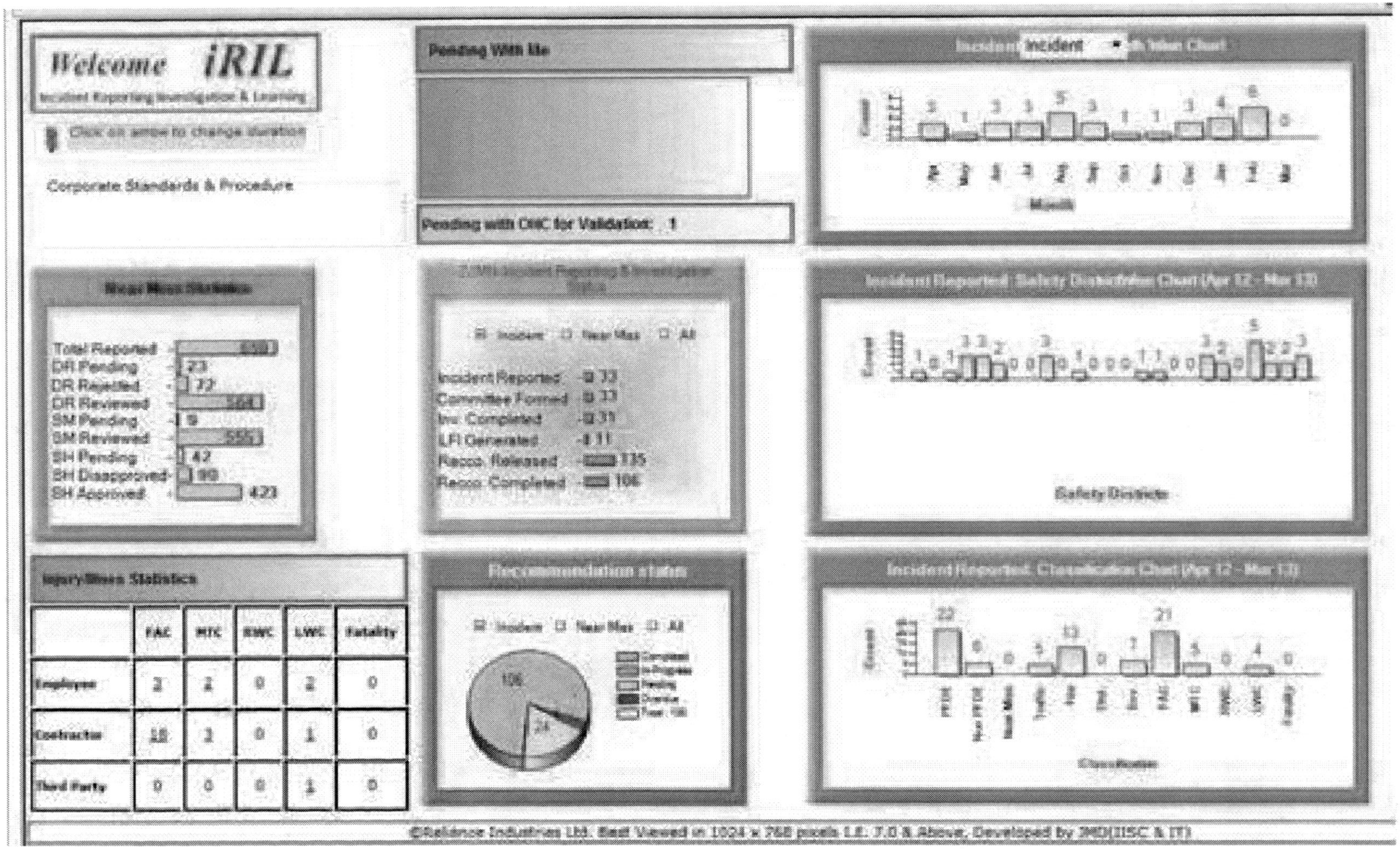

Fig1 : Snapshot of the Incident reporting system

While it is recognized that IT solutions alone cannot deliver high safety performance, RIL believes that the power of IT coupled with a strong subcommittee structure comprising members from line management, can deliver a superior functioning safety system in the long run for a large multi location organization.

4. Safety observation programs to influence behaviors at the workplace

RIL has taken an approach towards behavioral safety through a 'Safety Observation program'. The methodology of a safety observation program practiced at RIL follows the global approach of 'observe', 'discuss', 'comment', 'agree', 'discuss other safety issues' and 'acknowledge'.

The objective of the Reliance Safety Observation program is to reduce errors in working and manage defenses. By observing various unsafe acts and conditions, the operating manager, by virtue of his process knowledge, is able to create a situational awareness of what can go wrong in the surroundings, and by initiating a conversation with the employee at work and engaging the employee in the observation process, a behavioral change is initiated and reinforced. The situational awareness is a critical step practiced at RIL which is different from the conventional approach of a safety observation program. Findings of the Safety Observation Program are recorded electronically and analyzed as part of leading metrics.

While most organizations use the observation process primarily for workplace activities and behavior change, RIL is making efforts to dovetail PSM elements in the safety observation process to get maximum impact on the process safety related activities.

5. Procedures underpinned with technology

In pursuit of operating and safety excellence, RIL has evolved in synchronization with the global trend with respect to IT solutions. With technology, information inundation is quite possible. Therefore, to empower the operating managers with better decision making tools, support is taken from different software solutions.

Excessive alarms are a common issue faced by process operators with modern Distributed Control Systems (DCSs). This is a contributing factor for diminished capacity for the operator who has to attend to both critical and spurious alarms. These factors increase the likelihood that a critical alarm may be overlooked making it difficult to respond to a plant upset, and raising the chance of an unplanned shutdown or an incident.

To avoid such a situation, the alarm rationalization process was initiated to evaluate all potential or existing alarms against principles established through an alarm philosophy. This system minimizes spurious alarms and enables the operator to focus on the critical alarms to capture rationale such as cause, consequence and corrective action which can be used to guide operator response.

RIL is in the process of implementing an IT solution which will help the operating managers to more easily input as well as use all necessary information to arrive at informed decisions in the field quickly. Such application will also enable information sharing across all levels of the organization. The operations logbook feature of this application maintains an electronic log that provides shift summary reports, tracks daily operations and improves process and plant reliability. Operations monitoring helps users track operating performance against targets, systematically carry out data monitoring and highlight problem areas.

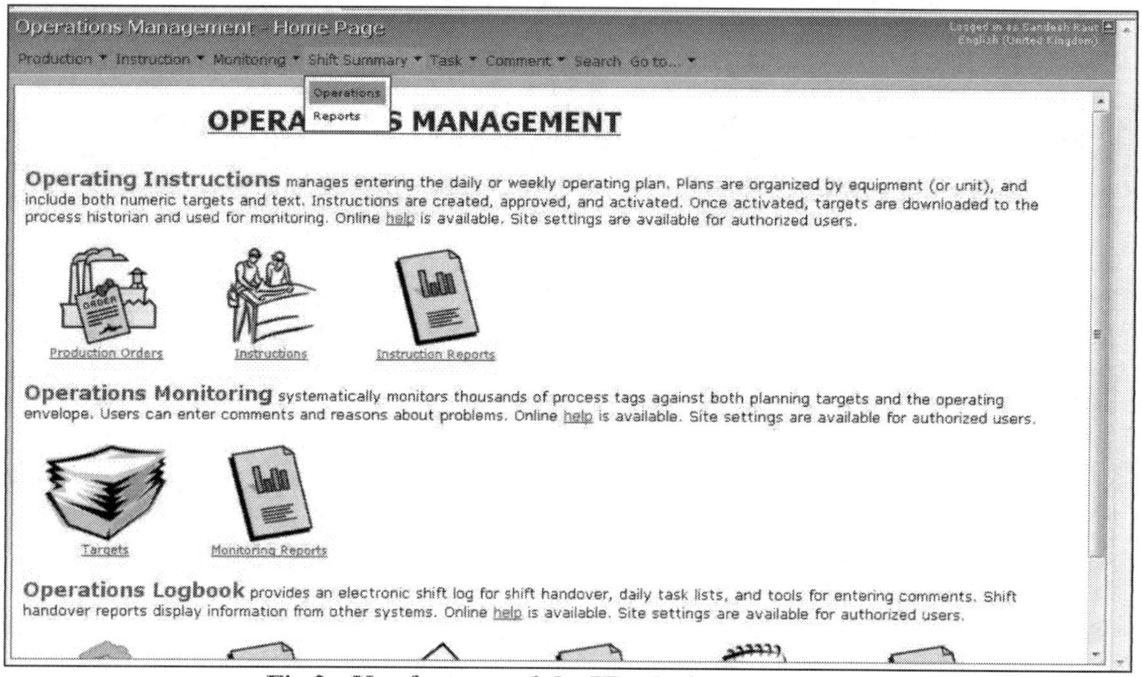

Fig 2 : Key features of the IT solution

Key advantages of the software solution are

- Integrated database to ensure that all functions (maintenance, operations, technology, safety) work on the same database
- Monitoring of deviations with respect to planning targets and standard operating conditions
- Knowledge repository of reasons for deviations
- Consistent and comprehensive e-enabled Interlock Bypass Management
- Complete and consistent record of hand over and take over activities carried out during the shift

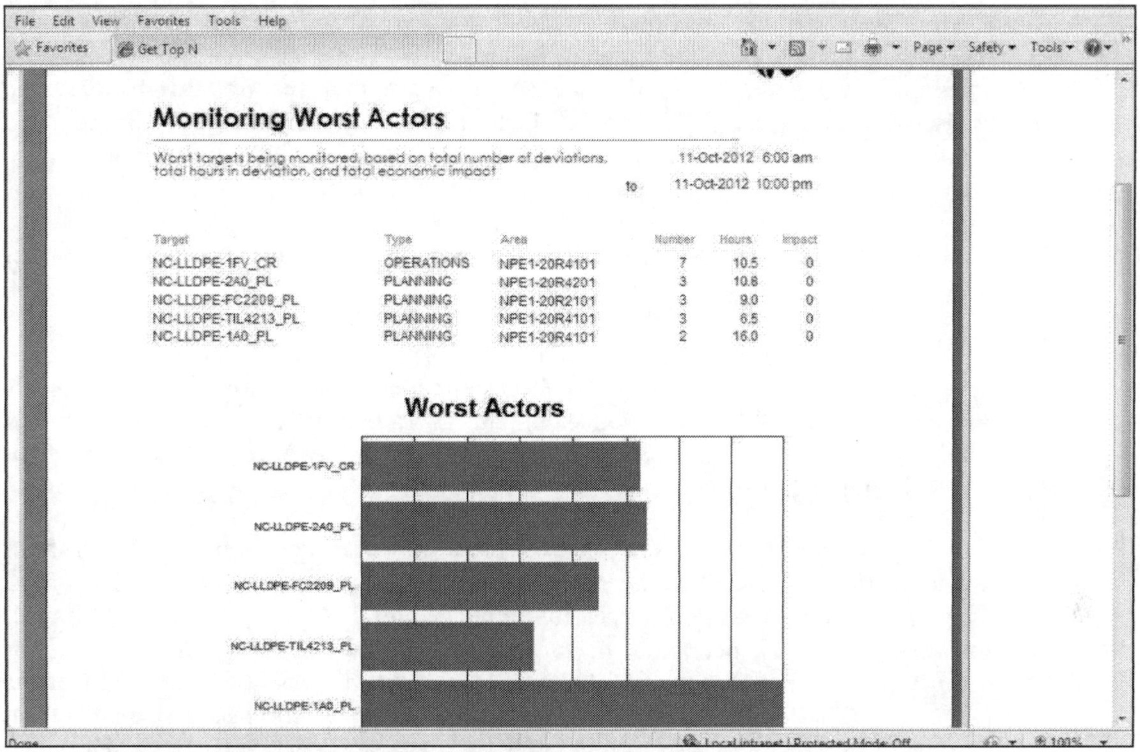

Fig 3: Typical example of the key process deviations monitored

This solution shall also enable the operating managers to establish risk-based inspection procedures, clarity on achieving production targets within the stipulated safe limits and reliability centered maintenance activities thereby resulting in increased health of the equipment and safe operations.

6. Developing understanding of risk-based approach during decision-making in case of emergencies

Risk is managed by identifying hazards, assessing consequences and probabilities, and evaluating and implementing preventive and mitigative measures. Periodic risk assessments are performed by qualified operating and maintenance personnel. Risk assessments are updated at specified intervals and as changes occur.

Risk perception has increased significantly among the operating personnel through their involvement in risk studies such as Process Hazard Analysis, Layer of Protection Analysis, and Quantitative Risk Analysis conducted for all the operating units as a part of the RIL PSM program. There has been an increased awareness among the operating group about the hazards that could be manifested in case of barrier failure. An important outcome of these risk studies is identification of barriers critical to process safety and maintenance of them. While preventive barriers prevent loss of primary containment and require periodic upkeep, mitigative barriers act as a last line of defence in case of an incident.

Having identified the maximum credible and worst case scenarios through risk assessment studies, operating managers are enabled to plan for an emergency by assessing the potentiality of damage and the resources required for handling an emergency. For clearly defined responsibilities and accountability during such conditions, to enable effective response and decision making during emergencies, systems like pre-incident planning and mock drills have been established at RIL.

7. Internal certification processes for increased competency of several levels of employees

A tiered competency certification method for process safety management has been designed taking in to account the level of competency an employee requires for his application during operations. Development of "Reliance Certified PSM Belts" has been initiated with the objective of bringing in the right level of PSM awareness for all levels of employees.

The target group for each level of certification may vary depending on their role in operations and the extent of the process safety applications they have to execute in their daily routine operations. The minimum required competency of the technical, functional skills of operating personnel also additionally defines the level of competency required that would be appropriate to the process safety demands for the job or role. The four levels of certification depend on the process safety demands on the functions of an employee, classified as the: awareness level, knowledge level, skill level, and mastery level of process safety knowledge. This certification process is also extended to all the employees – the maintenance managers, technical services, commercial team, and all levels of the management.

The first-level certified group of employees may not need to work extensively for process safety but need to have a basic understanding of the principles. The groups of personnel identified for the second belt certification are those who may have to apply PSM principles in their daily operational activities and contribute to increase the PSM trickle. Over a period of time it is expected that the third and fourth levels of certification will produce a group of PSM experts who will contribute to the continuous growth of PSM establishment in the organization by taking up coaching, and directing projects focusing on PSM and similar initiatives. The certification process is defined to complete with a formal validation in the system.

8. Conclusions

The journey to process safety excellence is a long one. Organizations all over the world are converging to some basic requirements to establish and improve process safety. However, best practices continue to evolve and proactive organizations continue to improve through internal experiences. The initiatives described above may in the overall sense not be very different from other organizations. However the finer differences are the ones which decide the extent and depth of success that each program can have in sensitizing as well as empowering the operating personnel in improving PSM. RIL continues to evolve systems in its continuous journey towards PSM excellence.

References

[1] RIL Group Process Safety Management Standard,- internal document

[2] *Guidelines for Process Safety Fundamentals in General Plan Operations,* New York: Center for Chemical Process Safety, American Institute of Chemical Engineers, 1993

[3] *Guidelines for Implementing Process Safety Management Systems,* New York: Center for Chemical Process Safety, American Institute of Chemical Engineers, 1994

[4] CCPS, "*Process Safety Leading and Lagging Metrics*", AICHE, New York, 2011.

9TH GLOBAL CONGRESS ON PROCESS SAFETY

Barrier Diagram (Bow Tie) Quality Issues for Operating Managers

Dr Robin Pitblado
DNV (USA) Inc.
1400 Ravello Drive, Katy TX 77449
Robin.pitblado@dnv.com

Pieter Weijand
Governors
Netherlands
Pieter.Weijand@gmail.com

Prepared for Presentation at
American Institute of Chemical Engineers
2013 Spring Meeting
9th Global Congress on Process Safety
San Antonio, Texas
April 28 – May 1, 2013

UNPUBLISHED

Barrier Diagram (Bow Tie) Quality Issues for Operating Managers

Dr Robin Pitblado
DNV (USA) Inc.
1400 Ravello Drive, Katy TX 77449
Robin.Pitblado@dnv.com

Pieter Weijand
Governors
Netherlands
Pieter.Weijand@gmail.com

Keywords: Barrier diagrams, Bow Ties, Quality, Operations

Abstract

The Oil & Gas (O&G) and petrochemical industries have been much more successful at reducing occupational safety events as compared to major process accidents over the past 20 years. A novel approach based on barrier diagrams, called Bow Ties, directly addresses major accidents and the focus on barriers and maintaining barrier functions has been shown to be effective in O&G operations in the North Sea. While not mandated in OSHA 1910.119, many companies are starting to apply barrier methods to improve operations and reduce major accidents. The Bow Tie approach is deceptively simple and many errors are made in Bow Tie diagrams. This is due to a lack of a formal standard or accepted methodology for the approach. This paper describes many common errors that appear in Bow Ties used for operational safety and how these might be rectified.

Since Operating Managers will rely on these risk diagrams as a key part of their process safety program, it is essential that they have good quality and not contain structural errors.

1. Introduction

The Oil & Gas and petrochemical industries have been very successful at enhancing occupational safety performance, achieving close to a factor of 10 reduction in OSHA lost time or recordable events, as reported in company annual sustainability reports. Conversely, major accident performance has hardly improved over the period, shown both by anecdotal evidence of major accident headlines, Chemical Safety Board lists of events, or Marsh data on 100 largest losses [1]. The relative performance between these two domains is shown in Figure 1.

While OSHA 1910 and EU Safety Cases have driven part of the improvement in occupational safety, they have been less successful for major accidents as shown in the figure. The authors believe this is due to a lack of focus on major hazards during operations. OSHA 1910 focuses more on design safety and typically the required PHA (Process Hazard Analysis) documents are not routinely used during operations. Conversely in the UK, the Piper Alpha accident in 1988 led to a detailed public inquiry chaired by Lord Cullen. That recommended new offshore safety case regulations that include a focus on barriers addressing major accidents. These were called Safety Critical Elements, and each required a Performance Standard (what it is expected to do, its reliability and its survivability) and a Written Scheme (the required inspection, maintenance and competence programs to keep the element functioning at its performance standard for the facility lifetime). This focus on barriers has led to close to a factor of 10 reduction in major leaks (which have the potential to lead to total platform loss) based on UK HSE (Health and Safety executive) statistics. Norway is broadly similar in barrier requirements. Statistically, there were 2 total loss events in the North Sea from 1980-1988, and none since then in the intervening 25 years. This supports a factor of 10 reduction – although the period is short.

The focus on barriers has led many companies to embrace a barrier approach to managing major accidents. The theory under-pinning this was developed by James Reason and Patrick Hudson, and is crystalized in the well-known Swiss cheese model of accident causation. The Bow Tie diagram was developed in the early 1990s, simultaneously in Australia and the Netherlands. This approach assists operations activities to consider major accidents and is considered the most effective of the various barrier methods as it provides both a clear summary of barriers deployed and a visual format that is easy to communicate. Unfortunately, there is no accepted international standard for Bow Tie diagrams and this leads to quality issues with many Bow Ties having fundamental errors that weaken the value of the method and hence the potential to reduce major accidents.

Figure 1. Comparison of performance - occupational vs major accident safety

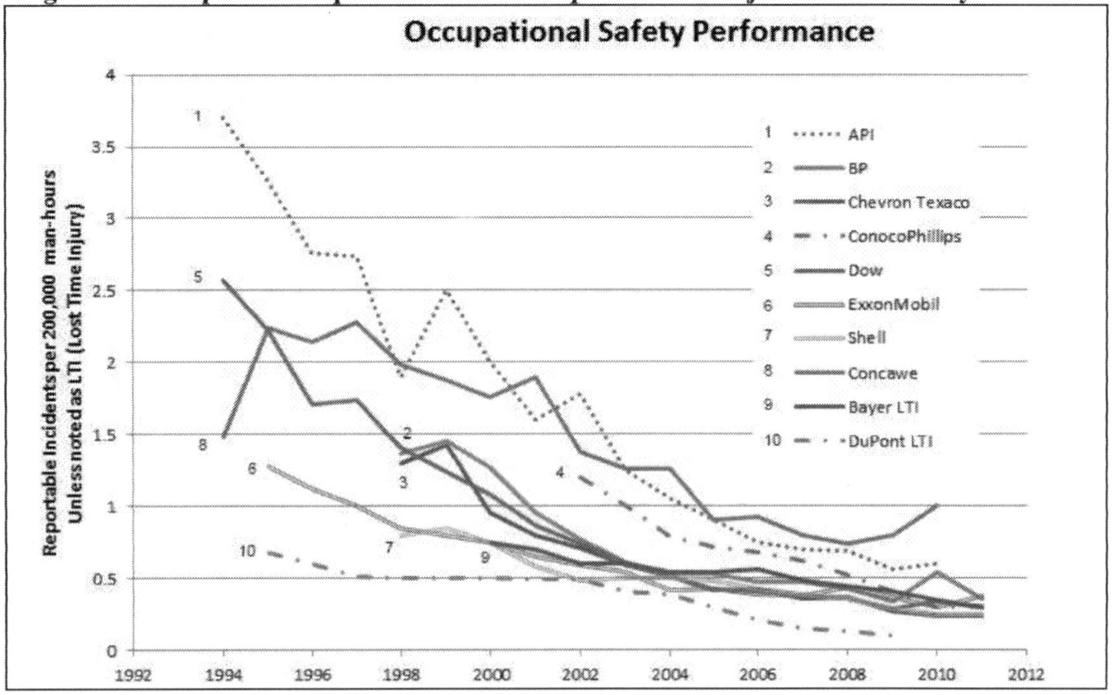

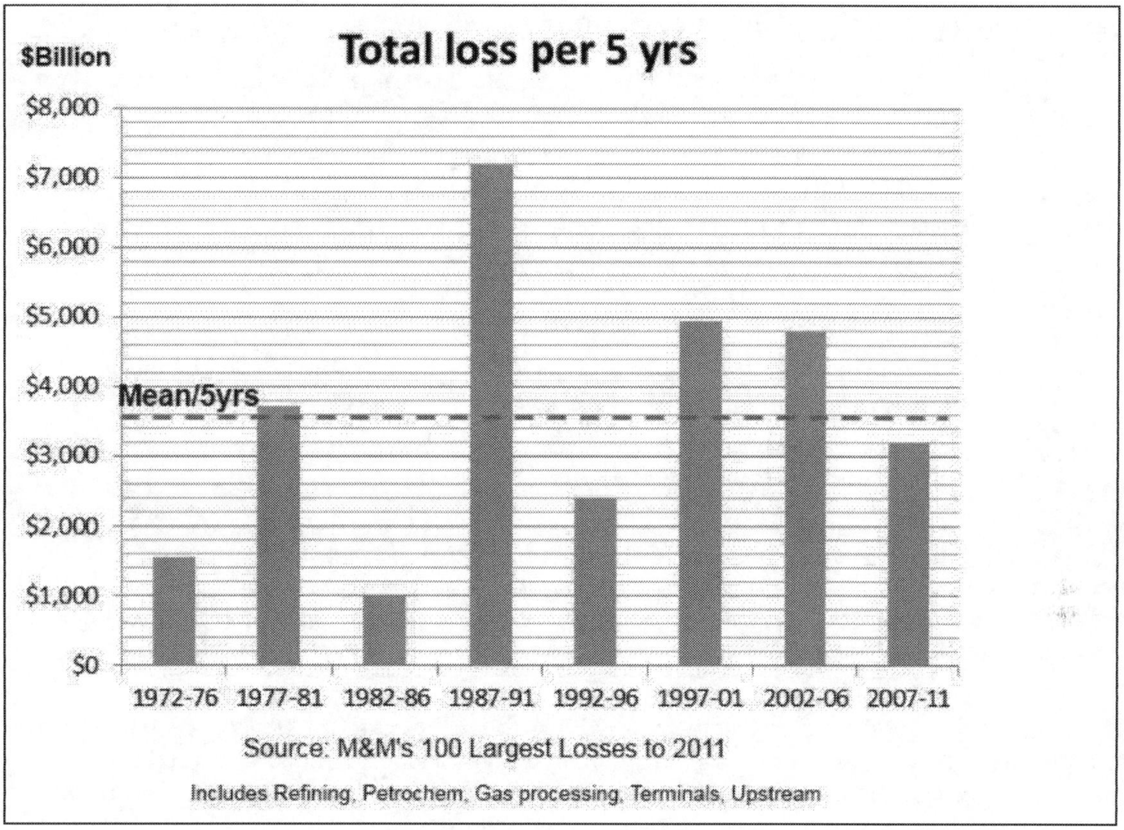

2. Bow Tie barrier diagram methodology

2.1 Method Overview

There are several descriptions of the Bow Tie method. These include a paper by Shell to the EU (Zuijderduijn [2]), the EU ARAMIS research project (Salvi and Debray [3]), the Norway BORA Project [4], and various company standards. The authors have worked with BP and Shell standards. While these are all good documents, they are not mutually consistent and none is accepted as a universal standard for method or terminology.

The methodology for Bow Tie diagrams appears straightforward and a typical diagram is shown in Figure 2.

Figure 2. Example Bow Tie diagram demonstrating method

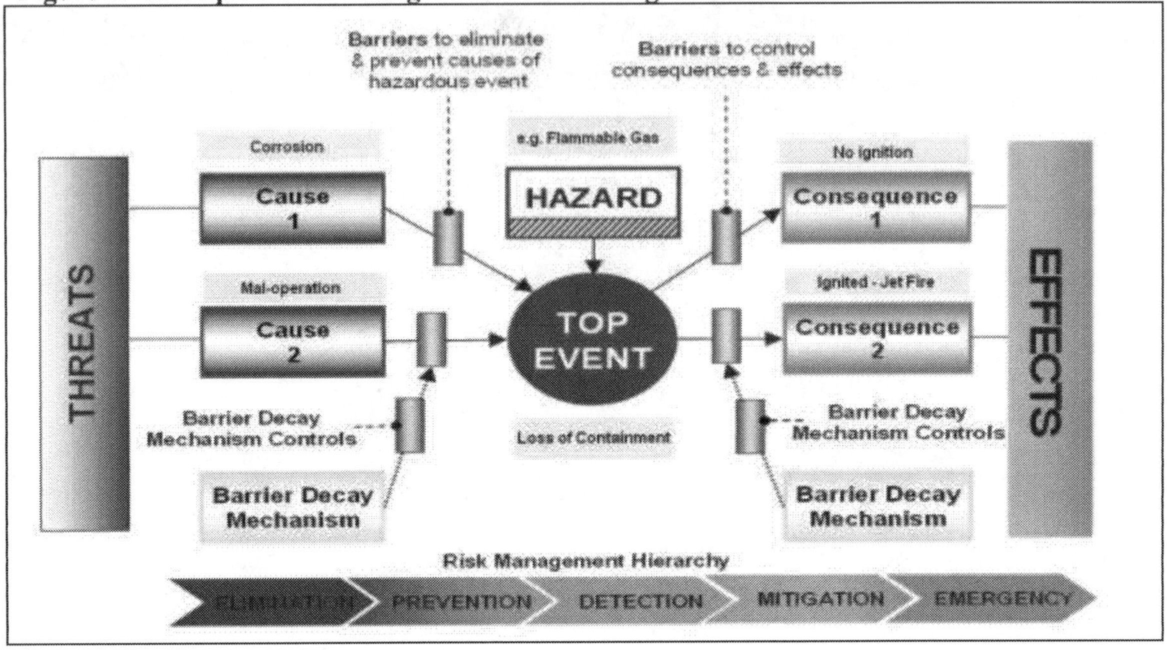

While Figure 2 is a good overview of the method, real Bow Ties can be extensive. A typical Bow Tie might have 8-10 threat causes and 3-5 outcome consequences. Each arm might have 3-5 barriers, but more are possible. Also several barriers will have Barrier Decay Mechanisms built-out.

A misconception by some is that the barriers, both prevention and mitigation, are simply marks on a threat pathway showing the existence of a barrier. This allows for good communications to showing all the barriers deployed and who is responsible for each. The specification of the barrier may be loosely defined and potentially inconsistent. However if the only intent is to list all the barriers on a figure, then this is acceptable and is simply a variant of providing a list.

While this is an important use of Bow Ties and enhances staff and contractor communications and understanding, it is only a partial use of the tool, and may contain structural errors that defeat more extensive and value-adding uses. The key is to understand the scientific basis for Bow Ties. The ARAMIS project provided a graphic (Figure 3) showing the Bow Tie as a combination of Fault Tree Analysis (FTA) on the left hand prevention side, and Event Tree Analysis (ETA) on the right hand mitigation side.

Figure 3. Bow Tie under-pinnings

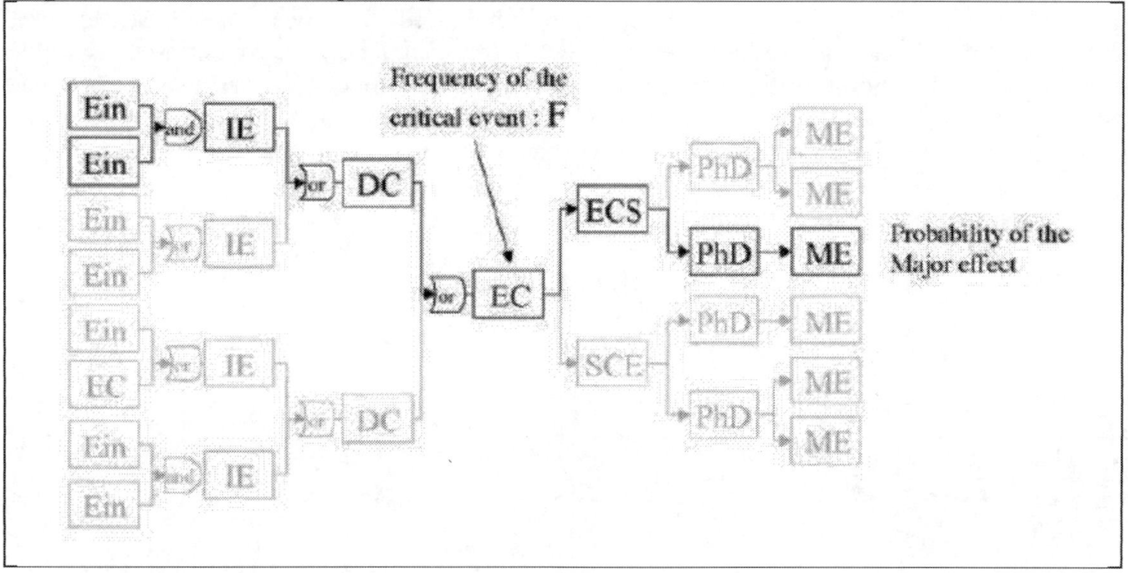

2.2 Fault Tree Basis

In order that the Bow Tie barrier method can be exploited for its maximum potential contribution to reducing major accidents, it is vital to link it conceptually to theoretically sound and more mature methodologies, which have formal rules for structure and for arrangement of events. A Bow Tie barrier is actually a fault tree AND-gate construct with a "demand" combined with "barrier fails" as shown in Figure 4.

Figure 4. Barrier is Fault Tree AND-gate

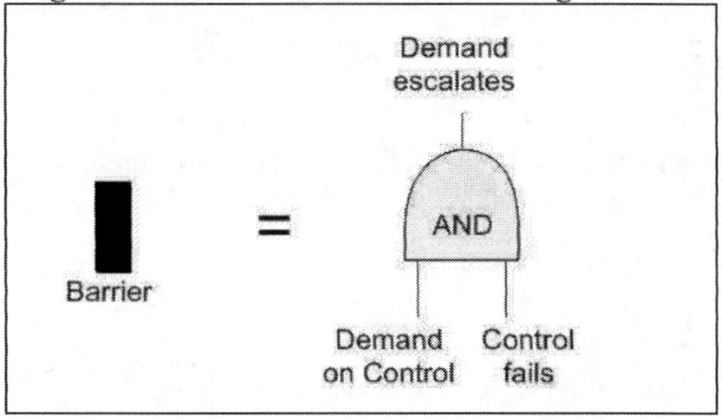

The different Threat arms coming into the Top Event are combined through a Fault tree OR-gate construct.

The Mitigation-side can be thought of as Event Tree Analysis as suggested in the ARAMIS project, but instead of the prevention side AND-gate format, the barrier on the mitigation side is an Event Tree branch with barrier fails and barrier succeeds as two outcomes. Since "barrier succeeds" terminates the event, that arm is discarded and the Bow Tie only builds-out the sequence of barrier failures until the ultimate consequence is reached. This explanation works at a first level, but it fails when Barrier Decay Mechanisms (escalation factors) are introduced. A Barrier Decay Mechanism is a fault tree build-out of how the main pathway barrier can degrade and what additional barriers are deployed to prevent this (Figure 5). This build-out is used equally on both sides and mitigation barriers can degrade just as prevention ones may. Also some barriers conceptually work after a consequence is realized, and the temporal order can be distorted for the sake of a 'clear' Bow_Tie diagram shape.

Figure 5. Barrier Decay Mechanism

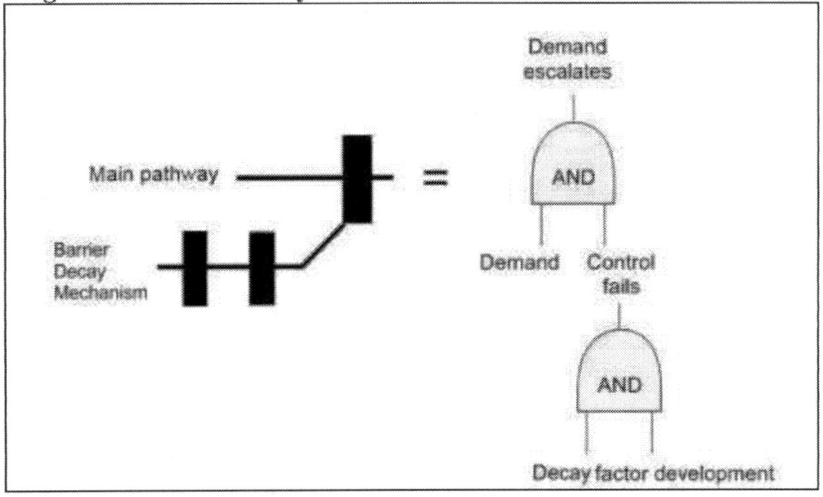

This build-out only applies to FTA, there is no equivalent form in ETA. Thus a Bow Tie uses fault tree methods on both sides, not just the prevention side. This is important as the aim is to reinforce the theoretical basis for Bow Ties, so that formal rules for construction and combination can apply, and fuller deliveries are possible from the Bow Tie. Examples of these include correct barrier counts between the Threat and the Top Event (Barrier Decay Mechanism barriers below the main pathway do not count in that total), a clarity in construction that places barriers in their correct locations, not simply plotted on a "useful graphic", an ability to draw correct conclusions knowing that the figure complies with Fault Tree systematics and that decisions taken on the basis of the Fault Tree are valid. A good overview of fault tree systematics and rule sets is provided by NASA [5].

2.3 Basic Definitions

The Bow Tie example shown in Figure 2 has several key elements. These can be defined as follows:

- The Hazard – this is the potential (material or activity) to lead to a Top Event
- Top Event – this is the event that can lead to undesired outcomes (usually loss of control or loss of containment) – it is named based on Fault Tree terminology
- Causes – these are the categories of challenges to safety systems that if not terminated will lead to the Top Event
- Outcomes or Consequences – these are the categories of undesired outcomes that can occur if the event starting from the Top Event is not terminated.

- Prevention barriers – these are the barriers / controls / safeguards that can terminate (or reduce the likelihood) of the cause before it reaches the Top Event
- Mitigation barriers – these are the barriers / controls / safeguards that can mitigate an event sequence before it reaches the undesired consequence
- Barrier Decay Mechanism or Escalation Factor – build-outs from the main pathway barrier for reasons it might degrade and the barriers deployed to prevent this.

There are several definitions for barriers; a good one is from Sklet [6]: "Safety barriers are physical and/or non-physical means planned to prevent, control, or mitigate undesired events or accidents". This definition allows for both hardware barriers (e.g. Emergency Shutdown Systems) and for non-physical barriers (e.g. inspection programs, training) and this allows the holistic deployment of barriers in a comprehensive major accident protection system to be properly mapped.

3. Problems observed in Bow Tie analysis

Given this understanding of the fundamental under-pinning of Bow Tie analysis by scientifically valid process safety methodologies, it is now possible to identify potential problems in Bow Ties produced without reference to such methods.

3.1 Hazard

The authors have observed several deficiencies with Hazards defined in Bow Tie analysis. A Hazard is a material, energy source or activity with a potential for an undesired outcome typically if there is a loss of containment or other form of loss of control. Some example valid Hazards and other incorrect hazards are shown in Table 1. The test for validity is that the Hazard should identify the source of (or damage to / disruption of) the problem and lead naturally to the Top Event, but should not be the top event itself. Note Hazards can have multiple Top Events, but there can only be one Bow Tie for each Top Event. There can be hazards which are correct for one Top Event but incorrect for another. Thus Collision may be a valid Hazard for a Top Event of "hull punctured", but not valid for a Top Event of "loss of steering" (where Collision would be a consequence).

Table 1. Valid and incorrect hazards

Valid Hazards	Incorrect Hazards
• Cyanide	• A leak of propane
• A 100t sphere of pressurized propane	• Ignition
• Driving a heavy vehicle at night	• Control system malfunctions
• Excess overtime being worked	

The significance of an error in defining the Hazard is that it might lead to a faulty Top Event and getting this right is core to success. For example if the Hazard is incorrectly defined as the Top Event (e.g. Leak of Propane in Table 1) then the resulting Top Event might be "large flammable cloud" and that would be incorrect. This is closer to an undesired outcome and this could lead to generation of incorrect barriers.

3.2 Top Event

The Top Event can be thought of as a point in time where there is some change in state that causes a loss of control over the hazard. Most often for major accidents this might be loss of containment or loss of control. Top Events are not the outcome; usually there are multiple barriers that can be deployed between the Top Event and the final consequence(s). Thus "explosion" would generally make a poor Top Event – it would be better to define this as a leak of hazardous material as this flows more naturally from the hazard. Then show how barriers can be deployed before a large flammable cloud forms that can explode: e.g. gas detection systems, emergency shutdown systems, ignition controls, and management of occupied spaces, etc. Some examples of good and poor Top Events are shown in Table 2. The Top Event should indicate how control is lost and / or give an indication of scale. The definition of a Top Event also depends on perspective of the analysis, for instance a Top Event defined from one perspective could be inappropriate from another perspective. For example for the Hazard: Reactive Chemicals a Top Event: Loss of Control of Reaction would make sense, but not a Top Event Incorrect Mixture. In this context Incorrect Mixture is a Threat, but it could be a valid Top Event in another context.

Table 2. Examples of good and poor Top Events

Good examples:	Poor or Incorrect Top Events

• Large leak of pressurized propane
• Small spill (<2kg) of eco-toxic material
• High speed turbine disintegration
• Control of exothermic reaction is lost
• Error in control of work
• Structural failure of equipment item
• Loss of intended separation
	• Explosion of propane cloud
	  o Outcome not Top Event
	• Ignition of flammable cloud
	  o Intermediate event in mitigation sequence
	• Crane dropped object
	  o Outcome, not loss of control
	• Corrosion
	  o Threat, not Top Event
	• Eco-toxic leak to environment
	  o Consequence, not Top Event

3.3 Cause and Effect

In the Bow Tie diagram it is important that there be a clear cause and effect between the various elements:

- Hazard should lead logically to the Top Event
- threats (which might have multiple separate causes) should lead to the Top Event
- the Top Event should lead logically to one or some set of Consequences
- Barrier Decay Mechanisms should lead logically to main pathway barrier failure

If the cause and effect linkage is absent or unclear, then it is likely that the Bow Tie will be incorrect, and this means that barriers may be missed or that new barriers are introduced that have no effect on major hazards.

3.4 Threats

Threats lead to the Top Event and are normally collected as a sequence of causes. These are often developed as a brainstorm of all possible causes. This might be 20+ causes. However, many of these causes (e.g. galvanic corrosion, thinning corrosion) may have similar controls and there is no value building out two pathways barriers – a single pathway would show all important barriers. Typical errors we see are too many causes identified, or conversely too few with important causes not enumerated. The suitability of barriers deployed against a major accident can only be assessed if all important threat causes are listed.

Not all threat causes have the same likelihood. For example, for a Top Event: "Large Leak of Propane", the threat corrosion is continuously present, whereas others like operator error or dropped object are high and low probability threats respectively. All should be treated similarly in terms of barrier build-out but there can be a different target for the number of barriers needed to manage the threat. The number of barriers required is discussed later. While it is a matter of judgment, some threats can be excluded (e.g. plane crash, earthquake, flood) although the proximity to airports, local earthquake category and flood susceptibility may change this assessment. In the last case – there is also not a direct link between flood and loss of containment.

Some general rules for defining threats are:

- Threats should address all significant risks leading to that Top Event (otherwise the Bow Tie would lead to misleading over-confidence in control)
- threats need to be direct (e.g. factors affecting human performance – such as drug and alcohol abuse – are not a direct cause but combine with other factors to affect Fitness for Duty, so Human Error might be the threat and alcohol and drug abuse would be a Barrier Decay Mechanism under Fitness for Duty)
- threats can be combined where the barriers deployed are the same
- all threats should lead logically to the Top Event and thereafter to every Consequence
- threats should be independent – two threats should not need to combine to lead to the Top Event (e.g. wet weather threat and dust on conductors threat leading to Top Event: short circuit – better to define a threat as Exposed Dirty Conductors)
- barrier failures are not threats (e.g. maintenance is a barrier, and failed maintenance is the barrier failing in a threat pathway and not a threat itself – referring back to the Fault Tree model: Threat AND Maintenance fails – so the Threat here will be corrosion or mechanical wear, etc.)

A good example of threats developed in the BORA Project [5] is provided in Table 3. This separates several threat classes from technical degradation of the system to external events. When building a Bow Tie the team needs to assess all these events to see if any can be combined or whether any can be eliminated as too unlikely.

Table 3. Example threats for offshore process safety barrier diagrams (from [4])

Initiating Event Type	Initiating Events
A. Technical degradation of system	1. Degradation of valve sealing 2. Degradation of flange gasket 3. Loss of bolt tensioning 4. Fatigue 5. Internal corrosion 6. External corrosion 7. Erosion 8. Other causes
B. Human intervention introduction latent error	1. Incorrect blinding/isolation 2. Incorrect fitting of flanges or bolts during maintenance 3. Valve(s) in incorrect position after maintenance 4. Erroneous choice or installations of sealing device 5. Maloperation of valve(s) during manual operation* 6. Maloperation of temporary hoses.
C. Human intervention causing immediate release	1. Break-down of isolation system during maintenance. 2. Maloperation of valve(s) during manual operation* 3. Work on wrong equipment, not known to be pressurised
D. Process disturbance	1. Overpressure 2. Overflow / overfilling
E. Inherent design errors	1. Design related failures
F. External events	1. Impact from falling object 2. Impact from bumping/collision

* This may lead to either introduction of a latent error or an immediate release

3.5 Consequences

Consequences are the logical outcome of all the mitigation barriers failing. They are usually grouped into 3-4 broad categories reflecting the range of mitigation measures deployed. Typical categories may include:

- Safety impact (Fatalities/injuries)
- environmental impact
- asset damage
- reputation impact

Consequences are events themselves and can be developed further. This aspect is most relevant if an actual event has occurred and the barrier diagram is being used for accident investigation (as with the BSCAT investigation approach [7] – Barrier-based systematic cause analysis technique). Thus a consequence can be a Top Event leading to further consequences. For example if the consequence is Environmental Impact, this can be converted to a Top Event leading to Consequences of short term Fish death, long term ecosystem damage, water treatment plant closure, etc. There will be prevention and mitigation barriers for this sequence as well.

For this reason consequences in the Bow Tie may not be the actual loss. Other factors may lead to larger or smaller losses.

3.6 Barriers

As noted earlier, a barrier as plotted in a Bow Tie is an AND-gate combining demand and barrier fails. If this happens then the next barrier in the sequence will be tested and so on until the Top Event is reached, and a similar process occurs on the Mitigation side until the Consequence is reached.

It may be seen that if there are 3 effective barriers on each side and each contributes a factor of 10 or more reduction (i.e. 90% effective) then 6 orders of magnitude reduction will occur from the original threat frequency. This is often very hard to achieve. However some Bow Tie diagrams show 10 or more barriers on both sides – and this is almost always incorrect as it implies a 20 order of magnitude reduction in frequency.

Some rule sets for barriers – these apply to both the Prevention and Mitigation side

- Barriers can be physical or non-physical
- barriers are real things that be tested or measured, non-physical should be systems
- barriers can be passive – pressure containment equipment, dikes
- barriers can be active – sprinklers, emergency shutdown systems
- barriers can be critical tasks – procedural systems, inspection programs
- barriers should be independent – this is based on fault tree rule sets
- sociotechnical aspect(s), barriers are mixtures of the above (frequently)

Some rule sets suggest that a barrier must be capable of completely terminating an event pathway. The authors believe this is too stringent. The basic concept being implemented here is the Swiss cheese model and this acknowledges that no barrier is 100% effective. This can be shown by way of example. There are important barriers such as water spray systems for toxic releases of ammonia. These have several issues affecting their effectiveness – their design basis (ammonia leak rate and water discharge rate and droplet size), their leak detection and actuation speed, and the spray deployment. It is common only to erect such sprays in the downwind direction towards significant population exposures. Thus such arrays, regardless of their technical efficiency can never protect leak events drifting in other directions and according to the rule that they have the potential to completely terminate the event – they would be excluded. This would not be correct – even partially effective barriers should be included. There is no real difference between partial effectiveness and reliability issues. Thus we would add a further rule

- Ideally barriers can terminate an accident pathway, but partially effective barriers should be included if these can make a material difference to the outcome

It is in the area of barriers where most errors are found in Bow Tie construction. Typical errors include:

- Too many similar barriers (Procedure 1 + Procedure 2 + Procedure 3, instead of a single barrier Procedural Control)
- inclusion of trivial or hardly effective barriers (e.g. sunglasses as a barrier to vehicle collision) *(Linked to scale of Top Event)*

- confusion of accident event sequence with barriers (e.g. Ignition – the correct barrier would be Ignition Controls, and if these fail then ignition may occur)
- inclusion of system element barriers on a main pathway rather than the system itself (e.g. escape and evacuation, escape lighting, muster area, etc., rather than Emergency Response System – the system details belong in the Barrier Decay Mechanism build-out as, if they fail, the Emergency Response System is degraded)

3.7 Barrier Decay Mechanism / Escalation Factor

This aspect of Bow Ties is poorly understood and many errors are seen here. The title that is most often used (Escalation Factor) contributes to this as there is no escalation of the event but there is an increase in failure likelihood. For this reason, the title Barrier Decay Mechanism is preferred as this clearly indicates the function of this pathway. The function is easily understood in fault tree terms (Figure 5). The main pathway barrier should be holistic procedural or administrative system or a complete hardware system. These systems have component elements and failures in these degrade the overall system and, thus, should properly be collected in the Barrier Decay mechanism.

Consider a critical shutdown procedure that is to be invoked if certain process parameters are exceeded. Key elements of such a procedure might be:

- Procedure correct and up-to-date
- staff all trained in the procedure
- staff competent to recognize the event and confident to act in a timely manner
- staff fit for duty
- staff supervision

It would be incorrect to include all of these on the main pathway. If simple order of magnitude reductions for each barrier are applied (i.e. 90% effective) this would imply that the elements of this procedure contribute to a 6 orders of magnitude risk reduction. This is wrong. This difference is more clearly shown in Figure 7 – located at end of paper.

A superficial inspection of part a) of the figure might satisfy the reader that the threat is very well protected – with 7 barriers. But many of these are dependent barriers and the correct manner to draw the barrier diagram is shown in part b) with a Barrier Decay Mechanism. This is titled Shutdown fails and it transposes all the main pathway shutdown barriers from part a) into the Barrier Decay Mechanism and leaves one overall Shutdown procedure. The diagram now correctly shows only 2 barriers and this is generally insufficient for an exothermic reactor hazard.

Some conclusions from this are that a Procedure Document is not a main pathway barrier – a piece of paper does not reduce risk. Similarly Training should not be a main pathway barrier. Training supports the real barrier, the Procedural Control.

3.8 Scoring Barriers

A key benefit of barrier diagrams is that they make explicit the critical safety barriers. An obvious benefit is to judge whether there are sufficient barriers in place for each threat. While there is no formally agreed method, Shell [3] outlined its approach that identifies needed barrier counts:

- For high risk threats – 3 barriers on both sides of the Bow Tie
- for medium risk threats – 2 barriers on each side
- for low risk threats – 1 barrier on each side

A surplus of barriers on one side cannot make up for a deficiency on the other side. That target still must be achieved. Also, since there are multiple threats, all with different likelihoods, the number of barriers on the Mitigation side will be determined by the highest risk threat.

Barriers can only be counted on a Bow Tie diagram if the structure is correct. This means that the rules identified above need to be implemented, and especially that Barrier Decay Mechanism barriers do not appear in the main pathway. When drawn correctly, it is often very difficult to achieve 3 barriers on the Prevention side (left side), although somewhat easier on the Mitigation side (right side).

3.9 Overall Complexity

Some Bow Tie diagrams can become very complex. One example is shown in Figure 6 – this is deliberately made small to show the complexity – it is not intended to be read. This may be correct, but it is too complex for good communication. However, without detailed examination, there appear to be some issues with this figure – on one Threat arm there are 9 barriers and on one Consequence arm there are 11 barriers. If this were true, then these pathways would have infinitesimal likelihood. More likely, this may mean that Barrier Decay Mechanism barriers have been located onto the main pathway. If the intent is purely to log barriers this is acceptable, but if the aim is to judge whether there are sufficient barriers for each threat – then this diagram may mislead and lead to incorrect safety expenditures.

Figure 6. Overly complex Bow Tie diagram

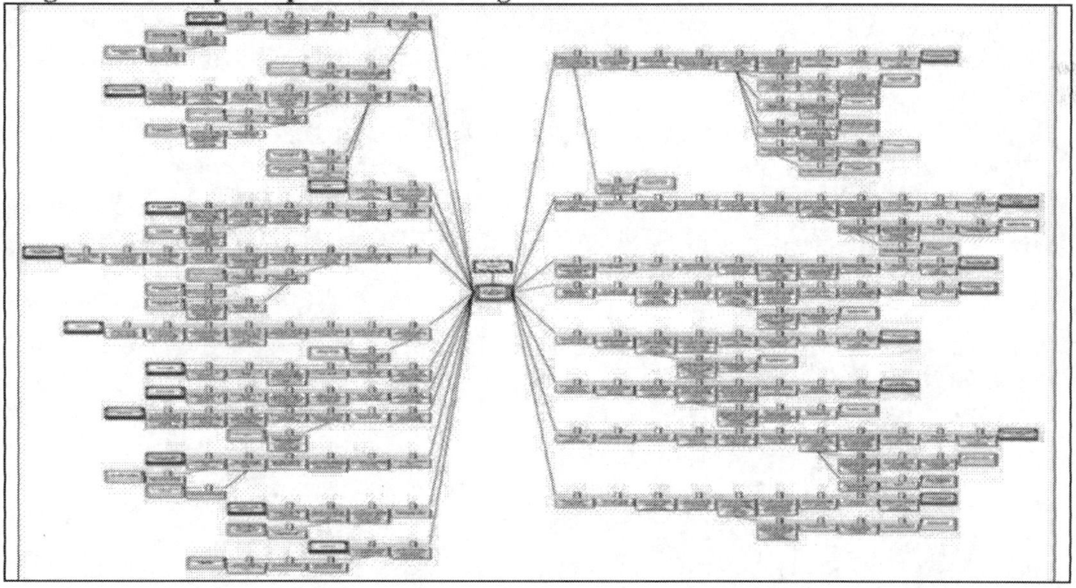

There are conflicting goals with Bow Tie diagrams. For facility managers and safety personnel – they need to know about all barriers. For regulators they may only need to know about main pathway barriers, not Barrier Decay Mechanisms. Individual operational or maintenance staff need clarity and simplicity as they might be confused if all detail for all staff is shown. They need an overview of the system of controls and all of the details for their contribution. Typically, software for Bow Tie diagrams provides for these different needs and diagrams can be customized from the same dataset to show only what is needed.

4. Conclusions

The Bow Tie barrier diagram method is growing rapidly in popularity – partly due to its sharp focus on major accidents, its compact format, and its ability to communicate critical safety barriers to staff and contractors. While several published documents exist describing the method, there is no globally accepted standard. This has led to many quality issues that detract from its use in an operational environment.

This paper has shown that the Bow Tie method should be carefully linked to its fault tree under-pinnings. FTA is the well proven and theoretically sound method that gives the Bow Tie diagram a real basis in safety science, much beyond being a convenient picture and aid for communication. One of the most important of these benefits is an ability to judge whether there are sufficient barriers for each identified threat.

The authors encourage the development of a standard document that would aid the industry in developing better quality diagrams without structural errors. Software can be used to increase or decrease the amount of detail shown to aid communication.

5. References

[1] Marsh (2011) 100 Largest Losses (1972–2011), Large Property Damage Losses In The Hydrocarbon Industry (22nd Ed), published by Marsh and McLennan Group of Companies.

[2] Zuijderduijn C., RISK MANAGEMENT BY SHELL REFINERY/CHEMICALS AT PERNIS, THE NETHERLANDS Implementation of SEVESO-II based on build up experiences, using a Hazards &Effects Management Process., EU Safety Conference Athens (2000)

[3] Salvi O., Debray B. (2006), A global view on ARAMIS, a risk assessment methodology for industries in the framework of the SEVESO II Directive, Journal of Hazardous Materials 130, 187–199

[4] BORA Handbook (2007) Barrier and Operational Risk Assessment, Rev 0, Norway Preventer Report for NFR / HSE / OLF.

[5] NASA (2002) Fault Tree Handbook with Aerospace Applications (Ver 1.1) Lead author William Vesely, available from www.nasa.gov

[6] Sklet S (2006) Safety barriers: Definition, classification, and performance. Journal of Loss Prevention in the Process Industries 19, 494–506.

[7] Pitblado R.M. and Fisher M., (2010), Novel Investigation Approach Linking Management System and Barrier Failure, European Refining Technical Conference (ERTC), Barcelona.

Figure 7. Comparison of incorrect and corrected barrier layouts

a)

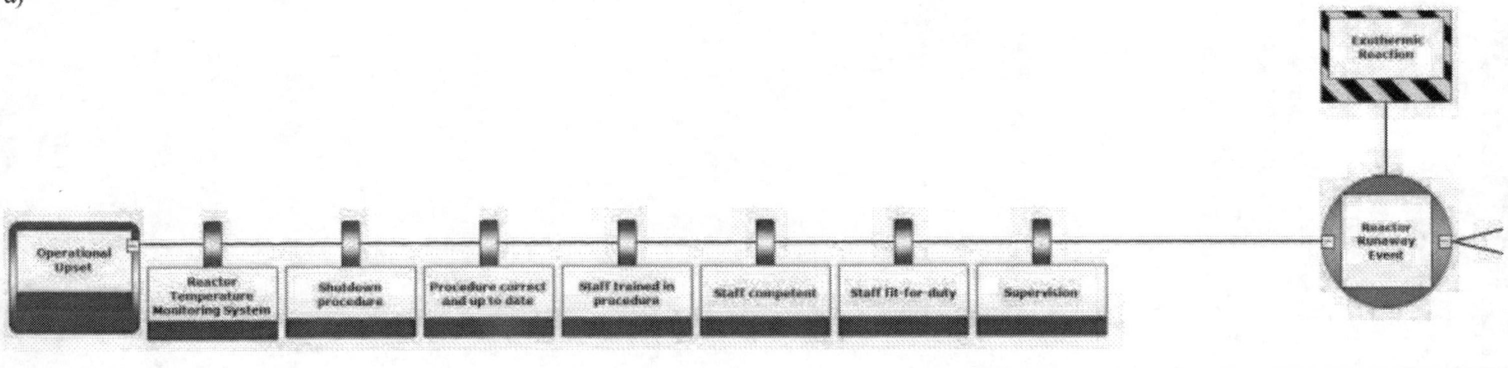

b)

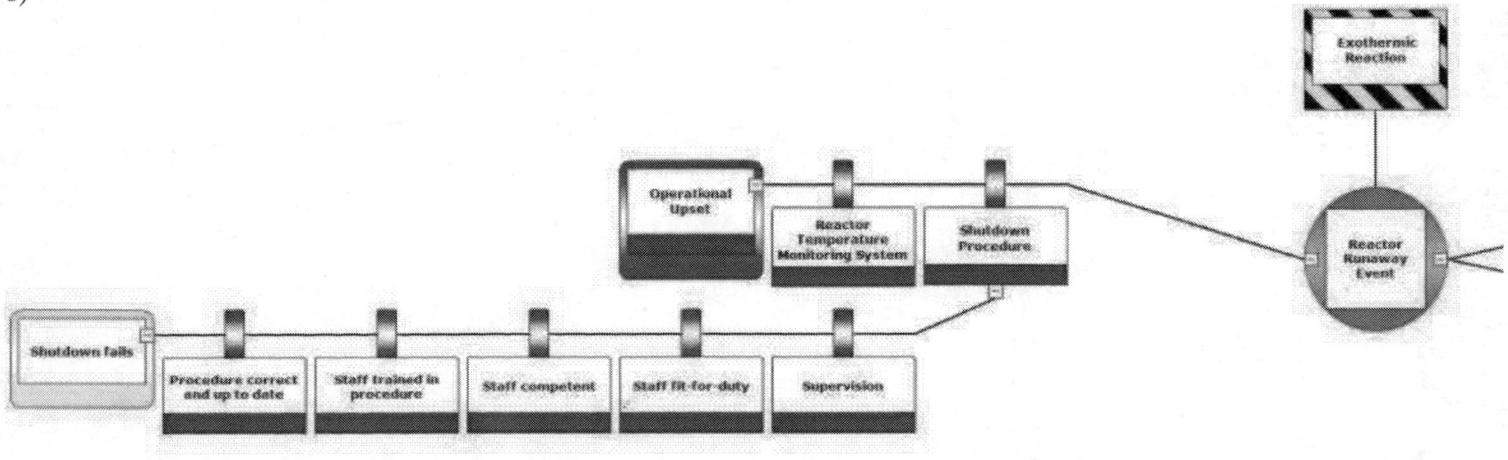

(45a) Pilot and Full-Scale Application of Technologies for Mercury Control

Noah Meeks, Research & Environmental Affairs, Southern Company Services, Inc., Birmingham, AL

With the final MATS rule going into effect in 2015, electric utilities must now consider mercury control as a needed technology in their emissions train. This talk wiill cover recent full-scale injections of activated carbon and alkali injection, as well as pilot-scale studies of non-halogenated carbons or non-carbon sorbents. We will also discuss recent mixed matrix membrane pilot studies. **(Most recent data will be included in the abstract closer to the meeting)

(45b) Acid Tunneling Vs Acidizing : A Case Study of Rudatain Oilfield-Kuwait

Adel Elsharkawy, Petroleum Engineering, Kuwait University, Kuwait, Kuwait and Mohamed Al-Hadher, Pet Eng, Kuwait Oil Company, Safat, Kuwait

Acid Tunneling Vs Acidizing : A Case Study of Rudatain Oilfield-Kuwait

By Mohammed Al-Hadher and Adel Elsharkawy

Kuwait University, College of Engineering & Petroleum

Po Box 5969 Safat 13060 Kuwait

Abstract

Acidizing is one of the most important techniques used in oil industry in order to enhance the reservoirs and increase oil production. There are three types of acidizing techniques which are matrix acid stimulation, acid fracturing, and acid tunneling. In this paper a comparison these types of acidizing is discussed. The case study of RA-0007 acid tunneling completed in Kuwait Oil Company is presented. It represents the first dendritic with extensive worm holes created by acid tunneling for maximum reservoir contact. The whole effect is aptly demonstrated by the change in PI of the well before and after the acid tunneling as mentioned before. Analysis of the well designed and implemented job showed that two tunnels were created successfully in MA-C formation through a two days rig-less operation. The results indicated an improvement in the well productivity index (PI) which had been increased from 1.3 to 3.87 bpd/Psi.

(45e) Sustained Casing Pressure Modeling and Well Integrity

Tony Rocha-Valadez, Mary Kay O'Connor Process Safety Center, Artie McFerrin Department of Chemical Engineering, Texas A&M University, College Station, TX, A. Rashid Hasan, Harold Vance Department of Petroleum Engineering, Texas A&M University, College Station, TX, C. Shah Kabir, EPT, Hess Corporation, Houston, TX and Sam M. Mannan, Artie McFerrin Department of Chemical Engineering, Mary Kay O'Connor Process Safety Center, Texas A&M University, College Station, TX

SUSTAINED CASING PRESSURE MODELING AND WELL INTEGRITY

T. Rocha-Valadez[1]*, A.R. Hasan[2], C.S. Kabir [3] and S.M. Mannan[1]

[1]Mary Kay O'Connor Process Safety Center, Artie McFerrin Department of Chemical Engineering, Texas A&M University, College Station, Texas 77843-3122, US., email: *tonyrochav@tamu.edu, mannan@tamu.edu.

[2] Harold Vance Department of Petroleum Engineering, Texas A&M University, College Station, Texas 77843-3122, US, email: rashid.hasan@pe.tamu.edu.

[3]Hess Corporation, 1501 McKinney Street, Houston, TX 77010, US, email: skabir@hess.com.

With an increasing demand for hydrocarbons and other sources of energy, the methods for delivering such energy needs have become complex. In the case of the oil and gas industry, exploration and production have gone to deeper waters and deeper horizons than ever before, resulting in drilling several miles into the earth's core. Deeper well penetration means encountering increasingly high formation pressures and temperatures. For example, in deepwater drilling, well construction and production management deal with significant heat transfer between the produced fluids and its surroundings. The high-rate production and the attendant heat transfer demand close attention in terms of operational safety and potential environmental impact. One phenomenon that occurs in producing wells is sustained casing pressure (SCP), which presents a risk to both process safety and the environment.

In oilfield terminology, the casing is a steel pipe cemented to the surrounding formation. Multiple concentric casings are involved in a typical well construction. The casing immediately surrounding the production tubing is known as the production casing, which is most vulnerable to thermal and mechanical stresses and fluid leakage. The cement sheaths provide zonal isolation, support the casing, and protect it from corrosive fluids. Ideally, pressures monitored at the wellhead for multiple casings should change minimally after being bled off through a needle valve at the surface. When the pressure rebuilds after being bled off, that casing is said to exhibit SCP [1].

SCP occurs because of stresses that the system experiences owing to fluids production up the tubing. Besides the normal source of SCP, a poor cement job may allow hydrocarbon migration from the producing or other horizons and accumulation into one or more of the casings. The risk of not attaining well integrity can range from the activation of rupture discs to the blowout of hydrocarbons. Because of the safety and environmental dangers of SCP, regulations require remedial treatments for a casing string that exhibits significant SCP problems. Unfortunately, available methodologies to test for SCP are often qualitative and empirical in nature [2, 3], rendering them inapplicable in many situations. No practical systematic testing procedure has emerged to quantify SCP.

In this paper, we present a model for SCP that is rooted in the transport processes of the system. The model assumes gas leaking through the cement barrier, and then migrating up the annular-liquid column. It accounts for the fluid-temperature profile along the wellbore and liquid compressibility. Some robust assumptions about the transport processes allow us to write the governing equation as a first-order, linear-differential equation that is solved with appropriate boundary conditions. The resulting algebraic expression for the casing pressure rise as a function of time paves the way for all applications. A comparative study of the model's performance with field data suggests excellent agreement for wells where the casing pressure increase is owes gas influx into the annulus. The proposed model could form the basis for standardization of SCP testing. This study also offers some test procedures to assess SCP-related safety concerns.

References

1. Bourgoyne Jr., A.T.J., S.L. Scott, and W. Manowski, ***A Review of Sustained Casing Pressure(SCP) Ocurring on the OCS***, 2000, Final Report Submiteted to MMS.

2. Xu, R. and A.K. Wojtanowicz, ***Diagnosis of Sustained Casing Pressure from Bleed-off/Buildup Testing Patterns.*** SPE, 2001: p. Productions and Operations Symposium.

3. Wojtanowicz, A.K., S. Nishikawa, and R. Xu, ***Diagnosis and Remediation of Sustained Casing Pressure in Wells***, July 2001, MMS.

(45f) Evaluation of Particle Growth in an Electrochemical Fluidised Bed Reactor

Emmanuel Ehirim, Rivers State University of Science and Technology, Port Harcourt, Nigeria

The process of electrodeposition within electrochemical reactor results in the increase of particle sizes, which increases the volume of the bed. This work entails a mathematical model to describe the effect of this phenomenon on the behaviour of a three-dimensional fluidised bed electrode operating under the condition of electron transfer. The model incorporates the pseudo-stationary behaviour in the reactor since the bed increases in volume due to the particle size growth. The Orthogonal Collocation method was used to solve the developed differential equations. The overpotential distribution and the reaction rate within the reactor have been calculated. The principal results are the effects of the electron transfer operation on the superficial displacement with time. The effects of total applied current, electrode geometry and the bed porosity on the behaviour of the system have been analysed. The result shows a very gradual increase in the bed height.

(45g) Debottlenecking Gas Dryers and Desiccant Adsorbers

Benjamin A. Schmitt, Hydrocarbon Processing Industry (HPI) - Business Unit, Johnson Screens, Inc., New Brighton, MN and Michael Ekholm, Engineering, Johnson Screens, Inc., New Brighton, MN

Debottlenecking Gas Dryers and Desiccant Adsorbers

Though Maximization of the Useable Bed Volume with Johnson Screens' SSG™

Benjamin Schmitt, PE

Senior Mechanical Engineer, Technology

Michael Ekholm, PE

Applications and Product Design Manager

Johnson Screens Inc. St. Paul, MN

AIChE Spring Meeting, April 2013

13th Topical Conference

San Antonio, TX

Introduction:

The following presentation of the Shaped Support Grid (SSG™) releases additional supporting data and updates to the initial presentation given by Robert G. Norell at the 2010 AIChE Spring Conference.

Limitations on ethylene dryers and desiccant adsorbers can create bottlenecks in the production process which can lead to a need for increased dryer and adsorber capacity. Capital intensive investments may be necessary in order to eliminate the bottleneck which may include the addition of an additional vessel. With limits on the plot plan, this may not be a workable solution. Adding to the problem is the amount of time consumed in engineering, planning and construction, which can have negative effects on the operation of the ethylene unit. Increased desiccant volume and operational efficiency can increase the unit's production with minimal capital expenditures. Johnson Screens', patent pending, SSG™ can add the increased process efficiency and cycle life necessary to extend the operating capability of the unit.

Extending the operation, though debottlenecking, can be achieved by maximizing the time before breakthrough of the media bed. There are three primary functions that maximize the breakthrough time:

Proper gas distribution in both normal operation and regeneration conditions which maximizes the bed utilization.

Increased molecular sieve or desiccant bed volume are achieved through the elimination of traditional bed supports comprised of support grids and beams. The support interface for the screen configuration also allows for thermal expansion typically present in vessels with a thermal cycle as part of the process. The design allows for thermal cycling without the risks of seal or structural failure inherent in other solutions.

Minimizing the bed volume consumed by the distribution and support system increases the net desiccant volume lost when utilizing alternative bed supports.

Johnson Screens ***SSG*** is designed to provide the industry with an alternative, cost effective, approach to increasing capacity. Supporting data and preliminary case studies will be depicted to quantify the benefits that are achievable through the implementation of the ***SSG***.

Conclusions:

The use of a Johnson Screens ***SSG*** provides a reliable immediate solution to reduce capacity bottlenecks. The ***SSG*** increases the bed volume, when compared to traditional bed support arrangements comprised of grids and beams. The increase in desiccant volume will allow for increases in feed flow rate and/or adsorbtion cycle time. The ***SSG*** maximizes the flow distribution during desiccant regeneration, maximizing the utilization of the complete cross section of the vessel. This solution can be implemented in both new and existing vessels, providing consistency in design between planned increases in performance of current vessels and future expansions.

The modular design of the ***SSG*** allows for easy installation and maintenance. Connection of the ***SSG*** to the vessel is performed within the outlet nozzle, eliminating all welding on the pressure vessel and eliminating any need for recertification of the pressure vessel after installation. The design allows for ease of dismantling for inspection of the vessel head during routine maintenance and inspections. The low profile will also reduce interference with any required media exchanges. The design package of the ***SSG*** provides the end user with a holistic solution that fulfills debottlenecking needs with limited capital expenditures.

Biography of Speaker:

Benjamin Schmitt is a Senior Mechanical Engineer in Johnson Screens' Hydrocarbon Processing Industry Technology Group. Ben graduated from North Dakota State University in 2004 with a Bachelor of Science degree in Mechanical Engineering. He currently working toward his Master's in Business Administration from Hamline University. Ben is member of AIChE and is a registered professional engineer in the state of Minnesota.

AMACS
PROCESS TOWER INTERNALS

AIChE 2013 Spring Meeting Poster board

Design Considerations for effective liquid removal In Suction Drums

Presented By:

Kanti Patel, PE

Compressor down again?

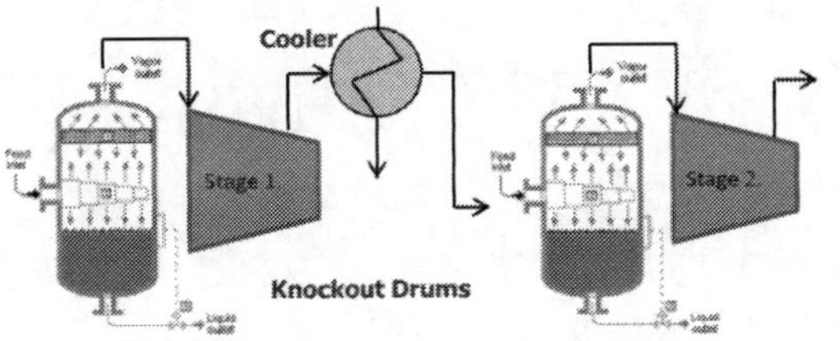

Is liquid carryover the culprit?

<u>We have heard it all before</u>:

-Premature compressor shutdown.

-Production time lost.

-High yearly compressor maintenance costs.

-Impeller damage due to pitting corrosion.

-Corrosion due to salt deposits.

AMACS
PROCESS TOWER INTERNALS

Common Design Problems with Suction Drums which promotes liquid carryover

1. Choosing the correct mist eliminator (M.E. equipment)
2. Knowing where to install the M.E. in a suction drum.
3. Consider a high mist loading in the gas when selecting your M.E.
4. Fouling considerations.
5. M.E. disengaging space, bad gas distribution & liquid slug in the feed.
6. One micron and great droplets, versus submicron mist removal.
7. High viscosity liquids.
8. Dislocated M.E. due to pressure surges.
9. Through put exceeds original design capacity.
10. Capacity range too large for suction drum.

AMACS
PROCESS TOWER INTERNALS

Choosing the correct Mist eliminator equipment

- Proper sizing

+ Use Souders-Brown equation to find optimum velocity. $$Vd = K \cdot \sqrt{\frac{Pl}{Pg} - 1}$$
+ Select the correct vapor velocity factor K. based on the operating pressure, and type of equipment.
    - Lower K-factor 0.01 for every 100 psi
    - K factor to select when using:

Mesh Pad	K= 0.1 - 0.55
Chevron Vane	K= 0.45 - 0.65
MultiPocket® Vane	K= 0.75 – 1.0

AMACS
PROCESS TOWER INTERNALS

Choosing the correct Mist eliminator equipment cont.

- Selection based on droplet size.

Table 2. Droplet sizes (water in air) typically captured with 99.9% efficiency by mist eliminator elements of various types

Element type	Size range microns
Fiber candles or panels	0.1 and larger
Mesh with co-knit yarn	2.0 μm and larger
0.006-inch knitted mesh	5.0 μm and larger
0.011-inch knitted mesh	10 μm and larger
AMACS MultiPocket ® Vane	10 μm and larger
Conventional vane arrays	15 μm and larger

AMACS
PROCESS TOWER INTERNALS

Where to install the M.E. in the suction drum

- Improperly installed M.E. can cause severe liquid carryover problems.
- General accepted spacing guidelines.

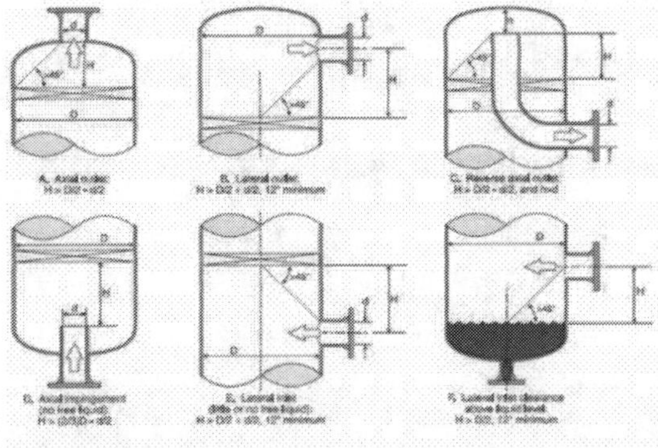

AMACS
PROCESS TOWER INTERNALS

Consider high mist loading when selecting your M.E. equipment

+ High mist loading in the gas stream can flood your M.E. equipment and cause liquid carryover.

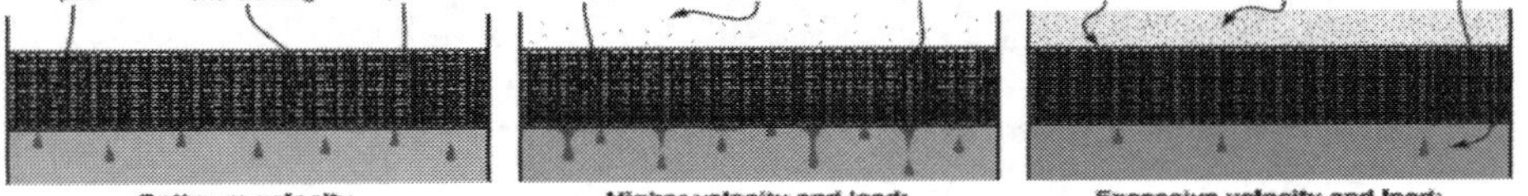

Figure 4. Envisioning mesh pad performance degradation as liquid and gas loads increase (vertical cross-sections)

AMACS
PROCESS TOWER INTERNALS

Consider high mist loading when selecting your M.E. equipment cont.

+ Limited drainage capabilities
  - mesh M.E. <0.5 gal/FT2/min
  - Vane type M.E. <2 gal/FT2/min

+ Select a combination of vane & mesh mist eliminators

AMACS
PROCESS TOWER INTERNALS

Fouling Applications

+ Fouling of mesh mist eliminator will cause liquid carryover.

+ For mild fouling install a water spray system below the mist eliminator for occasional inline cleaning (based on Delta-P)

+ For consistent fouling problems, install a Vane mist eliminator upstream of the mesh mist eliminator.
  - Vanes remove solids and bulk liquids therefore, protecting the downstream mesh mist eliminator

AMACS
PROCESS TOWER INTERNALS

Fouling Applications cont.

+ For consistent fouling problems where the solids are "sticky" install a water spray system below the vane mist eliminator.

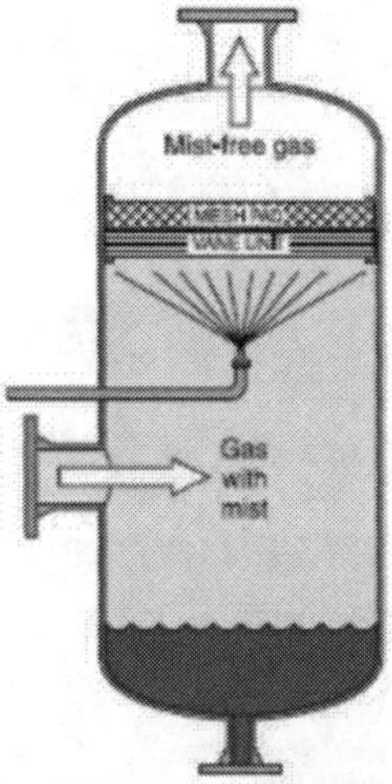

AMACS
PROCESS TOWER INTERNALS

Bad Gas distribution, limited disengaging space and liquid slugs.

+ Bad gas distribution, limited disengaging space in a suction drum can cause severe liquid carryover.

+ CFD analysis can assist in locating high velocity areas.

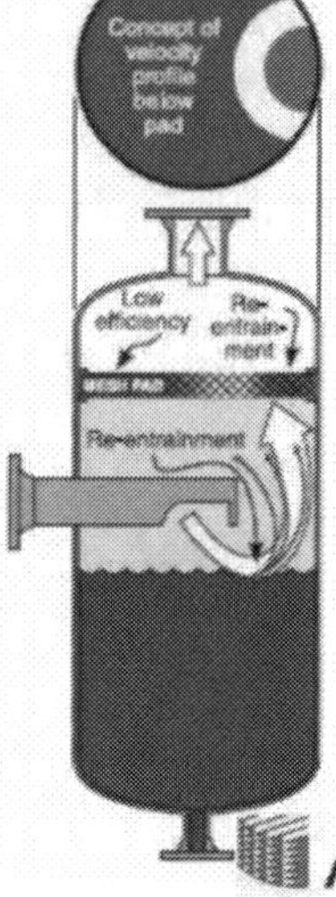

AMACS
PROCESS TOWER INTERNALS

Bad Gas distribution, limited disengaging space and liquid slugs cont.

+ Use inlet diffuser to force uniform gas distribution i suction drum.

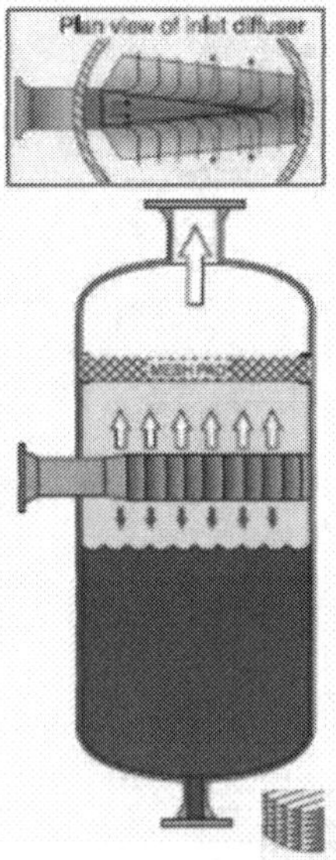

AMACS
PROCESS TOWER INTERNALS

Inlet Feed Device

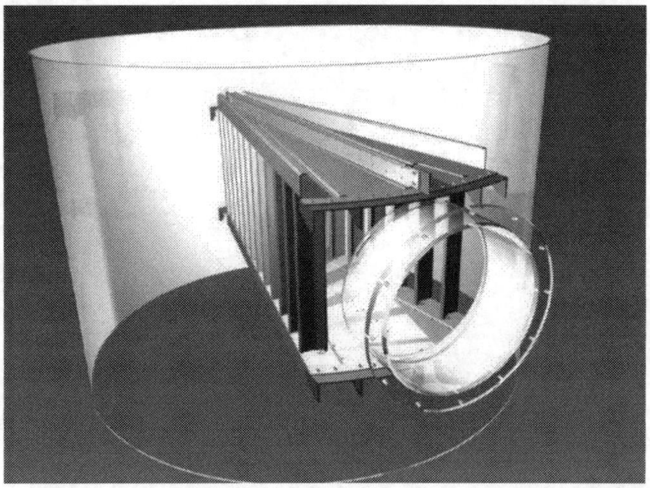

AMACS
PROCESS TOWER INTERNALS

Inlet Feed Device

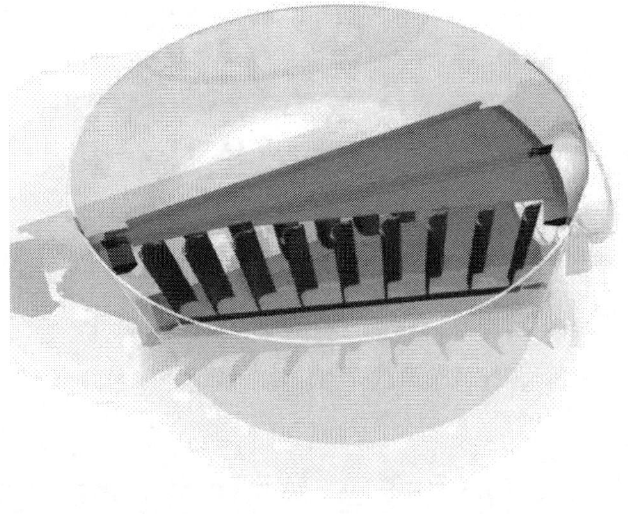

AMACS
PROCESS TOWER INTERNALS

Bad Gas distribution, limited disengaging space and liquid slugs cont.

+ Use a perforated distribution plate downstream of the M.E. for additional distribution assistance.

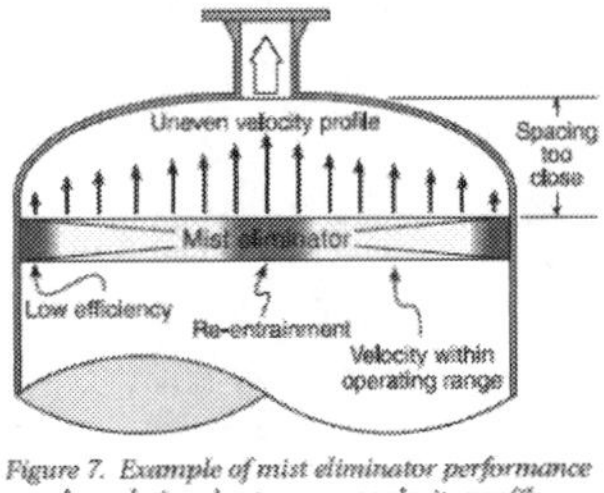

Figure 7. Example of mist eliminator performance degradation due to uneven velocity profile

+ Inlet diffusers also handle incoming liquid slug which otherwise can cause flooding of the Mist eliminator.

AMACS
PROCESS TOWER INTERNALS

Know what mist size droplets You are dealing with

+ Know if your mist is > 1 micron or submicron.
+ When gas travels from A to B, condensation in the pipeline will cause a fine mist (fog or submicron droplets)

  Examples:
  - Incoming gas from an offsite location
  - Fuel gas to furnaces/heaters with low NOx burners

AMACS
PROCESS TOWER INTERNALS

Know what mist size droplets You are dealing with cont.

+ Submicron mist is not captured by conventional M.E.

+ Submicron mist requires a coalescing filter, typically
installed in a two stage filter separator.

AMACS
PROCESS TOWER INTERNALS

Know what mist size droplets You are dealing with cont.

<u>1^{c} stage;</u> submicron droplets are coalesced to > 10 micron droplets.

<u>2^{c} stage;</u> greater than > 10 micron droplets are ejected from the coalescing filter and captured by a vane, which captures droplets > 8 micron.

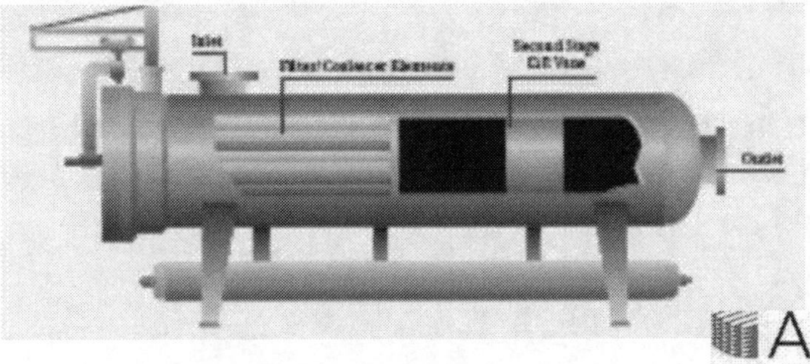

AMACS
PROCESS TOWER INTERNALS

High viscosity liquid as mist

+ Viscous liquids drain slower out of the M.E. and could cause localized flooding.
+ Use a mesh mist eliminator with an open drainage layer on the bottom to improve liquid drainage.
+ Use a lower K- value (lower face viscosity) for the M.E. To allow additional drainage time.
+ Use a vane type M.E. upstream of the mesh mist eliminator, to remove the bulk of the viscous liquid, so the M.E. can effectively handle the residual liquids.

Dislocated M.E. due to System pressure surges

+ M.E. sections can be dislodged into the downstream equipment, which then creates massive carryover.
+ Standard M.E. equipment often gets installed and secured to the support ring with tie wire.

Dislocated M.E. due to System pressure surges cont.

+ Where system pressure surges are common, there are more secure ways to install the M.E.
  - dual support rings
  - through bolts
  - install a pressure relief door

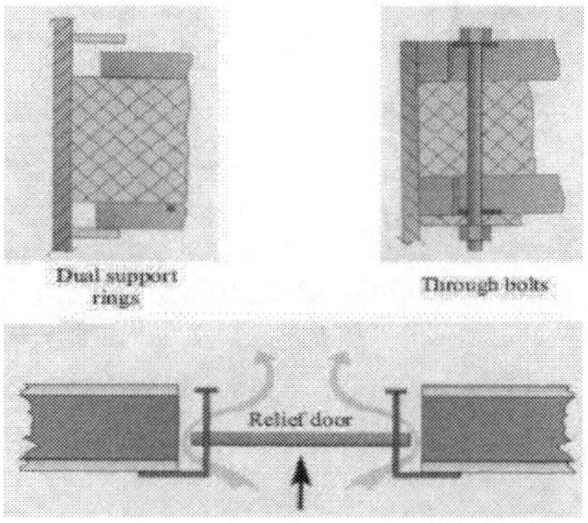

AMACS
PROCESS TOWER INTERNALS

When throughput exceeds original capacity

+ There are various solutions to increase capacity in an existing vessel.
  - Patented High capacit
    mist eliminator 0% - 1(

AMACS
PROCESS TOWER INTERNALS

When throughput exceeds original capacity cont.

-Patented MISTFIX™ cylindrical mist eliminator 0% - 25%

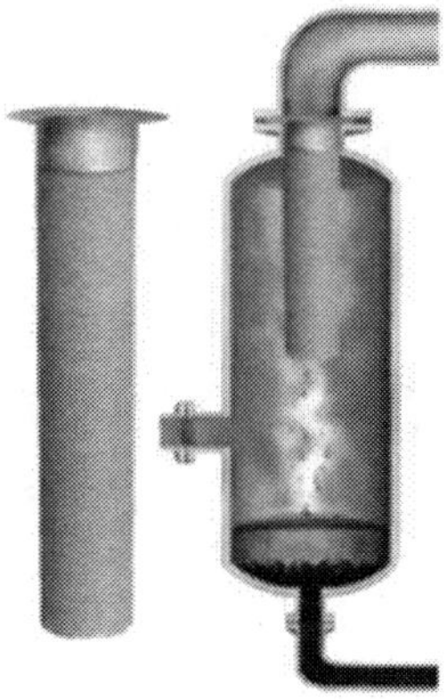

When throughput exceeds original capacity cont.

- High capacity Patented MultiPocket® vane. 0%-50%

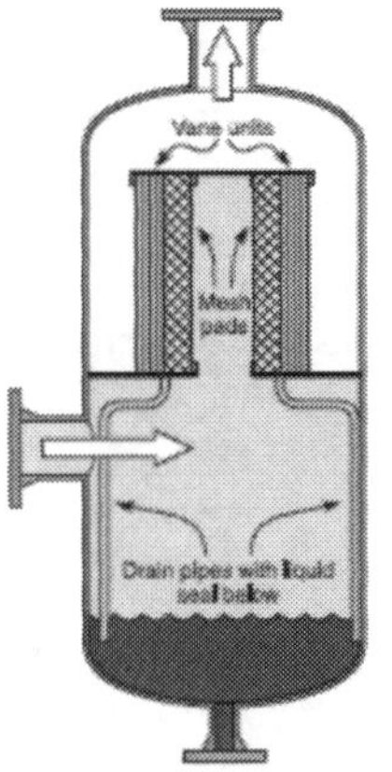

AMACS
PROCESS TOWER INTERNALS

When throughput exceeds original capacity cont.

+ MistFix™ is installed through the gas outlet, all other M.E. equipment comes in sections, installed through the manway.
+ With increased capacity comes increased bad gas distribution.
  - Use an inlet diffuser.
+ To avoid welding new support rings and housing for the high capacity new equipment, use double expansion rings, a compression fit.

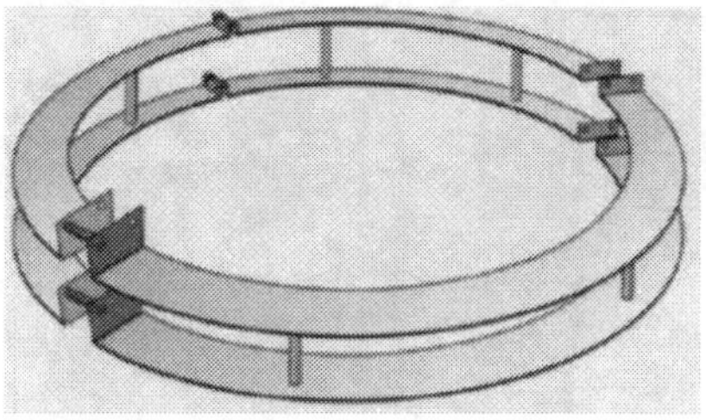

AMACS
PROCESS TOWER INTERNALS

Support Rings

AMACS
PROCESS TOWER INTERNALS

Support Rings

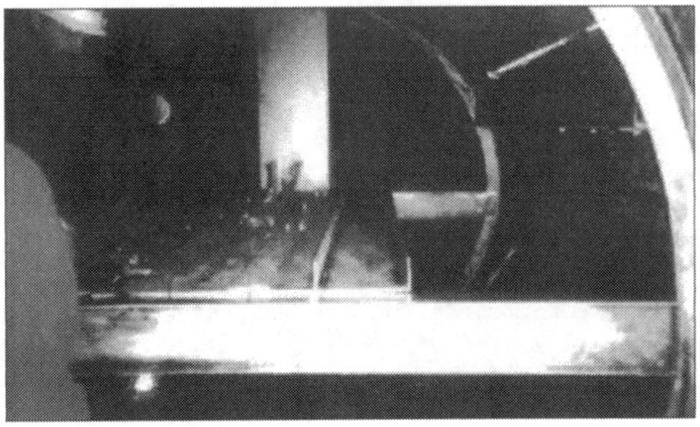

AMACS
PROCESS TOWER INTERNALS

When the capacity range is too large for a suction drum

+ Typical M.E. equipment will operate between
  30% – 110% capacity.

+ When a suction drum often operates below 30% liquid carryover may occur.

AMACS
PROCESS TOWER INTERNALS

When the capacity range is too large for a suction drum Cont.

To facilitate a capacity range of 10% - 110% consider a 2 phase M.E. in the same drum.

- one that operates 30% - 1
- second that operates between 10% - 30%

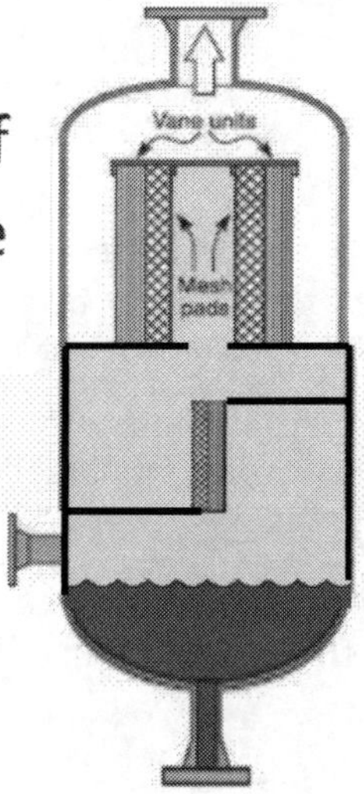

(45i) Renewable Feedstock for Steam Crackers: Catalytic Upgrading of Crude Tall Oil (CTO) Into Bio-Naphtha

Jinto Anthonykutty, Juha Linnekoski, Antero Laitinen and Ali Harlin, VTT Technical Research Centre of Finland, Espoo, Finland

Renewable feedstocks from bio resources are better alternatives to petroleum derived naphtha for the production of olefins and aromatics by steam cracking or reforming.[1, 2] Mostly, these renewable feedstocks (bio-naphtha) rich in n-alkanes are produced by catalytic upgrading methods via hydrotreating/hydroprocessing; in which oxygenates are removed from bio-oil by various deoxygenation processes. At present, hydrotreated vegetable (HVO) oil is produced in many countries on industrial scale using existing petroleum refinery methods[3]. However, employing higher priced vegetable oils could be a major concern in these processes and the main challenge is therefore finding an inexpensive and non-food chain affecting feedstock. Tall oil, obtained as a by-product from Kraft pulping process, is abundant and cheap;and hence, meets the criteria of a low cost starting material for upgrading[4]. Moreover, the utilization of tall oil in conventional oil refineries could serve as an eco-friendly method to convert bio-wastes into bio-naphtha[1].

In the present research approach, catalytic upgrading (hydrotreating) of crude tall oil (CTO) into bio-naphtha suitable for a steam cracker is discussed. CTO, which contains mainly fatty acids (42.9 wt. %) and resin acids (25.7 wt. %) was obtained from Stora-Enso Kaukopää (Finland) mill and used as received. A commercial NiMo catalyst was employed and the experiments were conducted in a continuous, down flow, fixed bed reactor at different process conditions (WHSV 1-3h^{-1}, Pressure 5 MPa and Temperature 350 - 450 °C). After the stabilization period (6 h.), the liquid sample was collected and separated into organic and aqueous phases.The product distribution in organic phase was analyzed with a Shimadzu GC/MS equipped with a Zebron MS 30m × 0.25 mm × 0.25μm column.

This current approach reports that CTO produces mainly paraffins, cycloalkanes and aromatics during upgrading process. Paraffinic hydrocarbons mainly comprise n-octadecane (20-25 wt. %) and n-heptadecane (15-20 wt. %), are obtained in maximum yield (55- 60 wt. %) in low temperature reactions. The yield of paraffinic hydrocarbons increased as a function of increasing space time in low temperature reactions. 18-Norabietane and other mono-, di- and tri- cycloalkanes are obtained from resin acids in CTO, and these cyclic structures are converted to aromatics such as naphthalene and phenanthrene in high temperature reactions (>400 °C). Stigmastane, apparently produced from sterols in CTO are also present as a major three ring component in the product stream. Additionally, gaseous products obtained and the yield increased as a function of temperature. The trend observed with the distribution of gaseous products reveal that decarboxylation and cracking reactions are favored at higher temperatures, while hydrogenation and hydrodeoxygenation (HDO) are suppressed.

The elemental composition of CTO shows that it contains relatively high sulfur content (1800 mg/kg), could be useful to keep the catalyst's activity with time on stream (TOS). However, the considerable amount (87 mg/kg) of alkali metals (Na, K, Ca and P) in CTO might be critical for the catalyst activity with TOS, will be discussed in this contribution.

The product composition obtained with CTO is assessed in comparison with compositions and characteristics of refinery naphtha feeds, and will be discussed in detail. Furthermore, this contribution will discuss the achieved deoxygenation rate and success criteria for the product composition of upgraded CTO.

References

1. Gornay, J.; Coniglio, L.; Billaud, F.; Wild, G. Steam Cracking and Steam Reforming of Waste Cooking Oil in a Tubular Stainless Steel Reactor with Wall Effects. ***energy&fuels***, **2009**, 23(11), 5663-5676.
2. Pyl, S.P.; Schietekat, C.M.; Reyniers, M.F.; Abhari, R.; Marin, G.B.; Van Geem, K.M. Biomass to olefins: Cracking of renewable naphtha. ***Chemical Engineering Journal***, **2011**, 176-177, 178-187.
3. Sunde, K.; Brekke, A.; Solberg, B. Environmental Impacts and Costs of Hydrotreated Vegetable Oils, Transesterified Lipids and Woody BTL- A Review. ***Energies***, **2011**, 4(6), 845-877.
4. Coll, R.; Udas, S.; Jacoby, W.A. Conversion of the Rosin Acid Fraction of Crude Tall Oil into Fuels and Chemicals. ***energy&fuels***, **2001**, 15, 1166-1172.

AIChE Paper Number 45j
High Efficiency Ethylene Cracking Furnaces

Rajaram Ramesh

Product Development Manager - Ethylene
Technip
Netherlands

Marco W. M. van Goethem, PhD

Product Development Manager – Modeling and Testing
Technip
Netherlands

25th Ethylene Producers Conference
AIChE 2013 Spring National Meeting
San Antonio, Texas, April 29 – May 3, 2013

Abstract

Environmental and economic pressures generate a driving force to produce products more efficiently. A direct effect of more efficient processes is that the emission of green house gases, particularly carbon dioxide decreases. In households nowadays "high efficiency furnaces" are available with efficiencies in the range of 94-97%. This paper describes the outcome of a feasibility study on applying the high efficiency furnace concept to steam cracking furnaces. The high efficiency is achieved by enhanced heat recovery from the flue gases before they are released to the atmosphere at ca. 65°C.

Technip studied the application of this concept for cracking furnaces based on gas as well as liquid feedstocks. Natural gas was used as basis for the energy price. Since energy prices have a strong impact on project economics, the evaluations were done for an EU average natural gas price and a US natural gas price.

Two alternative heat sinks have been considered: preheating of combustion air and preheating of cold hydrocarbon feed. Both are viable for liquid cracking furnaces. In the case of gas cracking furnaces feed preheating results in a pinch and this makes it not viable.

Capital costs are based on Technip's cost database and quotations from air preheater vendors. Special attention has been paid to the material selection of the affected convection section banks.

This study shows that high efficiency ethylene cracking furnaces are economically viable. For Europe the study indicates an attractive payback period. In the US the payback period is longer due to the lower energy prices. If energy prices increase in the future, the concept's economic viability will become better. Inclusion of CO_2 credits will make the payback period more attractive.

Furnace Efficiency

In this paper furnace efficiency has been calculated as net thermal efficiency as interpreted by API560[1]. Furnace efficiency is determined by drawing an envelope around the furnace and including in it the radiant section, convection section and any heat exchanger equipment that recover heat from the flue gas (air preheater, economizer, steam superheater, etc.). The efficiency is the ratio of the useful heat absorbed to the fired heat of the fuel and any other external heat inputs. Ambient temperature is a reference temperature for efficiency calculations and any heat inflow or outflow above this reference is taken into consideration as a heat input or heat loss.
Furnace efficiency can be determined as follows:

$$\frac{heat\ absorption\ in\ radiant\ section\ +\ convection\ section}{(firing\ rate\ \times\ fuel\ heating\ value) +\ sensible\ heat\ of\ fuel\ and\ air\ above\ ambient\ temp. +\ other\ external\ heat\ inputs}$$

Alternately, Furnace efficiency $= 100\% -$ Total heat loss, where:

$$\text{Total heat loss} = \frac{stack\ loss\ +\ radiation\ heat\ loss}{(firing\ rate\ \times\ fuel\ heating\ value) +\ sensible\ heat\ of\ fuel\ and\ air\ above\ ambient\ temp. +\ any\ external\ heat\ inputs}$$

In a furnace where fuel and combustion air enter the furnace at ambient temperature and there are no other external heat inputs, the first expression can be simplified to the following:

$$Furnace\ efficiency\ = \frac{heat\ absorption\ in\ radiant\ section\ +\ convection\ section}{fuel\ firing\ rate\ \times\ fuel\ heating\ value}$$

Please note that net thermal efficiency, as defined by API 560, is based on the lower heating value of the fuel.

Study objectives

Technip undertook this study with the goal to reduce fuel firing in cracking furnaces by at least two percent. Although this figure does not look very ambitious, it must be remembered that since efficiencies in current designs are already in the mid nineties (upper part of the technology S-curve), further improvements will not be easy and can only be incremental.

Since stack loss is the main source of inefficiency (radiation heat loss is comparatively smaller and more difficult / expensive to reduce) it was decided to make stack loss reduction as the main focus of the study.

By a relatively straight forward calculation it can be shown that the two percent efficiency rise target would require the flue gas to be cooled to approximately sixty or seventy degrees Celsius.

The main challenge in reducing stack temperature to this level is finding an appropriate heat sink for the recovered heat. In a cracking furnace there are two suitable heat sinks, combustion air and hydrocarbon feed, and both were studied. The following sections discuss the results for liquid furnaces and gas furnaces.

Basis of Study

We have done the study on a naphtha furnace and an ethane furnace. We have chosen these two feedstocks since a large number of furnaces operate on these two feeds and we can draw conclusions that can be applied approximately to other liquid and gas feeds. The study has been performed on the following bases:

- The reference liquid and gas furnace designs used in this study are based on actual furnaces from Technip's past projects. The furnaces were simulated using Technip's proprietary tool SPYRO which is used in the design and simulation of cracking furnaces.
- The budgetary capital cost need to achieve the target furnace efficiency was calculated as an incremental cost. It was estimated from vendor quotes and Technip's internal cost database.
- The cost estimates are for a new project, not a revamp, and are for a West European location.
- Operating costs are based on cost of base utilities and derived utilities. Base utilities are natural gas and demineralized water. These are taken from market data.
- Since natural gas prices are very different in different parts of the world, the operating costs and economic analysis have been done for European and United States gas prices.
- EU natural gas price was fixed at €25.1 /MWh of higher heating value (HHV) which is equivalent to €28.0 /MWh of lower heating value (LHV). The prices are based on EUROSTAT data for 1st half of 2010 and were obtained from the publications of the Department of Energy & Climate Change in the UK. They correspond to the EU-15 average for Large Industrial Consumers (i.e. 2.13 pence/kWh HHV).
- US Natural gas prices were obtained from the US Department of Energy. The average value was 6 \$/MM BTU (HHV) or 17 €/MWh (LHV) for 1st half of 2010. The average HHV used is 1029 BTU per cubic foot or 1.029 MM BTU per thousand cubic feet.
- Cost of demineralized water was assumed at € 2.00 /ton.
- The prices of other utilities (boiler feed water, steam and electricity) have been calculated from those of the two base utilities. Electricity was priced assuming 35% overall efficiency of power generation.
- The economic calculations do not include any CO2 credit for the fuel saved.
- Assumed exchange rate: 1 US dollar = 0.75 euros
- The plant is assumed to operate 8400 hours per year

The resultant utility prices used in the study are summarized below.

UTILITY COSTS

		EU basis	US basis	
FUEL as NATURAL GAS	Base	**28.0**	**17.0**	€ / MWh
DEMINERALISED WATER	Base	2.0	1.5	€ / ton
VHP STEAM (net value)	Derived	23.7	14.4	€ / ton
MP STEAM (net value)	Derived	13.9	8.5	€ / ton
LP STEAM	Derived	9.7	5.9	€ / ton
ELECTRICAL POWER	Derived	80.1	48.6	€ / MWh

Study Results

Naphtha Furnace using Air Preheat

Combustion air can be preheated in a plate type air preheater (APH). These are cross flow type exchangers and have low pressure drops. Combustion air is at a lower temperature than any of the process or utilities streams entering the convection section; therefore, from the standpoint of efficiency, APHs are ideally suited to be the last heat recovery section before the stack.

Fig 1 shows the Base Case Naphtha furnace while Fig 3 shows the same furnace with an APH. Please also refer to Table 1 for a performance comparison of the APH design with the Base Case naphtha furnace. The following points are of interest:

1) The overall efficiency of the furnace will improve from 94.4 to 97.0% for the APH option. This is mainly due to the lower stack temperature but also partly due to the smaller excess air, which is explained below.

2) This option will require a forced draft combustion air fan to overcome the pressure drop in the APH. As a result the furnace operation will change from induced draft mode to balanced draft. The availability of air in forced draft will enable more accurate control of the excess air and it will be possible to operate the furnace with lower excess air (5% instead of 8%). This leads to a lower flue gas flow, a lower stack loss and therefore better efficiency.

3) There will be a drop of 4.8% in fuel firing which is greater than the efficiency increase. This is because of the lower absorbed heat. The furnace is designed and controlled in such a way as to supply all the required process heat and to absorb the surplus heat into utilities to the extent possible. When the furnace operates more efficiently and the heat demanded by the process media remains unchanged, as in the APH case, there will be less surplus heat available for VHP steam. As a result the VHP steam generation will go down from 79.7 to 76.0 t/h.

4) The heat transfer surface in some parts of the APH will be below the water dew point of the flue gas (around 60°C). This can lead to water condensation and create two possible issues: 1) accumulation of water in the APH and 2) corrosion due to the presence of liquid water, oxygen and carbon dioxide.

   Two measures have been included in the design to counter these problems:
   - Placement of unit such that flue gas flows downward to aid easy draining
   - Use of corten steel to resist corrosion

5) We have evaluated the impact of the air preheat option on NOx emissions using API 535[2)]. The rise in air temperature and the drop in excess air will have opposing effects on the NOx production. The net effect is that the NOx concentration will be close to the Base Case.

   If a DeNOx unit is present, as is often the case in the US, the APH option may help to slightly extend the life of the DeNOx catalyst. This is because the total NOx load will be smaller than the Base Case due to the smaller flue gas flow.

Table 2 summarizes the results of the economic analysis of this option. We have used payback period to compare different options. Payback period is the ratio of the incremental capital cost above the Base Case CAPEX divided by the savings in yearly operating cost.

1) The capital cost (CAPEX) of the APH option includes the cost of the APH, the combustion air fan, extra ducting and structural steel. The operating cost (OPEX) savings is the savings in furnace fuel but this must be offset against the value of the lost VHP steam production (which notionally must be made up in a utility boiler elsewhere) and the additional electricity for the two fans.

2) It can be seen that the payback period is three years based on European gas prices and five years based on American gas prices.

Naphtha Furnace using Feed preheat

Naphtha is normally stored at ambient temperature and is preheated to 60°C using quench water before it is routed to the furnaces. Most cracking furnaces are equipped with a Feed Preheater (FPH) bank in the convection section. If flue gas is utilized to preheat naphtha from ambient conditions the stack temperature can be lowered to the desired range of sixty to seventy degrees Celsius. Till now designers and operators have not taken full advantage of this mainly because of the higher CAPEX and fear of dew point corrosion.

One of the consequences of this feature will be that the quench water duty previously absorbed by the naphtha will now shift to the air coolers in the quench water circuit. The additional electrical power that the air cooler fans will require will be about 10 KW per furnace.

Please refer to Table 3 for an overview of the performance of a naphtha furnace with and without an extended feed preheater. Please note that the Base Case naphtha furnace is the same as the one considered in the APH option.

1) The overall efficiency of the furnace can be improved from 94.4 to 96.0% for this option.

2) Unlike the APH option a combustion air fan will **not** be required. The furnace will continue to operate in induced draft mode.

3) The fuel firing and the VHP steam generation will drop marginally.

4) The upper tube rows of the FPH will be below the water dew point of the flue gas and some condensation will occur. As a precaution we have considered that the top four rows will be of stainless steel.

5) The flue gas NOx concentration will be unaffected by this option. However the total hourly NOx emissions and load on the DeNOx unit, if any, will decrease to the same extent as the fuel firing, which is around 0.6%.

Table 4 summarizes the results of the economic analysis of this option.

1) The capital cost (CAPEX) of this option covers the incremental cost of the extended FPH, extra ducting and structural steel. The OPEX savings is the reduction in furnace fuel after adjusting for the reduction in VHP steam production and the additional electricity for the ID fan and the Quench Water Air Cooler fans.

2) In the Base Case the naphtha feed piping between the external feed preheater and the convection section is insulated for heat conservation. In the FPH option naphtha will be fed at ambient temperature to the furnace and so these lines do not have to be insulated. This will yield some capital cost saving but this has not been taken into account in our analysis.

3) The payback period excluding CO2 credit will be three and a half years with European natural gas and five and a half years with US natural gas.

Ethane Furnace using Air Preheat

Fig 2 shows the Base Case Ethane furnace that was used in the study. The APH arrangement on an ethane furnace will be quite similar to the one on the naphtha furnace. Table 5 shows the impact that the APH would have on the furnace performance. The main points are as follows:

1) The overall efficiency of the furnace will improve from 91.7 to 96.5% for the APH option. This is mainly due to the lower stack temperature but also partly due to the smaller excess air.

2) In this furnace also the APH will require a forced draft combustion air fan. This will allow the excess air to be reduced from 8% to 5% which will result in lower stack losses and a better efficiency.

3) The fuel firing will go down by 11.0%, from 8.2 to 7.3 t/h. The absorbed heat will drop from 116.8 to 110.6 MW. As a result of these the VHP steam generation will go down from 97.0 to 88.8 t/h.

4) A portion of the APH will operate below the flue gas water dew point. Consequently the unit must be designed to prevent water accumulation and corrosion.

5) The impact on NOx was evaluated here also. The rise in air temperature will have a greater impact on NOx than the fall in excess air. We estimate that as a result the flue gas NOx concentration will increase slightly (around two percent). This can be a cause for concern if the furnace does not have a DeNOx unit and is operating close to the limits imposed by authorities. However since the total firing will go down, the hourly NOx production will drop. Plants can consider discussing with authorities to obtain a higher allowable NOx limit,

   If the furnace has a DeNOx unit, the chemical load on the unit will reduce, which should help to obtain a longer catalyst life.

The operating cost savings and payback for this option have been summarized in Table 6.

1) The capital cost includes the cost of the APH, the combustion air fan, extra ducting and structural steel. The operating cost savings is the savings in fuel firing with an adjustment for the value of the reduced VHP steam generation and the extra electricity for the ID fan and FD fan.

2) The payback period with natural gas priced at European rates will be one and a half years. The corresponding figure for US gas prices will be three years.

Ethane Furnace using Feed Preheat

Using feed at ambient temperature does not significantly reduce the stack temperature in a gas furnace. This is because the feed preheat curve has a steeper slope than the flue gas cooling curve, causing a pinch on the hot side of the Feed Preheater. Fig 4 shows the heating and cooling curves of ethane, boiler feed water and flue gas in the convection section. The hatched shape on the plot shows the maximum additional heat duty that can be recovered into feed, over and above the Base Case. The plot clearly shows that when the ethane feed temperature is lowered from the base value of 55°C to 30°C, the stack temperature will drop from 170°C to about 150°C. The flue gas cannot be cooled any more than this modest amount.

For this reason the feed preheat option was not investigated any further.

Conclusions and Summary

The key results from the different study options are listed together below.

Cracking furnace feed	Naphtha		Naphtha		Ethane	
Cold sink	Combustion air		Hydrocarbon feed		Combustion air	
	EU nat. gas	US nat. gas	EU nat. gas	US nat. gas	EU nat. gas	US nat. gas
OPEX savings	€ 0.60 mill	€ 0.36 mill	€ 0.17 mill	€ 0.10 mill	€ 1.27 mill	€ 0.77 mill
Payback time	3 years	5 years	3.5 years	5.5 years	1.5 years	3 years
Salient features of each option	Small number of additional equipment		No additional equipment		Small number of additional equipment	
	Higher CAPEX		Lower CAPEX		Higher CAPEX	
	More OPEX savings		Less OPEX savings		Best OPEX savings	
	Slight effect on NOx concentration		No effect on NOx		Around 2% higher NOx concentration	

All three scenarios have attractive payback periods and will be worth considering in high energy cost countries such as Europe and Asia. In US and other energy rich nations, the economic incentive will be much lower. In the US, where ethane cracking is common and increasing even more, improvement of furnace efficiency with air preheat can be worth considering as the payback time is less than 3 years.

If the options are compared against each other, in our opinion the Air preheating option has more potential, irrespective of feedstock.

The Feed Preheating option in liquid furnaces is cheaper but it also gives proportionally modest savings. Moreover, as earlier mentioned, the feed preheat duty is being shifted from one waste heat source (quench water) to another (flue gas). This option will not lead to a net improvement or cash savings in the plant as a whole unless there is a shortage of quench water and there are other, better uses for the saved quench water duty

Acknowledgements

The authors wish to express their thanks to their colleagues Mr Ashok Laghate, Ms Susan Cooper, Mr Sachin Singare and others in Technip Benelux who contributed to this study.

References

1) API 560: Fired Heaters for General Refinery Service, 4th edition

2) API 535: Burners for Fired Heaters in General Refinery Service, 2nd edition

Table 1
Naphtha Furnace with Combustion Air Preheating – Performance Summary

		Base Case	Air Preheat Case
Naphtha feed	t/h	60.8	60.8
Ambient conditions	-	15°C, 85% R.H.	15°C, 85% R.H.
Air temp. to burners	°C	15	77
Stack temperature	°C	108	61
Excess air	%	8	5
Absorbed heat	MW	113.2	110.4
Overall efficiency	%	94.4	97.0
Fuel flow	t/h	8.4	8.0
Fired heat	MW	119.9	113.9
VHP steam production	t/h	79.66	76.01
NOx concentration		Base	Close to Base

Table 2
Economics of Naphtha Furnace with Combustion Air Preheating

		Air Preheat Case (EU location)	Air Preheat Case (US location)
Savings in fuel firing	million €/yr	1.42	0.86
Cost of lost VHP steam capacity	million €/yr	0.73	0.44
Cost of additional electricity	million €/yr	0.10	0.06
Net OPEX saving	million €/yr	0.60	0.36
Payback	Years	3	5

Table 3
Naphtha Furnace with Feed Preheating – Performance Summary

		Base Case	Feed Preheat Case
Naphtha feed	t/h	60.8	60.8
Feed temperature	-	60°C	30°C
Combustion air temp.	°C	15	15
Stack temperature	°C	108	71
Excess air	%	8	8
Absorbed heat	MW	113.2	114.3
Overall efficiency	%	94.4	96.0
Fuel flow	t/h	8.4	8.35
Fired heat	MW	119.9	119.1
VHP steam production	t/h	79.66	79.63
NOx concentration		Base	Base

Table 4
Economics of Naphtha Furnace with Feed Preheating

		Feed Preheat Case (EU location)	Feed Preheat Case (US location)
Savings in fuel firing	million €/yr	0.19	0.12
Cost of lost VHP steam capacity	million €/yr	< 0.01	< 0.01
Cost of additional electricity	million €/yr	0.01	< 0.01
Net OPEX saving	million €/yr	0.17	0.10
Payback	Years	3.5	5.5

Table 5
Ethane Furnace with Air Preheating – Performance Summary

		Base Case	Air Preheat Case
Ethane feed	t/h	51.9	51.9
Ambient conditions	-	15°C, 85% R.H.	15°C, 85% R.H.
Air temp. to burners	°C	15	129
Stack temperature	°C	169	64
Excess air	%	10	5
Absorbed heat	MW	116.8	110.6
Overall efficiency	%	91.7	96.5
Fuel flow	t/h	8.2	7.3
Fired heat	MW	127.3	114.5
VHP steam production	t/h	97.0	88.75
NOx concentration		Base	Base + 2%

Table 6
Economics of Ethane Furnace with Air Preheating

		Air Preheat Case (EU location)	Air Preheat Case (US location)
Savings in fuel firing	million €/yr	3.01	1.83
Cost of lost VHP steam capacity	million €/yr	1,64	1.00
Cost of additional electricity	million €/yr	0.10	0.06
Net OPEX saving	million €/yr	1.27	0.77
Payback	Years	1.5	3

Fig 1
Naphtha cracking furnace

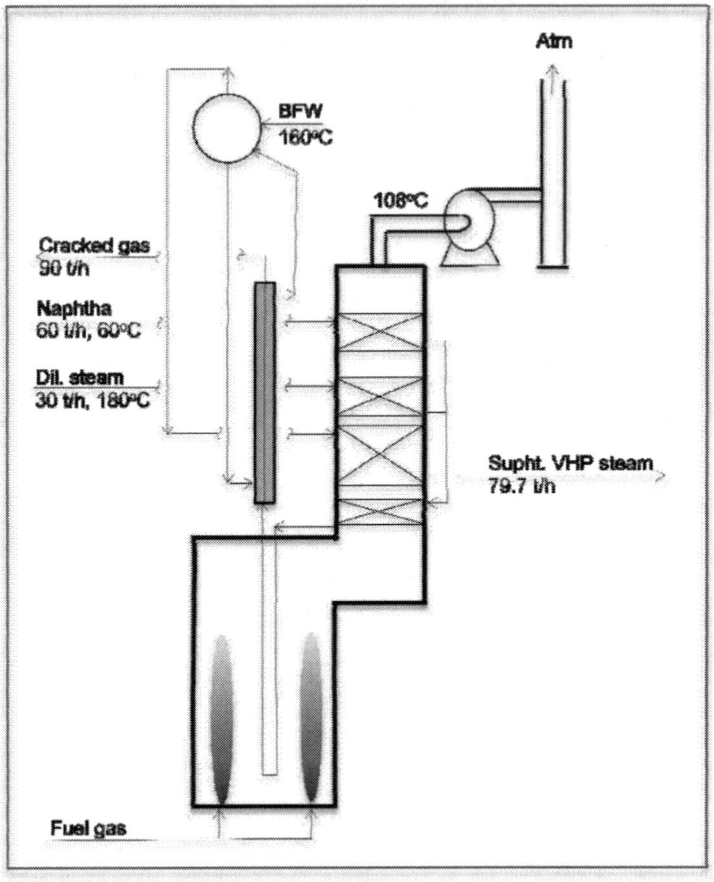

Fig 2
Ethane cracking furnace

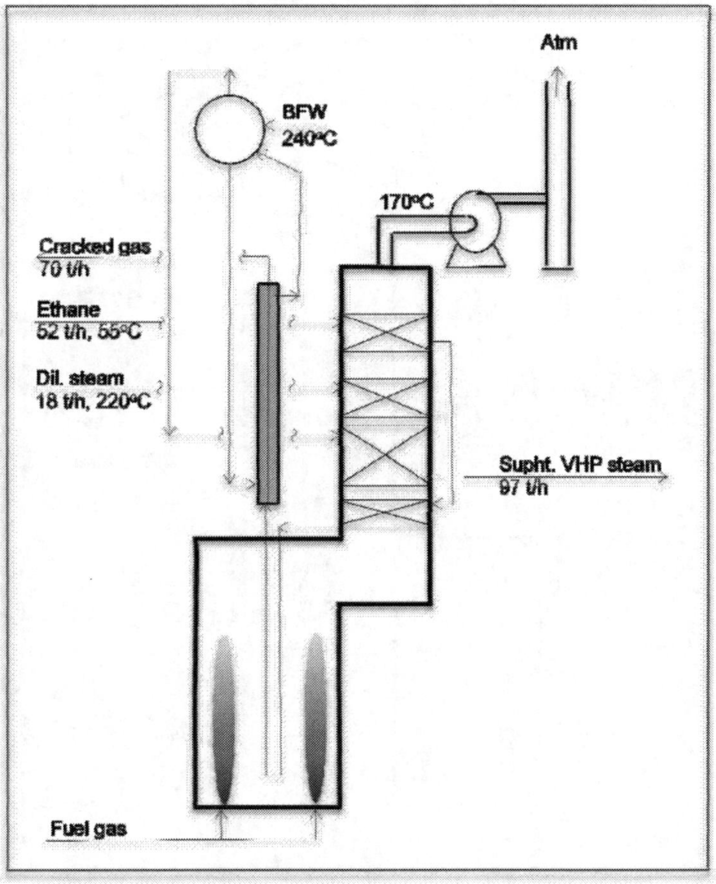

Fig 3
Naphtha cracking furnace – Air Preheater option

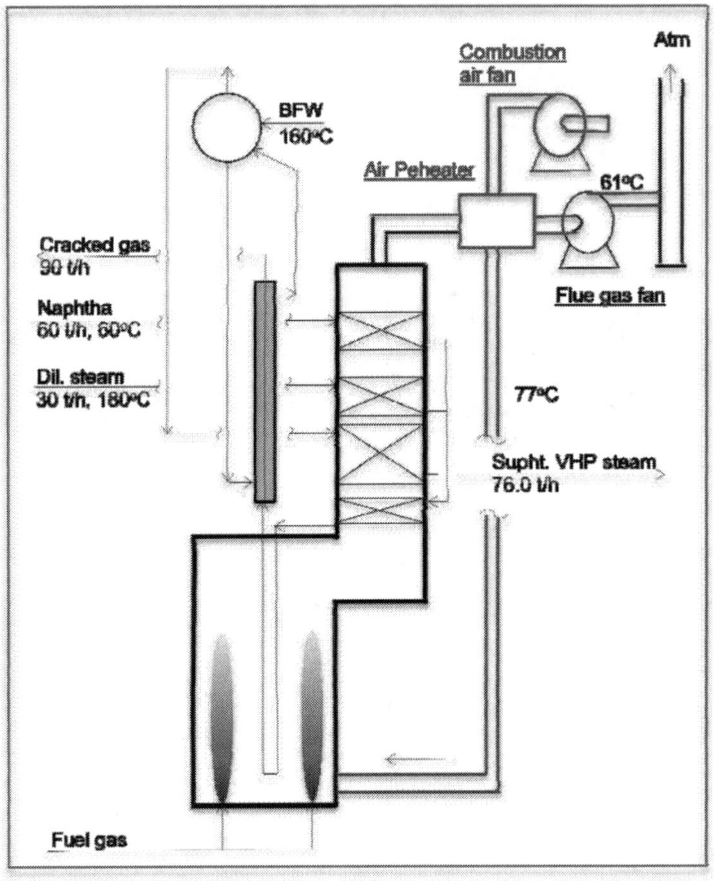

Fig 4
Ethane cracking furnace – Feed Preheater option

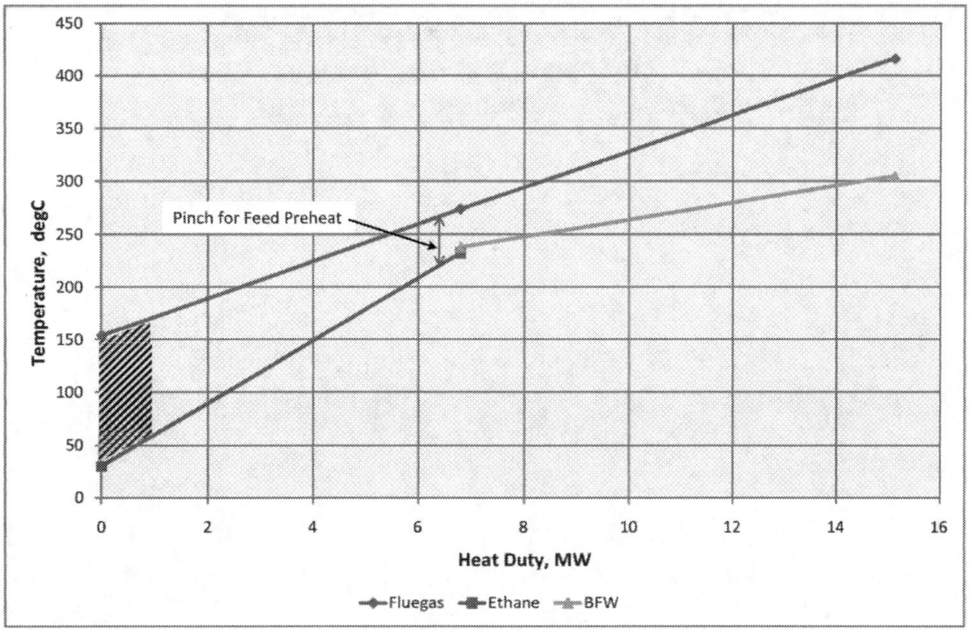

Modernize Your Control Functionality
Easy Steps to Start Process Optimization

13 AIChE
Spring Meeting, San Antonio, TX

Presenters

- Laurie Ben
- James Beall

EMERSON.
Process Management

Agenda

- Introduction
- Examples – Best Practices & Lessons Learned
  - Split Range Duty
  - Smith Predictor
  - Gain Scheduling/Adaptive Control
  - Signal Characterization
- Summary

Introduction

- Modernizing Control System – Decision Done Check!
- 'Do I' or 'Don't I' take advantage of new technology
  - Implemented older cumbersome techniques?
  - Complex custom strategies that handle automation challenges?
  - Hard to implement & maintain?
- Standard features that are field proven, practical, and easy to implement
  - Easier to implement & maintain
  - More efficient and effective control strategies
  - Achieve improvements in process performance
  - Lays the foundation for optimization

Agenda

- Introduction
- Examples – Best Practices & Lessons Learned
  - Split Range Duty
    - Control Challenge
    - Legacy Solution
    - Newer Technology Solution
    - Summary
  - Smith Predictor
  - Gain Scheduling/Adaptive Control
  - Signal Characterization
- Summary

Split Range Duty – Control Challenge

- Use steam and hot oil as energy source for distillation column for temperature controller.
- Hot oil is "free energy" and should be maximized
- Control Scenarios
  - Maximize hot oil and supplement with steam
  - Base load steam (if meter bad) and supplement with hot oil
  - Base load hot oil flow (if bad meter) and supplement with steam
- Constraints
  - Minimum and maximum hot oil valve position and flow rate
  - Minimum steam flow rate

Split Range Duty – Legacy Solution

– Old legacy system implemented this feature using a velocity PID with incremental output and a "micro expert rules system"

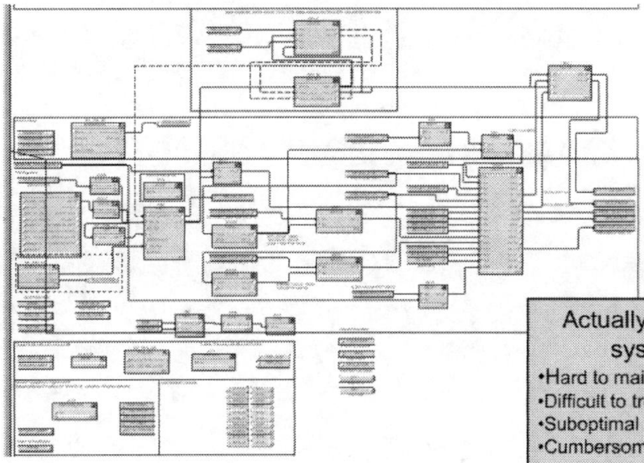

Actually Duplicated in New system, but still…

- Hard to maintain and modify
- Difficult to troubleshoot
- Suboptimal performance
- Cumbersome/Kludge/Ugly

Split Range Duty – New Technology Solution

Embedded MPC Function Block provides all control functionality! EASY!!!

Simple calculations to calculate limits

Split Range Duty – New Technology Solution

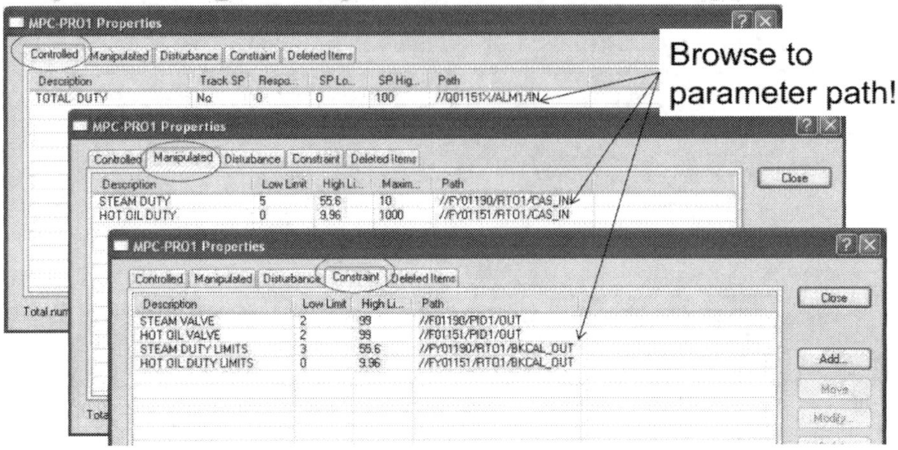

Split Range Duty – New Technology Solution

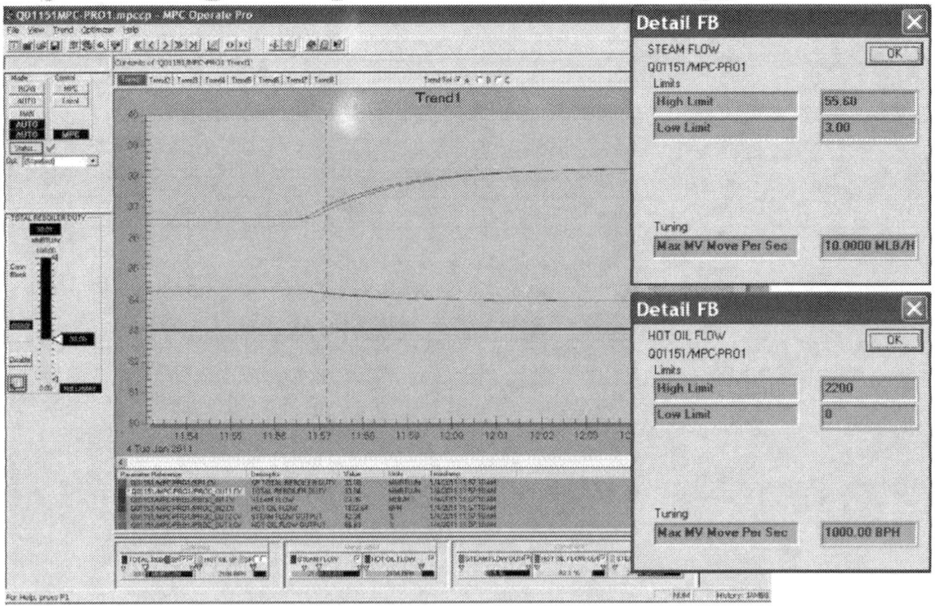

Summary Split Range Duty Control

Embedded MPC (New Technology) vs Legacy Solution

- Better performing solution
  - Bumpless transfers and constraint control
- Easier to Maintain
  - Change limits and settings
  - Simpler implementation
  - Easy to understand, train others, document
- Easily expands to handle future constraints
- Reduce energy costs!

Agenda

- Introduction
- Examples – Best Practices & Lessons Learned
  - Split Range Duty
  - Smith Predictor
    - Control Challenge
    - Legacy Solution
    - Newer Technology Solution
    - Summary
  - Gain Scheduling/Adaptive Control
  - Signal Characterization
- Summary

Smith-Predictor PID – Control Challenge

- Control of processes that are "dead time dominant" (dead time > time constant)
- Typically custom configured
  - some legacy control systems have preconfigured options, e.g. PROVOX
- Algorithm uses a process model
  - (gain, dead time and time constant)
- Challenge: as process dynamics change from model used, performance degrades rapidly

Smith-Predictor PID – Legacy Solution

- Traditional Smith-Predictor

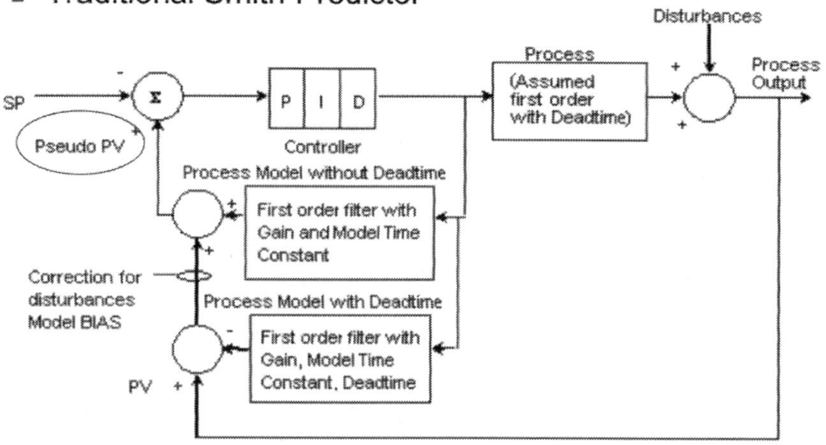

•High maintenance when process dynamics change
•Performance degrades rapidly as process dynamics change from configured model

Smith-Predictor PID – New Technology Solution #1

Easy to implement the Smith-Predictor but…

- High Maintenance when process dynamics change
- Performance degrades rapidly as process dynamics change from configured model

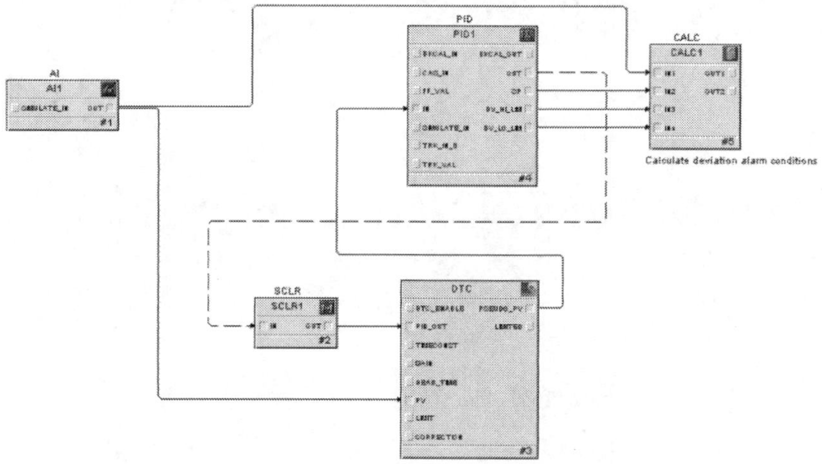

Smith-Predictor PID – New Technology Solution #2

- Embedded MPC Solution

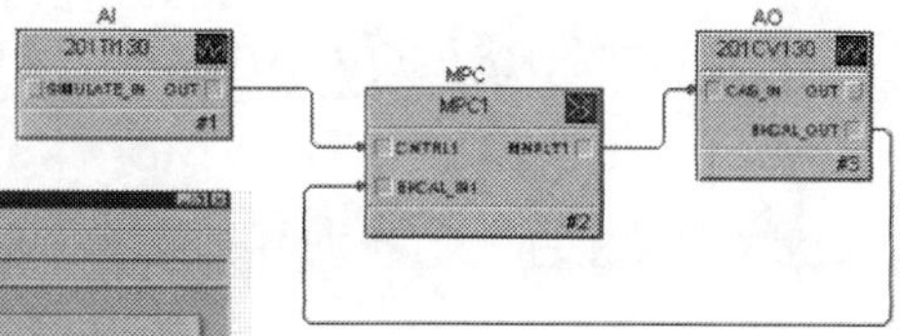

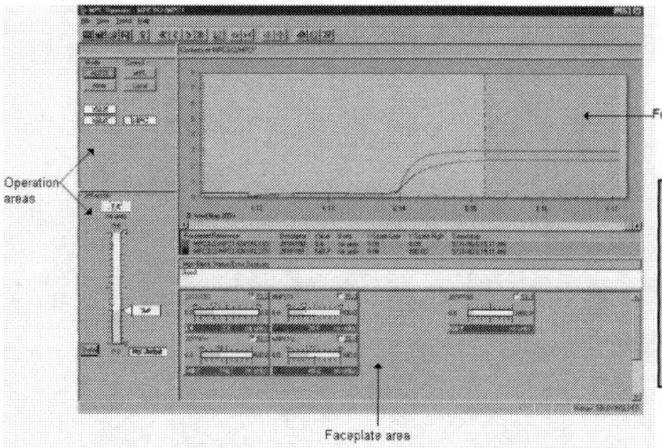

•Provides Future Prediction
•Provides MPC Faceplate – no configuration needed
•More robust under changing process dynamics

Smith-Predictor PID – New Technology Solution #2

- Embedded MPC Solution

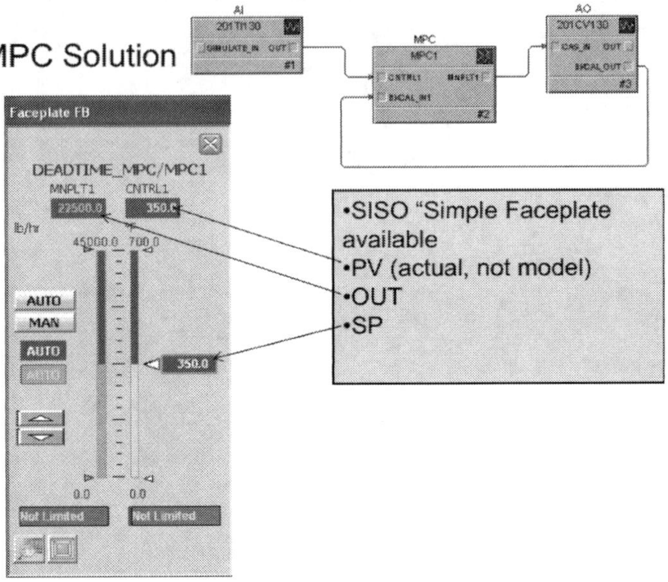

Summary Smith-Predictor PID

New Technology Solutions

- Standard Smith Predictor (#1) and Embedded MPC (#2)
- Both easy to implement and understand!

New Technology Solution #2 (Embedded MPC)

- More robust solution under a changing process model
- Provides visual display of the "future" PV
- Save money through better control performance

Agenda

- Introduction
- Examples – Best Practices & Lessons Learned
  - Split Range Duty
  - Smith Predictor
  - Gain Scheduling/Adaptive Control
    - Control Challenge
    - Legacy Solution
    - New Technology Solution
    - Summary
  - Signal Characterization
- Summary

Gain Scheduling – Control Challenge

- Legacy systems have limited capability to modify tuning based on process conditions
- Others had add-on application software, e.g. PROVOX "Gain Scheduler" system provided 3 sets of Gain, Reset, and Rate applied in 3 separate regions of a "state variable"
  - State variable = PV, OUT or "other"
  - Transition between tuning sets interpolated over "dead band"
- Beware: some suspend loop execution when tuning parameters change, so don't change the tuning every scan

Gain Scheduling – Legacy Solution

- Custom code
- Multiple points, registers, displays
- Restricted (slower) update rate

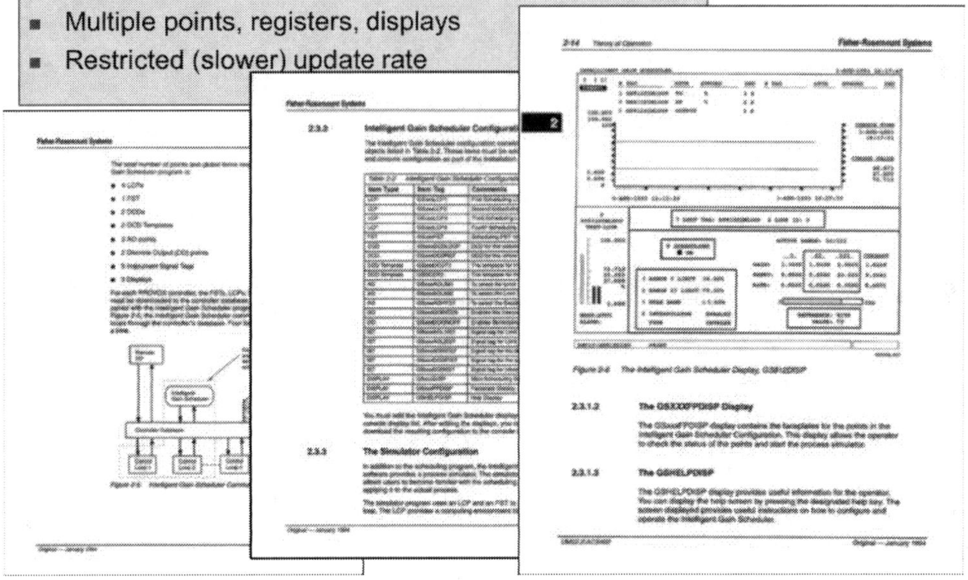

Gain Scheduling – New Technology Solution #1

Standard Modules

- Standard configuration
- Preconfigured faceplate & detailed display
- Easy to implement
- Tuning updates at loop execution

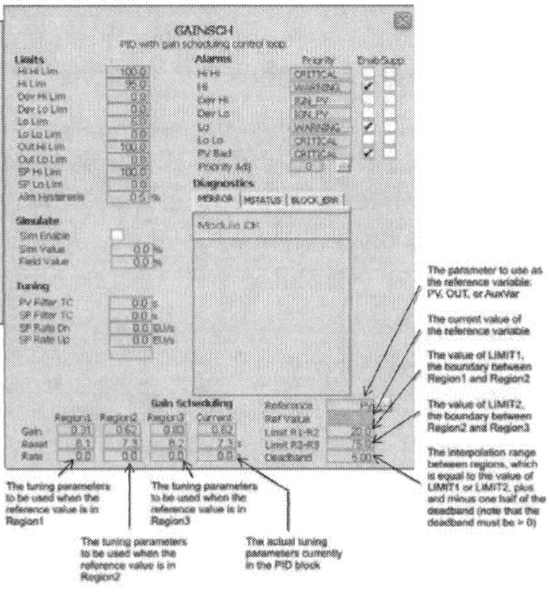

Gain Scheduling – New Technology Solution #2

New Technology "Adaptive Control"

- "Learns" process models from normal operating moves
- Allows up to 5 tuning regions of the "state variable"
- Partial Adaptive Control
  - Use "approved model" only & selected tuning rule in each region
  - Lambda PI or IMC PID; "Closed loop time constant" or "Lambda Factor".
- Optional Full Adaptive Control
  - continuous "Learning" of process model in each region
  - updates tuning while in each region per selected tuning rule
  - set allowable deviation (%) from approved model in each region

New Technology – Adaptive Control

Adaptive control adjusts tuning to improve control performance

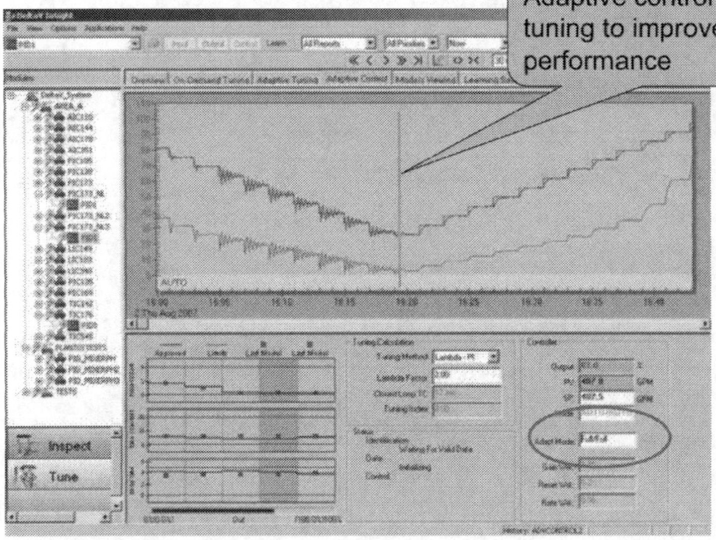

Summary Gain Scheduling

New Technology Solution(s) vs. Legacy Solution

- Gain Scheduling easy to implement and understand!
  - Standard configuration
  - Preconfigured faceplate & detailed display
  - Tuning updates at loop execution
- Adaptive Control provides additional benefit and ease of implementation
  - Learns models during normal operation – NO STEP TESTs!
  - More Regions – Up to 5
  - Automatic tuning updates when process model changes
- Improved control performance saves money!

Summary Gain Scheduling

New Technology Solution(s) vs. Legacy Solution

- Gain Scheduling easy to implement and understand!
  - Standard configuration
  - Preconfigured faceplate & detailed display
  - Tuning updates at loop execution
- Adaptive Control provides additional benefit and ease of implementation
  - Learns models during normal operation – NO STEP TESTs!
  - More Regions – Up to 5
  - Automatic tuning updates when process model changes
- Improved control performance saves money!

Agenda

- Introduction
- Examples – Best Practices & Lessons Learned
  - Split Range Duty
  - Smith Predictor
  - Gain Scheduling/Adaptive Control
  - Signal Characterization
    - Control Challenge
    - Legacy Solution
    - New Technology Solution
    - Summary
- Summary

Signal Characterization - Control Challenge

- Input Characterization (AI)
  - Combustion control – characterize fuel and air
  - Level – characterize pv (e.g. level to volume)
- Output characterization (AO)
  - modify installed flow characteristics of valves, dampers or other final control elements
  - Use to modify loop gain, e.g. output path to slave loop
- Legacy System Challenges
  - Lack of resolution (few points)
  - External black box characterizers may shift, lose documentation
  - Potential operator/maintenance confusion on signal going to final control element

Signal Characterization – Legacy Solution

- Legacy system using external analog characterizer

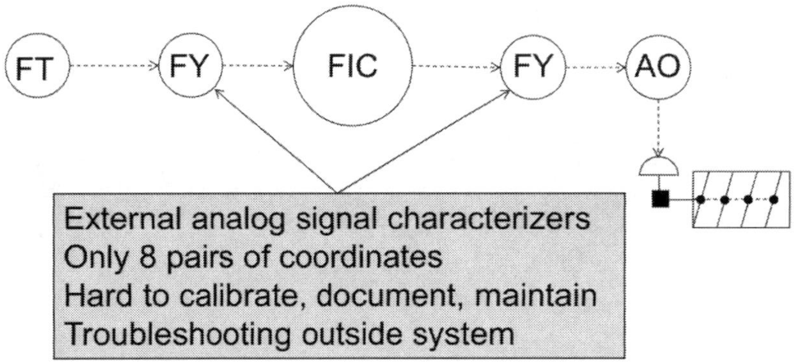

Signal Characterization – New Technology Solution #1

- Standard Module implementation provides 21 point signal characterizer function block (SGCR)

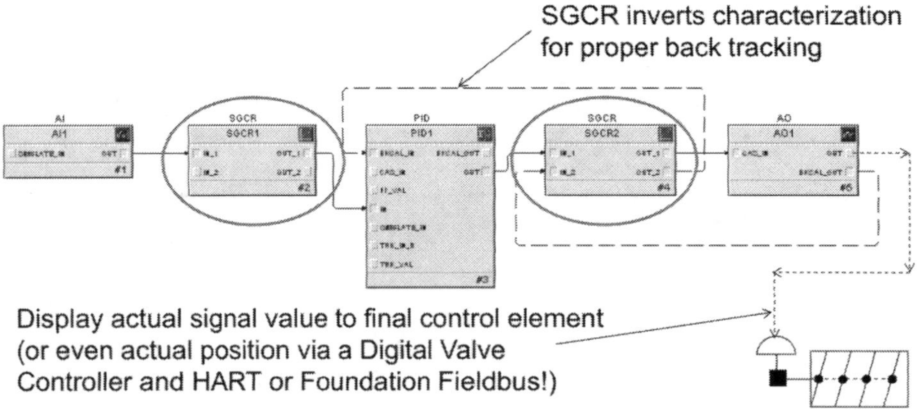

Signal Characterization – New Technology Solution #2

- Adaptive Control
  - Adjust tuning to compensate for final control element characteristics

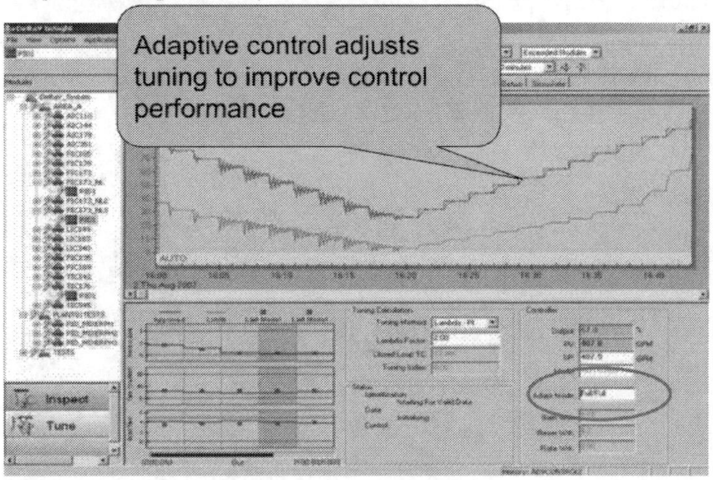

Summary Signal Characterization

- Newer Technology Signal Characterization function block
  - Better resolution (21 pairs of coordinates)
  - Self documenting (exists in control system)
  - No calibration required vs. 'external black box'
  - Provides "inverse" characterization for analog output characterizations
  - Reduce confusion: display signal being sent to final control element
- Applying Adaptive Tuning may provide a superior solution for AO characterization of final control element

Agenda

- Introduction
- Examples – Best Practices & Lessons Learned
  - Split Range Duty
  - Smith Predictor
  - Gain Scheduling/Adaptive Control
  - Controller Characterization
- Summary

Summary

- Practical application of control features that deal with common challenges
- Highlighted best practices & lessons learned
  - Split Range Duty
  - Smith Predictor
  - Gain Scheduling/Adaptive Control
  - Signal Characterization
- Do look at new functionality available when modernizing
  - Allows jump start to process optimization
  - Leads to improved process performance
  - Easier to implement and maintain over the life investment

Contacts & Questions?

James.Beall@Emerson.com
903-235-7935

Laurie.Ben@Emerson.com
512-834-7237

Paralleled Double-Effect Distillation: Simulative Case Studies

*Pei Wang, Leilei Dai, Ming Shi , Fan Wang, Hong Gao, Jinsheng Sun**
School of Chemical Engineering and Technology, Tianjin University, No. 92 Wei Jin Road, Tianjin 300072, P. R. China

Abstract

The increasing scarcity and growing demand for energy have led to an increasingly attention to the energy-efficient methods. Double-effect distillation is a common method to achieve remarkable energy conservation, which has three kinds of typical configurations including the feed splitting (FS), the light split/forward (LSF) and the light split/reverse (LSR). In spite of a good pathway for energy saving, better options are expected to cut down energy expenditure by a bigger margin than double-effect distillation does.

In this paper, case studies of paralleled or combined double-effect distillation schemes are proposed on the basis of current double-effect distillation. In order to verify the feasibility of this method, the improved methanol distillation scheme and organosilicon distillation scheme are optimized under this principle as examples. Exergy losses and the energy consumptions are measured as a tool to compare the two prevalent schemes with their corresponding optimized counterparts, respectively. The results of the total exergy loss and the energy consumption of these schemes show that the optimized schemes can cut the corner of the exergy loss by 22.8% and 40.41% and the energy consumption by 15.61% and 44.28%, respectively, in comparison to their popular cousins.

Keywords: double-effect distillation, paralleled double-effect distillation, methanol distillation, organosilicon monomer distillation, distillation process simulation, energy saving, exergy analysis

1. Introduction

Distillation, in chemical process and related industries, is by far the most widely applied for the separation of liquid mixture. In spite of its well-known benefit and widespread application, the major drawback is its high energy consumption. In classical distillation process, heat is transferred to the reboiler of distillation column. Along with the distillation column to separate the mixture, a large number of heat are directly discharge into the surrounding environment and cooling water. Araujo AB [1] has done some research to achieve that distillation has a relatively low thermodynamic efficiency and it requires the input of high quality energy in the reboiler to push the separation of liquid mixture. Therefore, the focus of energy conservation is to consider how to recover or reuse the discharged heat, or how to reduce the quantity of transferred heat to the column. In the early 20th century, many scholars have devoted to the study of saving energy in distillation, and now most of correlative techniques are widely used in distillation process [2,3]. These energy saving methods involve multi-effect distillation, heat pump distillation, heat integrated distillation and etc [3-8]. But without extra rotating machines by electricity or other power producers, multi-effect distillation has been more favored in research and application [9].

Many researches confirmed multi-effect distillation to be one of the effective methods to reduce the energy consumption [10-12]. The energy saving effect of those schemes used multi-effect distillation is associated with the number of the effects and the characteristics of separation system. Generally speaking, those schemes can offer energy savings around 30% with respect to the original or classical multicomponent distillation scheme [23, 27, 28]. Considering the interaction between investment of plants and energy saving effect, the double-effect distillation has been the most commonly used in not only research but also field application.

The use of the double-effect distillation technology for optimizing the conventional distillation schemes can cut a large amount of the energy requirement in comparison with the correspondingly prevalent schemes [13]. Although the system is very difficult to control in the sense that it is nonlinear, multivariable and interacting [14]. More and more published studies [15-19] have been focused on further questing the way of energy saving on the basis of the double-effect distillation schemes. However, the overhead vapor of one high pressured column only can integrate with the reboiler of one low pressured column in those published studies. In view of this, in this paper, the paralleled or combined double-effect distillation is proposed to broaden the thought for the application of the double-effect distillation, in which the overhead vapor from one of the high pressured column is divided into two or more streams to successively integrate with the reboiler of the other columns [9].

2. *Theory*

2.1 Double-effect or heat integrated distillation system

A double-effect distillation system is made up of two adjacent distillation columns running in different pressure, which can lead to a temperature difference between the top of the high pressured column and the bottom of the low pressured column. Through the thermodynamic connection, the reboiler of the low pressure column plays a role of the condenser for the high pressure column. According to the direction of the mass flow and the heat integration, the double-effect distillation can fall into three kinds of configuration: the feed splitting (FS), the light split/forward (LSF) and the light split/reverse (LSR), as showed in Figure. 1.

One of the configurations with a high thermodynamic efficiency is the feed splitting (FS), in which the feed is divided roughly equal between the two columns running in different pressure, and the products can be obtained from both of the top and bottom of the two columns, as illustrated in figure. 1 (a). The overhead vapor form the high pressured column is used to heat the reboiler from the low pressured column. There is only energy exchange between the two columns, so the heat duties of the feed splitting (FS) are approximately equal to the heat duty for single column with the same conditions divided by the number of effect.

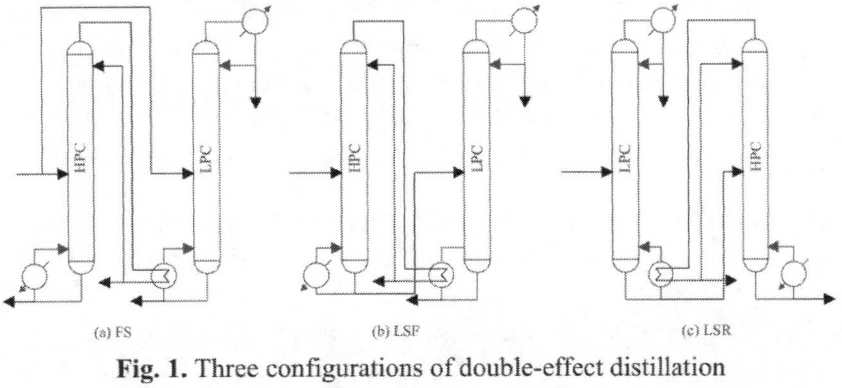

Fig. 1. Three configurations of double-effect distillation

Figure. 1 (b) and (c) respectively shows the configuration of the light split/forward heat integration (LSF) and the light split/reverse heat integration (LSR). Different from the feed splitting (FS), the total feed stream was sent to one column in the other two configurations. In the light split/forward (LSF), the total feed is pumped to the high-pressure column (HPC) and the bottom stream of HPC flows into the low-pressure column (LPC). In this way, all of the heavy components are from LPC and the light components from the top both of the two columns. Figure. 1 (b) indicates the direction of mass flow and heat integration is the same. The last configuration is the light split/reverse (LSR), as shown in figure. 1 (c), all of the feed is sent to the low-pressure column (LPC). It can remove the two kinds of light components from the top both of the two columns and one kind of heavy components from the bottom of HPC. Moreover, the direction of heat integration is in the reverse with the mass flow. These two kinds of configurations have not only energy exchange but also substantial exchange, therefore, when the operating parameters of one column are changed, the other column will also change.

Theoretically, the three kinds of configurations are empirically 30% less energy consumption than the single column distillation. As for choosing what kind of the configurations of the double-effect distillation, it depends on the properties of the multi-component mixture, the pressure of saturated steam, the temperature of cooling water, structure parameters of existing distillation column and so on.

2.2 Exergy analysis of the distillation process

According to the First Law of Thermodynamics, also called the Law of Energetic Balances, no matter what happens, the total amount of energy has not yet changed. However, when the various types of energy generate the mutual conversion, the quality of the energy is always decreased. The separation process in distillation is irreversible. In order to separate the mixture in distillation column, it needs the driving force which must consume a certain amount of external work, by this means, the decline of the quality of energy is inevitable. Therefore, we should choose a reasonable driving force in distillation process to avoid unnecessary loss of the quality of energy. It defined the concept of exergy (E_x) to compare the quality of the different types of energy. Exergy, which is a measure for evaluating the extent of available energy or the quality of energy, is based on the Second Law of Thermodynamics. Exergy analysis reveals the essence of the energy, and can be used as an important tool to evaluate the effect of energy saving. And it usually combines with process optimization studies, and as its guidance, to identify cost saving opportunities and amend the existing scheme.

Szargut [29] has summarized a comprehensive definition of the exergy, that is: 'Exergy is the amount of work obtainable when some matter is brought to a state thermodynamic equilibrium with the common components of the natural surroundings by means of reversible processes, involving interaction only with the above mentioned components of nature'. The state of one stream recorded as 1, according to the definition of exergy, the exergy of the stream can be expressed as:

$$E_x = (H_1 - H_0) - T_0(S_1 - S_0) \tag{2}$$

where, H_1 and S_1 is the enthalpy and the entropy of the stream 1, and T_0 is the temperature of the surrounding and H_0 and S_0 represent the enthalpy and the entropy of the surrounding.

The state of another stream recorded as 2, when the system stream change from state 1 (T_1, P_1) to state 2 (T_2, P_2), the variation of exergy (ΔE_x) is given by

$$\Delta E_x = E_{x,1} - E_{x,2} = \Delta H - T_0 \Delta S \tag{3}$$

In the conventional distillation process, it must input the heat or energy to drive the occurrence of separation process. The bottom of the column input the heat (Q_R) at this temperature (T_R), and the top of the column was sent to the heat (Q_C) at this temperature (T_C). The net work consumption (W) in distillation is

$$W = Q_R(1 - \frac{T_0}{T_R}) - Q_C(1 - \frac{T_0}{T_C}) \tag{4}$$

When the enthalpy difference between the product and the feed is very small, $Q_R=Q_C=Q$ can be approximately gotten, then Equation (4) is rewritten to be:

$$W = QT_0(\frac{1}{T_C} - \frac{1}{T_R}) \tag{5}$$

Then, the total exergy loss (Ex_{loss}) of the distillation column can be calculated by

$$Ex_{Loss} = W - \Delta E_x \tag{6}$$

3. *Prototype case: the prevalent distillation schemes*

3.1 The improved four-column methanol distillation scheme

In the methanol industry, it has gradually appeared two-column, three-column, four-column and five-column distillation scheme [20-23]. With the increase of the number of distillation column, the corresponding distillation scheme can not only increase the efficiency of the process, but also cut the energy consumption. At present, the four-column scheme has a dominant status to produce the purified methanol. However, the improved four-column distillation scheme can cut some more energy requirement with respect to the traditional four-column distillation scheme. The improved four-column distillation scheme is seen as the comparative reference in the paper.

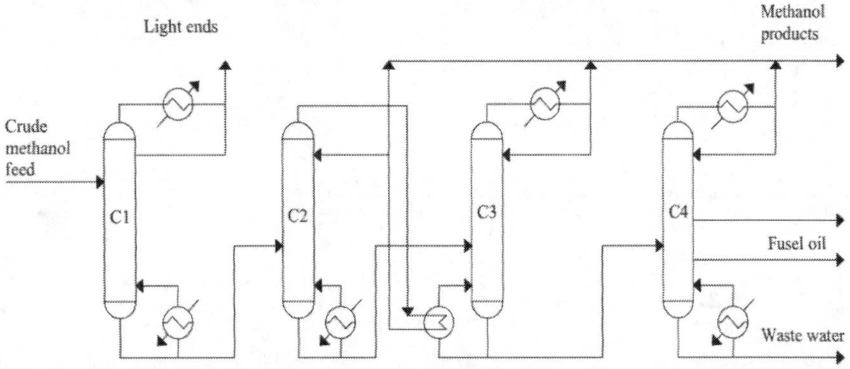

Fig. 2. The improved four-column double-effect methanol distillation scheme (C1, light ends column; C2, pressured column; C3, atmospheric column; C4, water column)

In the improved four-column distillation scheme, as shown in figure. 2 [24], the crude methanol flows into the light ends column (C1) in 87.5 ℃ for removing the light ends and the remnants of dissolved gas from the top of C1. Besides, in order to raise the methanol recovery rates, the light ends column (C1) designs a two-stage condenser system. The crude methanol was distillated successively from the pressured column (C2) and the atmospheric column (C3), and most of the methanol products can be received at the top both of C2 and C3. The bottom stream of C1, C2 and C3 is pumped respectively as the feed into C2, C3 and the water column (C4). Residual methanol, fusel oil and waste water are extracted from the corresponding position of the top, side stream and the bottom of the water column (C4).

In the improved scheme, the operation pressure of C2 is 0.9MPa, and the other three columns run under the atmospheric pressure. Due to the different of operating pressure, the overhead vapor of C2 can undertake thermal coupling connection with the reboiler of C3.

3.2 The current organosilicon monomer distillation scheme

According to the survey, there are several kinds of organosilicon distillation schemes. However, the 9-column organosilicon distillation scheme [9], in China, has an important role in the organosilicon industrial, as illustrated in Figure. 3, it is used as the base case.

Figure. 3 shows that the crude monomer is delivered to the heavies removal column (C1') and the bottom stream of C1' has a lot of the unsteady substances of high boiling points. These unsteady substances of high boiling points are pumped to the high boiling point column (C9') as the feed, and the products from the top of C9' continue to serve as the crude monomer by circumfluence. By this way, it can accomplish the separation between the unsteady substances of high boiling points and the useful substances at a relatively low temperature in the high boiling point column (C9') and avoid the unsteady substances of high boiling points from thermal cracking to generate many unknown substances which would greatly affect the subsequent separation of product. The low removal column (C2') provides trimethylchlorosilane (Me3), methyldichlorosilane (MeH), and other light components for lights removal column (C3'), MeH column (C4'), azeotrope removal column (C5') and Me3 column (C6'), which can finally achieve two products (MeH and Me3) and three wastes. It also provides methyltrichlorosilane (Me1) and dimethylchlorosilane (Me2) and other heavy components for Me1 column (C7') and Me2 column (C8'), which can finally get two products (Me1 and Me2).

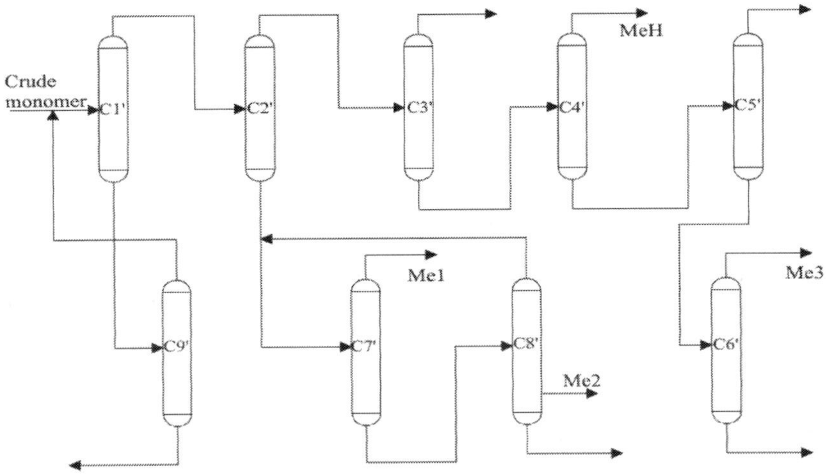

Fig. 3. The prevailing 9-column distillation scheme (C1', heavies removal column; C2', low removal column; C3', lights removal column; C4', MeH column; C5', azeotrope removal column; C6', Me3 column; C7', Me1 column; C8', Me2 column; C9', high boiling point column)

In this scheme, the operating pressure of all nine columns is relatively low, and the highest operating pressure is 260 kPa in the lights removal column (C3'). Although it has a different operating pressure between C3' and the other columns, but the temperature difference is not enough to make them use double-effect or heat integrated distillation.

4. *Simulation and analysis both of the prevalent schemes and the optimized schemes*

4.1 The novel five-column heat integrated methanol distillation scheme

Although the energy saving of the improved four-column methanol distillation scheme can reach 10% compared with the traditional four-column scheme, it still exist several shortages, which will be illustrated by the results of computer simulation to the improved four-column methanol distillation scheme. The flow rate of the crude methanol is 239230 kg/hr, and the temperature is 87.5 ℃, and the composition of the crude methanol is exhibited in table 1. The inlet and outlet temperature of the cooling water are 30 ℃ and 40 ℃.The methanol content is specified as 99.99 wt% in the methanol products and 10ppm in the wastewater of the water column. Table 2 shows the operating parameters of each column in the improved four-column scheme.

Table 1. Crude methanol feed data

Component	Mass fraction%	Component	Mass fraction%
H_2	0.0100	Pentanol	0.0338
N_2	0.0710	Formic acid	0.0030
Methane	0.0100	Acetic acid	0.0030
CO	0.0360	Methyl formate	0.0092
CO_2	0.6640	Acetone	0.0193
Methanol	95.0000	Methyl ethyl ketone	0.0030
Dimethyl ether	0.0309	Diisopropylether	0.0056
Trimethylamine	0.0010	Water	3.7100
Ethanol	0.0900	i-Pentane	0.0562
i-Propanol	0.0024	n-Pentane	0.0169
Propanol	0.0264	n-Hexane	0.0487
i-Butanol	0.0338	n-Heptane	0.0009
Butanol	0.0185	n-Octane	0.0050
3-Pentanol	0.1407	n-Nonane	0.0050

Table 2. The operating parameters of improved four-column scheme

The operating parameters	C1	C2	C3	C4
Top pressure (kPa)	150	900	110	110
Top temperature (℃)	81.0	132.6	67.7	67.7
Bottom temperature (℃)	83.8	139.1	96.5	104.5
Reflux ratio	0.58	2.44	1.68	11.57
Number of theoretical stage	33	52	58	61

The results of simulation indicate that the hot utility is still in a high level because the reboilers of C1, C2 and (C4) still need the saturated steam to transmit the heat. Table 2 shows the reflux ratio of C4 is 11.57 and this large reflux ratio can cause the increasing of the energy requirement of the water column (C4). Besides, according to the flowsheet of this scheme, the diameter of column is so large that the emergence of many urgent cases.

In view of above these defects in the improved four-column scheme, the novel five-column heat integrated methanol distillation scheme (shown as Fig. 4) is designed in this paper to reduce the impact of these factors as far as possible by using the paralleled or combined double-effect distillation. Juntao Zhang [23] and Jinsheng Sun [25] are also proposed a five-column methanol distillation scheme by using triple-effect distillation and double-effect distillation respectively, and the effect of energy saving of the two schemes is outstanding.

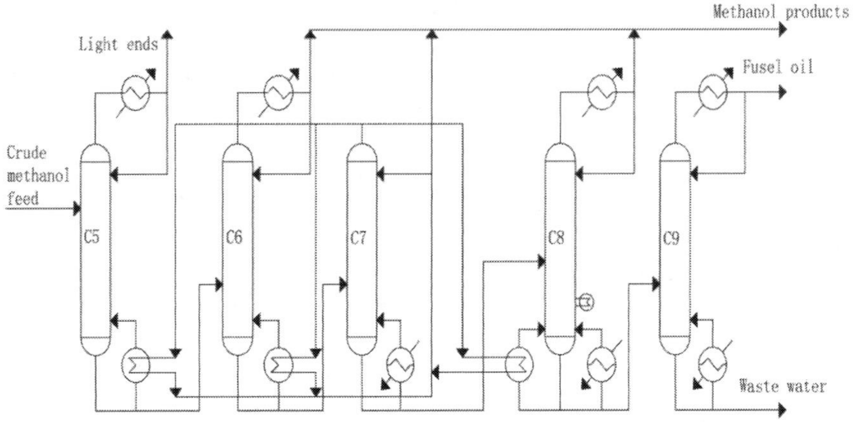

Fig. 4. The five-column heat integrated methanol distillation scheme (C5, light ends column; C6, prepositive atmospheric column; C7, pressured column; C8, postpositive atmospheric column; C9, water column)

According to the principle of the paralleled double-effect distillation, the four columns are unable to constitute the heat integration network in the improved four-column scheme. Through repeated verification, the prepositive atmospheric column (C6) is added after the light ends column (C5), meanwhile, the operating pressure of the pressured column (C7) is reduced to increase its throughput. In this way, the overhead vapor of C7 is split into three streams to respectively integrate with the light ends column (C5), the prepositive atmospheric column (C6) and the postpositive atmospheric column (C8). Due to use the paralleled double-effect distillation, the duties of the hot utility have decreased in a great margin. However, there are still part of heat from the background process is absorbed by the cooling water. So an inter-reboiler with a low temperature level is designed in the postpositive atmospheric column (C8) and it can share the heat with the background process streams, besides, we have proved the feasibility by using the column grand composite curve analysis of columns. Figure. 4 shows that the product of the top of the water column (C9) is fusel oil instead of purified methanol, which can not only reduce the reflux ratio of C9, but also make the diameter of column smaller to reduce the difficulty of operation.

Due to produce the fusel oil instead of purified methanol in the water column (C9) in the novel five-column distillation scheme, the methanol content of the feed for the water column have to limit at 0.016 wt%. In the five-column scheme, the operating pressure and the number of theoretical stage for each column are determined through the existing process, the onsite data and the requirement of the heat coupling. The proportion of the methanol products from the C6, C7 and C8 is determined by their own double-effect or heat integrated distillation system. The simulation results of the novel five-column scheme are listed in table 3.

Table 3. The operating parameters of five-column scheme

Column	C5	C6	C7	C8	C9
Top pressure (kPa)	160	110	400	110	110
Top temperature (℃)	83	66.6	104	66.6	86.5
Bottom temperature (℃)	85.6	84	118.8	104.5	103.2
Reflux ratio	1.46	1.49	2.23	1.87	4
Number of theoretical stage	33	54	54	65	29

4.2 Energy consumption and exergy loss analyses for the two methanol schemes

In the improved four-column scheme, the pressured column and the atmospheric column constitute the distillation system of heat coupling, there are three columns (C1, C2, C4) need the hot utility, and the entire duty of reboiler is 145.4×10^3 kw. As described above, the light ends column (C5) and the prepositive atmospheric column (C6) don't need the saturated steam in the five-column scheme and the total duty of reboiler is reduced to 122.7×10^3 kw. The detailed data of the reboiler duty as shown in table 4.

Table 4. The duty of reboiler in the two schemes

Column	C1	C2	C4	C7	C8	C9
Duty of reboiler (10^4 kw)	3.30	10.04	1.20	8.97	2.86	0.44

Obviously, the novel five-column scheme is much better than the improved four-column scheme in energy saving and the energy consumption of the novel scheme can be reduced by 15.61%. However, the recovery rate of methanol of the novel scheme is approximately equal with the improved four-column scheme.

Only the energy consumption analysis is not revealed the energy saving essence of the novel five-column scheme. Exergy analyses are used to further compare the two schemes, and the exergy flow rates are calculated through using the computer simulation data. The activity coefficient of the liquid is calculated using the Alcohol Package thermodynamic model. The calculation of the net work consumption of distillation column should pay more attention to the temperature of the cooling water. When the cooling water exchanges heat with other streams, the temperature of the cooling water is momentarily changed, so it is replaced the temperature of cooling water with the arithmetic average of inlet and outlet temperature of cooling water. Figure. 5 show the overall exergy loss of each column in the two methanol distillation schemes, in addition to the overall exergy loss of C6. Due to optimize the four-column scheme, the prepositive atmospheric column (C6) is added in the transformed scheme, and the overall exergy loss of C6 is 9157.6 kw. The exergy loss of each column, besides the pressured column (C7), in the novel five-column scheme is obviously less than their counterparts in the improved four-column scheme.

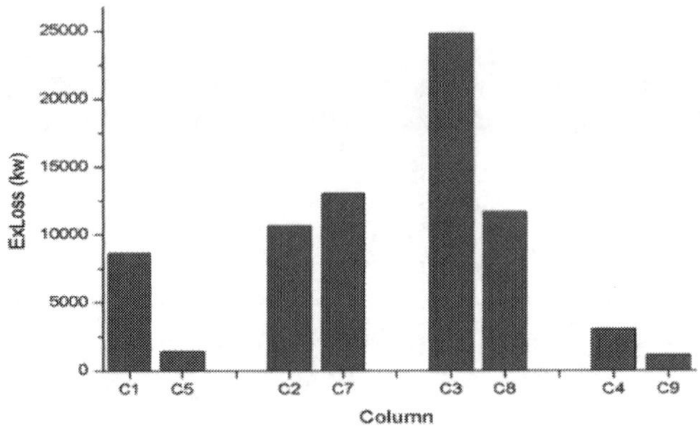

Fig. 5. The overall exergy loss analysis of each column in the two schemes

The total exergy loss of the improved four-column scheme and the transformed scheme is, respectively, 47100.1 kw and 36355.1 kw. The total exergy loss of the optimized scheme is dropped by 22.8%. Through the comparison between the exergy loss and energy consumption of the two schemes, it is indicated that the novel five-column scheme should be applied to produce the purified methanol in industry.

4.3 The novel heat coupling organosilicon distillation scheme

The primary source of the crude organosilicon monomers is the catalytic reaction products between chloromethane and silicon at high temperature and the table 5 displays the normal boiling points and mass percentage of the main components of the crude organosilicon monomers. Methyltrichlorosilane (Me1), dimethylchlorosilane (Me2), trimethylchlorosilane (Me3) and methyldichlorosilane (MeH) are the most valuable components in the crude organosilicon monomers. The purification requirements on organosilicon of these materials are shown in table 6.

Table 5. The crude organosilicon monomer feed data

Component	Normal boiling point at 1 atm/℃	Mass percentage %
LB		1
Methyldichlorosilane	41.8	4
Silicon tetrachloride	57.6	0.5
Trimethylchlorosilane	57.9	2.5
Methyltrichlorosilane	66.4	11
Dimethyldichlorosilane	70.2	80
HB		1
Temperature (℃)	40	
Pressure（kPa）	100	
Flow rate (kg/h)	26200	

Table 6. Purification demand on organosilicon

Monomer	Purification demand
MeH	99.5%
Me3	99.5%
Me1	99.5%
Me2	99.95%

The conventional nine-column scheme is simulated with the aid of existing process and onsite data to achieve the operating parameters of each column, as illustrated in table 7. The bottom temperature of the heavies removal column (C1') and the high boiling point column (C9') are obviously much bigger than that of the other seven columns in table 7. According to the principle of the double-effect distillation, the top temperature of the high pressured column is relatively high in comparison with the bottom temperature of the low pressured column, and the high pressured column can integrate with the low pressured column. Although the top temperature of C1' is less than the bottom temperature of the other seven columns, the overhead vapor of C1' still has a great opportunity to conduct heat with the other seven columns with the increasing of its top temperature.

Table 7. The simulation results of prevailing 9-column

Column	C1'	C2'	C3'	C4'	C5'	C6'	C7'	C8'	C9'
Top temperature (℃)	70.8	65.3	57.3	66.7	72.1	70.6	71.2	74.9	72.7
Bottom temperature (℃)	124.1	89.1	80.4	85.2	74.9	75.4	75.3	75.3	126.5
Reboiler duty (MW)	11.67	5.61	1.13	0.51	1.19	0.46	12.27	16.08	0.02
Condenser duty (MW)	-12.47	-5.53	-1.08	-0.53	-1.20	-0.47	-12.69	-16.12	-0.02
Top pressure (kPa)	110	180	260	220	170	150	110	110	110

According to the way of the paralleled double-effect distillation, the increasing operating pressure in the heavies removal column (C1') can make the top temperature of C1' ascending to integrate with the other seven columns. The idea is still in study to overcome some difficulties. However, we have got another feasible method, in which the operating pressure of C1' and the Me1 column (C7') is raised to 360 kPa and 280 kPa, respectively. In this way, the top temperature of C1' and C7' is increased to ensure they can integrate with the other columns, besides the high boiling point column, and the parallel double-effect or heat integrated organosilicon monomer distillation scheme is designed as shown in figure. 6 [26]. The feasibility of this heat coupling scheme has confirmed through three x-y diagrams including the heavies component and Me1, the heavies component and Me2, Me1 and Me2 [9].

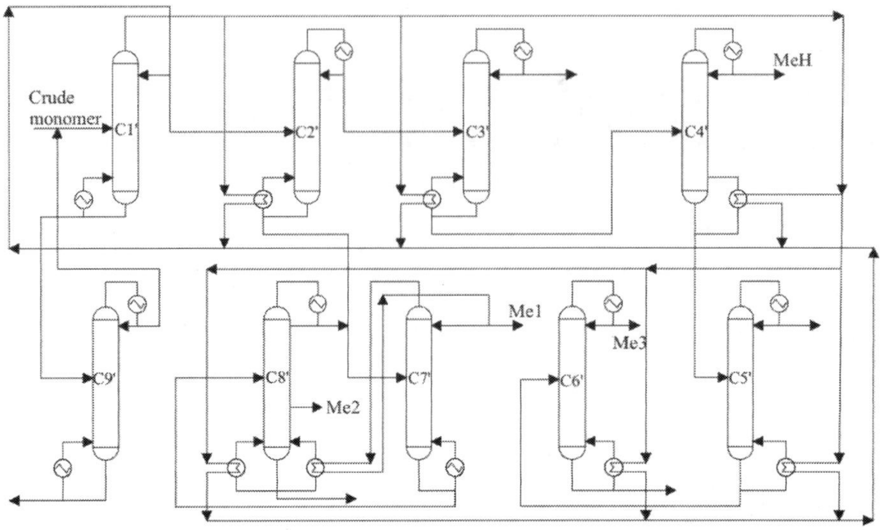

Fig. 6. The heat coupling 9-column scheme (C1', heavies removal column; C2', low removal column; C3', lights removal column; C4', MeH column; C5', azeotrope removal column; C6', Me3 column; C7', Me1 column; C8', Me2 column; C9', high boiling point column)

Figure. 6 shows the heavies removal column (C1') make the heat coupling with the low removal column (C2'), the lights removal column (C3'), the MeH column (C4'), the azeotrope removal column (C5'), the Me3 column (C6') and the Me2 column (C8'), successively. In order to gain a better energy saving effect, C8' is designed with two reboilers, the one integrated with C1' and the other integrated with the Me1 column (C7'). The operating parameters and the duty of reboiler and condenser of each column in the heat integrated 9-column are enumerated in table 8.

Table 8. The simulation results of heat integrated 9-column

Column	C1'	C2'	C3'	C4'	C5'	C6'	C7'	C8'	C9'
Top temperature (℃)	114.3	65.3	57.3	66.7	72.1	70.6	105.1	74.9	72.7
Bottom temperature (℃)	124.9	89.9	78.8	84.5	75.4	75.4	110.9	76.6	121.4
Reboiler duty (MW)	13.06	5.62	1.13	0.51	1.19	0.46	13.83	14.12	0.38
Condenser duty (MW)	-13.72	-5.54	-1.08	-0.53	-1.20	-0.47	-14.02	-14.43	-0.41
Top pressure (kPa)	360	180	260	220	170	150	280	110	110

4.4 Energy consumption and exergy loss analyses for the two organosilicon schemes

Figure. 4 indicate that each column of the prevalent nine-column scheme is not integrated with the other columns and the reboilers of all of these columns are required the saturated steam (600 kPa). According to the data from the table 7, the total duty of reboiler is 48.94 MW for the prevailing 9-column scheme. In the new nine-column scheme, C1' integrate with C2', C3', C4', C5', C6' and C8' successively and C7' integrate with C8', so C2', C3', C4', C5', C6' and C8' don't consume the saturated steam. The total duty of reboiler is 27.27 MW. Obviously, the heat integrated 9-column scheme has an absolute advantage over the prevailing 9-column scheme in energy saving, and the energy savings can reach 44.28%.

As the same with the methanol distillation scheme, the inlet and outlet temperature of cooling water are 30 ℃ and 40 ℃, and the total exergy loss is calculated through the results of computer simulation of the two schemes. The overall exergy loss comparison of each column in the two schemes is illustrated in figure. 7. In the normal scheme, the throughput of C9' is so small that the overall exergy loss of C9' is very near to zero, so it only have the exergy loss of C9' of the new scheme in figure. 7. The overall exergy loss of each column in the optimized scheme is declined in comparison with that of the corresponding column in the normal scheme except the high boiling point column (C9'). Because the crude organosilicon monomers contains some unstable components, in order to avoid these components happening thermal cracking in the heavies removal column (C1') in the novel scheme, more liquid from the bottom of C1' is pumped to C9' as the feed. That can cause the ascent of the throughput and the overall exergy loss of C9'.

The total exergy loss is 14035.5 kw and 8363.2 kw, correspondingly, in the conventional scheme and the novel scheme, and the exergy loss of the new scheme can be reduced by 40.41%. Through the above analysis and comparison, the heat integrated nine-column organosilicon distillation scheme should be a substitute for the prevalent nine-column scheme.

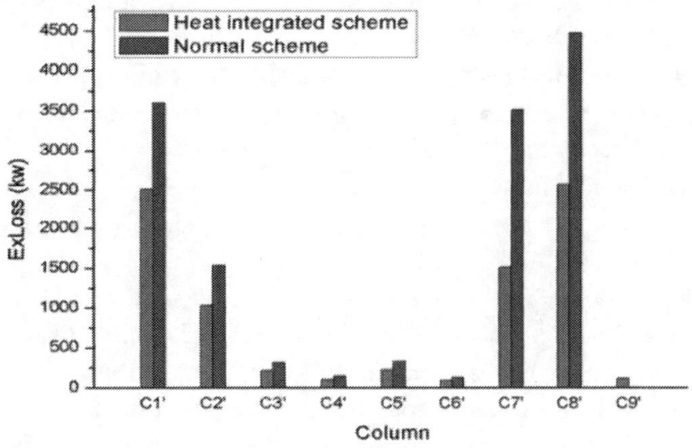

Fig. 7. The exergy loss comparison between the heat integrated scheme and the normal scheme

5. Conclusions

Paralleled double-effect or heat integrated distillation for energy saving is proposed, exampled by heat integrated five-column methanol distillation scheme and the heat coupling 9-column organosilicon monomer distillation scheme, with the aid of computer simulation. Besides, the optimized methanol distillation scheme and the transformed organosilicon scheme are comprehensively analyzed through the energy consumption and the exergy loss in comparison with the correspondingly prevalent scheme.

Adding a prepositive atmospheric column and decreasing the operating pressure of the pressured column in the novel five-column scheme can make the pressured column to simultaneously integrate with the light ends column, the prepositive atmospheric column and the postpositive atmospheric column. Besides, an inter-reboiler is added to integrate with the background process streams. The energy consumption and the exergy loss of the new scheme are 15.61% and 22.8% lower than the improved scheme.

Increasing the operating pressure of the heavies removal column and the Me1 column can make them to integrate with the other columns, in addition to the high boiling point column, to constitute a paralleled double-effect distillation system in the optimized nine-column scheme. The energy consumption and the exergy loss of the novel scheme are 44.28% and 40.41% less than the normal scheme.

The two cases expound that the key of paralleled double-effect distillation process lies in the reasonable adjustment of the energy distribution, the operating parameters and the configuration of some columns. It not only reduces the quantity of energy consumption, but also avoids the decrease of the quality of energy. For factories with a large numbers of distillation schemes of high energy requirement, the paralleled schemes lead to improve economic benefit.

References

[1] Araujo AB, Brito RP, Vasconcelos LS, "Exergetic analysis of distillation processes-a case study," Energy 2007;32:1185-93.

[2] Michele Mascia, Francesca Ferrara, Annalisa Vacca, Giuseppe Tola, Massimiliano Errico, "Design of heat integrated distillation systems for a light ends separation plant," Applied Thermal Engineering 2007;27:1205-1211.

[3] Amiya K. Jana, "Heat integrated distillation operation," Applied Energy 2010;87:1477-1494.

[4] Hilde K. Engelien, Sigurd Skogestad, "Multi-effect distillation applied to an industrial case study," Chemical Engineering and Processing 2005;44:819-826.

[5] James G. Gebbie, Michael K. Jensen, Piotr A. Domanski, "Experimental transient performance of a heat pump equipped with a distillation column," International Journal of Refrigeration 2007;30:499-505.

[6] Massimiliano Errico,Giuseppe Tola, Michele Mascia, "Energy saving in a crude distillation unit by a preflash implementation," Applied Thermal Engineering2009;29:1642-1647.

[7] Massimiliano Erricoa, Giuseppe Tola, Ben-Guang Rong, Daniele Demurtas, Ilkka Turunen, "Energy saving and capital cost evaluation in distillation column sequences with a divided wall column," chemical engineering research and design 2009;87:1649-1657.

[8] Ivar J. Halvorsen, Sigurd Skogestad, "Energy efficient distillation," Journal of Natural Gas Science and Engineering 2011;3:571-580.

[9] Jinsheng Sun, Fan Wang, Tingting Ma, Hong Gao, Yanzhen Liu, Fang Cai, "Exergy analysis of a parallel double-effect organosilicon monomer distillation scheme," Energy 2012;47:498-504.

[10] Hilde K. Engelien, Sigurd Skogestad, "Multi-effect distillation applied to an industrial case study," Chemical Engineering and Processing 2005;44:819-826.

[11] Kazuo Matsuda, Kenichi Kawazuishi, Yasuki Kansha, Chihiro Fushimi, Masaki Nagao, Hiroshi Kunikiyo, Fusao Masuda, Atsushi Tsutsumi, "Advanced energy saving in distillation process with self-heat recuperation technology," Energy 2011;36:4640-4645.

[12] M.A. Sharaf, A.S. Nafey, Lourdes García-Rodríguez, "Exergy and thermo-economic analyses of a combined solar organic cycle with multi-effect distillation (MED) desalination process," Desalination2011;272:135-147.

[13] Weiwen Wang, Wenjian Wang, "Energy Saving in Methylchlorosilane Distillation," International Conference on Computer Distributed Control and Intelligent Enviromental Monitoring 2012:290-293.

[14] Myungwan Han, Sunwon Park, "Multivariable control of double-effect distillation

configurations," J. Proc. Cont. 1996:247-253.
[15] Hilde K. Engelien, Sigurd Skogestad, "Multi-effect distillation applied to an industrial case study," Chemical Engineering and Processing2005;44:819-826.
[16] Debadrita Maiti, Amiya K. Jana, Amar Nath Samanta, "A novel heat integrated batch distillation scheme," Applied Energy2011;88:5221-5225.
[17] Anton A. Kiss, Servando J. Flores Landaeta, Carlos A. Infante Ferreira, "Towards energy efficient distillation technologies-Making the right choice," Energy2012;47:531-542.
[18] Diego C. Alarcón-Padilla, Lourdes García-Rodríguez, Julián Blanco-Gálvez, "Experimental assessment of connection of an absorption heat pump to a multi-effect distillation unit," Desalination2010;250:500-505.
[19] Diego C. Alarcón-Padilla, Lourdes García-Rodríguez, Julián Blanco-Gálvez, "Design recommendations for a multi-effect distillation plant connected to a double-effect absorption heat pump: A solar desalination case study," Desalination2010;262:11-14.
[20] Fiedler E, Grossman G. Methanol, "Ullmann's Encyclopedia of Industrial Chemistry," Weinheim : Wiley-VCH Verlag/Gmbh & Co; 2002.
[21] Meyers RA, "Handbook of synfuels Technology," New York : McGraw-Hill; 1984.
[22] Chu YZ, Qin LP, "Optimum design of methanol distillation process and columns," Chem Ind Eng Prog 2008;(27):1659-62.
[23] Juntao Zhang, Shengrong Liang, Xiao Feng, "A novel multi-effect methanol distillation process," Chem Eng Pro 2010;(49):1031-37.
[24] Jinsheng Sun, Fan Wang, Tingting Ma, Hong Gao, Peng Wu, Lili Liu, "Energy and exergy analysis of a five-column methanol distillation scheme," Energy2012;45:696-703.
[25] J. S. Sun, P. Wu, Y. H. Wang, H. Gao, L.L. Liu.CN Patent102039058 A. 2011.
[26] J. S. Sun, P. Wu, Y. H. Wang, Y. X. Su, H. Gao, J. Li, D. Yan. CN Patent 101798322 A. 2010.
[27] Michele Mascia, Francesca Ferrara, Annalisa Vacca, Giuseppe Tola, Massimiliano Errico, "Design of heat integrated distillation systems for a light ends separation plant," Applied Thermal Engineering 2007;27:1205-1211.
[28] Olga A. Flores, J. Carlos Ca´ rdenas, Salvador Herna´ ndez, Vicente Rico-Ramı´rez, "Thermodynamic Analysis of Thermally Coupled Distillation Sequences," Ind. Eng. Chem. Res. 2003;42:5940-5945.
[29] Szargut J, "Exergy Analysis of Thermal, Chemical and Metallurgical Processes," New York: Hemisphere Publishing Corporation;1988:55-152.

(45n) In Situ Transesterification of Jatropha Curcas Seed Particles with Alkaline Phase Transfer Catalyst

Subbarao Duvvuri[1], Shuhaimi Mahadzir[2] and Sintayehu Hailegiorgis[2], (1)Department of Chemical Engineering, Universiti Teknologi PETRONAS, Tronoh, Malaysia, (2)Chemical Engineering, Universiti Teknologi PETRONAS, Tronoh, Malaysia

Economic viability of biodiesel technology is at stake due to the high cost of edible oil feedstock and slow reaction rates due to limited solubility of oil in methanol. In situ transesterification of non edible oil seed particles can minimize the cost of feedstock. Microwave pretreatment of seeds can make oil molecules more reactive. Use of phase transfer catalysts can enhance the reaction rates. In the present work, these three concepts are utilized together to investigate alkaline in situ transesterification of microwave pretreated jatropha curcas seed particles in presence of benzyltrimethylammonium hydroxide (BTMAOH) as a phase transfer catalyst (PTC). It was observed that addition of PTC to alkaline catalyst has improved conversion and reduced reaction time. Microwave pretreatment of jatropha seed particles reduced reaction time drastically. The quality of biodiesel produced was within the limits of the international standard requirements of ATME D6751 and EN-14214. Thus, combined use of microwave pretreatment of seed particles, phase transfer catalysis in the presence of alkaline catalyst can increase the rate of in situ transesterification and product yield.

(45o) Catalyzed-Assisted Manufacture of Olefins (CAMOL): Updated for Use in Naphtha Service

Maximilian Walter, Petrochemicals, BASF SE, Ludwigshafen, Germany and **Suzanne Rech**, BASF Qtech, BASF Corporation, Iselin, NJ

Since the commercialization of the olefins plant, operators have searched for ways to be more competitive. One area always targeted includes keeping their furnaces in operation longer between decoking steps. Over the years, many anti-coking methods/technologies have become commercially available using different approaches to prevent coils from coking. One of the latest available technologies is the novel CAMOL™ (Catalyzed-assisted Manufacture of Olefins) Technology from BASF Qtech. As reported in the introduction of the technology to the EPC in 2008 & 2010, the CAMOL™technology differs from other available anti-coking technologies in that the technology not only eliminates the catalytic (or filamentous) coke make through a physical barrier, but also gasifies the thermal (or amorphous) coke for an even greater coking resistance. The coating process has generated a coating with almost identical thermo-mechanical properties as the base metal substrate (alloy). This allows bonding to the base metal similar to being welded with a minimum risk of spallation when going through thermal cycling. Commercially, the CAMOL Technology has been available since 2006 and is now in operation in furnaces of various designs.

With CAMOL™, operators have experienced extended run lengths, reduced energy requirement and GHG emissions, lower average TMT operating temperatures, reduced DMDS consumption, and less carburization. In addition, the technology has exhibited high operating temperature stability (>1130°C/ 2065°F).

At BASF SE in Ludwigshafen, Germany, the technology has been in operation since early 2010 in mostly naphtha service. After overcoming some initial challenges associated with a catalytic surface technology, such as reactivating the catalytic species after decoking, the coil is now successfully operating with run lengths significantly higher than uncoated coils. This paper will report on the results and learnings from the experiences gathered over the past three years along with the advancements made since the last report in 2010.

(45p) Extending Furnace Run Length and Coil Life by Optimizing Decoking

Vinod Mishra, Ingenero, Mumbai, India and **James Brigman**, Ingenero, Houston, TX

Abstract:

In the Ethylene manufacturing process, coke deposited in the radiant coils and Transfer Line Exchangers (TLEs) during normal operation is removed periodically by decoking the furnaces using a mixture of steam and air. There is a significant variation in operating parameters adopted by different plants for decoking. These are often either too conservative or aggressive, leading to production loss from increased off-stream time, TLE limitation and deterioration in coil life. This paper describes how the decoking procedure was modified based on furnace and TLE operating conditions and the decoking steps changed to obtain the best possible Ethylene yield, without adversely impacting coil life and TLE performance. This optimization has now been done in several major ethylene facilities and furnace run length has more than doubled and production has significantly increased.

SOLVING OPERATIONAL PROBLEMS IN A NAPHTHA SPLITTER

Flávio Weissheimer, GE Water & Process Technologies do Brasil Ltda, São Paulo/Brasil
Giorgio Franceschetti, Braskem, Santo André/Brasil
Rafael Henrique Camparin, GE Water & Process Technologies do Brasil Ltda, São Paulo/Brasil
Patrick Lucas, GE Water & Process Technologies, The Woodlands/US

Abstract

This paper sets out how the polymerization process of Reboilers of the (A-100) Crude Naphtha Fractionation Unit of Braskem UNIB 3 ABC can be controlled by applying polymerization inhibitors developed by GE Water & Process Technologies Ltda and Braskem´s engineering solutions.

These heat exchangers have a high degree of polymerization because of the content of olefinic compounds present in the feed composition which, when in contact with high temperatures in the heat exchange areas of reboilers, tend to polymerize. The joint solution of GE technology and Braskem engineering solutions applied to this system ensures regular operation of such equipment for periods of time greater than those hitherto recorded without manifesting loss of heat transfer efficiency.

In the 2012 shutdown while the opening of these reboilers was being inspected, it was observed that pipes had not been clogged nor was surface fouling present, which is a very positive state when compared to the historical records of previous campaigns.

The strategy adopted by GE/Braskem to ensure a longer campaign for these pieces of equipment was based on the quality control of feeding by adjusting dosages and monitoring performance using statistical analysis tools.

The use of statistical models based on Multivariable Regression Analysis (MRA) to evaluate performance, which is validated by inspecting reboilers internally, has been an efficient way to determine the degree to which the system has been fouled and also to schedule shut-downs for cleaning the mechanical parts.

Description of the Process

The (A-100) Naphta Fractionation Unit includes two distillation columns, the objective of which is to separate the light (C_6), medium (C_6 - C_8) and heavy (C_{9+}) cuts, included in the feed supply of crude naphtha.

The light cut and a portion of the heavy cut are sent to pyrolysis furnaces so as to produce ethylene. The medium cut is sent to the Hydrotreating and Catalytic Reforming Unit, with the objective of increasing the aromaticity of the system so as to produce BTX.

The A-100 processes 100 tons/h of crude naphtha, and approximately 35% of this current represents the average cut. Figure 1 below shows the schematic flow diagram of this fractionation process.

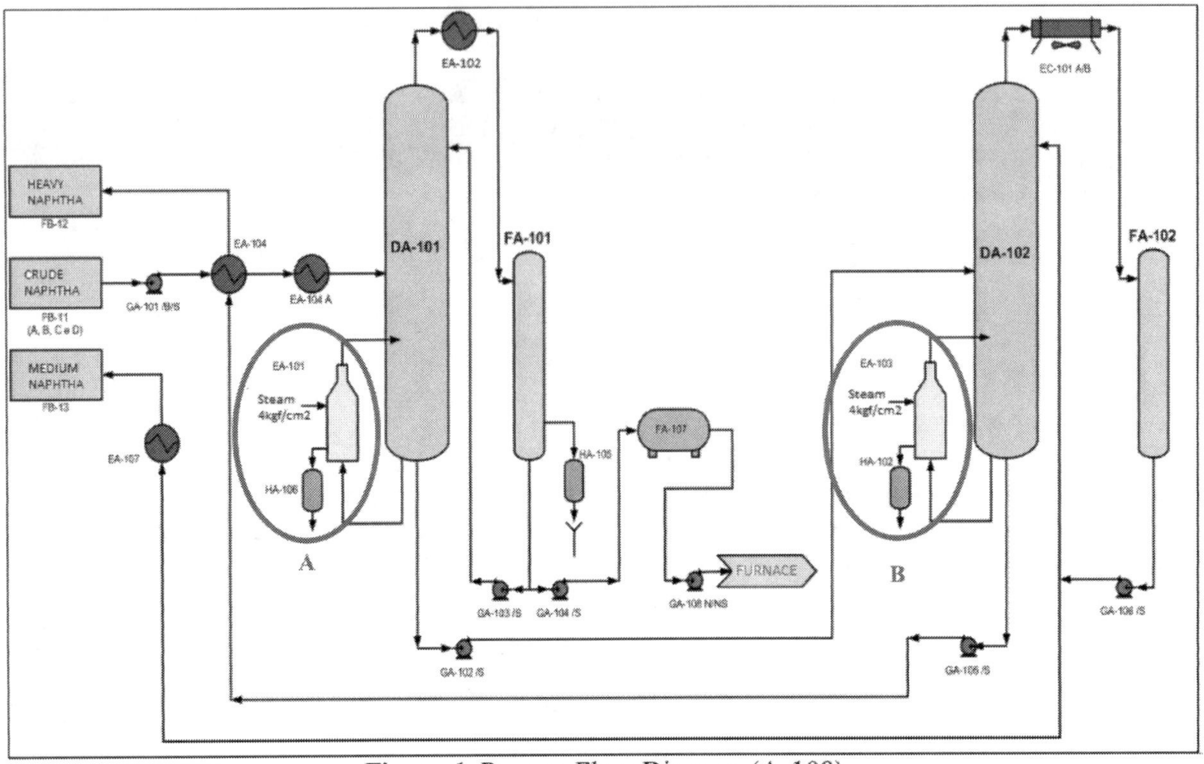

Figure 1. Process Flow Diagram (A-100)

The Reboilers, indicated by the letters A and B in Figure 1 above, operate with an injection of low steam (4kgf/cm2) to promote vaporization of the bottom stream of the A-100 distillation columns. These devices operate over a temperature range of about 150°C, which is sufficient to promote the polymerization of certain olefinic compounds present in the feed of crude naphtha.

Poor performance of these Reboilers can lead to production stoppages for maintenance and cleaning, which very often affects up to five days of production. The possibility of raising the times of campaigns and predicting the degree of fouling in downstream units helps to minimize production losses due to unscheduled stoppages.

This study will discuss the results of Reboiler A, which is the most critical equipment, with a view to synthesizing data, but gains in equipment B were also significant.

Evaluation System

Historical Record of the Campaign

Reboiler A was installed in 1972 but examination of the historical records did not reveal that the sheaf of tubes had ever been replaced. This exchanger has currently 11 tubes plugged into it.

The Campaign times of this equipment are shown in Table 1, with the last campaign being highlighted in blue so as to set it apart from the other ones. Table 1 gives a summary in the order of the timings of the campaigns, the average load processed in the period, the total load processed, the reason for operational stoppages etc.

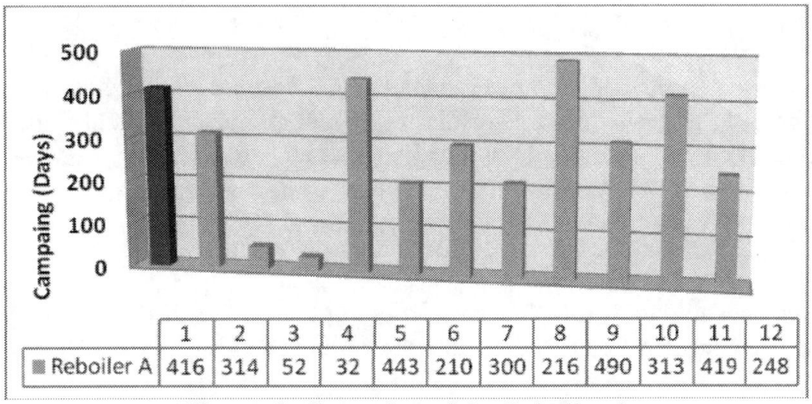

Figure 2. Campaign of the Reboiler A

Table 1. Historical Record of Reboiler A Campaigns

Date	Campaign *(days)*	Capacity *(ton/h)*	Capacity *(ton/campaign)*	Occurrence	Tubes Clogged *(%)*
13/06/12	416.0			Programmed Shutdown	0
17/03/11	314.0	93.1	701601.6	Heat exchange performance	30
03/05/10	52.0	90.8	113318.4	Stuck tube	-
11/03/10	32.0	71.1	54604.8	Heat exchange performance	50
09/02/10	443.0	72.0	765716.6	Heat exchange performance	20
16/11/08	210.0	98.1	494575.2	Heat exchange performance	-
16/04/08	300.0	104.0	748800.0	Programmed Shutdown	-
16/06/07	216.0	92.8	481075.2	Programmed Shutdown	-
10/11/06	490.0	-	-	Programmed Shutdown	-
30/06/05	313.0	-	-	Stuck tube	-
17/08/04	419.0	-	-	-	-
18/06/03	248.0	-	-	-	-
10/10/02	-	-	-	-	-

Conditions of Opening

In this review, a comparison is presented between two operational shut-downs of A-100, which occurred in March 2011 and May 2012, in order to observe the improvement in performance from the last campaign associated with the change in the GE Treatment strategy.

Campaign I: Production shutdown on 03/2011

Table 2 below summarizes the features observed during the inspection of the state of the inlet of this piece of this equipment during the shutdown of March 2011:

Table 2. Internal inspection of Reboiler A in 03/2011

Bottom Cover	Surface residue
Bottom Mirror	No deposits or internal blockage of the tubes were observed.
Top Mirror	Approximately 30% of the tubes were clogged, especially in the region of the vapor inlet, which has higher temperatures. Magnetic and non-magnetic deposits.

May/2011	**May/2011**

Figure 3. The mirror of Reboiler A before it was cleaned.	**Figure 4.** The mirror of Reboiler A at the vapor inlet before it was cleaned.

Note: It is difficult to carry out internal cleansing by Hydro jetting (pump pressure during cleaning is 20,000 psi). This reboiler has 11 tubes plugged into it, but none of them are plugged in during this shutdown.

Campaign II: Production shutdown on 03/2011

Table 3 below summarizes the features observed during the inspection of the state of the inlet of this piece of equipment during the shutdown in May 2012:

Table 3. Internal inspection of Reboiler A in 05/2012

Bottom Cover	Surface residue
Bottom Mirror	No deposits or internal blockage of the tubes were observed.
Top Mirror	Clogged tubes were not identified, not even in the region of the inlet vapor. There was only a thin layer of deposit that was easy to remove and which was evenly distributed over the upper surface of the mirror.
Spool	A thin deposit on the inner surface of the caisson.

May/2012	**May/2012**

Figure 5. The mirror of Reboiler A before it was cleaned.	**Figure 6.** The mirror of Reboiler A before it was cleaned.

Note: Hydro jetting was done quickly, without any obstructions due to deposits having been identified. This equipment has 11 tubes plugged into it, but none of them were during this shutdown.

Analysis of Deposits

The characterization of the deposits found in Reboiler A showed an 86.0% loss on ignition, thus indicating the strong presence of organic compounds in their composition, with the functional groups of carboxylic acid and ester standing out. The inorganic analysis of the sample indicated a high concentration of iron oxides and also sulfur oxides in the form of sulfides on the cover and spool of this reboiler. This is summarized in Table 4 below:

Table 4. Analysis of the Deposits of Reboiler A

Organic Compounds - LOI (%)	86.0
Water of hydration	Smaller
Sulfur compounds	Smaller
Carboxylic Acid and Ester	Greater
Dichloromethane soluble (%)	2.0
Elemental Sulfur	Greater
Dichloromethane insoluble (%)	98.0
Water of hydration	Smaller
Inorganics	Smaller
Carboxylic Acid and Ester	Greater
Composition of Dichloromethane insolubles (%)	
Carbon	56.0
Hydrogen	5.0
Nitrogen	3.0
Sulfur	8.0
Inorganic Compounds (%)	
Iron, $Fe_2O_3 + Fe_3O_4$	82.0
Sulfide, S^{2-}	16.0
Phosphate, P_2O_5	1.0
Sodium, Na_2O	1.0

Evaluation of Performance

The evaluation of performance of Reboiler A was prepared using the MRA tool, which is a GE good practice to predict fouling in critical systems. This statistical tool enables a dependent variable to be determined, in this case the percentage of opening the steam valve, by means of the numerical regression of the other operational variables of the column. This method generates a mathematical equation to predict the percentage of inlet steam when the equipment is clean, given that the equation obtained by regression includes data of this condition.

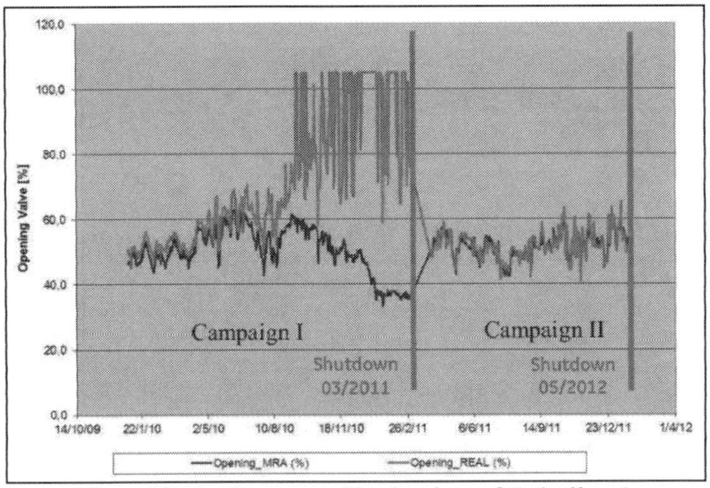

Figure 7. Performance Evaluation of Reboiler A

In Figure 7 the dark blue line represents the percentage of inlet steam when the equipment is clean, as per the equation defined by the MRA. The pink line represents the actual percentage of inlet steam. The sharp upward shift evidenced in the period between November 2010 and March 2011 shows the fouling of the system during Campaign I (prior to the 2011 operational shutdown).

After the 2011 shutdown, it can be seen that the two curves follow the same trend, without a shift, indicating that the unit has the equivalent performance of a clean piece of equipment. The change in the GE treatment strategy, taking the control of the dosages of the product Petroflo 20Y25 into account, was seen to have been effective in increasing the campaign time of this equipment.

Conclusion

The change of strategy in applying dispersant and controlling the dosages after the March 2011 shutdown, presented positive results, given the states of the inlet seen during the inspection of reboilers A and B in May 2012..

The low degree of fouling of these reboilers evidenced in the 2012 shutdown led to the production process being considered reliable, as the A-100 reboilers are no longer the pieces of equipment responsible for the loading limit of the unit. The downtime during the cleaning of the mechanical parts of the reboilers was about 10 times less than that observed in the previous shutdown, thus reducing Braskem´s maintenance costs.

In addition, the MRA statistical tool was able to produce a reliable representation of the degree of fouling of the system. This tool can be effective for scheduling shutdowns more than 03 months in advance.

References

1. Setal Lummus, "Manual de Operação da Unidade de Processamento de Nafta Bruta (Área 100)", Santo André, Brazil.
2. Ponce de Leon, Carolina; Weissheimer, Flavio and Camparin, Rafael H. (2011), "Proposta Técnica de Fornecimento de Especialidades Químicas, Equipamentos e Serviços para Tratamento Químico para a Unidade de Processamento de Nafta Bruta (Área 100)", Santo André, Brazil.
3. Kleinbaum, David G., Kupper, Lawrence L., Muller, Keith E. (2008) "Applied Regression Analysis and Other Multivariable Methods", Belmont, USA.

13AIChE
Spring Meeting, San Antonio, TX

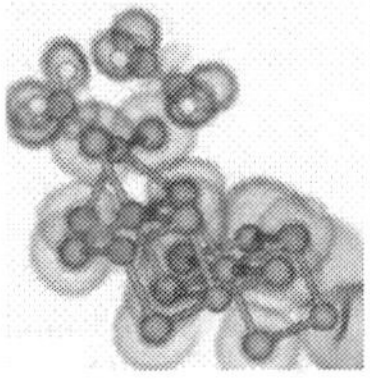

SmartProcess® Application Strategy Achieves Faster and More Efficient Control of an Ethylene Complex

Patrick Truesdale
Principal Consultant
Industry Solution Group
Emerson Process Management

April 29, 2013
San Antonio, TX

EMERSON.
Process Management

Speaker Introduction

- Principal Consultant
  Emerson Process Systems & Solutions
- BS Electrical Eng, PE
- 40+ Years experience with Refinery & Petrochem Process Control
- Joined Emerson in 2007
- Previous employers: duPont, Bonner & Moore, SETPOINT, AspenTech, Comcept Technology

Patrick
Truesdale

E-mail: patrick.truesdale@emerson.com

13AIChE
Spring Meeting, San Antonio, TX

EMERSON.
Process Management

Agenda

- Performance Package Review
- SmartProcess Ethylene Overview
  - Furnaces
  - Distillation
- Benefits

13 AIChE Spring Meeting, San Antonio, TX

EMERSON. Process Management

SmartProcess Ethylene Packages

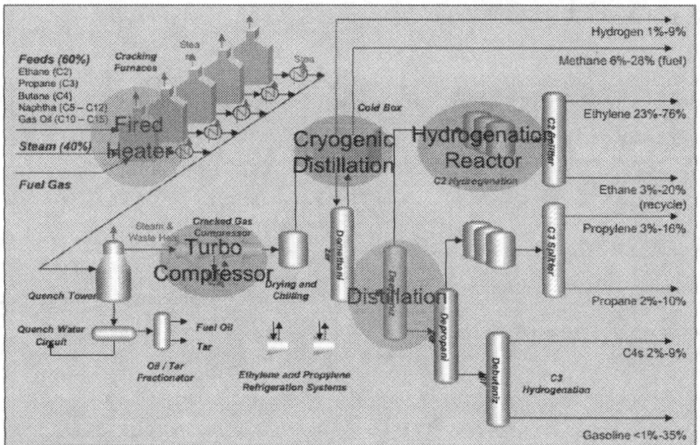

13 AIChE Spring Meeting, San Antonio, TX

EMERSON. Process Management

SmartProcess Scope

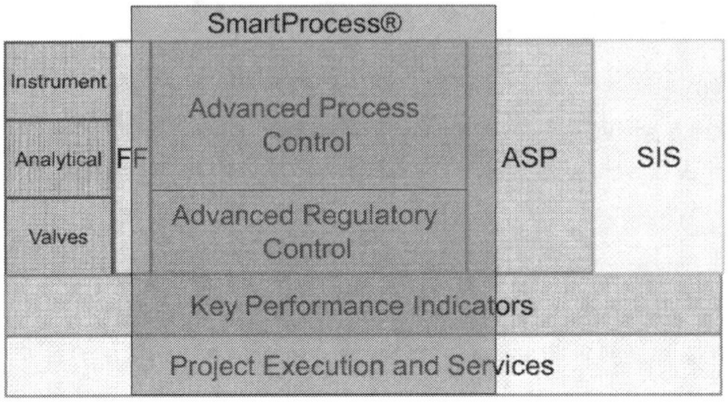

Basic Ethylene Furnace Variables

Key Operating Objectives:

- Constant severity
- Maintain HC/Stm ratios
- Min excess air
- Maintain within constraints
- Max charge (if needed)

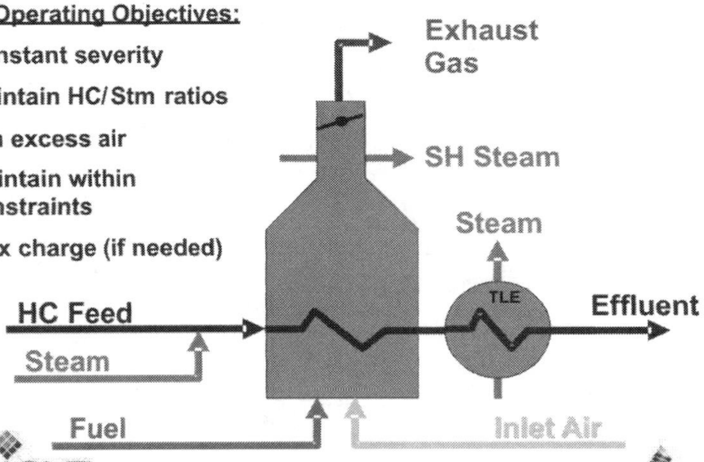

SmartProcess Furnace Module

- Design allows for different furnace configurations
  - Process Zones
  - Firing Zones
  - Multiple TLE analyzers
- Advanced Regulatory
  - Charge Control
  - Steam Control
  - Severity Calculation
  - Combustion Control
  - KPI's
- Advanced
  - Pass Balancing
  - Heat Balancing
  - Severity Control
  - O2 Control

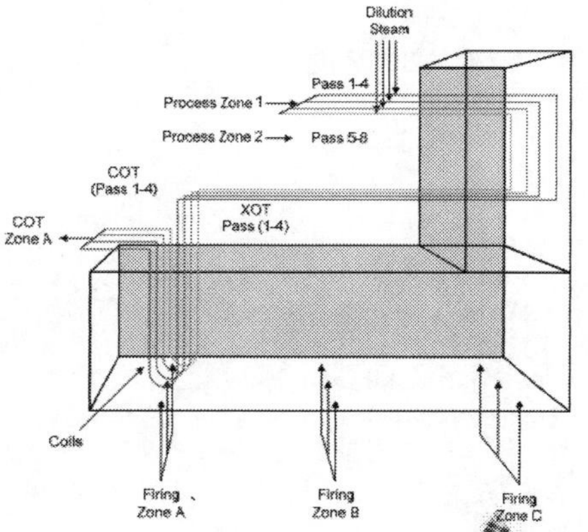

Standard Furnace Module Design

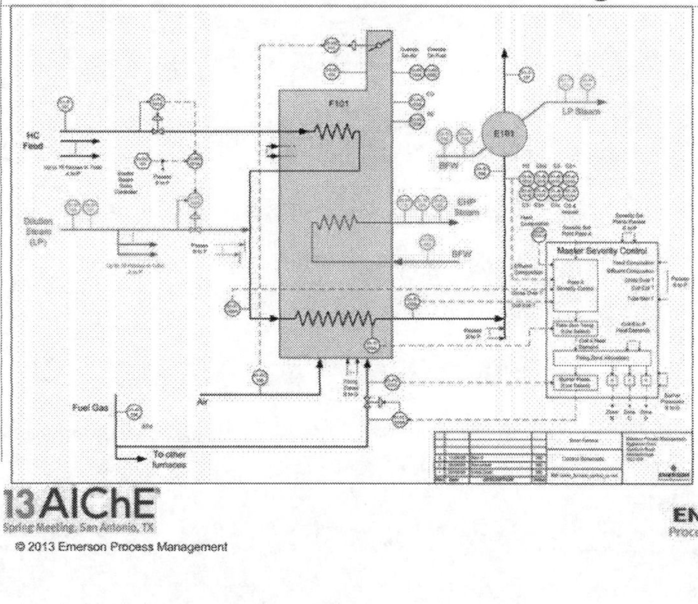

Charge Rate Control

Functions:
- Pass Balancing
- Constraints
- Feed Maximization

Functions:
- Ramping
- Limits & Alarms
- Multiple Feed Types
- Furnace Status
- Pass Offsets

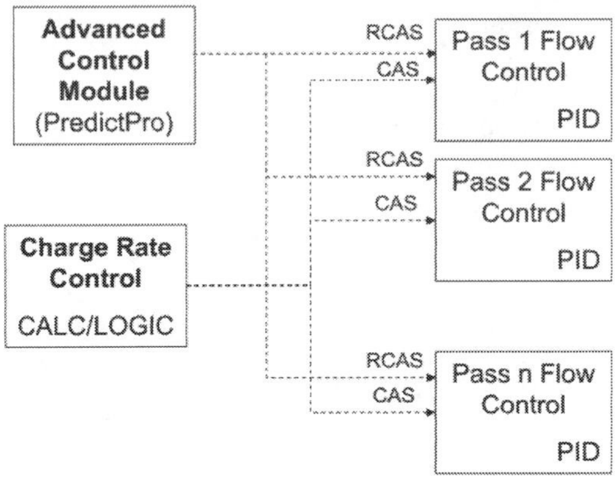

13 AIChE
Spring Meeting, San Antonio, TX

EMERSON
Process Management

Steam Ratio Control

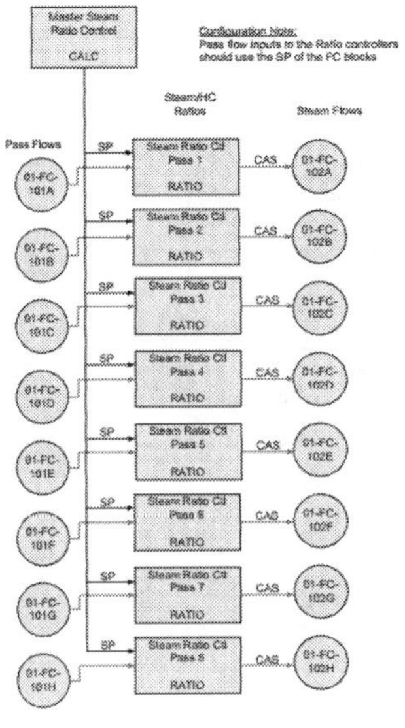

13 AIChE
Spring Meeting, San Antonio, TX

EMERSON
Process Management

Severity Control

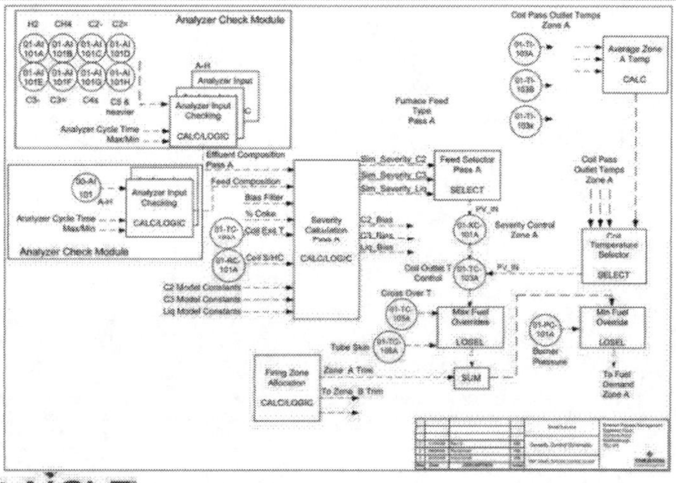

Automated Furnace Sequences

- Standard module templates to be used for Furnace Startup, Shutdown and Decoking sequences.
- Utilizes preconfigured Sequence Function Charts (SFC) running in DeltaV controllers or App Station
- Standard sequence functions can be customized to match site requirements
- Manual operator steps with confirmation can be mixed with automated steps
- Sequence execution control graphics and faceplate templates

Key Performance Calculations

- Furnace efficiency
  - Heat loss method
- Furnace Duty
- TLE Duty, Fouling factor
- Stack gas estimated composition
- Stack gas daily total tons exhaust for CO, CO2, S, N, etc.
- Stack gas NOx emissions (optional – based on CEMS)

Distillation - Classic Multivariable Problem

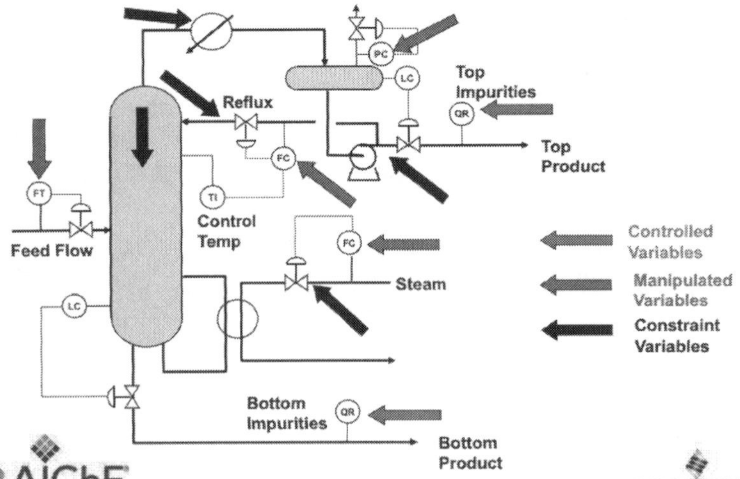

Distillation Control Module

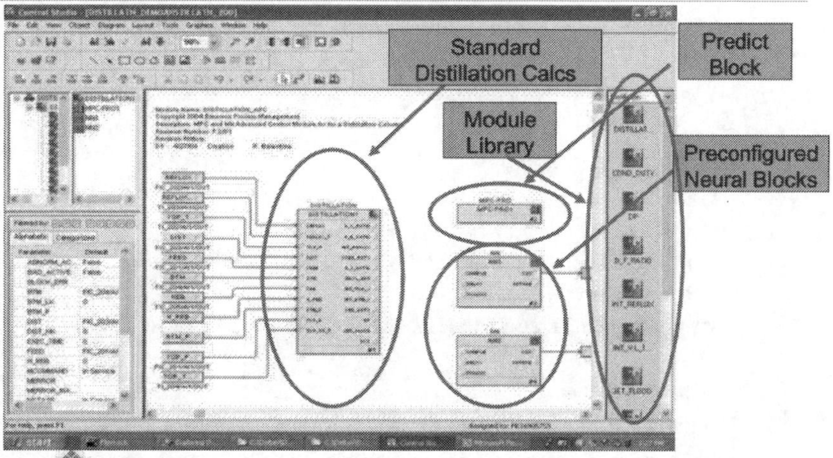

SmartProcess Distillation

- DeltaV PredictPro with up to 4 MV's
- DeltaV Neural for up to 2 product qualities
- Standard set of calculations & constraints
  - Internal reflux
  - Pressure compensated temperatures
  - Jet flooding
  - Reboiler & condenser duty
- Engineering Services to design, implement and commission

SmartProcess Applications

- Licensed industry applications for common processes that are:
  - Pre-engineered, tested, documented
  - Designed by APC Consultants with extensive practical experience
  - Supported & enhanced
- Bundled combination of:
  - DeltaV libraries: composite blocks, modules & templates
  - DeltaV simulation and demonstration example
  - APC tools (e.g. Predict, Neural)
  - Application code (e.g. VB, .Net)
  - Other databases and apps (e.g. SQL*Server, Excel, TankMaster)
  - Sample DeltaV Graphics
  - Documentation
- Templates to be used to configure for specific site requirements
- May be implemented by Emerson or Customer

13AIChE Spring Meeting, San Antonio, TX

EMERSON. Process Management

Benefits

13AIChE Spring Meeting, San Antonio, TX

EMERSON. Process Management

Ethylene Furnace High Temperature Imaging Cameras to Monitor Burner Flames ,Tube Temperature and Tube Failure Detection

Thomas M. Canty, P.E.
President
J.M. Canty Inc
Buffalo, NY USA

Abstract

The development of explosion proof, high temperature imaging cameras which operate in ethylene furnaces without cooling air now allows monitoring of ethylene furnaces to improve energy/operating efficiency along with providing important safety benefits. Imaging cameras work off the visual wavelength which provides 2 dimensional view and temperature measurement of the tube temperature which is more accurate than thermocouples and IR probes. In addition visual images of the furnace burners lets you detect flame detachment, smoking burners and flame shape in order to increase burner efficiency and save energy. The safety aspects of the Camera imaging systems is significant as it allows visual detection of pin hole leaks and tube failure in the furnaces.

Introduction

Ethylene furnaces, or any type of cracking unit, rely on thermal management in order to make product efficiently and to operate in a safe manner. Having 'eyes' in the furnace can help to identify problems in their early stages of development. High temperature camera technology is able to survive the environment and give a wide angle view of the tubes from which temperature and tube blockage may be determined. Achievable benefits from a camera system include:

1. Temperature analysis capacity for all areas of the image, not just single point. Multiple zone capability for user customization of analysis.
2. Flame detection and size measurement
3. Detection of tube movement and ability for software to automatically move measurement zones as tubes distorts.

Imaging Fundamentals

Basics of an imaging system include a high resolution sensor, proper illumination, appropriate hardware for the process and a robust software analysis package. For this particular application illumination is created by the process and need only be managed optimally for best results. A high density pixel array yields resolution of the image and measurement as well. The hardware and software are the more challenging aspects of the application and will be discussed in some detail.

Hardware

The design of the camera system is unique in that it's internally exposed lens assemblies are **designed to survive without cooling air**. When installed properly, the combination of system materials and heat transfer coefficients at the camera tip limit heat transfer to the system and allow it to run hot without chance of damage. This feature, when system is installed in a clean, gas burning environment, does not disrupt the temperature or air balances of the furnace because no cooling air is being introduced to the furnace. In addition, the design includes a disposable, quarts shield which protects the process exposed lenses from direct contact with the internal atmosphere. Should any particulate contact the system, the quartz shield is there to protect the optical lenses. Replacement of the protective quartz shield is inexpensive. Replacement of damaged lenses is very expensive and so the Canty design guards against that possibility.

Internally, the system is a two piece assembly consisting of the internal insertion lens and the external camera package. As mentioned earlier in this document, Canty employs a fused glass window as part of the electronic housing package. This has a dual purpose; Firstly, the camera housing assembly is a separate entity which carries WP and EXP ratings. Secondly, the camera housing bolts to the internal lens system and transmits the optical rays through the fused glass window to the CCD. This mounting also creates a hermetic barrier between the process and the electronic housing thereby preventing migration of hot, process gasses outside the pressure boundary of the system.

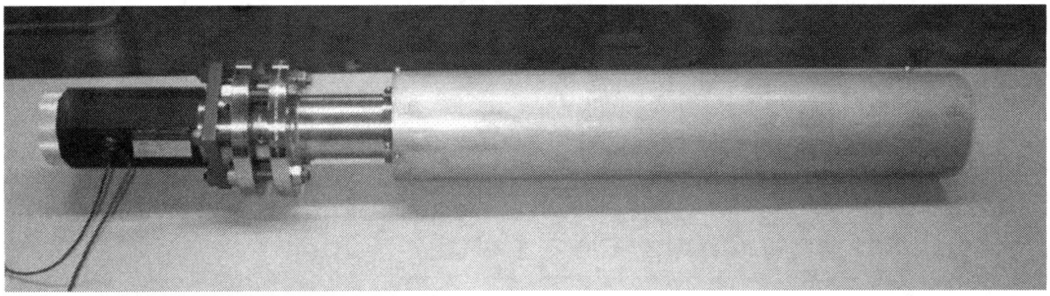

Figure 1. High Temp Camera Assembly

The camera electronics come with a full array of controls that allow it to be optimized to the software function to achieve accurate, consistent data output.

As previously mentioned, the installation is an important feature of the design. When installed correctly, the camera sits inside the refractory (see Figure 2) where its exposure to the process environment is minimized thus minimizing heat transfer into the internal lens.

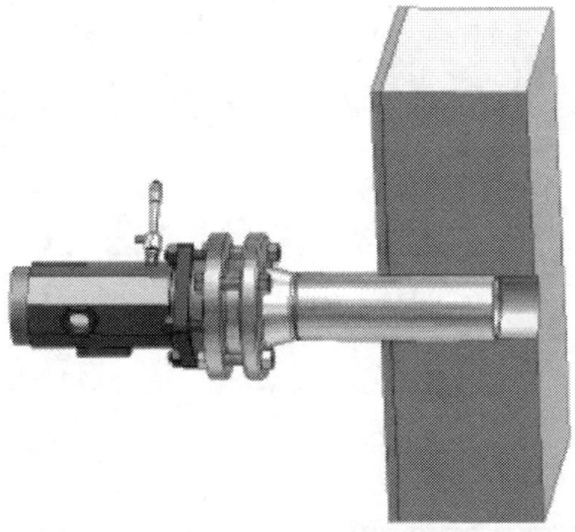

Figure 2. Camera Installation Schematic

Software

With respect to software design, the Canty software is fully flexible in analyzing any portion of the captured image. Multiple analysis zones can be drawn to monitor discreet areas of the view and each zone can run its own particular algorithm depending on need. The benefit here is that multiple tubes can be continually surveyed for changes and trends. The analysis data is also not single point, but is comprised of all the data points in the analysis zone. By choosing proper software filters various information can be obtained from the full array of data such as hot or cold spots. This data can also be archived in total or as snap shots at determined increments of time. The on screen data output can also be updated in the short or near term and can thus be used to judge what is happening in the furnace over time with just a glance at the screen. For instance, using the histogram feature with a 1x daily output a long term trend can be presented on screen in real time.

The basic theory of temperature measurement is based on the radiative emissions of a body at temperature – as the body temperature increases, the radiation curve shifts towards the shorter wavelengths and that is why as a body heats up the eye cannot detect it until it reaches 1000F which is the approximate visible point where emitted radiation is dark red in color. As the temperature increases, the color graduates to bright red, then orange then toward white. A pyrometer functions by measuring a narrow band of wavelength intensity and comparing it to the intensity of a black body to determine the temperature. From the graph below it can be seen that if a single wavelength is monitored the intensity of that measurement changes as the body temperature changes (curve shifts) which then corresponds to a different temperature on the calibration scale.

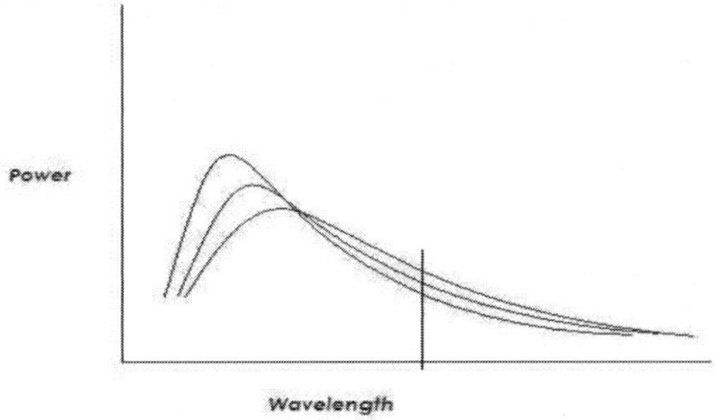

Figure 3. Power Curve at Varying Temperatures

The same is true of a camera system measuring the increasing intensity of an object as it radiates from approx 1000 F (dim, dark red color) on up the temperature scale to 3000+ F for instance which would be a bright white. From the graph above one can see that as color goes from red toward white the peak intensity is always increasing and it is the measure of this intensity which indicates the object temperature. Using an RGB camera the measurement can also be made by monitoring the individual color components of the signal impinging on the chip. For industrial uses, the measurement of intensity usually suffices.

Standard instrumentation used in industry include thermocouple type devices (contact) and thermal imaging and pyrometer systems (non contact). In severe temperatures and atmospheres, the contact devices deteriorate quickly and provide gradually less reliable information leaving the operator running blind. They also create a discontinuity on the surface they are welded to which, in the case of a furnace tube, can lead to premature tube failure.

Non contact systems have difficulties with calibration drift and emissivity adjustments to account for the fact that many items being measured are not black body emitters. In addition, these technologies are single point measurement systems and are best used as hand held devices which can be difficult to use and provide subjective measurement results based on operator preferences for targeting.

Vision technology offers several benefits these other technologies do not. The visual aspect of the system allows the operator to view the process and confirm the readings received. It also allows flexibility to set up measurement zones on screen at any time to assess any or all areas of the process in view should a change be required. Since vision systems are generally fixed installations, an added benefit is that the information is taken from the same process points and features within a given measurement zone at all times which means there is no error in sighting of the instrument.

The vision system is also unique in that it does not suffer calibration drift as almost all instruments do. Once the calibration is set the readings repeat themselves within a couple degrees F. The calibration is based on thermocouple or pyrometer readings initially, however after that calibration is finished it does not have to be redone in the future as long as the system remains in its location.

Process history and performance can also be best documented by a vision system. First off, the camera can analyze any area of process that is within the systems view. This affords the operator the capability of viewing many points of interest instead of just single point temperature readings. As an example, if the system is monitoring a tube bank it can assess the temperature of the tubes for the full length of view. If there is a local flow problem in a tube it will become overheated in one section. The camera can locate that section and indicate its temperature rise over the rest of the tube or adjacent tubes. This information allows the tube location to be identified as well as the position in the tube that has encountered the upset condition. Over time this historical data, recorded in video or still image format by the control software, may be significant in improving the process. In addition to the ability to monitor the full area of the view, the full area of view can also be segmented into multiple zones of measurement, i.e. each tube having its own measurement zone for automated process control capabilities. This capability allows the tube to be operated at optimum efficiency since the addition of excess heat can be avoided thereby reducing energy draw to operate the furnace.

Thermal Analysis

Tube Temperature

The following figures illustrate the temperature analysis capability for tubes in a live test situation. Measurement zones are placed over the tubes (figure 4) as the furnace is heated and cooled. The temperature plots are shown in figures 5 and 6. The results are as expected and uniform from tube to tube, although a small number are included in this demonstration.

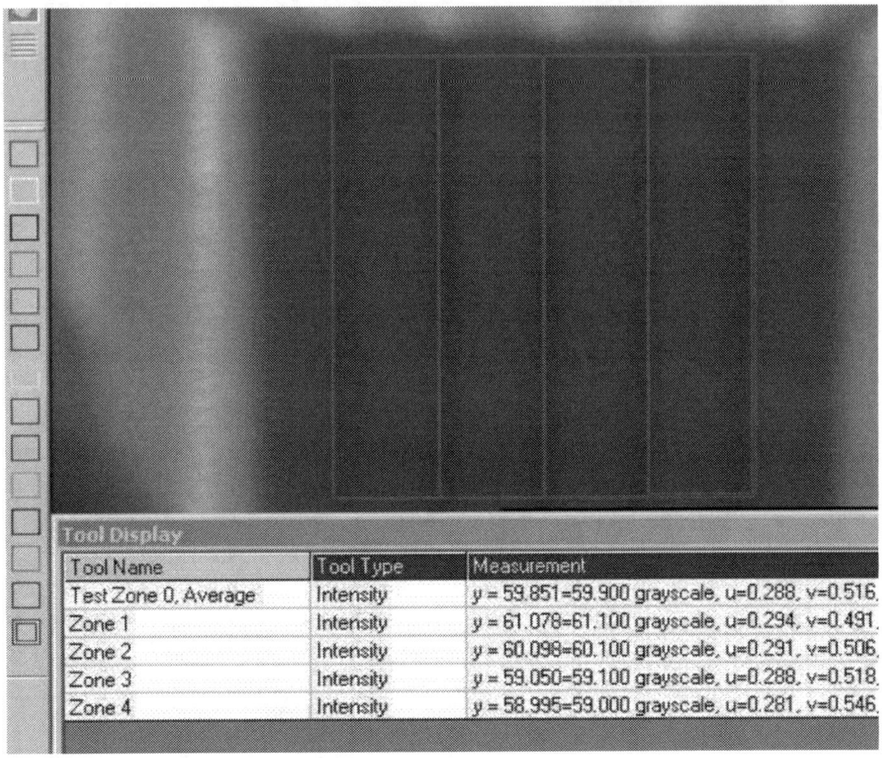

Figure 4. Software Screen Shot

As thermal loads are fluctuated up and down the following graphs track those changes. The cooling cycles are left to natural convection and thus are not linear in nature.

For the following curves, the vertical scale is gray scale intensity which is measured 0 – 255 (pitch black to bright white). The hotter the tube, the higher, or more intense, the gray scale reading. This value in process would be calibrated to the temperature. The horizontal scale is the image number, or time.

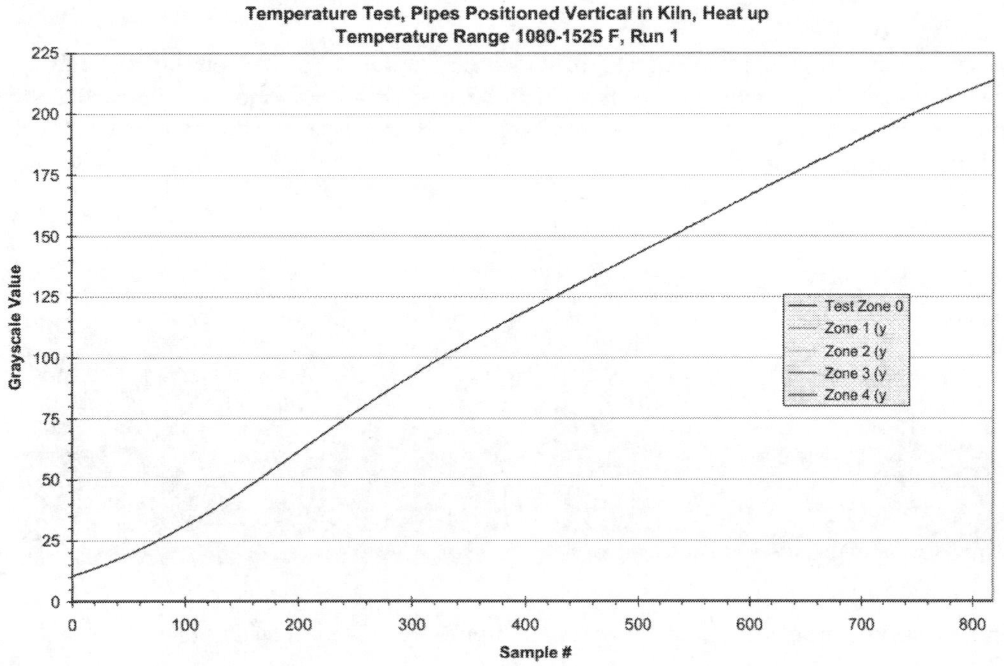

Figure 5. Heating Cycle

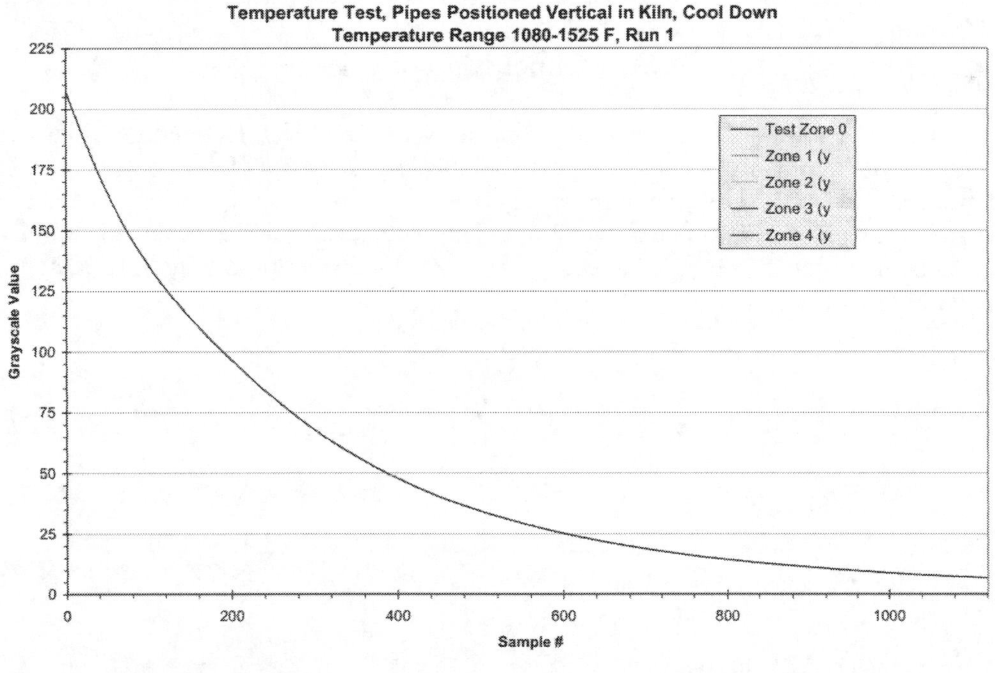

Figure 6. Cooling Cycle

Flame Measurement

Flame temperature can be monitored in the same way as described previously, however the flame size, or presence, can also be monitored. Figure 7 shows a live flame and Figure 8 shows a digital capture. Computing the flame area, or shape, can be done to track its changes over time.

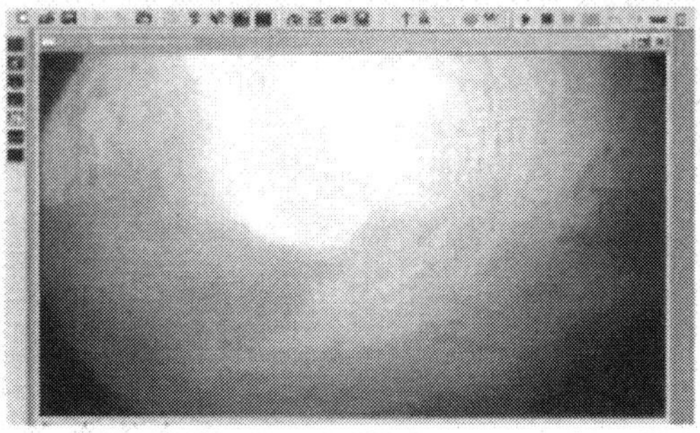

Figure 7. Kiln Flame (Live)

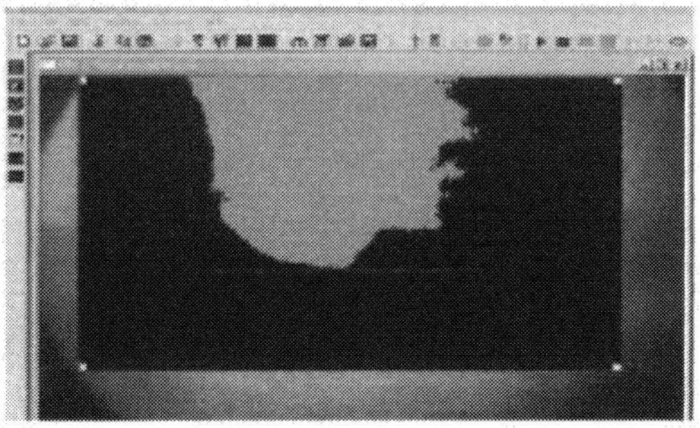

Figure 8. Kiln Flame (Digital)

Conclusion

Vision Technology presents an opportunity to increase effectiveness of furnace monitoring through hardware and software capabilities. The thermal design at the core of the hardware enables the system to provide the operator a view into the furnace and feed back on its thermal performance over time. This data can assist in detecting building upset conditions, but also provide a record for study to assist in improving the operational method of the furnace.

(45v) Straighten up! Non-Linear Compensation Using Feed Forward Control

Barbara Hamilton, Process Systems and Solutions, Emerson Process Management, Hampden , ME

Non Linear processes occur throughout the petrochemical plant. While the PID loop is surprisingly robust for most situations, there are cases where compensation of these non-linearity's is critical for stable control. Similarly, disturbances can cause major disruptions in control stability.

A method of feed forward control, using data based models, can provide robust control throughout the process range. Historical data are used to build models to drive the output toward the solution with PID control providing the final drive to zero error.

The model is also used to develop dynamic tuning. The result is a simple to configure, robust solution to manage non-linear loop and/or disturbance variations.

(45w) Adsorption of CO2 by Alkaline Activated Ca-Al2O3-SiO2 Material

Naim Faqir, King Fahd University of Petroleum & Minerals, Dhahran, Saudi Arabia and Reyad Shawabkeh, Chemical Engineering, King Fahd University of Petroleum & Minerals, Dhahran, Saudi Arabia

Natural gas contains a slight level of carbon dioxide which leads to several industrial and domestic heating problems. In addition, it is one of main greenhouse gases that contribute for global warming, where its pollution is a wide-spread problem arising from a wide range of industries, leading to complex contaminant in atmosphere. Given the increasing pressures worldwide to minimize the emission of greenhouse gases into atmosphere, CO_2 must be remediated efficiently and cost effectively.

In this short abstract a chemically treated limestone and alumino-silicate material were utilized for removal of CO_2 from natural gas using a combined adsorption-chemical reaction processes. This functionalized Ca-Al_2O_3-SiO_2 material was characterized according to its crystalline phase, surface area, pore size distribution and surface morphology. Then it was tested against different concentrations of CO_2 in CO_2-N_2 gas streams.

Pressure Gauges: Hidden Danger or Visible Safety

by Jeff Placek, WIKA Instrument, LP, Lawrenceville, GA

Introduction

In recent years, refining plants, petrochemical plants and chemical plants have been required to increase capacity, but have been provided with very few upgrades to existing equipment to accommodate the increased capacity requirements. This has caused plants to have increased stress and strain on their systems and operating units. For example, higher pulsating pressures and heavy mechanically-induced vibrations are translating to all areas of the processing equipment, especially pressure instrumentation.

Pressure gauges are one of the most important, yet most overlooked or ignored keys to ensuring the continued safety, productivity and profitability of process plants. Process pressure gauges are used to read pressure on transfer lines, discharge pumps, and other places where it is important to be able to read pressures.

Problems Associated with Misapplied, Faulty or Ignored Gauges

Incorrect, misapplied, damaged, faulty or ignored gauges can lead to a number of issues, including safety problems, downtime, and costly regulatory agency fines and other penalties.

Process Safety:

One group of problems relates to process safety issues, such as loss of containment that leads to fugitive emissions, fires and explosions. Bourdon tubes in malfunctioning gauges can rupture, causing the release of media, such as gasoline, oil or gas. Because this media is often volatile and corrosive, it can cause fires, explosions, or other serious safety issues, such as a toxic gas releases or acids spraying onto workers.

Media releases are common results of misapplied, faulty or ignored gauges. Fires and explosions are less common but they do occur. When reviewing plant instrumentation, findings have revealed instances where the wrong gauges have been installed in locations on pumps. In one instance, a gauge vibrated so severely when the pump came out of alignment that it caused the Bourdon tube in the gauge to crack. The media coming out of the tube leaked onto the pump, which was excessively hot because of age and misalignment, causing a fire.

Expensive Downtime:

A second group of problems relates to expensive downtime, which can result from the breakdown of process safety just mentioned, or simply from having to shut down a process in order to replace incorrect, misapplied, broken or otherwise damaged gauges.

1 - Safety-Related Downtime:

As a result of the problems identified above (emissions, fires, explosions, etc.), there can be any number of causes of subsequent downtime, including fire suppression activities, released media containment, line shutdown, personnel evacuation, and then eventually having to rebuild or repair the lines, replace equipment, etc. In addition, if there is a media release with no injuries or fatalities, the plant is required to notify the EPA. However, if there is a media release that does cause injuries or deaths, the plant is required to notify OSHA. These can lead to very long process shutdowns, anywhere from a day to a month or more. (More on OSHA and the EPA below.)

2 - Non-Safety-Related Downtime:

When a line fails, causing a gauge to fail, even if there is no release or safety incident, downtime can still occur as a result of the process needing to be shut down in order to replace the gauge. Gauge failure rarely causes other parts of the process to fail, although gauge failure is usually an alert that there is a problem. In most cases, a plant worker will simply replace the pressure gauge and move on. When another failure occurs, the worker will do the same, until the root cause of the failure, such as a misaligned pump, causes the plant to shut down the line, resulting in more costly, unplanned downtime and a major maintenance time.

Regulatory Agency Fines and Penalties:

A third group of problems relates to the potential for significant fines and other penalties from federal regulatory agencies (primarily OSHA and the EPA), as well as from state and local jurisdictions, that result from compromised process safety.

1 - OSHA:

OSHA addresses many process-related safety issues in 1910 Subpart H, "Hazardous Materials." More specific is 1910.119, "Process safety management of highly hazardous chemicals." In its Process Safety Management section, OSHA notes that, "Unexpected releases of toxic, reactive, or flammable liquids and gases in processes involving highly hazardous chemicals have been reported for many years in various industries that use chemicals with such properties. Regardless of the industry that uses these hazardous chemicals, there is a potential for an accidental release at times they are not properly controlled, creating the possibility of disaster."

OSHA goes on to note that: "To help ensure safe and healthful workforces, OSHA has issued the Process Safety Management of Highly Hazardous Chemicals standard (29 CFR 1910.119), which contains requirements for the management of hazards associated with processes using highly hazardous chemicals."

OSHA's standard emphasizes the management of hazards associated with highly hazardous chemicals and establishes a comprehensive management program that integrates technologies, procedures, and management practices.

While OSHA does not monitor gauges in specifically, they will enforce guidelines for worker safety. OSHA investigates injuries and deaths, and may levy fines or other penalties.

2 - EPA:

According to Greg Neneman, director of business development solutions for Valin Corporation (San Jose, Calif.), fugitive emission regulations from the Federal level have tightened up over the past several years. In many states, such as California, Oregon and Washington, the regulations are even tighter.

"One challenge companies face is ensuring compliance with emission regulations when they upgrade or expand their processes," points out Neneman. "They are not allowed to produce additional emissions beyond the original limit set by regulations."

From the standpoint of federal EPA regulations, each company that produces or emits volatile organic compounds (VOCs) is required to come up with a leak detection and repair program. "In some areas, county and local requirements are even tighter than federal requirements," continues Neneman.

In certain areas, companies that produce or emit VOCs are required by regulation to do audits on valves, flanges, gauges, and all other process connections on a quarterly basis. "And in some jurisdictions, the regulatory agencies can give plants 24 hours notice before they descend on that plant to check all of their records and conduct other investigations," adds Neneman.

In May 2012, the U.S. Environmental Protection Agency (EPA) levied a $1 million civil penalty on a plastics company for alleged violations of the Clean Air Act at two of its chemical manufacturing plants. In addition to paying the penalty, the company agreed to implement a comprehensive program to reduce fugitive emissions of hazardous air pollutants (HAPs) from leaking equipment such as valves and pumps. The settlement also required the company to implement enhanced work practices, including more frequent leak monitoring, better repair practices, and new efforts designed to prevent leaks. After being fined by the EPA, the company realized that instrumentation was one of the areas where they having fugitive emissions. Until that time, workers had been replacing gauges left and right, because they weren't aware of the problems that could occur. The gauges weren't a big expense to them, so they considered the replacement activity to be more of a nuisance than anything. It's when they had the rupture that they realized there was a real problem.

An audit of the plant, which involved reviewing at all of the instrumentation, found that in many applications, they were using the wrong instrumentation. To correct the issues, the right instrumentation was specified for them, and the correct gauges were ordered. The plant now has its instrumentation under control.

While conforming to EPA regulations has always been a challenge for process facilities, it is going to become even more challenging in the future as a result of EPA's recent introduction of new requirements related to fugitive emissions and leak detection and repair (LDAR) strategies. On April 17, 2012, the EPA Administrator signed a suite of final air quality regulations that affect both upstream and midstream oil and gas sectors - the production/processing and transmission/storage sectors of the oil and natural gas industry.

For the first time, EPA will regulate air emissions from natural gas wells that are hydraulically fractured, as well as other emission sources associated with exploration, production, processing, and transportation of oil and natural gas.

Promulgation of the final rules will result in new control equipment requirements, as well as increased requirements for monitoring, recordkeeping, and reporting for owners, operators, contractors and vendors.

The final rules require the use of emission controls, as well as work practices, through two different air regulatory programs. One is the New Source Performance Standards (NSPS). The other is the National Emissions Standards for Hazardous Air Pollutants (NESHAP).

EPA finalized a new NSPS at 40 CFR Part 60, Subpart OOOO, which regulates emissions of volatile organic compounds (VOC) and sulfur dioxide (SO_2) from a variety of oil and gas exploration, production, processing, and transportation facilities, including natural gas wells, certain compressors and pneumatic controllers, storage vessels, and sweetening units at onshore natural gas processing plants. Subpart OOOO also includes startup, shutdown and malfunction provisions.

The EPA has also made it clear that oil and gas production and processing sources will be the focus of its National Enforcement Initiative (NEI) for fiscal years 2012 and 2013. With the new NSPS and revised NESHAPs in place, EPA now has the tools it needs to enforce in a comprehensive manner.

As a result of these new regulations, plants should look at everything more closely, including gauges.

How Problems Can Occur With Gauges

Technical problems can occur with process pressure gauges in a number of ways, which can be conveniently categorized into seven broad categories with specialties or subsets. The most common reason for gauge problems is that workers often install gauges without knowing how they should be installed or accessorized. As a result, the gauges can fail because they were never designed to function in those specific situations. They can also fail because they were installed incorrectly, or because they were not properly maintained.

One way to eliminate or minimize these problems is to consider specific pressure gauge accessories. These can prove to be the extra step needed to ensure a successful, proactive preventive maintenance program, and for extending the life of the gauges. Accessories can provide additional protection for the gauges during their lifetimes.

1 - Mechanical Vibration.

This is oscillating, reciprocating or other periodic motion of a rigid or elastic body or medium forced from a position or state of equilibrium. Typically, vibration is the main cause of pressure gauge failure in process plants. Pump applications produce most of this vibration.

Gauges that aren't designed to withstand vibration may be installed in applications where vibration occurs, causing the gauges to come out of accuracy. Even more serious, a gauge can vibrate so much that the Bourdon tube inside it can crack, releasing media."

One way to prevent vibration-related problems is to install liquid-filled gauges, which help to dampen the vibration and also protect the internal components. The liquid case filling also lubricates the internals and prevents corrosion.

2 - Pulsation.

When media rapidly cycles through the pressure system of the gauge, pulsation occurs. If the pointer pulsation increments are greater than 5 percent of the full scale value, you must intervene to prevent damage to the gauge.

A restrictor or throttle screw can solve most pulsation problems. It will slow down the media by reducing the size of the orifice and averaging out pressure fluctuations. For extreme pulsation, such as spiking and surging, a snubber works best. It allows more control over the amount of snubbing and the flow of media into the pressure system.

3 - Temperature:

If a gauge is exposed to a media with very high temperature, the metal alloy in the gauge may not be able to handle the temperature, causing the Bourdon tube to fail and release media.

Long pipes that separate the pressure gauge from the process, or small cooling elements with radiator fins, protect the pressure instrument from high process temperatures by allowing ambient air to flow over the surface area. In environments with steam, using a siphon with the pressure gauge cools the steam down and drives it back to its liquid state before it gets to the gauge. With very high-temperature media, a diaphragm seal with a cooling element will isolate and prevent the hot media from reaching the gauge.

4 - Spikes or Overpressure.

In some cases, process media may be under extreme pressure. Pumps also can cause spikes and over-pressure, both of which shorten gauge life. Gauges can fail as a result of these pressure spikes. When workers turn pumps on or off, or open or close valves, a surge of media or oil goes through the pipe and hits the pressure gauge, which causes a huge spike. For example, if a line is pumping 100 psi of oil, someone may install a 100 psi gauge, not realizing that this leaves no allowance for any overpressure, surging, or spiking. When pressure does increase, it causes overpressure on the gauge, sometimes to the point of where the Bourdon tube ruptures and leaks media.

One way to prevent this problem is to use a different type of gauge, such as 200 psi gauge for a 100 psi line. Another option is to accessorize the gauge with an over-pressure protector, which would shut off the pressure if there were a surge or spike. Overpressure protectors are designed to avert spikes and the resulting overpressure damage. These protectors provide protection for pressure gauges against the effects of pressures exceeding their maximum pressure ratings.

5 - Corrosion.

Corrosion problems can occur with gauges if the process media is not compatible with the pressure gauges' wetted parts material. Media can sometimes be so corrosive that it corrodes through the Bourdon tube and leaks onto the ground.

When the media is not compatible with the wetted parts material of the pressure gauge, a diaphragm seal is required to isolate the gauge from the media. The benefit of a diaphragm seal is that it can be made of special alloys that are designed to withstand corrosion. It is essential to select a diaphragm material that is compatible to the process media.

6 - Clogging.

In some processes, the media may eventually crystallize or may otherwise harden. If a standard gauge is used in such an application, it will eventually get clogged and will not be able to read the pressure accurately. For example, if a gauge gets clogged and ends up reading zero pressure, when in fact it is under tremendous pressure, there will be an unexpected and dangerous release of media if and when the piping system is opened. Diaphragm seals protect the pressure gauge from solids, particles and viscous media by creating a barrier between the process and the pressure inlet of the gauge. There is also an option to equip the diaphragm seal with flushing ports to constantly flush material out from the diaphragm surface.

7 - Mishandling and Misapplication.

Gauges can also fail prematurely if they are misapplied. For example, a gauge may have been ideal for the application when it was installed. However, if there are equipment modifications or more components subsequently added, this can change process parameters, resulting in the gauge not longer being right for that application. Misapplication can lead to serious problems associated with failure and subsequent media release.

It is also possible to accidentally damage gauges during installation. Gauges should be installed using the proper sized wrench on the flats. Gauges should never be installed by turning the case. After installation, gauges can also fail if they are mishandled or used improperly.

Why Problems Can Occur With Gauges

In a typical process plant, about 25 percent of pressure gauges have failed or are about to fail.[1]
A lack of management focus and worker education and experience are key reasons plants experience issues with gauges.

Management Focus.

Gauges are often viewed as inexpensive commodities, with management failing to realize that they are very critical parts of the processes. Often, plants will replace failed gauges with the same replacement gauges, which will also fail, only to have to be replaced again. Eventually, their luck may run out and a serious failure may occur because the incorrect gauge has been installed.

Why do gauges tend to receive such little attention from management? Management tends to focus on the expensive equipment - $100,000 or million dollars items. They often don't understand the importance of focusing attention on $100 gauges. This is not the fault of management. The problem is that no one has explained to them just how important gauges are.

As noted earlier, though, ignoring instrumentation can cause serious safety and financial problems. Plant process gauges can be thought of in the same way as vehicle gauges. If the temperature gauge on an automobile is defective, it is possible to destroy the engine as a result of overheating that is not registered by the defective temperature gauge. Similarly, if the oil pressure gauge on an automobile is defective, it is again possible to destroy the engine as a result of seized components resulting from loss of oil that is not indicated by the defective oil pressure gauge.

Worker Education and Experience.

A lack of worker education and experience also contributes to gauge issues. In many instances, when maintenance technicians look at gauges, they don't pay much attention to them unless they leaking or look as if they are going to explode. They don't realize all of the information that that gauge can provide. In addition, technicians may not realize that a gauge has failed or is about to fail. Again, as long as it is not leaking media, they will often just keep walking. This is rarely the fault of the technicians. Many have not been properly trained to understand process pressure gauge performance.

Along the same lines, the industry has been experiencing a "brain drain" for several years, with large numbers of experienced technical workers and engineers retiring and being replaced by younger people who lack the education and experience to understand what gauges are telling them, or, worse, not being replaced at all.

According to research conducted by Cambridge Energy Research Associates in 2007, more than half of all oil-field professionals will reach retirement age during the next decade. The Interstate Oil & Gas Compact Commission found that the average age of production and service company employees is 46 to 49 years, and the average retirement age in the oil and gas industry is 55 years. [2]

Historically, plants had specialists whose jobs focused exclusively on instrumentation. They spec'd the gauges, maintained them, and made sure they were operating correctly. However, as plants got leaner in recent years, they started letting these people go. Plants first shifted gauge responsibility to the pipefitter whose main responsibility was to make sure the pipes were put together correctly and not leaking.

[1]WIKA data aggregated from more than 250 instrument audits.
[2]Peter Peek, Jocelyn Fenard, Patrick Gantes, Christine Theiler. December 2008, "Skills Shortage in the Global Oil and Gas Industry: How to close the gap" Part I. Pages 14-15.

Then, as plants began letting pipefitters go, they shifted the responsibility to the unit operators or technicians, who now install the gauges. These people usually don't have any experience or knowledge of how to install instrumentation. They usually just replace old gauges with similar gauges. The problem is that the gauges are usually the wrong ones to begin with. The result is that, these days, few plants have people who have much if any education and experience with instrumentation.

Taking Stock of Process Pressure Gauges

When it comes to improving the safety, productivity and profitability of process plants through an effective gauge management program, companies can take a couple of steps. First, an understanding of the vital importance of gauges must be established. Second, workers must understand what kind of gauges will be effective.

Focus on Gauges:

World-class process plants maintain their assets proactively. In such plants, pressure gauges receive as much attention as pumps and control valves. Management understands the importance of proactively maintaining their assets, and they realize that one of the best ways to maintain their most expensive assets is to properly source and maintain the gauges that protect these assets. That is, they understand the value of spending $100 for a gauge that can subsequently prevent damage or destruction to a $100,000 or $500,000 piece of equipment and the resulting negative safety, downtime and regulatory implications.

There is an interest in predictive maintenance these days in process plants, which is a good thing. A lot of this, though, involves expensive software. In addition, when something is out of kilter, the system may be set up such that technicians may have to wait for it to get far enough out of the normal range before it will generate an alert on the computer. However, properly-maintained pressure gauges, studied by people who know what to look for, can be very inexpensive but reliable and useful mechanical predictive maintenance tools.

Gauge Procurement Considerations:

Over time, as noted earlier, the status of pressure gauges has unfortunately diminished to that of commodities in many process plants. In order to ensure process plant safety, productivity and profitability, it is essential that plant professionals understand the importance of selecting the right gauges for applications.

It is also important that these be high-quality gauges, not "commodity" gauges. That is, not all process gauges are created equal. The so-called "lowest price" gauges appear attractive at first, but are rarely the "lowest cost" in the long run, because they usually have to be replaced more often, leading to excessive downtime and inconvenience. Worse, they may fail unexpectedly, leading to catastrophic results.

Consider these 10 steps to select the right gauges that will help to ensure safe, productive and profitable plant processes.

1 - Know what is in the pipe. This way, you can select a gauge alloy that will be compatible with the process media.

2 - Next, know the temperature of the media. If it is very hot, for example, it will be unsafe to use a standard pressure gauge. A long pipe, siphon or diaphragm seal assembly may be needed to cool the media sufficiently before it reaches the gauge internals.

3 – Third, know the ambient conditions surrounding the gauge. For example, if it is corrosive around the gauge, you need to make sure that the external components of the gauge such as the case and window can handle the environmental conditions.

4 - Next, know the pressure in the piping that the gauge is meant to measure. A best practice is using a gauge that is designed for two times the normal operating pressure. For example, if normal pressure is 100 psi, use a 200 psi gauge.

5 – Know the conditions that affect gauge wear. These could include mechanical vibration, pulsation, pressure spikes, etc. For example, if vibration exists, a gauge with a liquid-filled case is recommended. If there is pulsation, use a throttle plug or restrictor accessory.

6 - Select gauges that are designed for a particular service. Process type gauges are designed with strong Bourdon tubes and other heavy duty components. If a gauge is misapplied, the internals of the gauge may not be rugged enough to resist failure and subsequent media release in these applications.

7 - The gauge case must be rugged and have the correct safety features. In some case designs, if there is a rupture of the Bourdon tube, media will be released out the front of the gauge, which can spray workers. Select a gauge where media would release out from the back of the gauge.

8 - Accuracy is important. In most process plants, +/-0.5% accuracy is the norm, and +/-1% is acceptable.

9 - Size is also important. The gauge dial needs to be large enough that operators can read it from a distance – four inches or larger.

10 - Finally, the gauge should be mounted properly. A gauge is usually mounted by its stem, but it can also be mounted in a panel or against a flat surface.

Looking Outside for Assistance

As noted earlier, it can be difficult for plant management to devote enough time and attention to an effective process pressure gauge program. Similarly, it can be difficult for technical personnel to get up-to-speed in terms of education and experience related to process pressure gauges.

There has been a growing trend over the last 30 years in virtually all industries for companies to outsource services that are not considered "core competencies." That is, companies have come to realize that it makes sense to focus their attention on their missions and what they do best, and leave secondary and peripheral activities to third parties that specialize in these other activities and services. Becoming experts in pressure gauges is certainly not a core competency for refining plants, petrochemical, and chemical plants, so this is an area that can easily be outsourced to an experienced third-party audit team.

If a plant elects not to gain the experience that it needs on its own, free assistance is available. To ensure that gauges are working properly and are able to protect the valuable assets that they were designed to protect, a five-phase instrument review is recommended. A best practices review identifies gauges with unwanted conditions and recommends corrective actions to ensure optimum performance. The review identifies issues that affect reliability, longevity and safety. Such a review can uncover evidence of pulsation, vibration, pressure spikes, excess process or ambient temperature, installation damage, clogging, chemical incompatibility, and other problems that can lead to serious problems.

Phase 1 - Interview. This involves discussions between the third-party audit team and plant personnel. During the interview, plant personnel learn the purpose of the instrument audit, how it will be conducted, and how they will benefit from the audit. The interview also helps determine the best way to match the audit to the needs of that specific plant.

Phase 2 - Pre-Meeting. This involves determining the units or sections of the plant to audit, as well as outline roles and responsibilities. It also identifies prohibited and exempt areas of the plant. The third-party team reviews problematic instrument installations and discusses safety procedures with plant personnel.

Phase 3 - Instrument Application Audit. This involves having key plant personnel lead the third-party team on an actual plant walk-through. The audit team reviews the target gauge installations, identifies gauges that demonstrate unwanted conditions, as well as those that demonstrate installation best practices.

Phase 4 - Post-Meeting. This involves explaining the instrument audit results to key plant personnel, specifically noting gauge installations that pose immediate safety concerns. Together, the audit team and plant personnel determine which installations to investigate immediately and which to block out. The audit team and plant personnel then set schedules for the follow-up meeting.

Phase 5 - Follow-up Meeting and Presentation. The third-party team presents its general findings and the executive summary resulting from the instrument audit. The team also presents a spreadsheet that lists details about the gauges from the instrument review, color-coded by severity. The team also provides recommendations for improvements based on experience and best practices. Together, the audit team and plant personnel develop a strategy for moving forward.

The plant should also consider leveraging the audit team in other ways. For example, over time a plant's storeroom inventory can get out of hand. A qualified third-party audit team can help plant personnel rationalize, consolidate and manage their inventory of pressure gauges. For example, if a plant stocks spares for 10 or more ranges, the audit team may be able to help reduce that number to five or six ranges.

Essentially, when used properly, gauges play a significant role in plant safety, productivity and profitability. They act like silent alarms, indicating when something is seriously wrong or helping to diagnose deeper issues within systems and processes that can eventually lead to serious safety issues. To ensure safety and reliability at your plant, make certain the proper gauges are in place by reviewing your processes, gauges and your inventory. Everything is at stake.

REFERENCES

1. WIKA data aggregated from more than 250 instrument audits.

2. Peter Peek, Jocelyn Fenard, Patrick Gantes, Christine Theiler. December 2008, "Skills Shortage in the Global Oil and Gas Industry: How to close the gap" Part I. Pages 14-15.

9TH GLOBAL CONGRESS ON PROCESS SAFETY

RAGAGEP 101

Lisa Long
James Lay, P.E.
Michael Marshall, P.E.
Jeffrey Wanko, P.E., C.S.P.
US Occupational Safety and Health Administration
200 Constitution Avenue NW
Washington, DC 20037
long.lisa@dol.gov

Prepared for Presentation at

American Institute of Chemical Engineers

2013 Spring Meeting

9th Global Congress on Process Safety

San Antonio, Texas

April 28 – May 1, 2013

UNPUBLISHED

RAGAGEP 101

Lisa Long
James Lay, P.E.
Michael Marshall, P.E.
Jeffrey Wanko, P.E., C.S.P.
US Occupational Safety and Health Administration
200 Constitution Avenue NW
Washington, DC 20037
long.lisa@dol.gov

Keywords: Recognized and generally accepted good engineering practice, RAGAGEP, OSHA, process safety management, PSM, consensus codes and standards

Abstract

Recognized and generally accepted good engineering practices (RAGAGEPs) play an important role in the Occupational Safety and Health Administration's Process Safety Management standard (29 CFR 1910.119). Since 2007, OSHA issued numerous RAGAGEP-related citations under the Petroleum Refinery Process Safety Management and PSM Covered Chemical Facility National Emphasis Programs. Even though the PSM standard became effective over 20 years ago, noncompliance with PSM's RAGAGEP requirements remains an issue.

This paper reviews RAGAGEP basics from OSHA's perspective, using examples from national emphasis program inspections to illustrate common RAGAGEP compliance problems.

1. Introduction

Good engineering practice (GEP) has always been part of the foundation of safe operations in the chemical industry. GEP is a part of all aspects of chemical processing including the design of equipment, the overall process, and the man-machine interface.

This paper on recognized and generally accepted good engineering practice (RAGAGEP):

- Provides background information;
- Describes how RAGAGEP evolved from early industry descriptions, the PSM promulgation process, Occupational Safety and Health Administration (OSHA) interpretations, and later industry publications;
- Illustrates how other federal agencies use GEP and RAGAGEP;
- Describes OSHA's enforcement of RAGAGEP in the Petroleum Refinery Process Safety Management National Emphasis Program (Refinery NEP) and Process Safety Management Covered Chemical Facility National Emphasis Program (ChemNEP);
- Documents examples of OSHA citations related to RAGAGEP; and

Provides incidents in which RAGAGEP noncompliance was a causal factor.

2. Background

Prior to the enactment of OSHA's Process Safety Management of Highly Hazardous Chemicals Standard (PSM) [1], OSHA primarily enforced GEP through specific consensus standards incorporated by reference into existing OSHA standards. For example, OSHA's 29 CFR 1910.106 standard, *Flammable and Combustible Liquids*, incorporates the American Society of Mechanical Engineers (ASME) *Boiler and Pressure Vessel Code*, Section VIII - 1968.

However, with the promulgation of PSM in 1992, OSHA began enforcing regulations related to GEP differently. Notably, the PSM standard does not incorporate any GEP. Rather, the PSM standard includes nonspecific requirements that employers comply with RAGAGEP in equipment design and in inspection and testing.

A phrase that is common in the industry relating to management system performance is "Say what you do and do what you say." OSHA's PSM standard parallels this philosophy in that it is largely a performance standard. The PSM standard sets forth a management system framework upon which employers build. The design and maintenance of the process is an essential part of the standard in which RAGAGEP plays a critical role.

The PSM standard is a set of requirements for employers with processes containing highly hazardous chemicals intended to prevent releases. The PSM standard is not prescriptive because of the number and types of industries, facilities, chemicals, and technologies covered by the standard. Employers have latitude to develop and initiate the required management systems in a manner that is flexible and takes into account the attributes of their covered processes and equipment while still achieving safe operations.

Specifically for RAGAGEP, the PSM standard:

1. Allows employers to select appropriate RAGAGEP for their processes;
2. Requires employers to document the RAGAGEP to which they comply ("Say what you do"); and

Requires employers perform ("do what you say") to the provisions of the chosen RAGAGEP and document compliance with the chosen RAGAGEP.

3. Early RAGAGEP Recognition

3.1 American Institute of Chemical Engineers

One of the first authoritative references on the use of GEP as a guideline in the chemical processing industry was by the American Institute of Chemical Engineers' (AIChE) Center for Chemical Process Safety (CCPS) in 1989. In the *Guidelines for Technical Management of Chemical Process Safety* [2], CCPS provided guidelines on how to use external and internal GEP to achieve safe facilities. Some of the guidance by CCPS related to GEP included:

- Management of standards, codes, and regulations as a necessary part of chemical process safety activities;
- Application of external codes, standards, and regulations to the layout, design, and operation of chemical plants and the protection of equipment, employees, and the public;
- Means of deciding which codes, standards, and guidelines the facility follows;
- A table listing of some example external codes/regulations including ASME - Boiler and Pressure Vessel Code and Code for Pressure Piping (B31), American Petroleum Institute (API) - RP 520/521 Pressure Relieving Systems, AIChE – Design Institute for Emergency Relief Systems Procedure for Emergency Venting Sizing, OSHA – 29 CFR 1910, etc.;
- Clear documentation of the chosen codes, standards and guidelines so that those involved in design and maintenance know which RAGAGEP to follow.
- Means of documenting the reason for deviating from a code or standard. The facility seeking to use an alternative approach should demonstrate that the approach is at least as safe as the chosen code or standard;
- Development of internal standards to complement external codes and design practices;
- Developing and instituting internal standards to achieve consistency in design, where appropriate, and to safely operate and maintain the facilities; and
- Management commitment to following all chosen RAGAGEP whether internally- or externally-developed.

3.2 American Petroleum Institute

In January 1990, API issued Recommended Practice (RP) 750, *Management of Process Hazards* [3]. RP 750 states the following with respect to GEP:

- The mechanical design information should include design codes employed (Section 2.3.1);

- The mechanical design should be consistent with the applicable consensus codes and standards in effect at the time the design is prepared or, in the absence of such codes and standards, recognized and generally accepted engineering practices (Section 2.3.2);
- When the mechanical design is not consistent with applicable consensus codes and standards, the deviation and its design basis should be documented (Section 2.3.2);
- When a process hazards analysis or other review reveals that existing equipment is designed and constructed in accordance with consensus codes, standards, or practices that are no longer in general use, procedures should be implemented to ensure that the equipment is fit for service (Section 2.3.2);
- Facilities should establish inspection and testing programs for critical equipment. Such programs should include testing and inspection procedures in accordance with common standards and codes, such as API 510 or the API Guide for the Inspection of Refinery Equipment. (Section 8.5(a)).

3.3 OSHA Proposed PSM Rule

In 1990, OSHA issued a Notice of Proposed Rulemaking (NPRM) for the PSM rule [4]. The NPRM proposed regulations including some related to RAGAGEP, however, the NPRM contained no background information on the genesis or intent of the specific RAGAGEP requirements. The NPRM contained the following language on RAGAGEP:

- 1910.119(d)(3)(i)(F) - The employer shall develop and maintain a compilation of written safety information…information pertaining to the equipment in the process shall include….Design codes employed;
- 1910.119(d)(3)(ii) - The employer shall document that equipment complies with applicable codes and standards, such as those published by the ASME, API, AIChE, ANSI, ASTM and NFPA, where they exist; or recognized and generally accepted good engineering practices;
- 1910.119(d)(3)(iii) - For existing equipment designed and constructed in accordance with codes, standards, or practices that are no longer in general use, the employer shall determine and document that the equipment is designed, maintained, inspected, tested, and operating in a safe manner;
- 1910.119(j)(3)(ii) - Inspection and testing procedures shall follow applicable codes and standards, such as those published by ASME, API, AIChE, ANSI, ASTM, and NFPA, where they exist; or recognized and generally accepted good engineering practices; and
- 1910.119(j)(3)(iii) - The frequency of inspections and tests of process equipment shall be consistent with applicable codes and standards, or, more frequently if determined to be necessary by prior operating experience.

4. OSHA Final Rule on PSM and RAGAGEP

In 1992, OSHA promulgated its PSM regulation containing some, but not all, of the RAGAGEP provisions listed in the NPRM. The preamble to the PSM rule explains that the phrase "recognized and generally accepted good engineering practices" included both appropriate internal standards and applicable published codes and standards.

The preamble to the final PSM standard [5] describes how and why RAGAGEP became part of the regulation:

- (57 FR 6375) As part of the process safety information the employer is required to document that the process equipment being used complies with applicable consensus codes and standards, where they exist; or must be consistent with recognized and generally accepted engineering practices.
- OSHA modified the paragraph 1910.119(d)(3)(ii) from the NPRM by eliminating the specific list of codes and standards producing organizations.
- OSHA stated that the discussion in paragraph (j), *Mechanical Integrity*, section of the preamble describes the reasons for the change from the specific list of required codes and standards.
- (57 FR 6390 – 6391) OSHA explained the intent of proposed paragraph (j)(3)(ii) was an effort to assure that inspections and tests are performed properly.

Many commenters to the NPRM, though, objected to OSHA's inclusion of a list of consensus documents in the regulation. While OSHA believed the lists in the proposed (d)(3)(ii) and (j)(3)(ii) were not exhaustive nor attempts at incorporation by reference, OSHA published the final rule without lists.

The following "RAGAGEP rules" show the final text of the existing RAGAGEP standards. Bolded text represents language added to the final standards following the comment period and the struck-through portion represents language removed:

- 1910.119(d)(3)(i)(F) - The employer shall develop and maintain a compilation of written safety information…information pertaining to the equipment in the process shall include….Design codes **and standards** employed;
- 1910.119(d)(3)(ii) - The employer shall document that equipment complies with **recognized and generally accepted good engineering practices.** ~~applicable codes and standards, such as those published by the ASME, API, AIChE, ANSI, ASTM and NFPA, where they exist; or recognized and generally accepted good engineering practices~~;
- 1910.119(d)(3)(iii) - For existing equipment designed and constructed in accordance with codes, standards, or practices that are no longer in general use, the employer shall determine and document that the equipment is designed, maintained, inspected, tested, and operating in a safe manner;
- 1910.119(j~~)(3)~~**(4)**(ii) - Inspection and testing procedures shall follow **recognized and generally accepted good engineering practices.** ~~applicable codes and standards, such as those published by ASME, API, AIChE, ANSI, ASTM, and NFPA, where they exist; or recognized and generally accepted good engineering practices~~; and

- 1910.119(j)~~(3)~~(4)(iii) - The frequency of inspections and tests of process equipment shall be consistent with applicable **manufacturers' recommendations and good engineering practices, and** ~~codes and standards, or,~~ more frequently if determined to be necessary by prior operating experience.

4.1 Appendix C to the PSM Final Rule

Appendix C to the PSM Final Rule, *Compliance Guidelines and Recommendations for Process Safety Management (Non-Mandatory),* contains information for the regulated community related to RAGAGEP. Appendix C states the following with respect to the design of equipment:

> The information pertaining to process equipment design must be documented. In other words, what were the codes and standards relied on to establish good engineering practice. These codes and standards are published by such organizations as the American Society of Mechanical Engineers, American Petroleum Institute, American National Standards Institute, National Fire Protection Association, American Society for Testing and Materials, National Board of Boiler and Pressure Vessel Inspectors, National Association of Corrosion Engineers, American Society of Exchange Manufacturers Association, and model building code groups.
>
> In addition, various engineering societies issue technical reports which impact process design. For example, the American Institute of Chemical Engineers has published technical reports on topics such as two phase flow for venting devices. This type of technically recognized report would constitute good engineering practice.

Note, the codes and standards listed above are examples and not mandatory – the employer selects the appropriate RAGAGEP with which to comply.

The following excerpt from Appendix C discusses RAGAGEP as it relates to mechanical integrity inspections and tests:

> Also, applicable codes and standards such as the National Board Inspection Code, or those from the American Society for Testing and Material, American Petroleum Institute, National Fire Protection Association, American National Standards Institute, American Society of Mechanical Engineers, and other groups, provide information to help establish an effective testing and inspection frequency, as well as appropriate methodologies.
>
> The applicable codes and standards provide criteria for external inspections for such items as foundation and supports, anchor bolts, concrete or steel supports, guy wires, nozzles and sprinklers, pipe hangers, grounding connections, protective coatings and insulation, and external metal surfaces of piping and vessels, etc. These codes and standards also provide information on methodologies for internal inspection, and a frequency formula based on the corrosion rate of the materials of construction. Also, erosion both internal and

external needs to be considered along with corrosion effects for piping and valves. Where the corrosion rate is not known, a maximum inspection frequency is recommended, and methods of developing the corrosion rate are available in the codes. Internal inspections need to cover items such as vessel shell, bottom and head; metallic linings; nonmetallic linings; thickness measurements for vessels and piping; inspection for erosion, corrosion, cracking and bulges; internal equipment like trays, baffles, sensors and screens for erosion, corrosion or cracking and other deficiencies. Some of these inspections may be performed by state or local government inspectors under state and local statutes. However, each employer needs to develop procedures to ensure that tests and inspections are conducted properly and that consistency is maintained even where different employees may be involved. Appropriate training is to be provided to maintenance personnel to ensure that they understand the preventive maintenance program procedures, safe practices, and the proper use and application of special equipment or unique tools that may be required. This training is part of the overall training program called for in the standard.

5. Further Guidance on RAGAGEP

5.1 Center for Chemical Process Safety

In 2006, CCPS published, *Guidelines for Mechanical Integrity Systems* [6] which discusses RAGAGEP extensively - what it is, how it is used, application to different classes of equipment, design, fabrication, construction/installation, inspection, testing, and maintenance activities. The text contains the following information on RAGAGEP:

- A description of RAGAGEP with many examples;
- Facility-develop internal standards provided by the manufacturer or licensor of a process and often these are based on RAGAGEP;
- Recognition that to effectively use RAGAGEP, facility management must determine which practices are available and then assess the applicability of each practice to their facility; and
- Several chapters that address the applicability and use of RAGAGEP for detailed operations, e.g., vessel design and fabrications, in-service piping inspections, RAGAGEP for instrumentation and controls, RAGAGEP for prime movers, etc.

5.2 Incorporation of Consensus Standards

In federal regulation, use of the terms recognized and generally accepted good engineering practices and RAGAGEP originated with OSHA. However, the Federal government has long encouraged the use of voluntary consensus standards in regulatory affairs. The National Technology Transfer and Advancement Act of 1995 (NTTA), as interpreted by the Office of Management and Budget's (OMB's) Circular Number A-119 [7], instructs agencies to "use voluntary consensus standards in lieu of government-unique standards except where inconsistent with law or otherwise impractical." In addition, OMB A-119 states that "many voluntary consensus standards are appropriate or adaptable for the government's purposes."

OMB A-119 defines "standards" and "technical standards" as:

1. Common and repeated use of rules, conditions, guidelines or characteristics for products or related processes and production methods, and related management systems practices.

2. The definition of terms; classification of components; delineation of procedures; specification of dimensions, materials, performance, designs, or operations; measurement of quality and quantity in describing materials, processes, products, systems, services, or practices; test methods and sampling procedures; or descriptions of fit and measurements.

OMB further defines "voluntary consensus standards" as standards developed or adopted by voluntary consensus standards bodies, both domestic and international. These bodies must have attributes including openness, balance of interest, due process, an appeals process, and consensus, defined as general agreement, but not necessarily unanimity, with a process for attempting to resolve objections by interested parties. Circular A-119 also defines other types of standards as "non-consensus standards", "industry standards", "Company standards", or "de facto standards" developed in the private sector, but without the full consensus process.

Government agencies regulating highly hazardous chemicals (HHCs) use voluntary consensus standards in a variety of ways. For example, Section 6(a) of the Occupational Safety and Health Act of 1970 [8] gave the nascent OSHA a two-year window in which to "by rule promulgate as an occupational safety or health standard any national consensus standard, and any established Federal standard…" The 6(a) provision enabled OSHA to quickly establish a basic suite of protective standards. National consensus standards incorporated by reference into regulations under this provision of the OSH act are listed in 29 CFR 1910.6 *Incorporation by reference.* Examples include the ASME *Boiler and Pressure Vessel Code*, Section VIII *Unfired Pressure Vessels, Including Addenda* (1969), incorporation by reference approved for OSHA standards §1910.261, 1910. 262, and 1910.263(i)(24)(ii); and API 650 (1966) *Welded Steel Tanks for Oil Storage, 3rd Ed.*, incorporation by reference approved for §1910.106(b)(1)(iii)(a)(2). Since OSHA's incorporation by reference provision expired in 1974, the agency has followed the rule making process referenced in section 6(b) of the OSHA Act, including for updating references to more recent versions of national consensus standards. Incorporation of consensus standards by reference has benefits to both OSHA and the regulated community. However, OSHA must incorporate a specific version of a standard. When the consensus standard is updated, OSHA's incorporated standard becomes dated and can only be updated through the rulemaking process. Commenters on OSHA's ANPRM realized this, and this resulted in a different approach, the introduction of the RAGAGEP concept.

6. OSHA RAGAGEP Enforcement

6.1 OSHA RAGAGEP Policy

In the past, OSHA issued enforcement policy related to RAGAGEP. The first milestone enforcement policy related to RAGAGEP was a 2000 OSHA Letter of Interpretation (LOI) to the Instrument Society of America [9]. In this LOI, OSHA stated the following regarding the agency's RAGAGEP policy:

- 1910.119(d)(3)(i)(F) requires an employer to document which design codes and standards are used for SIS as well as all covered equipment;
- 1910.119(d)(3)(ii) requires an employer to document that SIS comply with RAGAGEP;
- 1910.119(j)(4) requires an employer to conduct inspection and testing on process equipment, including SIS, in accordance with RAGAGEP;
- If an employer documents compliance with ISA S84.01 for SIS in accordance with 1910.119(d)(3)(i)(F) and it meets all S84.01 and other OSHA PSM requirements related to SIS, OSHA considers the employer in compliance with PSM requirements for SIS.

OSHA issued a second important RAGAGEP enforcement policy LOI in 2010 [10]. In this LOI related to blast resistant structures, OSHA reiterated its policy for enforcing the RAGAGEP sections of the PSM standard.

6.2 OSHA RAGAGEP Enforcement

Because Process Safety information and, specifically, compliance with RAGAGEP is such an important part of a strong PSM program, OSHA emphasizes this in both the Refinery and Chem NEPs. Both NEPs rely on focused questions and compliance guidance that Compliance Officers can ask to help assess compliance with the standard. The Refinery NEP contains both static and dynamic (or changing) questions. OSHA published the static questions in the NEP while the dynamic questions changed periodically and limited to distribution within OSHA.

The Refinery NEP static questions focus on the following areas with respect to RAGAGEP:

- Relief Systems / Blowdowns
    - Does the PSI contain information on the RAGAGEP used?
    - Does the PSI contain the RAGAGEP's required design parameters?
    - Do the relief valves comply with the RAGAGEP?
    - Are intervening valves appropriate and adequately controlled?
    - Do the valves / blowdowns relieve to a safe location?
    - Are blowdown alarm systems in accordance with RAGAGEP?
    - Is blowdown quench water system in accordance with RAGAGEP?

- Does the relief testing program comply with RAGAGEP?
    - Frequency
    - Personnel qualifications
    - Testing procedures
    - Testing blowdown alarms
- Facility Siting
- Pressure vessels/Piping
  - Does the PSI contain information on the RAGAGEP?
  - Are the safeguards properly designed?
  - Does the testing and inspection comply with RAGAGEP
    - Bonded liners
    - Thickness measurement locations
    - Corrosion under insulation
    - Testing frequency
    - Personnel qualifications
    - Testing basis
    - Injection points
    - Appropriate testing locations

It is important to note that OSHA also uses RAGAGEP compliance questions in it dynamic list of inspection priority items in both its Refinery and Chem NEPs.

OSHA analyzed citation data from the Refinery and Chem NEPs to determine the most frequently cited standards and any trends. In both enforcement efforts, the most frequently cited PSM sub-element is 1910.119(d)(3)(ii)—the requirement that employers document that equipment complies with RAGAGEP. Of these citations, most common areas cited involved valves and facility siting.

The following list outlines some of the valve-related issues cited while enforcing the Refinery NEP:

- Valves between pressure vessel and relief device not locked open or controlled.
- Relief device appropriate back pressure not maintained.
- Relief device inlet line pressure drop or chattering problems.
- Relief devices not installed on pressure vessels.
- Relief devices not appropriately sized for the circumstances.
- Valves inaccessible.
- Valves inoperable.

The following list outlines to two most frequent concerns regarding facility siting:

- Facilities not compliant with RAGAGEP chosen by the facility, and
- Structures either unprotected by adequate separation or building construction from consequences of HHC release such as fire, explosion, or toxic releases.

In the case of facility siting, citations often reference multiple on-site structures including control rooms and other buildings. In many cases, siting studies identified issues that the facility had not resolved in a timely manner.

OSHA also commonly cites the failure to document RAGAGEP in Chemical NEP inspections. However, due to the larger variety of facility types inspected, it is difficult to determine trends. The following list is a sample of RAGAGEP violations found during Chemical NEP inspections:

- Safety interlocks do not follow RAGAGEP
- RAGAGEP not documented for equipment (glass lined vessels, piping, rupture disks, chemical hoses, pumps)
- RAGAGEP not documented for ventilation systems design
- Ammonia system labeling not in accordance with RAGAGEP
- Chlorine alarm systems not designed in accordance with RAGAGEP
- Piping supported did not follow RAGAGEP
- Corrosion rates not calculated in accordance with RAGAGEP
- Chlorine did not have shut-off valves in accordance with RAGAGEP
- Relief devices not designed in accordance with RAGAGEP

PSM explicitly requires employers to document that the equipment complies with RAGAGEP at the time the process was originally built. However, employers should stay abreast of current engineering codes, standards, and practices. Changes in codes, standards, and practices often result when industry has discovered new hazards or new means of hazard control. If a the PHA team finds the process design does not adequately protect against a newly discovered or newly recognized hazard, employers should consider compliance with a more recent code or additional safeguards to ensure employees are protected.

7. Chemical Process Safety Incidents and RAGAGEP

Process safety incidents in the chemical and refining industries occur for many reasons. Investigators find multiple broken barriers that result in a process safety incident. Barriers such as engineering and administrative controls often play a role in the prevention of chemical process incidents. Barrier design, operation, and maintenance are frequently based on RAGAGEP.

Highlights of incidents reveal many employer failures regarding RAGAGEP allowing incident trajectories to continue to harmful outcomes.

Bellwood, Illinois - June 14, 2006

The incident occurred when a mixture of flammable liquids boiled in an open-top kettle. Vapors escaped the kettle and flowed into the processing room and into an adjoining warehouse/general purpose room where it ignited. The flash fire and deflagration killed one and injured two.

The facility installed the chemical processing area in direct communication with the general use area not in accordance with NFPA 30. In addition, the ventilation system was designed with only overhead point pick-up and not general ventilation with floor sweeps.

Middleton, Massachusetts - March 13, 2011

The release of acetone vapor into a congested workspace caused a deflagration that nearly destroyed the building. The ventilation system design was not designed to sweep or remove flammable vapors from the room. The facility did not use NFPA 30 in its ventilation system design. The inadequately designed ventilation allowed flammable vapors to accumulate.

Houston, Texas - April 12, 2004

The employer modified a large pressure vessel for its operation. The modification involved cutting and subsequent replacing a large coupon in the vessel wall. The modification did not follow RAGAGEP for performing welding on a pressure vessel. After the facility placed the modified vessel into service, it failed at the welded coupon, ejected its contents, rocketed into a neighboring building, and caused a large fire.

Illiopolis, Illinois - April 23, 2004

An explosion and fire killed five and injured three at this polyvinyl chloride facility. Many factors combined to cause a release of vinyl chloride monomer into the congested workspace. However, the facility had not designed and maintained its safety interlocks in accordance with RAGAGEP as they were easily disabled or bypassed. When needed to prevent the release, the interlock was bypassed and the release occurred.

Two of the most basic backbones of any process safety program are to design equipment in accordance with RAGAGEP and to maintain, inspect, and test equipment in accordance with RAGAGEP. Given these two fundamental tenets of process safety, there is little doubt that contributing to most incidents are failures to adhere to these requirements.

8. Conclusion

The RAGAGEP concept as used by OSHA comprises hundreds of industry-published codes and standards and, possibly, many thousands of internally-produced practices and protocols. When promulgated, the PSM standard made mandatory that facilities adopt code and standards published for their industry or produce internal standards that meet similar requirements. The management system that OSHA envisioned with the PSM standard is no different from any quality cycle (plan-do-check-act) and RAGAGEP plays a critical role.

Because OSHA is not in a position to be expert in any industry, use of RAGAGEP is critical to the success of a performance standard such as PSM. Using the collective wisdom and experience of industry experts, RAGAGEP helps achieve safe operations throughout industries using highly hazardous chemicals.

9. References

[1] "Process Safety Management of Highly Hazardous Chemicals," 29 CFR 1910.119, 57 FR 6356, Feb. 24, 1992

[2] "Guidelines for Technical Management of Chemical Process Safety," (pp 123 – 127), Center for Chemical Process Safety, 1989

[3] "Management of Process Hazards," American Petroleum Institute Recommended Practice 750, 1990

[4] 55 FR 29150 – 29173, Occupational Safety and Health Administration, July 27, 1990

[5] 57 FR 6356, Preamble to OSHA PSM Final Rule, 2/24/1992

[6] "Guidelines for Mechanical Integrity Systems," Center for Chemical Process Safety, 2006

[7] "Federal Participation in the Development and Use of Voluntary Consensus Standards and in Conformity Assessment Activities," OMB Circular A-119, Revised, February 10, 1998, http://www.whitehouse.gov/omb/circulars_a119/ (accessed February 13, 2013)

[8] "Occupational Safety and Health Act of 1970," 29 USC 651 – 678, December 1970

[9] "Compliance with PSM and ANSI/ISA-S84.01 for safety instrumented systems," Occupation Safety and Health Administration Letter of Interpretation, March 23, 2000, http://www.osha.gov/pls/oshaweb/owadisp.show_document?p_table=INTERPRETATIONS&p_id=23722

[10] "Applicability of the PSM Standard's Mechanical Integrity Requirements to Refinery Structures." Occupational Safety and Health Administration Letter of Interpretation, February 1, 2010
http://www.osha.gov/pls/oshaweb/owadisp.show_document?p_table=INTERPRETATIONS&p_id=27443

Disclaimer

This paper represents the views of the authors. Although is the authors based the paper on OSHA programs, it does not represent official OSHA policy.

Summary of Expert Forum on the Use of Performance-Based Regulatory Models in the U.S. Oil and Gas Industry, Offshore and Onshore

M. Sam Mannan*
Mary Kay O'Connor Process Safety Center, Artie McFerrin Department of Chemical Engineering, Texas A & M University, College Station, TX 77843, USA
Email: mannan@tamu.edu

Xiaodan Gao
Mary Kay O'Connor Process Safety Center, Artie McFerrin Department of Chemical Engineering, Texas A & M University, College Station, TX 77843, USA
Email: xiaodan.gao@tamu.edu

Prepared for Presentation at
American Institute of Chemical Engineers
2013 Spring Meeting
9th Global Congress on Process Safety
San Antonio, Texas
April 28 – May 1, 2013

UNPUBLISHED

Summary of Expert Forum on the Use of Performance-Based Regulatory Models in the U.S. Oil and Gas Industry, Offshore and Onshore

M. Sam Mannan, PhD, PE, CSP DHC*
Mary Kay O'Connor Process Safety Center, Artie McFerrin Department of Chemical Engineering, Texas A & M University, College Station, Texas 77843, USA
Email: mannan@tamu.edu

Xiaodan Gao
Mary Kay O'Connor Process Safety Center, Artie McFerrin Department of Chemical Engineering, Texas A & M University, College Station, Texas 77843, USA
Email: xiaodan.gao@tamu.edu

Keywords: Performance-based and prescriptive regulatory regime; Federal agency oversight; Safety case

Abstract

Over the recent decade, catastrophic incidents such as the BP Texas City Explosion and the Deepwater Horizon oil spill have grabbed much public and media attention. It is important now to discuss the effectiveness of the current safety regulatory regime in the US. Recommendations have been made to move forward to performance-based regulation. To gain opinions from different parties and the public, agencies including OSHA, BSEE, USCG and EPA initiated a stakeholder meeting to discuss several important issues of a performance-based regulatory regime. Discussions included: advantages and disadvantages of performance-based and prescriptive regulatory approaches, ways to advance performance-based regulations and standards in the oil and gas industry, and effective oversight of performance-based regulatory models. This paper will summarize the standpoints of different parties such as CCPS, API, Steel Workers and Contra Costa Accidental Release Prevention, and then provide detailed evidence to address the topics based on a comprehensive literature review.

1. Introduction

As presented by May, [1] a regulatory regime must provide a framework with a perfect equilibrium between control, accountability for results, and flexibility to achieve the results. A prescriptive regulatory system [2] "sets specific technical or procedural requirements that must be complied with by regulated entities". A performance-based regulatory (PBR) system [2] which is also referred to as goal-based, "identifies functions or outcomes for regulated entities but allows them considerable flexibility to determine how they will undertake the functions and achieve the outcomes".

After a series of accidents that occurred in the US, such as the 2005 Texas City Explosion and the 2010 Deepwater Horizon oil spill accident, recommendations have been made to adopt thorough performance-based regulatory regimes and develop a proactive and risk-based performance approach, taking the example of Safety Case by UK Health and Safety Executive (UK HSE).

To obtain diverse opinions about this important issue, multiple US agencies including Occupational Safety & Health Administration (OSHA), Bureau of Safety and Environmental Enforcement (BSEE), United States Coast Guard (USCG), and United States Environmental Protection Agency (EPA) invited interested parties to participate in a co-sponsored stakeholder meeting, and to submit comments on the use and implementation of performance-based regulatory models for enhanced safety and environmental performance in the United States oil and gas industry. The agencies proposed specific topics about different aspects of performance-based regulatory regime, which should address: 1) advantages and disadvantages of performance-based and prescriptive regulatory approaches; 2) whether these methods could create synergies between multiple agencies; 3) what types of models or combinations could result in long-term economic benefits; 4) ways to advance performance-based regulations and standards in oil and gas industry; and 5) effective oversight of performance-based regulatory models.

Mary Kay O'Connor Process Safety Center (MKOPSC) was invited to interpret our standpoints as a representative from the academic field. This paper will summarize the standpoints of different parties such as Center for Chemical Process Safety (CCPS), American Petroleum Institute (API), United Steel Workers and Contra Costa Accidental Release Prevention Program, and then provide detailed evidence to address all the topics based on a comprehensive literature review.

2. Standpoints of Federal Agencies and Other Parties

2.1 Federal Agencies

The speeches from the federal agencies explained their expectations and motivations to hold this meeting, including specific examples that came from their experience. Each of their presentations is briefly summarized as followed:

2.1.1 Occupational Safety & Health Administration (OSHA)

The representative from OSHA stated that federal agencies are reorganizing their work in ways that best serve competitive America, which is the spirit to organize this stakeholder meeting. In such a collaboration manner, the differences from agencies could be carefully considered. Much progress will be made through this opportunity to move beyond the tired, old cat-and-mouse relationship between regulators. Given the complexity of the oil and gas industry, it is important in our society to maintain and improve employee and public safety and protect the environment.

2.1.2 Environmental Protection Agency (EPA)

EPA stated that it is a very big challenge for all of the regulators to come up with programs that provide the necessary flexibility for all different facilities to comply with and actually meet the requirement. EPA is trying to develop programs that are with high flexibility for companies to comply with and take their major responsibility. The companies should always try their best to protect the public health, the environment and the workers that they have on their facilities.

2.1.3 White House Office of Information and Regulatory Affairs (OIRA)

President Obama has issued a series of policies, statements, and directives aiming at regulatory reform, and the most principal of which is Executive Order 13563 titled "Improving Regulation and Regulatory Review." This executive order outlines a series of core principles and procedures to direct agency regulatory activity with the aim of promoting predictability and avoiding uncertainty in the marketplace, and in the process, eliminating unnecessary or unjustified burdens on the private sectors. The government does not always have all the answers and we are working with agencies to consider alternatives to regulatory options before they issued those regulations. The representative introduced the six key provisions in this executive order: 1) Predictability and Certainty, 2) The best available science, 3) Public participation, 4) Use of the least burdensome tools, 5) Considering costs and benefits (both quantitative and qualitative), and 6) seeking to improve actual results.

2.1.4 Bureau of Safety and Environmental Enforcement (BSEE)

The responsibility of BSEE is to protect the environment, and conserve offshore resources through regulatory oversight and enforcement. The representative provided a regulatory update of BSEE, including Safety and Environmental Management Systems (SEMS) I&II, and summarized the BSEE way ahead with the following strategies:

- Blend prescriptive and performance-based approaches
- Compel safety, emergency preparedness, environmental responsibility, and appropriate development and conservation
- Build and sustain this capacity within and across BSEE's key functions

2.1.5 United States Coast Guard (USCG)

The USCG representative stated that they have a blend of both prescriptive and performance-based regulations. They utilize international standards, guidance, recommended practices, classification society rules and industry standards. In general, they regulate US vessels that are involved in oil and gas under Outer Continental Shelf Activities (OCSLA) rules.

2.1.6 Pipeline and Hazardous Material Safety Administration (PHMSA)

It was mentioned that PHMSA has been experimenting with risk-based, management-based, and performance-based regulations for about two decades. They made some progress, but still there are a lot of things that they have yet to learn. When there is a major accident, there is a demand for the regulatory agency to fix the problem. Generally people want performance-based regulatory approaches only until we give it to them because of some disadvantages. Federal agencies have limited resources to obtain a perfect regulatory regime. It will be the operator's responsibility to identify the risks for their facilities if a performance-based scheme will be mandated. It will not be easy to conform to performance-based regulations.

2.2 Viewpoints from Other Parties

The American Petroleum Institute (API) supports a performance based regulation to encourage innovation in designs, techniques, processes and materials. API has developed around 600 standards covering oil and natural gas industry, which have been cited in multiple regulations by agencies such as OSHA, EPA, DOT and BSEE. For example, API RP 75 has been cited by BSEE, and API 2610 was cited by four agencies covering regulations in design, construction, operation, inspection, maintenance of a petroleum terminal.

Opinions from the US Steelworkers Union reflect the perspective of workers working in the refining sector. They recommended a combination of prescriptive-based and performance-based regulation. The challenges would come from comprehensive database/lessons learned from industrial accidents. We should also learn lessons from incidents with less severe consequences than catastrophic incidents. They believed that agency duplications should be reduced to increase efficiency.

Dr. Andrew Hopkins from Australia argued that we should move toward a safety case regime. He pointed out that regulators and industry should work in collaboration to prevent accidents in the safety case. The elements of a successful safety case include:

- Hazard management framework, like PSM or SEMS, with PHA as the heart of the safety case
- Case made to the regulator, which forces the operator to choose the best standard for their purposes and justify it to regulators
- Competent and well-resourced regulator
- General duty of care, from which the regulated body must genuinely want to improve safety

The Center for Chemical Process Safety (CCPS) stated that a risk-based process safety regulatory regime is necessary to thoroughly understand potential consequences, apply appropriate systems to achieve desired risk and ensure systems are functioning as expected. CCPS has guidelines that are similar to OSHA PSM, but with several added elements such as process safety culture, measures and metrics, and stakeholder outreach. CCPS stressed the use of metrics and finding root causes to problems, all within a risk-based approach.

The California Accidental Release Prevention Program (CalARP) indicated that they have been mandating an Industrial Safety Ordinance (ISO) since 1999, which is the most stringent process safety regulation in the United States. It is similar to OSHA PSM, but contains additional requirements, such as Human factors, Management of Organizational Changes, Inherently Safer Design, Root Cause Analysis Investigations, and Safety Culture Assessments. The actions to ensure oversight include:

- Conducting unannounced inspections of facilities
- Performing audits of Safety Programs at the Facilities
- Detailed on-site inspections/audits at least every three years (4-5 weeks per audit)
- Screening process before each audit to focus on problem areas
- Audit follow-up on previous issues
- Facilities are required to develop process safety metrics

3. MKOPSC perspective

3.1 Advantages and Limits of Performance-Based Regulatory Approaches

Prescriptive regulatory approaches emphasize control and accountability, and performance-based regulatory approaches promote flexibility with accountability for results. These features must be acknowledged as integral parts of the discussion regarding the advantages and limits for adapting performance-based regulations. The debate must consider two main points: the possibility for a long-term hazard reduction (Health, Safety and Environmental) and the utilization of resources.

3.1.1 Advantages of Performance-Based Regulatory Approaches

Numerous authors have provided discussions concerning the advantages of performance-based regulatory regimes. The key points raised can be summarized as follows:

- *Innovation and adoption of new technologies* [3-7]

  The flexibility that accompanies performance-based regulations allows for a continuous improvement by means of the implementation of new technologies as they become available. This is a desired feature in order to keep up with a complex and continuously changing industry and market, where unique situations often occur. Moreover, contrary to common belief, opportunities for innovation arise not only for the regulated entities, but also for the regulatory agencies. According to EPA [8], regulatory agencies are granted discretion to set performance targets within broad policy areas and to experiment with means to achieve them.

- *Goal-oriented risk management system* [7-8]

  The implementation of a performance-based regulatory system is accompanied by a goal-oriented risk management system. There is a clearer quantification of the performance, and any decision must then be based on the assessment of the hazards implied. Moreover, this management system will be common to all participating companies, thus allowing for unification in the objectives, communication, planning, and response to HSE-related issues.

- *Safety ownership, responsibility and pro-activity* [7-12]

By means of the joint responsibility generated, traditional distinctions between regulated entities, regulatory agencies and regulation beneficiaries become fuzzier [12]. It results in a further understanding of the involvement and support necessary, and a wider use of the industry knowledge [11, 13-14]. The ownership of risk and the generated responsibility switch the focus to company's HSE objectives, abolishing dedication limitations. This provides plenty of opportunities for independent assessment and actions that go beyond minimum performance objectives.

- *Potential for more efficient utilization of resources* [1, 7-8, 11, 14-15]

As a consequence of the flexibility to achieve results, the cost-effectiveness of the compliance strategies is stimulated. Companies are encouraged to innovate with cost-effective solutions [15] and proven successful alternatives that have been applied in different locations of the world [7]. In the long-term, this will not only benefit the companies, but also their workers and the public in general.

3.1.2 Limits of Performance-Based Regulatory Approaches

The Regulatory Policy Program of the Center for Business and Government at Harvard University organized a workshop in May 2002 to discuss performance-based regulations. Several government agencies and researchers from a variety of areas participated in the workshop. The workshop concluded that some of the most important criteria to consider when determining or selecting the regulatory approach are the effectiveness, efficiency, equity, clarity, and the ease and accuracy of enforcement [18]. The discussion provided in this section follows the aforementioned report and incorporates other literature sources.

- *Cost increase pitfalls for companies and regulatory agencies* [11, 13,14,16, 17]

Apart from the obviously overwhelming task that the application of performance-based regulations implies, and the associated economic burden, several possible pitfalls can be identified for both the regulators and the industry. Several authors assure that an increased cost to regulators is unavoidable due to the number of trained personnel that a more active regulator will demand. Some authors even point out monitoring compliance being too costly as a factor for failure [11].

- *Lack of suitable metrics to ensure compliance*

One of the most widely debated issues about performance-based regulations is the lack of performance indicators for both regulators and regulated entities to measure the outcomes. Dave Drummond from the University of Wisconsin-Madison states that as long as the outcome cannot be measured, performance-based regulation should not be used [19]. However, even if the outcome can be measured, the measurements or indicators must be adequately reliable in order to provide appropriate feedback. For this purpose, performance-based regulation may be accompanied by appropriate prescriptive requirements for measuring, reporting, recordkeeping and data management [20].

- *Performance may be difficult to measure directly for rare and catastrophic events*

Performance-based regulations might be difficult to implement when the consequences of poor performance are very serious. In that case, the regulation must be formulated very carefully. Youngblood and Kim [20] recommend the use of an objectives hierarchy for the selection of performance measurements and establishment of performance criteria. These measurements and criteria should focus on levels of performance in which the failure to meet a criterion or a given target will not lead to severe consequences or immediate safety concerns [20].

Performance-based regulations are less desirable when used to address rare catastrophic events, such as big fires, explosions or toxic releases, as it is almost impossible to measure performance for these types of events. Instead, performance must be predicted, which causes uncertainty and difficulties in the implementation [18].

- *Small companies having limited expertise or capacity for innovation*

  With regard to the size of the company, performance-based regulation is usually more difficult to implement in small companies. Medium and large companies have a larger availability of resources to create customized programs to achieve the desired outcomes. They also have the capacity of developing innovative, efficient and cost-effective solutions to meet performance goals [21]. Smaller companies may not have the resources to develop new programs or processes or invest in innovation [21]. Some small companies would rather be told the process they have to follow to achieve the desired outcome instead of creating one on their own [18].

- *There is no detailed understanding of the problem*

  Performance-based regulations can be either tightly or loosely specified, depending on the level of discretion provided to the regulator and the regulated entity [18]. For tightly specified performance-based standards or those where a quantitative threshold is provided, it is necessary to understand in detail the cause-effect or dose-response relationships between the sources and the final objective in order to establish optimal thresholds [18]. Lack of understanding of the problem or loosely specified requirements, including words such as "adequate" or "effective" [22], cause difficulty in the implementation and enforcement of the regulation.

  It is worth pointing out that the workshop at Harvard University [18] was conducted 10 years ago. However, the challenges and difficulties to implement performance-based regulation remain more or less the same. More research is needed to obtain data on the effectiveness of performance-based regulation as compared to prescriptive regulation in order to determine the best approach for each industry/situation.

  Overall, it is desirable to use a combination of performance-based and prescriptive regulations in order to address a specific issue or problem [21].

3.2 Advancement of the Use of Performance-Based Regulation in U.S.

In order to advance the use of performance-based regulations in the US Oil and Gas Industry, several criteria must first be met. The main factors in whether or not the programs are acceptable and successful include the following:

- There must be convincing statistical evidence that the performance-based measures will outperform the prescriptive-based measures they are replacing.
- There must be a fair way to implement these performance-based measures in relation to the prescriptive-based measures (not overregulated, not too much burden on companies to comply too quickly) as well as in relation to the differences in situations that the companies currently have (large vs. small, no one taking advantage of rules).
- There must be an economic benefit to the switch from prescriptive measures to performance-based measures, where the financial sunk costs and operating cost increase is counterbalanced by a tangible economic benefit, whether that benefit is decrease in incidents/fatalities or a decrease in cost through more efficient use of resources and innovation.

These concepts are graphically represented in the following Figure 1.

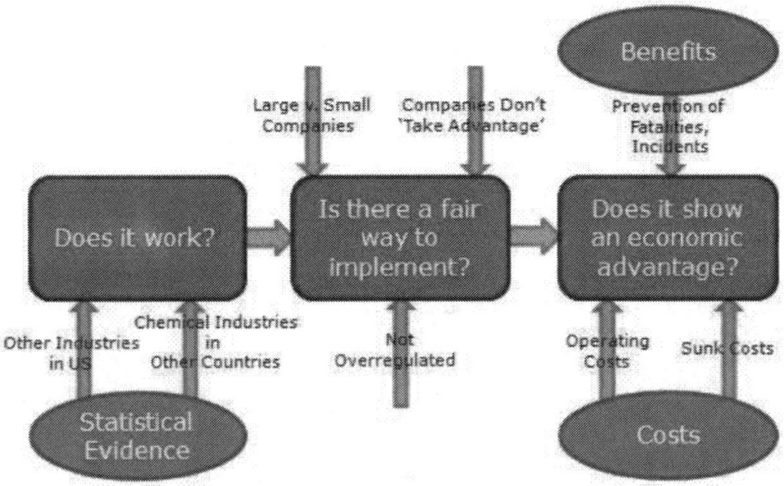

Figure 1. Advancement of performance-based regulation

These points are further discussed in the following paragraphs.

3.2.1 Statistical evidence

It is simple human nature to resist change if the change proposed does not fulfill the need it is intended to satisfy. Therefore, it is natural to expect resistance to change from the current types of regulation if it cannot be proven that the safety performance stands a significant chance of improving. There is precedent for the advancement of performance-based regulation, both in other industries in the US (nuclear industry and their Performance-Based and-Risk Informed regulation), and in the oil and gas industry in Europe (UK and Norway).

The question becomes, then, have these industries seen any tangible increase in safety statistics by either PSM or Safety Case? Unfortunately, the answer to this question is that it is hard to tell for several reasons (Figure 2 and Figure 3). Based upon the statistics from the UK HSE [23] and US API [24], it is difficult to tell whether the statistics show any real upward or downward trend.

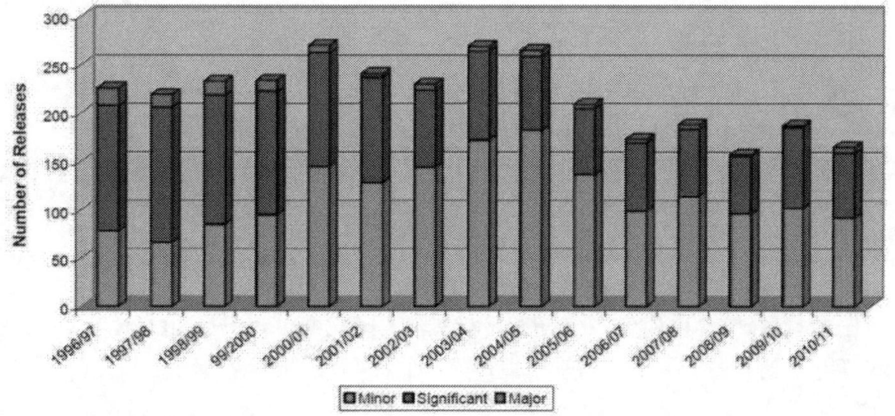

Figure 2: HSE data for offshore hydrocarbon releases in the period from 1996-97 to 2010-11[23]

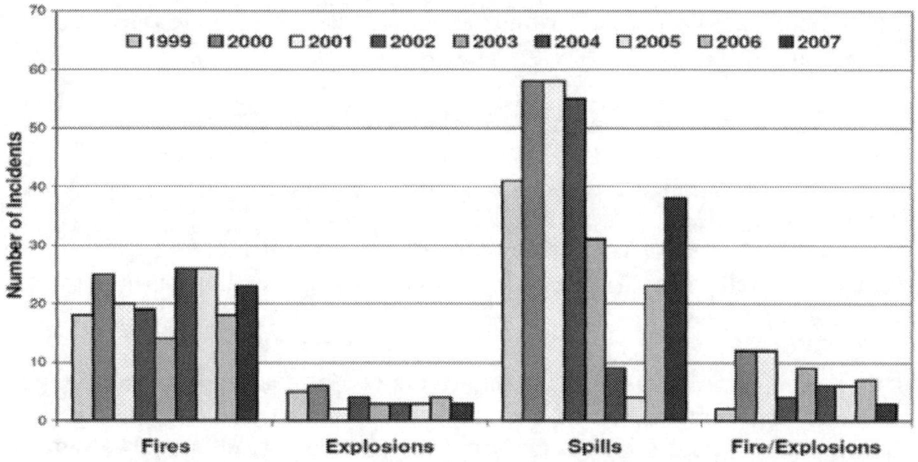

Figure 3: API Process Safety Performance Measurement Report 1999-2007[24]

If there is a trend, either upward or downward, it would also be very difficult to tell whether it was due to the regulatory regime or whether it was a function of some other factor (*e.g.*, an increase in minor accidents due to better reporting or a decrease in major accidents due to more of them being reported as significant). As such, the regulatory bodies would have to prove that these gains that other industries and other countries are making are not just "smoke and mirrors" based on other influencing factors.

Finally, it is difficult to say that any trends can be taken from a mere 15 years of data, especially when the goal of the regulation is to decrease significant events that may happen once every 10, 25, or even 50 years. Furthermore, an extended amount of time may be needed to get the programs fully implemented and working properly before results can be seen. Thus, many years of data collecting may be needed before one can conclusively state that the program is working as intended.

3.2.2 Is there a fair way to implement?

When PSM was first implemented and RMP was on the horizon, industry had several doubts and fears for the two seemingly overlapping regimes with new procedures for compliance. In the two decades since, many of the same fears would apply to the advancement of new regulation.

Not overregulated [25]: When PSM started and RMP was around the corner, one of the main concerns that industry had was whether or not they would be overregulated by the dual regulations. Since PSM and RMP have different aims (occupational vs. environmental), there was concern that each would be quite different although aimed at basically the same end. Thus, Andrews, speaking for industry through his paper, made several key requests:

- EPA chemicals be a subset equal to or less than the chemicals in OSHA PSM
- EPA threshold quantities be greater than or equal to the threshold quantities of PSM, arguing that if they were less, it would effectively extend PSM to operations that OSHA did not deem it necessary for PSM to be extended to
- Exclusions for certain (unmanned low-hazard) facilities
- Recognition that compliance with PSM would constitute compliance with RMP

This shows the level of concern with what industry saw as "fairness" in overregulation. They wanted two mutually inclusive regulations that would not increase their burden, even though the regulations had different goals.

Companies do not take advantage [26]: Companies under a performance-based regime, since it is not as concrete as a prescriptive-based regime, may attempt to take advantage of the added flexibility (not in an innovative way, but in a way such that they would be 'cheating' the system).

May and Koski [26] suggest that extra education could be a helpful way to keep this from happening, basically arguing that it will lead to the owners, operators, and engineers 'owning' safety, thus making it so that they will not be morally capable of cheating. However, they also note that due to differing capabilities and willingness to learn, this may not be the most effective method.

The second method suggested was to hold engineers (and whomever does safety reviews, tests, *etc.*) responsible for the accuracy of the results, saying that if it is found that anything is forged or unreal they can be held legally responsible. They also take it one step further and say that engineers could be held liable for performance (whether they did a good job complying with the regulation or not). They admit that this may cause insurance costs for liability to go too high to afford, causing engineers and consultants to charge more, possibly pricing them out of a job.

The third method mentioned was similar to a peer-review process where documents are reviewed by others for adequacy. The quality of results would depend on the quality of the reviewer and how enthusiastic they are about making suggestions and questioning safety cases presented.

Large vs. Small Companies [18]: Small companies face their own specific set of challenges in the move to performance-based regulation. Coglianese, Nash, and Olmstead [18] posit that small firms may not have the resources to look for a way to achieve a performance-based goal and that it is oftentimes easier for them to follow orders rather than innovate for the simple reason that they may not have the money, manpower, or time to find a better way to reach the moving target. They also point out that it is the job of the government and industry to work together so that the government can understand exactly how difficult it could be for a small firm to implement a performance-based regulation and their possible preference for a prescriptive-based regulation. In such a way, the large companies must help to protect the smaller companies from unduly strenuous regulation by an open communication with those creating the regulation.

3.2.3 Does it show an economic advantage? [25]

It is simple economics that, above all else, a company (whose motive is to maximize economic return, without which they will not stay in business) needs to see an economic benefit in order to accept new regulations. Andrews specifically mentions that many in industry believe that OSHA underestimated the cost of compliance with the new PSM system (OSHA said $810 million per year for the first five years, while Andrews argued it would be five times that). He also argues that though there will obviously be some economic benefit through lower fatality rates and incidents and increased reliability, industry was still concerned about trying to gain a competitive advantage (being able to gain a competitive advantage means being able to maximize economic benefit) in the face of such a large resource investment.

3.3 Multiple Agency Jurisdiction Issues

3.3.1 Duplication of regulations

Reduction of regulatory overlaps is a theoretically attractive way to streamline operations and simplify compliance for industry where there may currently be situations in which similar hazards are managed in seemingly different ways. The following paragraphs explain the origin of some overlaps, advantages that could possibly be gained, and the disadvantages that may be created in the elimination of regulatory overlap.

Origins:

Aagaard [27] has presented several ways in which regulatory overlap could occur. Overlap can occur organically, as a function of the growth of a regulatory body or as an attempt to gain more power, prestige, or oversight for a body, either because they want the extra power or because they believe they can do a better job or fill some gaps in the regulation. It can happen arbitrarily, when two regimes' missions somewhat overlap, giving them both the power to regulate similar areas. It is possible that overlap may arise intentionally, such as with OSHA PSM and EPA RMP, where Congress specifically mandated that both be developed in conjunction with one another; or unintentionally when the delegating bodies are unsure which regulatory body should be responsible for what, or if there is political conflict. It could also arise from overlapping legal fields, which is difficult to avoid and nearly impossible to change in hindsight. For example, OSHA regulates employee health while EPA regulates environmental and public health. Their legal fields necessarily overlap because in many cases what is dangerous to a worker is dangerous to the public or environment, so the legal overlap could not have been avoided, nor can it be eliminated.

In some ways, this overlap could be used to the advantage of the regulators, specifically when the overlap is intentional and is meant to fill gaps in between regulatory bodies that may have similar fields but differing missions. In other ways, the overlap could be destructive, particularly when it is inadvertent and causes conflict in how to deal with compliance in industry. Thus, when handled properly, overlap can be a powerful tool in making sure that regulations cover the largest span possible while causing minimum inconvenience to industry, who must comply, report, and improve.

Disadvantages:
It is obvious that disadvantages would arise from overlap, mostly because of efficiency issues. The extent of these disadvantages depends mainly on the coordination between the regulatory bodies and the degree of the overlap. Four categories were identified by Aagaard [27] as the classical arguments against overlap:

- *Duplication* – a waste of resources if the functions can be consolidated or eliminated from one of the overlapping bodies, though it is not a complete waste if there is benefit to the duplication
- *Conflict* - opportunities for conflicting regulations, whether directly conflicting (cannot satisfy both) or indirectly conflicting (can comply with both with increased effort)
- *Coordination* – necessity of increased coordination between the regulatory bodies
- *Complexity* – difficult to monitor who is in charge of what, or assign blame for failures, *etc.*

Rectanus and Natalicchio [28] further posit that overlap may have harmful effects during regulatory functions such as incident investigation, stating that regulatory bodies may prefer to independently work on investigations so as not to get delayed by other bodies. They state that it may be difficult or time-consuming to have to coordinated investigations and in the end the regulators may not be able to obtain all of the necessary information or work on the timeline of another regulator. Thus, they trade the inconvenience to the operator (having multiple investigations, procuring the same documents multiple times, *etc.*) for the convenience of being able to work on their own terms. Other overlaps exist, such as creation and administration of training materials, development of emergency response procedures, and record-keeping.

Industry itself has also expressed concern over some of the disadvantages of duplicative and inconsistent regulation. For example, Andrews [25] stated in his paper on industry views of regulation in the early days of PSM, that if a regulation (RMP) was implemented after and in conjunction with PSM and included PSM controlled chemicals at lower than PSM threshold values, it would effectively be extending PSM to sites that OSHA did not deem appropriate. This is an example where, with poor coordination, industry could have been severely affected by an overlap in regulation. Thus, care must be taken in the implementation of closely-related regulations so that a regulation is not inappropriately extended into a realm where it is not necessary.

Advantages

Perhaps counter-intuitively, there are also advantages that can be gained by these overlaps, particularly when they are intentional, well-thought out, and allow opportunity for collaboration between regulators that work in similar areas, but have differing missions and expertise.

Rectanus and Natalicchio [28] found that advantages could be gained in precisely these areas through overlap. They state that it is often reasonable to have some degree of overlap in such closely related fields, and that it is not necessarily a bad thing if it allows for competition between agencies, better service delivery, or emergency backup. They do, however, state that overlap must be properly managed in order to retain these advantages. Aagaard [27] further states that overlap can increase reliability by reducing the number of errors in action (regulators do not act when they should have acted) and can foster policy innovation where regulators have a heightened sense of what other bodies are doing and can thus act in a more informed manner, such as making better policy decisions in conjunction with one another, and learning from the successes and mistakes of others.

Should Overlap Be Eliminated?

It is intuitive to want to eliminate as much overlap as possible, especially from the industry standpoint where this elimination could possibly result in less work, greater efficiency, and less confusion over compliance. However, care must be taken to avoid eliminating overlap that was put in place intentionally to gain the possible benefits of competition and collaboration. Thus, as much as one would perform a risk analysis in a hazardous process if they were considering making a change, if overlap is to be eliminated, it must be as a result of careful analysis of the risk of the elimination.

3.3.2 Implement PBR uniformly by multiple agencies

Initially, it is important for every agency to determine how they are going to implement performance-based regulations individually and then uniformly across the industry. Primarily, agencies will have to develop strategic plans that cover the functions and operations of the agencies; then, they will have to identify the strategic goals and objectives. Further, a detailed description of the required resources and processes needed to fulfill the goals and objectives should be completed. Later on, the agencies will have to identify which are the factors that could affect the achievement of the goals and objectives suggested earlier by the agencies. Finally the agencies will have to describe how the performance goals included in annual performance plans will be related to the agencies goals and objectives [1, 29 - 30].

Once all the elements mentioned in the previous paragraph have been addressed, the agencies must receive feedback from the US Congress, the Board of Directors of the companies, the public and the stakeholders (as shown in Figure 4).

Congress will have to review the plan for every agency to achieve the corresponding goals and objectives and analyze the processes and resources required to meet such goals. This is necessary information for Congress to assign the appropriate budget for the agencies to fulfill their objectives. However, for this to be completed a performance plan has to be analyzed and discussed between the agency and Congress. Some points that must be addressed in this plan are:

- Describe the processes and resources required to meet the performance goals;
- Determine which are the performance indicators to be used in measuring or assessing the relevant outputs and outcomes of each activity;
- Provide a comparison basis of the actual program results to the established performance goals [29].

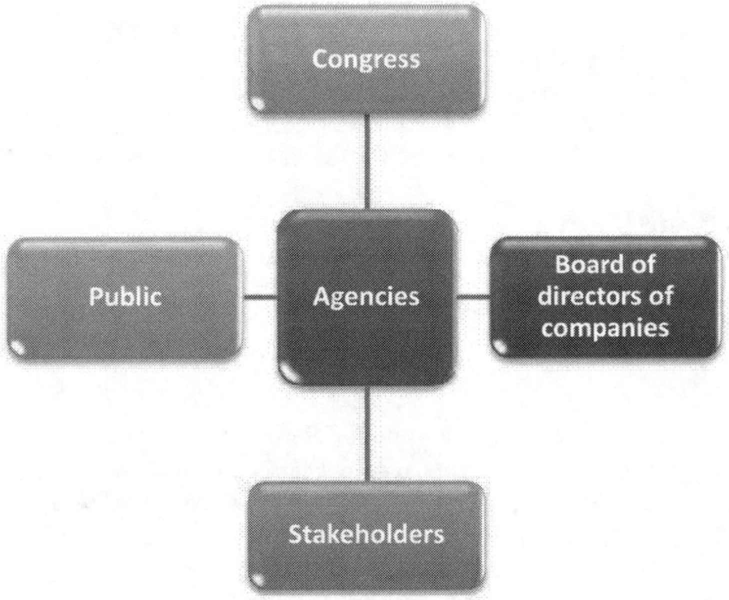

Figure 4. Feedback path for the agencies

Furthermore, before this plan is given to Congress, several perspectives have to be considered. For instance, the expertise of the Board of Directors of the companies, stakeholders and the public will strengthen the performance regulations by making them more achievable and realistic. Wholey *et al.* [29] proposed a criteria methodology to grant the political, senior management and analytical support for the successful implantation of performance-based regulations. These criteria are:

- Agreement on goals and strategies by using the professional judgment of experts.
- Performance measurement systems of sufficient quality. In this respect, the national database system recommended and discussed earlier could be used to develop the performance measurement systems.
- Performance-based management systems to build the communication of the performance information to the relevant agency.
- Accountability, demonstrating effective or improved performance, and supporting policy decision making, establishing communication between the agencies with the stakeholders and the public.

Moreover, it is recommended that a proper inspection methodology for agencies to enforce regulations is established, as well as a uniform terminology across agencies and industries. Finally, a training program has to be implemented. This training should include the political and bureaucratic context for performance-based management and should focus mainly on the following issues [29]:

- Measuring, evaluating, and reporting on performance to analyze the implications of alternative policies and program approaches;
- Using performance information to support resource allocation and other policy decision making;
- Reinforcing performance-based management and building expertise in performance-based management.

4. Conclusions

For all the various combinations of agencies and regulatory programs, it is very difficult to judge which regulatory regime is more effective. To answer this question, performance data should be collected adequately and uniformly under consistent incident reporting system. Otherwise, biased data collection will always lead to incorrect conclusions.

Secondly, it has been discussed that a "one size fits all" regulatory approach is not appropriate due to the diversity of existing industries, processes and company sizes. It is therefore necessary to determine the most adequate regulatory approach to implement according to the situation. Some situations in which prescriptive regulation is preferred are mentioned below:

- High risk situations, particularly, when there are technologies known to work well. For example, in the case of the airline industry where the consequences of an accident are severe, there are prescriptive methods to achieve the required level of safety, which have shown successful results.
- When companies have limited capabilities to develop their own approach to achieve performance targets, such as small companies. These companies would benefit from prescriptive regulations or prescriptive components given as optional guidance to complement performance-based regulations.
- Standard procedures that have little need for innovation.
- In situations where experience exists, or there is little room for innovation, a prescriptive regulation should be preferred. Instead, when the rate of change of technology or innovation is an important consideration, performance-based regulations might be desirable.
- Businesses could benefit from prescriptive components as a complement to performance standards to provide more concrete guidance.

Finally, to answer the question that if we should consider further changes in the regulatory framework, we should figure out how to improve our existing program instead of changing the whole system and allow time for existing regulations (*e.g.*, PSM) to be implemented properly. We could increase training and reinforce the regulation compliance, gradually add regulatory reform and efficiencies to the existing system, and implement overall quantitative risk assessments for plants. We should put in place a well-planned national database system to collect appropriate and statistically significant data on incidents to be able to evaluate the performance of regulatory agencies as well as the regulated entities.

References

[1] May P., "Performance-Based Regulatory Regimes", 13th World Conference on Earthquake Engineering, Vancouver, B.C., Canada August 1-6, 2004 Paper No. 3254.

[2] Jennifer Dagg, Peggy Holroyd, Nathan Lemphers, Randy Lucas and Benjamin Thibault, "Comparing the Offshore Drilling Regulatory Regimes of the Canadian Arctic, the U.S., the U.K., Greenland and Norway", 2011.

[3] C. J. Burke and J. A. Veil., "Potential environmental benefits from regulatory consideration of synthetic drilling muds", Environmental Assessment Division, Argonne National Laboratory 9700 South Cass Avenue, Argonne, Illinois 60439.

[4] Office of Technology Assessment, U.S. Congress. "Environmental Policy Tools: A User's Guide. OTA-EVN-634", Washington DC: U.S. Government Printing Office, 1995.

[5] John, D. "Civic environmentalism: Alternatives to regulation in states and communities", Washington, DC: CQ Press, 1994.

[6] Hopkins Andrew, Professor, Australian National University & consultant to the Chemical Safety Board Deepwater Horizon team.

[7] National Academy of Engineering and National Research Council, "Macondo Well - Deepwater Horizon Blowout: Lessons for Improving Offshore Drilling Safety", The National Academy Press, Washington, D.C., 2011.

[8] EPA, "Reinventing Environmental Regulation".

[9] Coglianese C, Lazar D. "Management-Based Regulation: Prescribing Private Management to Achieve Public Goals." Law and Society Review, 37(4): 691-730, 2003.

[10] Gunningham N, Johnstone R., "Regulating Workplace Safety, Systems and Sanctions", Oxford: Oxford University Press, 1999.

[11] Queensland Goverment, Department of State Development, "Guidelines on alternatives to prescriptive regulations".

[12] Karkkainen, Bradley C., Archon Fung, and Charles F. Sabel, "After backyard environmentalism: Toward a performance-based regime of environmental regulation", American Behavioral Scientist, 44(4): 692-711, 2000.

[13] DNV, "Key Aspects of An effective U.S. offshore Safety regime", Whitepaper.

[14] U.S. Regulatory Council. "Performance Standards, A Practical Guide to the Use of Performance Standards as a Regulatory Alternative. Report of the Project on Regulatory Alternatives", Washington DC: Administrative Conference of the United States, 1981.

[15] Yates Report. "Weathertightness of Buildings in New Zealand. Report of the Government Administration Committee's Inquiry into the Weathertightness of Buildings in New Zealand", Wellington, NZ: House of Representatives, 2003.

[16] Arnold Ken, Senior Technical Advisor, Worley Parsons.

[17] Coglianese C, Nash J, Olmstead T. "Performance-Based Regulation: Prospects and Limitations in Health, Safety, and Environmental Protection." Administrative Law Review, 55: 705-729, 2003.

[18] Coglianese, C., J. Nash, and T. Olmstead, "Performance-Based Regulation: Prospects and Limitations in Health, Safety and Environmental Protection", Regulatory Policy Program Report No. RPP-03. John F. Kennedy School of Government. Harvard University, 2002.

[19] Drummond, D. "Regulatory strategies: Building performance oriented chemical waste rules", Available from: http://www2.fpm.wisc.edu/chemwasteinfo/PerformOrientedRules.htm.

[20] Youngblood, R. and I.S. Kim, "Issues in Formulating Performance Based approaches to regulatory oversight of nuclear poweder plants", Nuclear Engineering and Technology, 37(3): p. 231-244,2005.

[21] Queensland Goverment. Department of State Development, "Guidelines on alternatives to prescriptive regulation", Available from: https://www.wbginvestmentclimate.org/uploads/23.Queensland.pdf.

[22] Regulatory Uncertainty in Performance-Based Regulations, "Analysis of the Uncertainty in STUK Guide YVL 4.3. Fire Protection at Nuclear Facilities in Finland". Available from: http://www.google.com/url?sa=t&rct=j&q=&esrc=s&source=web&cd=2&cad=rja&ved=0CEYQFjAB&url=http%3A%2F%2Fwww.neimagazine.com%2FJournals%2FPower%2FNEI%2FMarch_2004%2Fattachments%2FTrubatchUncertain.doc&ei=xVMxUJ2tE-K02gWlloHACQ&usg=AFQjCNGtK1cILOPvH5ek8Qweg7l9eMKTlQ.

[23] "Offshore Injury, Health, and Incident Statistics 2010/2011," UK Health and Safety Executive, Merseyside 2011.

[24] American Petroleum Institute, "Process Safety Performance Measurement Report", American Petroleum Institute Statistics Department, June 2008.

[25] D. C. Andrews, "Industry views on chemical process safety," Process Safety Progress, vol. 12, pp. 104-105, 1993.

[26] P. J. May and C. Koski, "Performance-Based Regulation and Regulatory Regimes," Pacific Earthquake Engineering Research Center, University of California, Berkeley, Berkeley, CA September 2004.

[27] T. S. Aagaard, "Regulatory Overlap, Overlapping Legal Fields, and Statutory Discontinuities," Virginia Environmental Law Journal, vol. 29, 2011.

[28] L. Rectanus and J. Natalicchio, "Better Coordination Can Improve Safety at Hazardous Material Facilities," United States General Accounting Office, Washington, D.C. 2000.

[29] Wholey, J. "Performance-Based Management: Responding to the Challenges", M.E. Sharpe, Inc., Vol. 22, No. 3, pp. 288-307, 1999.

[30] Altmayer, C. "Moving to performance-based management", Government finance review, 2006.

A Guide to the Legal Framework of the PSM Standard for Engineers

Christopher Cunio, Esq.
Cooley Manion Jones LLP
21 Custom House Street
Boston, MA 02110
ccunio@cmjlaw.com

Georges Melhem, Ph.D.
ioMosaic Corporation
93 Stiles Road
Salem, NH 03079
melhem@iomosaic.com

Prepared for Presentation at
American Institute of Chemical Engineers
2013 Spring Meeting
9th Global Congress on Process Safety
San Antonio, Texas
April 28 – May 1, 2013

Keywords: *acceptable limits, audit, citation, compliance, corrective action, coverage, deficiency, fine, flammable, General Duty Clause, highly hazardous chemical, inspection, timely manner, training, mechanical integrity, National Emphasis Program, non-serious, performance based, process, process safety information, process hazards analysis, RAGAGEP, serious, violation, and willful.*

Abstract: *Compliance with the PSM Standard is challenging for even the most sophisticated operators because of the broad scope and highly technical nature of the 14 PSM elements. This article provides guidance on how to comply with the three elements most frequently cited by OSHA—process safety information, process hazards analysis, and mechanical integrity—and the consequences of a failure to do so.*

The Process Safety Management (PSM) Standard is a U.S. regulation that was issued by the Secretary of Labor in 1992. Its enactment was motivated by a series of chemical disasters in the 1980s and early 1990s, including the 1984 Bhopal, India gas leak disaster, which resulted in more than 2000 fatalities, and deadly chemical explosions in the U.S. facilities of Phillips 66, ARCO and BASF.[1]

The primary purpose of the PSM Standard is to prevent, or minimize the consequences of, the release of highly hazardous chemicals into locations that could expose people to serious bodily injury and deadly harm. To achieve its purpose, the Standard requires regulated facilities to implement a "PSM program," which is a systematic approach to proactively review chemical processes and identify, evaluate, and prevent or mitigate chemical releases. PSM programs must contain 14 components (referred to as "elements"): (1) employee participation, (2) process safety information, (3) process hazard analysis, (4) operating procedures, (5) training, (6) contractors, (7) pre-startup safety review, (8) mechanical integrity, (9) hot work permit, (10) management of change, (11) incident investigation, (12) emergency planning and response, (13) compliance audits, and (14) trade secrets.[2]

Compliance with the PSM Standard is challenging for even the most sophisticated operators because of the broad scope and highly technical nature of these 14 elements. This article focuses on the elements of the PSM program—specifically process safety information, process hazard analysis, and mechanical integrity—that have resulted in the majority of OSHA citations under the Petroleum Refinery (>50%) and Chemical (>60%) PSM National Emphasis Programs.[3]

The goal of this article is to provide in-house engineers with guidance on how to comply with these elements and when to consult external resources, such as engineering and/or legal PSM experts. The number of citations issued by OSHA in connection with the National Emphasis Programs demonstrates that there is still considerable misunderstanding of the PSM Standard, which is now over 20 years old. Consultation and collaboration with external PSM experts can facilitate the creation and maintenance of a robust, compliant PSM program. It can also be beneficial when addressing PSM violations identified through audits (1st, 2nd, and 3rd party) and OSHA inspections to ensure efficient and effective corrective action. Finally, it can be beneficial during due diligence performed in connection with mergers and acquisitions. External PSM expertise can help identify PSM violations and quantify corrective action prior to closing, which can substantially impact purchase price and/or the scope of indemnification rights under a purchase agreement.

Performance Based

The PSM Standard is "performance-based," which means that the regulator, in this case the Secretary of Labor, dictates "what to do" (e.g., to implement a PSM program that contains the 14 elements), and the operators determine "how to do it" (e.g. through the adoption of internal policies, external standards, or a hybrid/combination approach) based on their knowledge, industry practice, facility needs and conditions, and economic considerations. A site's PSM program can be a single document or series of documents; the only requirement being that it contains all 14 elements.

Operators should consider consulting PSM experts to determine whether a site's PSM program appropriately incorporates all 14 elements. In addition, they should consider consulting PSM experts to review the content of any existing PSM programs because OSHA can issue citations for a site's failure to follow its own PSM program, even when the site's program is arguably more stringent than industry practice. For example, OSHA recently cited a petroleum refinery for allowing the wall of a flare line to fall below minimum acceptable thickness identified in its internal inspection manual. OSHA v. Wynnewood[4] In short, operators must not only develop compliant PSM programs, but they must be understood, implemented, and followed.

PSM Coverage

The threshold inquiry for PSM compliance is to determine what areas of a facility must be covered by a PSM program. Under the PSM Standard, a PSM program must cover any "process" that involves "highly hazardous chemicals." The Standard broadly defines "process" as any activity involving a highly hazardous chemical, such as the use, storage, manufacturing, or handling of a chemical. This broad definition includes all vessels interconnected within an activity, as well as vessels that are separated from an activity but could be involved in a hazardous release.[5]

The PSM Standard defines "highly hazardous chemicals" in two ways. First, it lists 137 chemicals in Appendix A to the Standard as "high hazardous chemicals" when they are present at a facility in excess of certain threshold quantities (TQ). OSHA has determined that these chemicals have a potential for catastrophic release—*i.e.*, a major uncontrolled emission, fire, or explosion that puts employees in serious danger—when present above their TQs. Second, the Standard defines highly hazardous chemicals as *flammable gases* (as defined by 29 C.F.R. 1910.1200) and *flammable liquids* (with a flashpoint below 100 °F) that are on site in one location in a quantity of 10,000 pounds or more. This definition, however, excludes hydrocarbon fuels used solely for workplace consumption, and flammable liquids stored in atmospheric tanks or transferred that are kept below their normal boiling point without chilling or refrigeration.[6]

The reasonable interpretation of these definitions is that all equipment involved in, or that could have an impact on, any activity related to highly hazardous chemicals must be covered by the PSM program unless it falls under a limited exception.[7] Operators should consider consulting PSM experts in connection with defining the boundaries of PSM coverage. An expert's demarcation of what activities constitute "process" and what chemicals rise to the level of "highly hazardous" will help a facility develop and maintain an effective, compliant PSM program. OSHA has recently issued citations related to the improper delineation of the boundaries of PSM coverage. See, e.g., OSHA v. Delek[8] OSHA cited Delek, a petroleum refinery, for its failure to inspect a ventilation system for a control room on the basis that it was "process equipment" involved in the "handling" of chemicals.

Where a majority of processes at a facility are PSM covered, operators should consider applying PSM requirements to the non-covered processes as well to avoid, among other things, having two different process safety programs in place at the same facility. Indeed, many PSM elements may already be applied facility-wide (e.g., management of change, pre-startup safety review, hot work permits, contractor safety, emergency planning and response, and incident investigation).

Process Safety Information

Process safety information (PSI) is critical data regarding the highly hazardous chemicals, technology, and equipment involved in the chemical process. The PSI element of the PSM Standard requires operators to have the following types of PSI in a complete, accurate, up-to-date and accessible format: (1) information pertaining to chemical hazards, which shall consist of, at a minimum, toxicity information, permissible exposure limits, physical data, reactivity data, corrosivity data, thermal and chemical stability data, and hazardous effects of inadvertent mixing of different materials that could foreseeably occur; [9](2) information pertaining to the technology, which shall consist of, at a minimum, block flow or process flow diagrams similar to exemplar diagrams contained in Appendix B to the Standard, process chemistry, maximum intended inventory, safe upper and lower limits for temperatures, pressures, flows or compositions, and an evaluation of the consequences of deviations, including those affecting the safety and health of employees;[10] and (3) information pertaining to equipment, which shall include materials of construction, piping and instrument diagrams (P&IDs), electrical classification, relief system design and design basis, ventilation system design, design codes and standards employed, material and energy balances for processes built after the Standard was enacted (May 26, 1992), and safety systems (e.g., interlocks, detection, or suppression systems).[11]

The amount of information needed to meet the PSI element can be considerable. There is no requirement that all PSI be compiled in a single document, or that it be located in a single file. Where it is contained in various documents and/or locations, good practice is to compile an index of the PSI and/or locations.

Compliant PSI—including P&IDs, electrical classifications, equipment files, and relief system documentation—is essential to the safe operation of a facility. P&IDs are critical because they depict the connectivity of equipment and the instrumentation used to control the equipment. They are also required to, among other things, conduct effective process hazards analysis (discussed below) and relief system evaluation. Electrical classifications are critical because improperly classified electrical equipment can become a source of ignition and can cause an explosion. Equipment files are critical because operators must be able to document that all equipment complies with current recognized and generally accepted good engineering practices (RAGAGEP), or is designed, maintained, inspected, tested, and operating in a safe manner. Relief system documentation is critical because relief systems are the last line of defense for protecting equipment from scenarios that can cause catastrophic equipment failure and explosions, fires, and toxicity impacts. OSHA has recently issued citations for incomplete and inaccurate PSI. See, e.g., OSHA v. Hunter Panels L.L.C.[12] OSHA cited Hunter, a manufacturer of insulation paneling, because its PSI did not contain specific corrosivity data ("process safety information did not include corrosivity data for resistance of Type 321 Stainless Steel … to n-Pentane").

Operators should consider consulting PSM experts to determine whether a site's PSI is compliant and, if not, how to close the compliance gap. Common errors include assuming that all chemical hazard information is located on material safety data sheets; however, many MSDSs do not contain complete information regarding reactivity, corrosivity, stability and the hazardous effects of inadvertent mixing. This gap can be addressed by creating hazardous chemical interaction matrices.

Process Hazard Analysis

Process hazard analysis (PHA) is an organized set of assessments designed to identify and analyze the significance of potential hazards associated with the processing or handling of highly hazardous chemicals. PHAs help facilities determine how to control such hazards. The PHA element of the PSM Standard requires operators to have conducted all initial PHAs by May 26, 1997, and to conduct follow-up PHAs at regular intervals not to exceed five years.[13] In order to conduct an effective PHA, all PSI should be complete, accurate, and up-to-date. Any deficiencies in PSI must be identified in the PHA and listed as a recommendation.

In addition, PHAs must be conducted by a team that includes individuals with expertise in engineering and process operations using one of the following methodologies: What-If, Checklist, What-If/Checklist, Hazard and Operability Study, Failure Mode and Effects Analysis, Fault Tree Analysis, or an appropriate equivalent method.[14] A popular new methodology being used is the Layer of Protection Analysis (LOPA). LOPA is used to identify safety integrity level requirements for safety instrumented systems as well as to provide better quantification of the frequencies of scenarios identified in PHAs.

PHAs must address the hazards of the process, any previous incidents with potentially catastrophic consequences, engineering and administrative controls, and the consequences of failure of the controls. PHAs must also address facility siting, human factors and failure of controls.[15] Although many companies use checklists to address facility siting issues during PHAs, good practice is to also conduct separate site-wide facility studies focused on occupied buildings according to industry standards such as API 752.

Upon their completion, PHAs may contain recommendations, which must be resolved (i.e., accepted, rejected, or withdrawn) in "a timely manner." Corrective action pursuant to recommendations must be completed "as soon as possible."[16] Although the PSM Standard does not define these terms, government publications indicate, for example, that OSHA expects employers to complete corrective action within a one or two year timeframe absent unusual circumstances.[17] In order to obtain consistent results from PHAs across a facility or an entire company, it is important to establish a preferred PHA methodology and other supporting tools such as risk ranking matrices, and recommendation tracking software.

Operators should consider consulting PSM experts to help determine whether PHAs are being conducted in compliance with the Standard. OSHA has recently issued citations for failure to properly conduct PHAs. See, e.g., OSHA v. Westlake Vinyls Co.[18] OSHA cited Westlake, a manufacturer of chemicals and plastic products, for failing to address hazards of the process, administrative controls, facility siting and human factors in its PHAs. PSM experts can also elucidate unfixed compliance terms like "a timely manner" and "as soon as possible" to assist a facility's compliance effort. OSHA has recently issued citations for failure to address recommendations within an appropriate time-scale. See, e.g., OSHA v. Heritage-WTI, Inc.[19] OSHA cited Heritage-WTI, an environmental services company that incinerates hazardous wastes and other materials, for failing to complete certain PHA recommendations for over 5 years. OSHA has also recently cited companies for failure to complete corrective actions in shorter time periods. See, e.g., OSHA v. Cargill Meat Solutions Co. (recommendations from a 2 year old PHA revalidation were not documented as closed)[20] and OSHA v. Hunter Panels L.L.C. (operating company cited in August 2012 for failure to "document the resolution of action items from the 2012 Process Hazard Analysis" related to providing a spill kit for small releases at a tank area).

Mechanical Integrity

Mechanical Integrity (MI) is the system of assuring that process equipment is in satisfactory condition to safely and reliably perform its intended design function and operate properly within the limits of established PSI. Process equipment is equipment that can control, prevent, or mitigate catastrophic chemical releases. It includes pressure vessels, storage tanks, piping systems (including piping components such as valves), relief and vent systems and devices, emergency shutdown systems, controls (including monitoring devices, sensors, alarms, and interlocks), pumps, and utility systems.[21]

The MI element of the PSM Standard requires operators to establish and implement an MI program to maintain the ongoing integrity of process equipment.[22] A critical first step in developing an MI program is to identify the process equipment and instrumentation that needs to be included. A good practice is to look at the PHAs to determine if failure of equipment could result in the release of hazardous chemicals. Also, any safeguards listed for PHA scenarios should be included.

Additionally, MI requires written procedures by which an operator shall: (1) conduct MI training; (2) perform inspections and testing of process equipment following RAGAGEP; and (3) correct deficiencies in process equipment that are outside of acceptable limits (defined by the PSI, or RAGAGEP) before any further use, or in a timely manner while necessary means are taken to assure the safe operation of the equipment during an interim period.[23]

The term RAGAGEP includes "national consensus" codes and standards like those published by National Fire Protection Association, American Society for Testing and Materials, American National Standards Institute, and American Petroleum Institute.[24] Examples of deficiencies outside of "acceptable limits" include equipment that was not designed, fabricated, constructed or installed per RAGAGEP; equipment with mechanical defects that interfere with its capability to function as intended; a degradation in the equipment exceeding acceptable limits (including retirement thickness); operating equipment outside its normal limits; equipment leaks; and by-passed equipment.[25] For example, equipment that is operated below retirement thickness is outside acceptable limits until it is repaired or replaced to meet PSI requirements or it is determined fit-for-service "as is" under API 579, which establishes a new acceptable limit.

Individuals who perform inspections and testing may need to have certifications, such as for inspecting pressure vessels and welding. Also, the testing and inspections can be costly, particularly if equipment must be taken out of service to perform the test or inspection. Implementing risk-based inspection programs such as outlined in API 580/581 may allow resources to be more effectively utilized.

Operators should consider consulting PSM experts concerning MI to help determine whether the scope of an existing MI program is adequate, whether inspections are being properly conducted, and whether deficiencies are being addressed in compliance with the Standard. OSHA has recently issued citations for an MI program's failure to correct process equipment deficiencies in a safe and timely manner. See, e.g., OSHA v. Arkema Inc.[26] OSHA cited Arkema, a chemical company, for operating with compromised rupture discs in relief systems.

PSM Violations

Deficient PSI, PHAs and/or MI violate the PSM Standard, which could result in an OSHA citation. There are three categories of PSM violations: (1) serious; (2) non-serious; and (3) willful. A violation is "serious" if death or serious physical harm (i.e., a substantial impairment to bodily function) could result from the violation. Conversely, a violation is "non-serious" if death or serious physical harm could not result from the violation. OSHA assesses civil penalties of up to $7,000 for every serious and non-serious violation. A "willful violation" is one that is committed with either intentional disregard or plain indifference to the requirements of the PSM Standard. OSHA assesses civil penalties between $5,000 and $70,000 for every willful violation. Willful violations that cause death to an employee are subject to criminal sanctions, including imprisonment of up to 1 year.[27]

The severity of each OSHA penalty is determined by the gravity of the violation. Two factors largely determine the gravity of a violation: (1) the severity of the injury that could occur from the violation (i.e., high, medium, or low); and (2) the probability that the injury could result from the violation (i.e., greater probability and lesser probability). Other factors OSHA may consider include the size of the operator, the operator's good faith, and the prior history of violations at the site. These factors, however, are not defenses to the underlying violations.[28]

When confronted with a potential PSM violation (e.g., inadequate PSI, PHAs and/or MI), OSHA may issue a citation under the PSM Standard or, in the alternative, under the OSHA General Duty Clause in the event that the PSM Standard does not apply. The General Duty Clause is a "catchall provision" that imposes an independent duty on operators to provide a safe work environment. Specifically, it requires an operator to provide a place of employment that is free from "recognized hazards" that are currently "causing, or are likely to cause [should they occur] death or serious physical harm to . . . employees." A hazard is "recognized" where: (a) the employer has identified it; (b) it is known in the industry; or (c) it is blatantly obvious. Penalties under the General Duty Clause are the same as those issued under the PSM Standard.[29]

Conclusion

Personnel who collaborate with PSM experts will understand the complexities of today's PSM standard. This understanding will make their facilities better equipped to maintain PSM compliance, help avoid costly OSHA citations, better control cost of corrective action, and foster a safe and reliable work environment for all employees.

References

[1] 29 CFR 1910.119 (OSHA PSM Standard); 57 FR 6356-01 (Preamble and Final PSM Rule); Public Law 101-549 §304 (Clean Air Act Amendments 1990).
[2] 29 CFR 1910.119(c-p).
[3] See CPL 03-00-010 (Petroleum Refinery NEP 2009); CPL 03-00-014 (Chemical NEP 2011).
[4] OSHRC Docket No. 07-0609 (8/4/08); 340 Fed. Appx. 462 (10th Cir. 2009).
[5] 29 CFR 1910.119(a)(1)(i)-(ii) and 29 CFR 1910.119(b).
[6] 29 CFR 1910.119(a)(1)(i)(Appendix A) and 29 CFR 1910.119(a)(1)(ii).
[7] The PSM Standard lists a few, limited exceptions where the PSM program will not apply: at retail facilities, oil or gas well drilling or servicing operations, or normally unoccupied remote facilities that are geographically remote and have no permanent employees (29 CFR 1910.119(a)(2)(i)-(iii)).
[8] OSHRC Docket No. 08-1386 (4/27/11).
[9] 29 CFR 1910.119(c)(3), (d), (l), Appendix C, par. 3(complete, accurate, up-to-date and accessible requirements). Material Safety Data Sheets that meet the requirements of 29 C.F.R. 1910.1200(g) can be used to comply so long as they contain the requisite information.
[10] Where original technical information no longer exists, operators may develop such information in conjunction with process hazard analysis.
[11] 29 CFR 1910.119(d)(1) - (3).
[12] Inspection #544919 (citation issued: 1/25/13).
[13] 29 CFR 1910.119(e).
[14] Id.
[15] Id.
[16] 29 CFR 1910.119(e)(5).
[17] 57 FR 6356-01, p. 6379.
[18] Inspection #329260 (citation issued: 9/18/12).
[19] Inspection #189909 (citation issued: 8/6/12).
[20] Inspection #315 121 848 (citation issued: 12/6/11).
[21] 29 CFR 1910.119(j).
[22] 20 CFR 1910.119(j)(2).
[23] 29 CFR 1910.119(j)(3) – (5).
[24] 57 FR 6356-01, p. 6390.
[25] CPL 03-00-010, pp. 6-7.
[26] Inspection No. 316 217 835 (citation issued: 7/10/12).
[27] 29 USC §666 (Penalties); Trinity Indus., Inc., 504 F.3d 397 (3d Cir. 2007) (serious); OSHA Field Operations Manual Ch. 4-II-C-3(non-serious); J.A. Jones Constr. Co., 15 OSH Cas. (BNA) par. 2201(willful). In 2009, a bill known as the "Protecting America's Workers Act," was proposed in both the U.S. Senate (S.1580) and House of Representatives (H.R.2067) to increase OSHA civil penalties to $250,000 (maximum) per violation, and criminal penalties to up to 20 years for repeat deaths.
[28] 29 USC §666(j) (criteria); OSHA Field Operations Manual Ch. 6-III-A (gravity).
[29] 29 USC §654 (OSHA General Duty Clause); Occupational Health & Safety Law; Compliance and Practice §§7.24-7.25.

New CCPS Guideline Book: Guidelines for Enabling Conditions and Conditional Modifiers in LOPA

J. Wayne Chastain
Plant Protection Technical Services, Eastman Chemical Company
200 South Wilcox Drive
Kingsport, TN 37660
chastain@eastman.com

Stanley A. Urbanik
DuPont Corporation
Wilmington, DE
Stanley.A.Urbanik@dupont.com

Robert W. Johnson
Unwin Company
Columbus, OH
rjohnson@unwin-co.com

John F. Murphy
Process Safety Services
Punta Gorda, FL
hamjfm@embarqmail.com

Prepared for Presentation at
American Institute of Chemical Engineers
2013 Spring Meeting
9th Global Congress on Process Safety
San Antonio, Texas
April 28 – May 1, 2013

UNPUBLISHED

New CCPS Guideline Book: Guidelines for Enabling Conditions and Conditional Modifiers in LOPA

J. Wayne Chastain
Plant Protection Technical Services, Eastman Chemical Company
200 South Wilcox Drive
Kingsport, TN 37660
chastain@eastman.com

Stanley A. Urbanik
DuPont Corporation

Robert W. Johnson
Unwin Company

John F. Murphy
Process Safety Services

Keywords: Layers of Protection Analysis, LOPA, conditional modifier, enabling condition

Abstract

This paper outlines the content of a new CCPS book entitled *Guidelines for Enabling Conditions and Conditional Modifiers in LOPA*. An enabling condition is a condition that is not a failure, error or a protection layer but makes it possible for an incident sequence to proceed to a consequence of concern. It consists of an operating phase or condition that does not directly cause the scenario, but that must be present or active in order for the scenario to proceed to a loss event; expressed as a dimensionless probability. A conditional modifier is one of several probabilities included in scenario risk calculations, generally when risk criteria endpoints are expressed in impact terms (e.g., fatalities) instead of in primary loss event (e.g., release, vessel rupture). Conditional modifiers include, but are not limited to: probability of a hazardous atmosphere, probability of ignition, probability of explosion, probability of personnel presence, probability of injury or fatality, and probability of equipment damage or other financial impact. These *Guidelines for Enabling Conditions and Conditional Modifiers in LOPA* contain information useful to the experienced analyst and accomplished practitioner. In this book, various kinds of enabling conditions and conditional modifiers are defined and illustrated as they might be used in Layer of Protection Analyses (LOPAs). There is a discussion on how enabling conditions and conditional modifiers can be used in other risk analysis applications. The appendices discuss "double jeopardy" and simultaneous failures, peak risk, and example rules for enabling conditions and conditional modifiers.

1. Introduction

Enabling conditions and conditional modifiers are two types of factors that have long been used in quantitative risk analysis. Both of these factors were introduced for use in LOPA in *Layer of Protection Analysis: Simplified Process Risk Assessment [1]*. Enabling conditions are conditions that generally influence the initiating events for scenarios. They are conditions that do not directly cause the scenario, but which must be active in order for the scenario to proceed to a loss event. Enabling conditions are generally included in a LOPA independent of the type of endpoint used in the analysis. Conditional modifiers are probabilities included in LOPA when the risk criteria endpoints are expressed in terms of impact as opposed to primary loss event terms. Neither enabling conditions nor conditional modifiers are independent protection layers and do not typically meet all of the criteria used to qualify an independent protection layer (IPL). The development of enabling conditions and conditional modifiers in *Layer of Protection Analysis: Simplified Process Risk Assessment* was sparse given that the major foci of the book were the overall methodology, initiating events, and independent protection layers [1].

Guidelines for Enabling Conditions and Conditional Modifiers in LOPA provides a more in-depth treatment of these two important factors used in LOPA. Although these factors are used in other types of risk analysis, the focus of this text is on their use in LOPA. The text provides suggested values or means of obtaining appropriate values for these factors, and also provides guidance on when the values should and should not be used. Misuse of these factors can result in an analysis that can significantly under-predict or over-predict the actual risk of a scenario.

2. LOPA Overview

Layer of Protection Analysis is a simplified form of quantitative risk analysis that uses order-of-magnitude categories for initiating event frequency, consequence severity and probability of failure of independent protection layers (IPLs) to analyze and assess the risk of one or more incident scenarios. LOPA is a risk analysis method that was developed to answer questions such as "what layers of protection are needed to meet our risk goals?" and "how much risk reduction does each layer provide or need to provide?" LOPA can help to answer these questions with less time and effort than more complex methods such as quantitative risk analysis.

LOPAs are typically conducted after a qualitative hazard evaluation has been used to identify the hazard scenarios of the process. Often only a subset of the identified hazard scenarios from a qualitative study is evaluated using LOPA. Likewise, LOPA can be used as an endpoint for risk analysis or can be used as a screening tool to determine scenarios that should be evaluated using more detailed quantitative risk analysis techniques.

LOPA is applied to a selected incident scenario. For the scenario being evaluated the initiating event and the frequency associated with that initiating event must be determined. The safeguards that are present to prevent the scenario are then evaluated to determine if they meet the requirements to be used as an independent protection layer. Each safeguard should meet seven key attributes in order to qualify as an IPL. These attributes are:

- Independence
- Functionality
- Integrity
- Reliability
- Validation, maintenance, and auditing
- Access security
- Management of change

For each IPL, the probability of failure on demand must be determined.

By combining the initiating event frequency with the probability of failure on demand for each IPL, the scenario frequency can be determined. This predicted frequency can be compared to an established risk criterion.

Enabling conditions and conditional modifiers are additional factors that are included by some organizations in their LOPAs. Enabling conditions can be required for the evaluation of some scenarios in order to obtain an accurate evaluation of the risk of a scenario. Their exclusion from a LOPA can result in overestimation of the risk of an event.

Conditional modifiers are used when the risk criteria associated with the LOPA process have been selected based on their use in the analysis. When risk criteria used in the analysis are based on releases from primary containment, conditional modifiers are typically not applicable. However, if the risk criteria selected are based on the potential ultimate consequences associated with an event, the use of conditional modifiers may be warranted.

Enabling conditions and conditional modifiers have associated probabilities, similar to IPLs.

3. Enabling Conditions

Enabling conditions make the initiation of a scenario possible. Enabling conditions are not initiating events, but for a given scenario, if the enabling condition is not true, the scenario will not progress from the initiating event to the hazardous outcome under consideration. Mitigating factors such as probability of personnel presence and probability of ignition, which come into play after the primary loss event, are not enabling conditions but conditional modifiers. Enabling conditions generally fall into one of several categories.

3.1 Time At Risk

Time-at-risk enabling conditions come into play when an incident sequence may only be realized a certain fraction of time when the conditions are right for the event sequence to proceed to a loss event. One of the most important considerations about time-at-risk enabling conditions is that they only apply to revealed failures. A revealed failure is one that becomes apparent and will be corrected after its occurrence. In contrast, an unrevealed failure can occur at any time, but only becomes apparent when a specific event occurs that reveals it or a test reveals that the failure has occurred.

3.1.1 Seasonal Effects

One group of common time-at-risk enabling conditions are various seasonal effects. For example, low ambient temperatures that can cause freezing of process or utility lines may only be present for a certain amount of time during the year. Likewise, sufficiently high ambient temperatures to effect cooling capacity or cause excessive vapor pressure of some process components may only be present during a portion of the year. Seasonal enabling conditions can typically be characterized by climatological data available for the site. When the seasonal effect itself is the initiating event, then it is not also an enabling condition.

3.1.2 Process State Risks

Another type of time-at-risk enabling condition is process state risks that express the probability that the process is in a certain state. These enabling conditions apply when an initiating failure only has the undesired impact when the process is in the particular state. Process state time-at-risk can apply to batch processes where the risk being evaluated is only present during certain periods of the batch cycle, or to scenarios that are only applicable during the startup, shutdown, or clean-out of the process.

3.1.3 Situations Where Time-at-Risk Should Not be Used as an Enabling Conditions

Time-at-risk enabling conditions are not appropriate in all situations. Some examples where time-at-risk enabling conditions are not appropriate include:

- Continuous operations where hazards are constant over time
- Infrequent short-duration operating modes with high potential severity
- Procedure-based operations involving human error in the execution of a stepwise procedure
- Continuous or procedure-based operations involving any combination of factors that are failure or fault conditions, regardless of whether these conditions are revealed or latent
- Incident sequences where time-at-risk applies to the severity of consequences of the loss event rather than its likelihood of occurrence

3.1.4 Campaign Enabling Conditions

Campaign enabling conditions reflect different conditions that are present due to changes in processes from batch to batch or from time to time. In a production facility where some batches pose certain hazards while other batches do not because of the presence of certain raw materials, final products, catalysts, or operating conditions, campaign enabling conditions may be appropriate in a LOPA. Campaign enabling conditions may also apply to operations with certain modes that are present for limited periods of time such as recycle or a final control element being down in an effluent handling system or a process that only operates a limited portion of the year due to low demand.

Campaign enabling conditions may potentially provide high levels of risk reduction. When evaluating the use of campaign enabling conditions several factors must be considered.

Campaign enabling conditions may change over time as the production needs of a facility change. Campaign enabling conditions are not appropriate to be used in situations where there is the potential for an unrevealed failure that only becomes apparent with the initiation of the loss event sequence when the condition is true.

An example of an unrevealed failure that would prevent the use of a campaign enabling condition is a level control loop on a weigh tank that is only used during a particular campaign. If the level control loop failure could occur at any point in time but would only be revealed during the hazardous campaign, the campaign enabling condition could not be applied. If the level transmitter is checked prior to the campaign being started and the failure can only occur during the campaign, then the campaign enabling condition could be properly applied.

Because of the high levels of risk reduction that may be claimed with a campaign enabling condition, caution must be used. In certain cases, the application of an enabling condition can result in a risk that meets the criteria being evaluated, but for brief periods of time exposes personnel to very high risks. This concept is called "peak risk" and is evaluated in the text as a special case.

4. Conditional Modifiers

Conditional modifiers are probabilities included in a LOPA or other risk analysis when risk criteria are expressed in impact terms (e.g. fatalities) as opposed to loss of primary containment. These modifiers include:

- Probability of a hazardous atmosphere
- Probability of ignition or initiation
- Probability of explosion
- Probability of personnel presence
- Probability of injury or fatality
- Probability of equipment damage or other financial impact.

Although conditional modifiers have the same effect on the predicted frequency for an event in a LOPA as independent protection layers, they differ significantly since IPLs are engineered or administrative controls that are capable of responding to a process deviation to either avoid or mitigate a loss event. Conditional modifiers, by contrast, are not response actions but relate to the condition that the facility is likely to be in at a particular point or period of time in the incident sequence. The conditional modifier indicates that probability that the condition is true at any time. Even though conditional modifiers are not engineered and administrative controls, the probability that a conditional modifier is true can be impacted by the way that the facility is managed. As an example, access control procedures and training can have an impact on the probability of personnel presence.

Conditional modifiers, like IPLs, need to meet the basic requirement of independence; i.e. independent of other considerations and factors in the scenario risk calculations. Particular care needs to be taken to ensure that conditional modifiers are independent of the severity of the event.

The improper use of conditional modifiers in a LOPA can result in an underestimation of the risk of an event. Because of this, their use needs to be intentional, disciplined, and consistent. To ensure appropriate application, a good approach is to develop a standardized baseline set of conditional modifiers and the rules for their use. Common pitfalls that exist in the use of conditional modifiers are:

- Using more conditional modifiers than an incident scenario warrants
- Being overly optimistic in estimating conditional modifier probabilities
- Including more additional modifiers until a desired risk level is reached
- Using conditional modifiers to justify avoiding a standard practice
- Using conditional modifiers in situations where the risk criteria are based solely on loss of primary containment or implicitly assume conditional modifiers

When selecting a conditional modifier for use in a LOPA, the value ought to be expected to remain relatively constant over time. A change to the process that results in a change to probability of a conditional modifier can invalidate the risk decision made using LOPA unless such a change is properly accounted. A company's management of change process may need to be expanded to capture the potential impacts to conditional modifiers that a process change may entail.

4.1 Probability of a Hazardous Atmosphere

Probability of a hazardous atmosphere is a conditional modifier that may be applied to loss of primary containment events with the potential to cause harm, but may not depending on the specific situation. A loss-of-containment event of an ignitable material may form a flammable atmosphere or not depending on the ambient temperature. When evaluating the use of a probability of hazardous atmosphere, care needs to be taken to ensure that the conditional modifier is independent of other aspects of the LOPA that may reflect a similar effect such as time-at-risk or procedural controls that may be counted as IPLs.

4.2 Probability of Ignition or Initiation

When a flammable atmosphere is present, there is a potential that it will be ignited. The probability that ignition will occur is a affected by a broad range of factors that have made detailed treatment of this topic beyond the scope of *Guidelines for Enabling Conditions and Conditional Modifiers in LOPA*; however, several pages are devoted to providing initial guidance for the LOPA practitioner in dealing with this important conditional modifier. A closely related event is the initiation of an uncontrolled reaction in a reactive material mixture.

There are three basic scenarios in which probability of ignition needs to be determined if it is going to be included in a LOPA.

- Ignition of a flammable atmosphere inside process equipment
- Ignition of a flammable vapor, gas, or dust cloud external to process equipment
- Initiation of uncontrolled reaction in a reactive material or mixture.

Each of these is given some level of attention in the text.

Probability of ignition inside process equipment is applicable whenever a flammable or combustible atmosphere exists normally or can be formed as the result of a process upset condition in process equipment. Each situation needs to be evaluated carefully, taking into account a broad range of process parameters, details about the equipment and its operating and upset conditions, the properties of the materials being processed, historical data, and other applicable data in reaching the determination of the value to be used for conditional modifier.

Many of the same considerations that are pertinent to the probability of ignition inside process equipment also apply when determining an appropriate value for the probability of ignition of a flammable or combustible atmosphere outside process equipment. Other ignition mechanisms that are not applicable to flammable and combustible atmospheres inside process equipment must also need to be considered such as traffic, hot work, personnel with cell phones or other non-classified equipment, personnel smoking, etc. The size of the flammable cloud outside equipment and the duration of release are also significant factors in determining the probability of ignition. The potential for immediate vs. delayed ignition is a factor to be considered, since delayed ignition can potentially pose a more hazardous situation than immediate ignition.

The probability of initiation of an uncontrolled reaction is a function of the different mechanisms that are available for imparting the necessary energy to begin the reaction. These mechanisms may include shock or impact, friction, electrostatic discharge, and heat. The effect and probability of contamination of the process material needs to be accounted for, since contamination can result in significantly lower levels of energy required to cause the reaction to begin.

4.3 Probability of Explosion

Several different conditional modifiers can be applied to situations in which an explosion could potentially occur. Some of the conditional modifiers related to probability of explosions that are evaluated in the text are the probability of dust explosion, probability of a vessel rupture, probability of a vapor cloud explosion, and probability of a deflagration-to-detonation transition.

4.4 Probability of Personnel Presence

Probability of personnel presence is one of the most commonly used conditional modifiers within LOPA. This conditional modifier reflects the probability that one or more people are in the affected area when a loss event occurs. Several key considerations for using (or not using) probability of personnel presence as a conditional modifier include:

- When using the conditional modifier in conjunction with the probability of injury, the LOPA practitioner needs to ensure that the factors are not "double-counted"
- Using probability of personnel presence may require a more detailed evaluation of the incident sequence which may require source term calculations and dispersion modeling

- Probability of personnel presence needs to take into account all potentially impacted personnel including not only operations personnel, but also maintenance workers, and any other personnel that may be in the area
- The probability of personnel presence needs to be independent of the scenario being evaluated which will often not be the case when operator response to an alarm will take personnel into the field
- Probability of personnel presence should not be credited when the initiating event is an operator error that immediately results in the loss of containment
- When personnel presence is embedded in the determination of the severity of consequences, a conditional modifier should not be used
- Probability of personnel presence should not be credited when the corporate risk criteria are not based on personnel presence

An alternate approach to using probability of personnel presence as a conditional modifier is to use access control as independent protection layer. Administrative access control should only be used as an IPL when it will effectively limit access and the control meets all of the requirements of an IPL. When administrative controls are used as an IPL to limit the access of personnel to an area, it would be incorrect to also use probability of personnel presence as a conditional modifier.

The effect area determination is an important part of the probability of personnel presence. Effect areas can be determined from dispersion modeling, the determination of toxic release effect areas, fire and explosion effect areas. Once the effect area has been determined, the sum of the time that personnel are going to be in the effect area can be determined. In the event that more than one person can be present in the effect area, corporate risk criteria may require that a different severity category be used in the analysis.

4.5 Probability of Injury or Fatality

The probability of injury or fatality is a conditional modifier that reflects the probability that personnel that are present in the effect area will suffer the consequence being evaluated. Four approaches are presented in the text for addressing the probability of injury or fatality.

- Approach #1: Use P=1 for the probability of injury or fatality conditional modifier, because of how the effect boundary is defined.
- Approach #2: Apply the same probability of injury or fatality conditional modifier ($P<1$) to everyone in the effect area.
- Approach #3: Estimate the probability of serious injury or fatality conditional modifier for individual(s) inside the effect area.
- Approach #4: Use of a combined approach of the first three in the evaluation.

There are several notes and cautions when using the probability of injury or fatality.

- More than one consequence may need to be evaluated.
- The conditional modifier needs to apply to the entire effect area, not just at the outside boundary of the effect area.

- Controls that limit personnel from entering a dangerous area that has been created can be addressed as independent protection layers.
- The characteristics of the particular chemical involved in the release may impact the probability of injury or fatality.

Probability of equipment damage or other financial impact as a possible conditional modifier is also discussed in the text.

5. Application to Other Methods

Guidelines for Enabling Conditions and Conditional Modifiers in LOPA briefly addresses the use of these factors in other types of risk analyses. The applicability and use of these factors in the various types of quantitative risk analysis is discussed included their use in fault tree analysis, event tree analysis, and quantitative impact analysis. The use of the factors is also discussed as they might apply to scenario identification methods such as Hazard and Operability Studies and What-If Analysis. Finally the text includes a brief discussion of the applicability to barrier analysis and diagrams.

6. Appendices

The first two appendices of the text address two common issues that arise in LOPA when using enabling conditions or conditional modifiers; namely, simultaneous failures and the concept of "double jeopardy" and peak risk concepts. The last appendix in the text provides an example rule set for LOPA enabling conditions.

6.1 Simultaneous Failure and "Double Jeopardy"

The first appendix describes what is meant by "double jeopardy" in the scope of risk analysis. Specifically "double jeopardy" is the concurrent incidence of two independent initiating events or other revealed failures. Per this definition, the text points out that the true incidence of "double jeopardy" events is very small. It is important to note that the text draws a firm distinction between revealed and unrevealed failures. If any of the failures in the "double jeopardy" case being evaluated are unrevealed failures, then the scenario is not a case of double jeopardy.

6.2 Peak Risk

The second appendix deals with the concept of peak risk. Peak risk was developed to help address situations in which a relatively high risk is present in a facility for a very short period of time. When this type of risk is annualized, as would be done when an enabling condition is added to the scenario LOPA indicating the amount of time the process is in the state where the event can occur, the risk can be shown to be low enough. Peak risk is a means of setting a limit on the risk that personnel may be exposed to without consideration being given to the duration of the event.

6.3 Example Rule Set

The final appendix in the text provides an example rule set for LOPA enabling conditions. This example is not meant to provide recommended guidance, but simply a possible path forward for what guidance on the topic might look like.

7. Conclusions

Guidelines for Enabling Conditions and Conditional Modifiers in LOPA is a text that will move the understanding and consistent application of these important LOPA factors forward in the industry. The warnings and caveats given in the text will help to prevent the misuse of these factors that can result in under-prediction of the risk being evaluated in LOPA. The text gives an overview of the major types of enabling conditions and conditional modifiers being used by LOPA practitioners.

8. References

[1] *Layer of Protection Analysis: Simplified Process Risk Assessment*; American Institute of Chemical Engineers, Center for Chemical Process Safety: New York, New York, 2001.

9TH GLOBAL CONGRESS ON PROCESS SAFETY

Implementing LOPA Recommendations into Design of Instrumented Protective Systems

Rajeev Limaye PE, CFSE
Director - Control Systems & Instrumentation
Praxair, Inc.
1585 Sawdust Road
The Woodlands Texas 77380
rajeev_limaye@praxair.com

Prepared for Presentation at
American Institute of Chemical Engineers
2013 Spring Meeting
9th Global Congress on Process Safety
San Antonio, Texas
April 28 – May 1, 2013

UNPUBLISHED

Implementing LOPA Recommendations into Design of Instrumented Protective Systems

Rajeev Limaye PE, CFSE
Director - Control Systems & Instrumentation
Praxair, Inc.
1585 Sawdust Road
The Woodlands, Texas 77380
rajeev_limaye@praxair.com

Keywords: ISA84, SIS, Instrumented Protective System, IPS, SIL, IEC61511, IEC61508, OSHA, PSM, LOPA, SIF, ISS, ISF, Process Safety Time, Risk Reduction, PHA, HAZOP

Abstract

When Independent Protection Layers (IPLs) are identified during Layer of Protection Analysis (LOPA), there may not be enough time to verify the validity of each IPL. In functional safety lifecycle, the next steps after LOPA are preparation of Safety Requirement Specification (SRS) and conceptual design of the Instrumented Protective System (IPS) which are typically performed by control system engineer(s). If certain protection layers are found to be inadequate, an iterative approach to revisit the LOPA is required to ensure the required risk reduction is achieved by the IPLs.

In most cases, multiple Instrumented Protective Functions (IPFs) and control functions require the same process measurement. For example, an alarm, a trip and a control loop may require the same process value. Adequate instrumentation must be provided to meet the independency criteria of IPL. Various scenarios are discussed on how and when to share the process signals between an IPS and the Basic Process Control System (BPCS). Good engineering practices to achieve safety as well as reliability of the system by means of different fault tolerant configurations are discussed. Typical Piping & Instrumentation Diagram (P&ID) representation of some of the common scenarios is also presented.

If operator response to an alarm is one of the IPLs, then some additional requirements need to be taken into consideration such as operator response time, human factors, etc.

1. Introduction

Inherent process risk is reduced to a tolerable level by implementing protective functions. Each organization has to define the tolerable risk level for safety, environmental and commercial hazards. Each protective function reduces the risk by a certain order of magnitude. LOPA is one of the most widely used semi-quantitative methods of analyzing and documenting protective functions. An important outcome of LOPA is identification of IPLs essential for required risk reduction. The required Safety Integrity

Level (SIL) of each IPL is also determined during LOPA. SIL defines the target performance level of an IPL in terms of a range of average probability of failure on demand (PFD_{avg}).
Safety Instrumented System (SIS) design, implementation, maintenance and operation is covered by ISA84 standard as functional safety lifecycle. ISA84 is endorsed by the Occupational Safety and Health Administration (OSHA) as a Recognized And Generally Accepted Good Engineering Practice (RAGAGEP). If an employer documents that it will comply with ISA84 and meets all ISA84 requirements, the employer will be considered in compliance with OSHA Process Safety Management (PSM) requirements for the SIS.

2. IPS classification

Over the last decade, several terms and definitions were introduced through ISA84, IEC61508 and IEC61511 standards as well as CCPS books and other publications. It is important to get clear understanding of these terms and acronyms to clearly define the scope of this discussion.

A safeguard is any device, system or action that would likely prevent an undesirable process incident triggered by an initiating event [1]. The safeguards could be non-instrumented (e.g. pressure safety valve) or instrumented (e.g. trip). In this paper, the focus is only on the instrumented safeguards.

A safeguard can be classified using following three attributes

Purpose	• Personnel Safety • Environmental Protection (releases) • Asset Protection (commercial)
Performance (Integrity Level)	The amount of risk reduction offered by the safeguard is measured in terms of Risk Reduction Factor (RRF). The RRF is also represented in terms of average Probability of Failure on Demand (PFD_{avg}). $RRF = 1/PFD_{avg}$
System	System in which the safeguard is implemented (BPCS, PLC, SIS, BMS, HIPPS etc.)

Fig. 1 shows the visual representation of the classification of safeguards and IPSs.

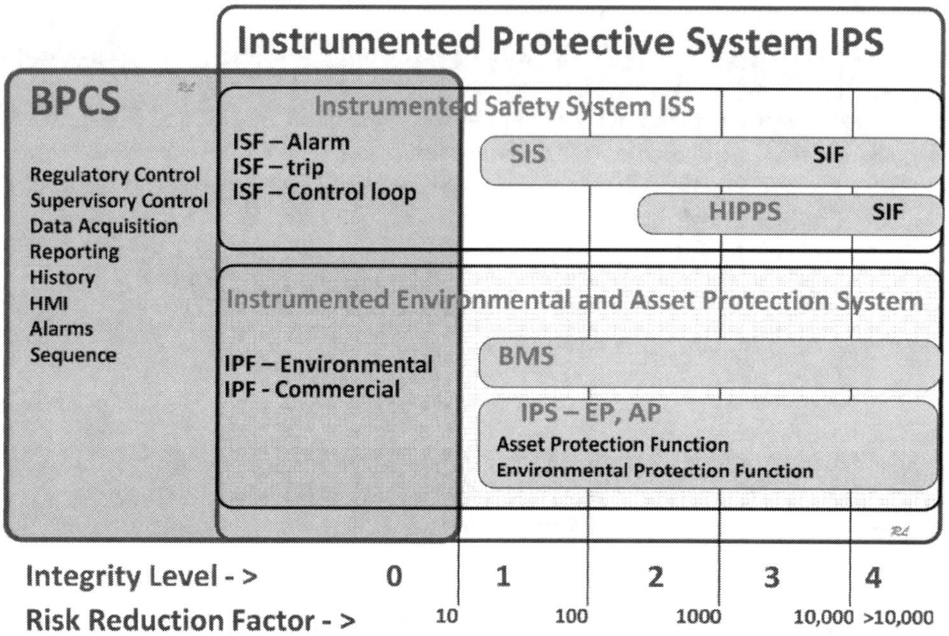

Fig. 1. Classification of safeguards and instrumented protective systems

Instrumented Protective System (IPS) refers to all systems that implement instrumented protective functions (IPFs) for the safety, environmental protection and asset protection. BPCS could be used to implement some of the IPFs.

Instrumented Safety Function (ISF) is a type of IPF for safety.

Instrumented Safety System (ISS) is a type of IPS intended for implementing ISF [2]. Some of the ISFs could be implemented in BPCS.

Safety Instrumented System (SIS) is an ISS that requires compliance with ISA84/IEC61511 standard for all safety lifecycle steps [3].

Basic Process Control System (BPCS) can be used to implement the IPF including ISFs. As per the ISA84, risk reduction credit cannot be more than 10 for the ISF implemented in BPCS.

Safety Instrumented Function (SIF) is a type of ISF that is implemented in SIS and must follow safety lifecycle as per ISA84/IEC61511 standard

High Integrity Pressure Protection System (HIPPS) is a type of SIS that offers higher risk reduction factor and must comply with ISA84/IEC61511 standard. Standards like API 521 and ASME Section VIII, Division 1 & 2 allows the use of HIPPS in lieu of a pressure relief device, as long as the HIPPS meets or exceeds the protection provided by the pressure relief device [4].

Burner Management System (BMS) implements furnace burner startup / shutdown logic as per the standards like NFPA 85.

Independent Protection Layer (IPL) is a type of IPF that is credited in LOPA for reducing the risk of undesired event and meets the criteria of being independent, specific, auditable and dependable defined in ISA84/IEC61511 Part 3 [5].

3. IPS design

While the BPCS is independent from the rest of the IPSs, often it is not practical to physically separate the BMS and other IPSs from the SIS. There is no requirement that these protective functions need to be implemented in separate systems. It may not be economical to do so. Whenever possible, SIFs should be separated from non-safety related IPFs. Most likely the IPFs with RRF of up to 10 are implemented in BPCS. The IPFs requiring higher level of risk reduction are implemented in SIS. ISA84 clause 11.2.2 states that "Where the SIS is to implement both safety and non-safety instrumented functions then all the hardware and software that can negatively affect any SIF under normal and fault conditions shall be treated as part of SIS and comply with the requirements of the highest SIL" [3].

In most discussions and literature, the focus of LOPA is identifying the layers of protection for personnel safety. However, safety cannot be addressed in isolation. It is also important to design the systems for asset protection and to prevent the environmental releases. Similar to the risk tolerance for safety incidents, each organization should develop the risk tolerance for commercial and environmental risks. The same LOPA approach should be used for analyzing and identifying the protection layers for safety, environmental and commercial risks. The scenarios should be properly classified during LOPA to indicate the type of risk it is addressing. The IPLs should be classified for protection against safety, environmental or commercial risk. An IPL may offer protection against combination of safety, environmental and commercial risks. Depending on the tolerable risk of consequence in each category, the integrity level requirement of the IPL may be different for safety, environmental and commercial risks. The highest integrity level of all three categories should be selected for the IPL.

Usually, the control systems engineer is responsible for allocating the IPFs to various IPSs and determining the number and type of systems needed. A comprehensive list of safeguards that are credited for risk reduction in LOPA with the classification using three attributes explained above is very handy. It is usually called an "IPL List". Each IPL should be given a unique ID which can be referred to on the P&IDs and the rest of the safety lifecycle documentation and planning including the SRS.

If most IPLs are SIL1 with an exception of just a few that require higher SIL rating, it is worth evaluating the cost of implementing just those IPLs in a separate small High Integrity Protection System, while selecting relatively less expensive SIS hardware for the rest of the IPLs that constitute the majority of the inputs and outputs.

4. LOPA issued for design

One of the main outcomes of the LOPA is identification of IPLs to mitigate the frequency of an undesired process event below the tolerable level. There have been discussions and theories on how to effectively conduct LOPA. Some recommend that it be done by a smaller group of people after the HAZOP for the scenarios with risk ranking above certain predetermined level. With this approach, a focused attention can be placed on the

scenarios identified for LOPA. This approach adds another activity on the project schedule and is perceived to take a longer time. There is another concept of performing LOPA during the HAZOP while the captive audience is present in the room. Usually HAZOP is performed by a team of much larger size compared to the LOPA team. There are pros and cons of each method. There is no one method better than the other. The setting in which LOPA can be effectively done depends on each organization. The quality of LOPA outcome depends on the team members' familiarity with LOPA concepts, nature and amount of experience, familiarity with other ISA84 safety lifecycle steps as well as the enthusiasm. If the team performing HAZOP is tired due to its long duration and schedule pressure, there is a tendency to cut corners and wrap up the LOPA scenarios quickly. Usually in such cases, the documentation of IPLs is inadequate and poorly done. There is not enough time during LOPA to analyze in detail the validity of each IPL.
Once the initial LOPA report is published, the next step is the SRS development and conceptual design of the IPS which is typically performed by control system engineer/s. It is often necessary to revisit the LOPA during the conceptual design of IPS. Some of the most commonly found reasons to revisit the LOPA are as follows:

- Improve the description of the IPLs, consequence and initiating event for better clarity.
- There is not enough time during LOPA to verify details such as process safety time, and the safety function response times.
- Availability of instruments and/or the process taps to meet the independency criteria. If identified ISFs do not meet the criteria of IPL, additional means of mitigating risk need to be identified.
- IPL listed is also part of the initiating event.
- Consideration of ISFs in BPCS (either excessive or ignored).
- During HAZOP the items that need further investigation are documented as HAZOP action items. These items are resolved after the HAZOP and proposed modifications may impact the ISFs.

Once the LOPA is revisited to incorporate the changes or improvements described above, it is typically referred to as LOPA issued for design (IFD). It is important to go through this iteration at least once to ensure the correct documentation of LOPA – which is the basis of any IPS design. During the verification step of each milestone in safety lifecycle, the functionality and performance of each ISF is verified against the LOPA documentation.

5. Process example

To illustrate these concepts, consider a process example consisting of high pressure knock out drum as shown in Fig 2.

The product gas at high pressure and temperature exiting the reformer is cooled in a process gas cooler, which results in condensation of water in the gas. The condensate water is removed in the knockout drum V-100. The level of the liquid in the knockout drum is controlled by a level control loop LC-100. The normal operating range of 35-

50% has been established to maintain a liquid level blanket in the knockout drum. If the level goes high, the separation will not occur as the liquid will get carried over in the product stream. If the level drops too low, there is a risk of the high pressure gas entering low pressure system downstream of the control valve LV-100. Maintaining the liquid level blanket to avoid this hazardous event is very important.

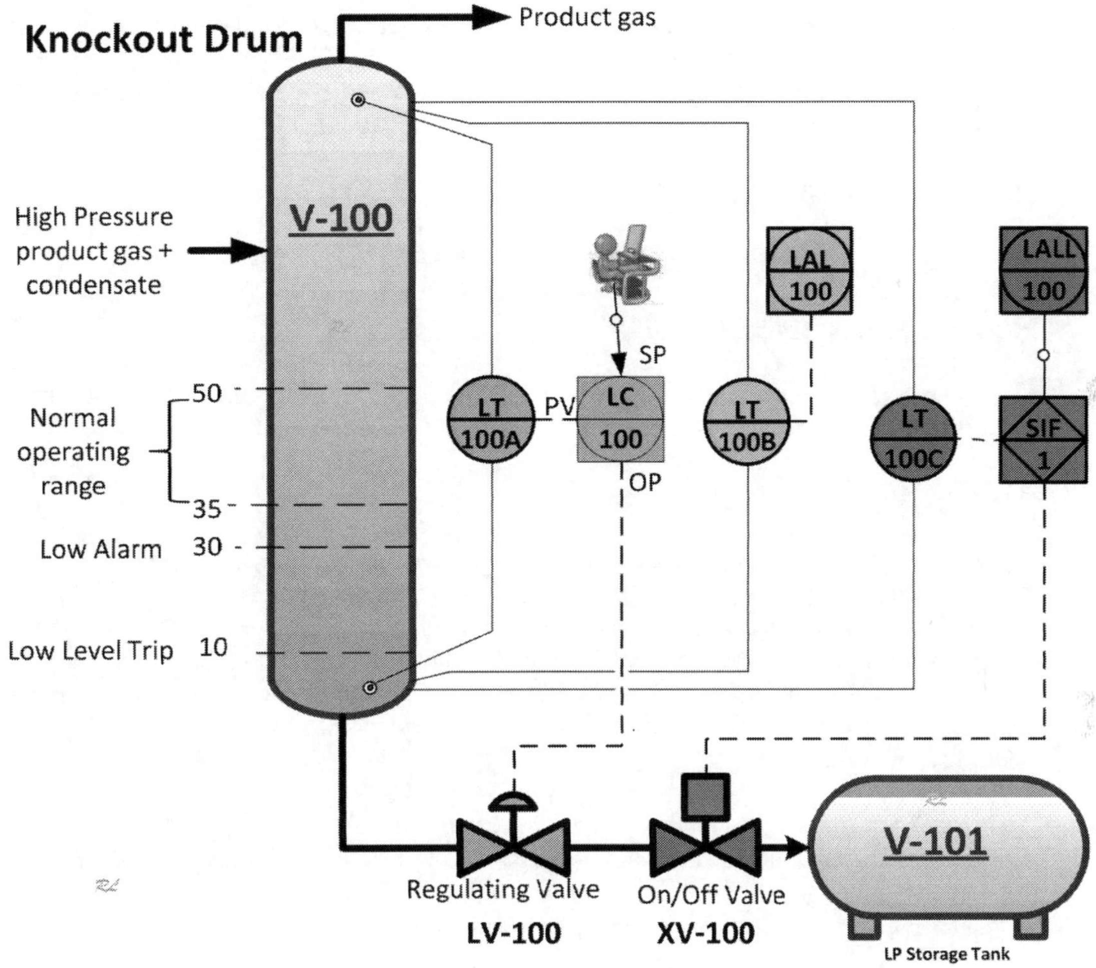

Fig 2. Knock out drum process example

Two independent protection layers are identified in the LOPA for this scenario to mitigate the risk to tolerable level.

The first protection layer is the operator response to an alarm LAL-100 implemented in BPCS with a Risk Reduction Factor (RRF) of 10. The low alarm limit for this pre-trip Safety Related Alarm is set to 30%.
$PFD_{avg} = 1/RRF = 1/10 = 1 \text{ X } 10^{-1}$

As per ISA84, the maximum risk reduction than can be assigned to operator response to an alarm implemented in BPCS is 10. It is commonly referred to as SIL0. An alarm is called as Safety Related Alarm (SRA) when operator response to the alarm is used as a protection layer and the risk reduction credit of one order of magnitude is claimed in LOPA.

The second protection layer to prevent the high pressure gas from entering low pressure system is SIF-1 with an RRF greater than 100. This SIL2 SIF is implemented in the SIS. If the level falls below the trip limit of 10%, then the on-off valve XV-100 closes. SIF is a combination of sensor, logic solver and final control element.

6. Time considerations

The response time of ISF must be less than the Process Safety Time.

6.1 Process Safety Time (PST)

Process Safety Time is the difference between the time at which the unacceptable condition occurs ($T_{CONDITION}$) and the time where unwanted event occurs (T_{EVENT}) [1].

Process Safety Time = $T_{EVENT} - T_{CONDITION}$

In the above example, there are two protection layers. The first ISF is operator response to an alarm.
The process safety time for an alarm is the time when level reaches 30% till it goes to the trip set point of 10%. The time can be calculated by dividing the volume of the knockout drum for the 20% of instrument range (difference between the alarm set point and trip set point) by the worst case flowrate of condensate when the level control valve LV-100 stays wide open.

The process safety time for the SIF can be calculated in a similar manner. It will be the time when the level reaches trip point of 10% till it goes to 0% at maximum flowrate.
Typically the Process Safety Time calculation is done by process engineers. Such calculations become easy when the process model is available.

6.2 Alarm Response Time

Alarm Response Time is the difference between the time at which the alarm condition occurs and the time when process starts responding in the direction to correct the alarm condition. It includes the sensor lag, BPCS lag, operator response time and any process lag. Process deadtime is the amount of time it takes for the process to begin reacting after corrective action.

Alarm Response Time = Sensor delay + BPCS delay + Operator response time + process deadtime

The Process Safety Time for alarm has to be greater than the Alarm Response Time. These different time elements are shown in Fig. 3.

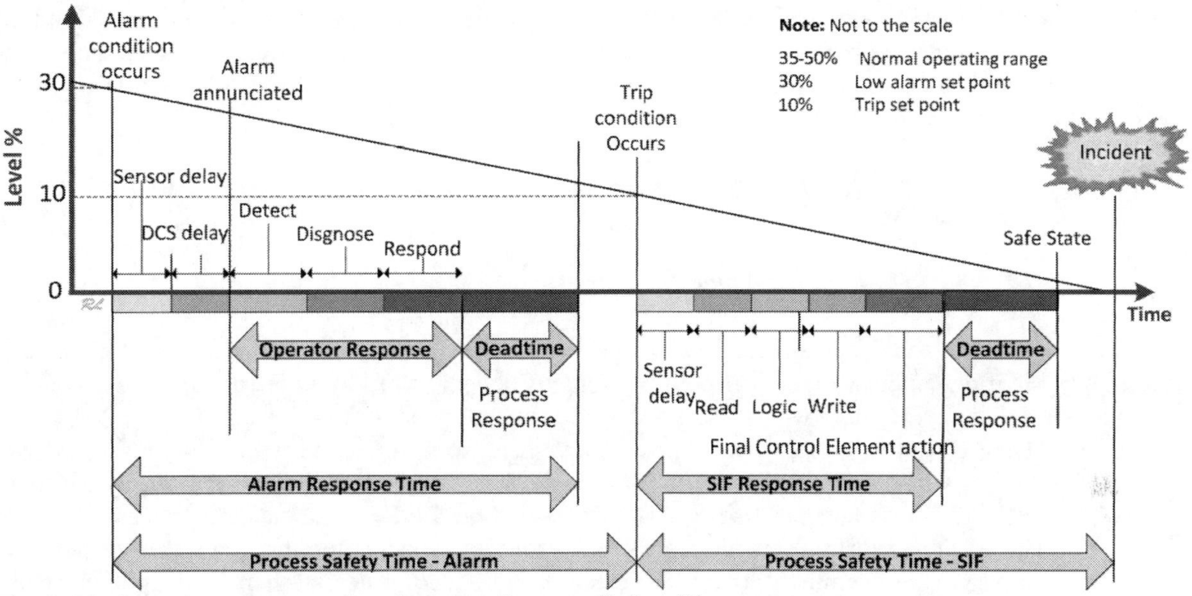

Fig. 3. Various time elements in relation to Process Safety Time

6.3 Operator Response Time

Operator response time is impacted by human factors, ergonomics, training, etc. collectively called as performance shaping factors. As per ISA18.2 feedback model of operator process interaction, the operator response time constitutes the following human interactions.

Detect: The operator becomes aware of the deviation from the desired condition. The design of the alarm system and the operator interface impact detection of deviation.

Diagnose: The operator uses knowledge and skills to interpret the information and diagnose the situation and determine the corrective action to take in response.

Respond: The operator takes corrective action in response to the deviation.

6.4 Minimum Time To Respond

This is defined in the alarm philosophy document. Each organization or plant site should develop an alarm philosophy document as defined in ISA18.2 [6]. Minimum Time To Respond is the quickest possible time to allow operator to go through detect – diagnose – respond steps. It is not physically practical to take necessary corrective actions in less than this time. Three to ten (3 - 10) minutes is most commonly used value as a minimum time to respond.

If the required operator response based on the PST for alarm is less than the Minimum Time To Respond, then no credit can be taken for the operator response to alarm as protection layer. This requirement is applicable to not just a Safety Related Alarm, but any alarm configured in the system. In such situations, various options should be reviewed to allow sufficient time of operator response. In the above process example, the simplest option is to check if the low alarm set point (LAL-100) can be increased to get more Process Safety Time.

If the alarm set point cannot be increased, using a restriction orifice to limit the maximum flow could be another option.

6.5 SIF Response Time

SIF response time includes the sensor delay, analog input card scan time, logic execution, writing output to the final control element and the time it takes for the valve to close. Typically valve closing time is most significant in these time elements. The logic may have ON or OFF delays. All these delays need to be included while calculating the SIF response time. SIF response time should be less than half of the Process Safety Time for SIF. If that is not the case, various options should be considered. The easiest option is increasing the trip limit if it is possible. Often times, adjusting trip settings does not have much impact as much faster SIF response is needed. The first step is to understand the most significant contributor to the SIF response time. Are there any logic elements with excessive delays? Can those be safely reduced? If the valve is the biggest component, are there options to install quick exhaust and/or volume boosters to improve the response? If none of the solutions are possible, selecting altogether a different valve with fast response may be necessary.

7. Instrument selection

The independency requirement of IPL greatly influences the selection of instruments and the process vessel design as well. The brownfield vs. greenfield project may also have an impact on deciding the instrumentation.

7.1 Level measurement

If there is a requirement of multiple level measurements on the vessel, and if the process taps are not enough, it is easier to influence the vessel design to add the required number of taps on a greenfield project as most likely, the vessel design is still not issued for fabrication when the initial LOPA is performed.

On a brownfield project, if LOPA determines the need for additional independent level measurements, it is unlikely that the new taps will be drilled into the vessel that is already in service. In such situation, different measurement technologies and options need to be explored. For example, if the vessel has a flange on the top, a radar level gauge or Guided Wave Radar (GWR) could be selected. Another option that requires no changes to the vessel integrity is a nuclear level gauge. Although it is expensive and prone to high maintenance and may require permits, it might be the best alternative of all the available options.

If the same process tap is to be shared by two level instruments, then common mode failure analysis using Quantitative Risk Analysis (QRA) techniques need to be performed. For clean service, if the common mode failure analysis determines that it may be possible to share the process taps, it is highly advisable to choose transmitters with advanced diagnostic features such as detection of plugged impulse lines. The type of process fluids involved and the process conditions play a big role in deciding whether a pair of common taps can be shared by two instruments. The process taps should not be shared if there is a history of impulse line plugging or possibility of deposit formation or the impulse line freezing.

7.2 Flow measurement

Orifice plate is the most common flow element used to measure the flow using differential pressure transmitters. Most often LOPA identifies multiple IPLs requiring the same flow measurement. A triple tap flange is most commonly used to provide three independent taps when independent flow measurements are required for the same flow. On a brownfield project, where only one process tap for the flow measurement is available, replacing the orifice flange is an option. Typically the whole pipe spool upstream and downstream of orifice plate is replaced. The new pipe spools with triple tap flanges could be fabricated and pressure tested in advance. It is relatively easy to replace the pipe spools during turn around.

If replacing orifice flanges is not an option, there are various other ways of additional independent flow measurement such as sonic flow meters, magnetic meters, anemometers etc. Depending on the process, an appropriate method should be selected.

7.3 Temperature measurement

When independent temperature measurements are required, it is often possible to use the measurements upstream or downstream of the vessel that are indicative of the problem. When no other measurements are available, it is possible to drill thermowells on process piping or use a new pipe spool with required number of thermowells. Using a bundle of thermocouples in the same thermowell and connecting the thermocouples to different temperature transmitters introduces a common cause failure of measurement from the same thermowell and needs to be taken into account during PFD_{avg} calculation of the SIF.

7.4 Pressure measurement

Independent pressure measurement of a vessel is not as complicated as independent level measurement as, adequate pressure taps usually exist on most vessels. It is also possible to use pressure taps on the downstream pipeline as long as there are no isolation valves between the pressure tap and the vessel.

7.5 *Final control element*

In most of the cases, final control element is the most significant contributor to the PFD_{avg}. SIL verification calculations will confirm if the selected device and test frequency can meet the target SIL of the SIF.

Increasing the proof test frequency and/or partial stroke testing options should be considered to improve the PFD_{avg} if required.

If the final control element involves solenoid valves to cut the instrument air to the regulatory control valve, it must be verified that the regulatory loop is not part of the initiating event. Other considerations include the tight shutoff requirements, time taken to complete the fail safe action and metallurgy.

7.6 *Failure rate data requirements*

The components used in SIF should be suitable for safety application with appropriate SIL rating. It is important to verify the Failure Modes, Effects, and Diagnostic Analysis (FMEDA) data sheets from the manufacturers while selecting and purchasing the instrumentation for SIF. This information is required during SIL verification calculations.

8. IPL being part of initiating event

Many times the IPL listed is part of the initiating event. If the LOPA team is not careful or lacks experience, one may find such IPLs listed in the LOPA. In the above process example, the initiating event is failure of regulatory loop to control the level. This failure may be due to a valve malfunction, frozen transmitter, impulse line plugging or human error. The operator response to alarm is used as the first IPL. If the transmitter used to generate the alarm is the same as the one used in the regulatory loop (LT-100A), then it is not a valid IPL. It is important that the alarm is generated from an independent transmitter.

9. Sharing of devices between BPCS and SIS

ISA84 clause 11.2.10 states that “A device used to perform part of SIF shall not be used for BPCS where the failure of that device results in failure of basic control function which causes a demand on SIF”.
Therefore, the same sensor used for generating alarm cannot be shared with the control function in BPCS and cannot be shared with the SIF implemented in SIS. When independent sensors are available for each function, they can be configured in fault tolerant mode to achieve higher reliability and to increase the diagnostic coverage. Some of the most commonly encountered scenarios are discussed below. These examples are for illustration purpose only. Each owner operator company is responsible for doing analysis to ensure their configurations are valid and can satisfy the independency criteria of ISA84.

9.1 Standalone SRA

When the operator response to an alarm is identified in LOPA as a protection layer, and there are no other ISFs or control functions associated with the measurement, the transmitter is wired to BPCS as shown in Fig. 4.

Fig 4. Standalone SRA P&ID representation

9.2 SRA and control function

Fig 5a shows a scenario where SRA and control function (such as PID loop) is implemented in BPCS. When the same process measurement is required for both the functions, separate transmitters should be used for alarm and control as shown in the Fig 5a. Each transmitter should be wired to separate cards in the controller or preferably separate controllers of BPCS.

The two transmitters could be configured in a BPCS as shown in Fig. 5b to improve availability, facilitate maintenance and improve diagnostic coverage. As shown in Fig. 5b, a software switch HS-100x is provided for the operator to manually change the source of input for the alarm as well as for the control function. This allows taking one of the transmitters out of service for maintenance or proof testing. Depending on the functionality in BPCS, a deviation alarm should be configured for the maintenance technician. If the difference between the readings of two transmitters is more than a pre-set threshold, a low priority deviation alarm is generated. The operator action for this alarm is typically to generate the maintenance work order for instrument technician to check the transmitters and correct the situation.

When the hand switch is used to switch the input to another source, a timer KS-100x should be started with alarm KAH-100x. If the time in switched input mode exceeds preconfigured limit, an alarm should be generated. The preconfigured timeout limit to generate warning alarm should be less than the Mean Time to Repair (MTTR) of the transmitter. If the time in switched state exceeds MTTR, then the SIF should initiate the action to put the process in safe state.

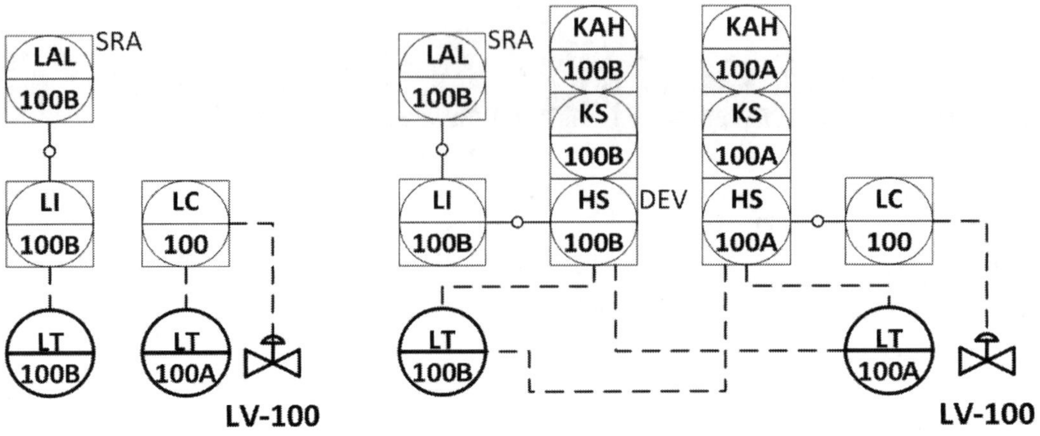

Fig 5a SRA & Control

Fig 5b Arrangement for easy maintenance and higher diagnostic coverage

Fig 5: Safety Related Alarm and control function both implemented in BPCS

9.3 SRA, control function and SIF

Fig. 6 shows a scenario where the SRA, control and SIF are all using the same process measurement. The SRA and control function are implemented in the BPCS, while the SIF is implemented in the SIS. When the available instrumentation is adequate to meet the independency criteria of each function, it is beneficial to wire it in fault tolerant configuration to improve reliability.

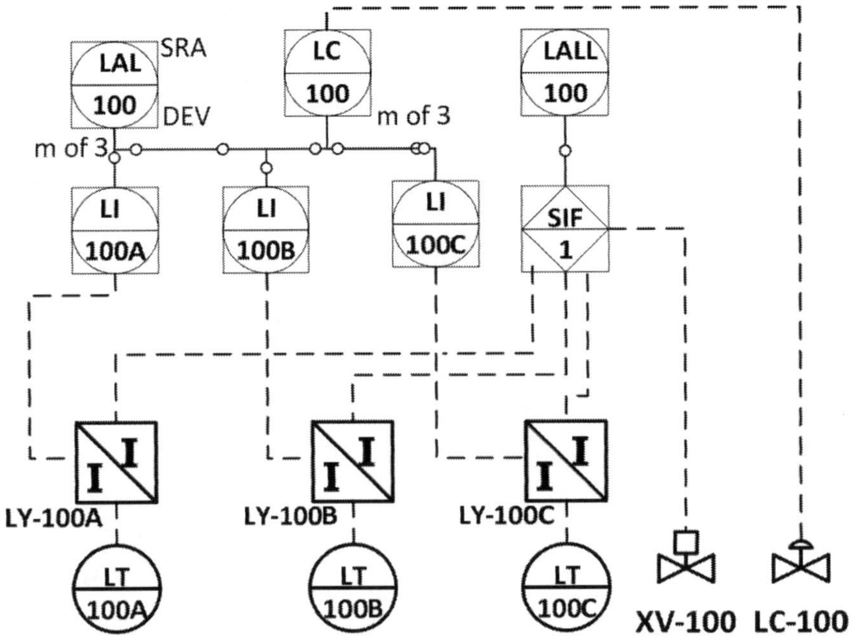

Fig 6. SRA, SIF and control function implementation example to achieve high reliability

In this example the three level transmitters using independent taps for process connection are wired to SIS through the safety certified current loop isolator and repeater. The transmitters using Highway Addressable Remote Transducer (HART) communication protocol are powered by SIS and the isolators have capability to pass through the HART signals on each channel.

SIL 1 or SIL2 SIF is implemented in SIS with 2oo3 voting logic for the level input signals. Depending on the BPCS, the actual implementation may differ. Fig 4 shows a generic representation where a middle of 3 selector block is used in BPCS which has SRA configured. Some BPCS have a standard 2oo3 block which can be used as well. Output of another middle of 3 selector block is used as a PV for the control function. When sharing the transmitters between BPCS and SIS, the following considerations should be taken into account [7].

- The failure of any hardware or software outside the SIS should not prevent any SIF from operating correctly.
- The failure of a BPCS component does not result in the initiating cause for the process hazard and the failure (or defeat/bypass) of the SIF that protects against the specific scenario under evaluation.
- The probability of common mode, common cause or dependent failures, such as plugged impulse lines, maintenance activity including bypasses, incorrectly operated line isolation valves, etc., has been adequately evaluated and determined to be sufficiently low. It is often recommended to use diverse measurement technology to reduce the common cause failure problems. A combination of differential pressure and GWR transmitter is an example of using diverse technologies to measure the same process value.
- The shared components are managed according to ISA84, including proof testing, access security and management of change.
- The sensor (e.g., transmitter, analyzer, switch) is powered by the SIS. The signal is transmitted to the BPCS by an optical isolator or other means to ensure that no failure of the BPCS affects the functionality of the SIS.

10. ISFs in BPCS

BPCS is normally the first line of defense against process excursions. Typically, three types of ISFs are implemented in BPCS:

- Regulatory control
- Alarm
- Trip (or BPCS interlock)

Risk reduction credit can be taken in LOPA for the functions implanted in BPCS as long as following guidelines are observed:

- ISF is independent of the initiating event.
- Credit for each ISF in BPCS is no more than RRF of 10 (PFD_{avg} >= 1×10^{-1}).
- No more than two ISFs are credited for the same scenario (combined RRF of two BPCS functions is no more than 100). A detailed analysis recommended in CCPS

LOPA book should be performed if credit is taken for two ISFs implemented in BPCS [8].
- Separate measurement devices and final control elements are used for each ISF. For example, the transmitters and valves used for regulatory control should be separate from those used for the trip action. Alarm should be generated from transmitter that is not used in regulatory control or trip. (They could be wired in fault tolerant configuration as shown in Fig. 5 and 6).
- The devices used in ISF for same scenario are wired to different IO cards of BPCS.
- Procedures for BPCS maintenance exist and are followed. Good Management of Change (MOC) procedures are followed for changes to BPCS logic and parameters, all changes are auditable and well documented, appropriate access control to BPCS is in place.

11. Instrument index and P&ID updates

Each IPL is evaluated in detail for its validity as part of the conceptual IPS design. It may result in modifications such as addition or deletion of instruments and sensor elements, tag name changes, logic changes and re-assignment of signals to BPCS and other IPSs. It is important to maintain and update the IO database and with appropriate comments that could prove very useful in later stages of the lifecycle. For example, tag assignment to IO cards of IPS is done at the beginning of the detail design phase of IPS. A comment to allocate the transmitters used in the same scenario to different IO cards in the system is very handy during IO allocation step. Maintaining and updating the IO index is an essential step for project change control. IO count of each IPS is an important key quantity in any project. Reporting the changes to this key quantity to project management on a routine basis is an absolute must to keep all stakeholders informed of the scope changes which may impact the project cost and schedule.

Any updates to P&ID as a result of changes to IPL should be marked on the master P&ID set and be reviewed and approved per the P&ID MOC procedures of the organization.

12. Conclusion

LOPA should address the mitigation of risks in all three categories – safety, environmental and asset protection. An IPL list should be maintained with at least the three attributes to classify the safeguards - integrity level, system and the purpose. An IPL list is very useful in allocating them to appropriate IPS.

Once the initial LOPA report is published, the next step in the functional safety lifecycle involves preparation of SRS and conceptual design of IPS. During the conceptual design, some IPLs may be found invalid. Some of the most common reasons for invalid IPLs are: process safety time, independency criteria, common mode failures and poor documentation. It is important to revisit the LOPA during the conceptual design to correct all discrepancies and publish the LOPA as issued for design. Usually it requires at least one revision to LOPA to update the findings from the conceptual design phase.

The greenfield vs. brownfield project could influence the selection of instrumentation. When independent sensors are available for each IPF, they can be configured in fault tolerant mode to achieve higher reliability and to increase the diagnostic coverage.

13. References

[1] Guidelines for Safe and Reliable Instrumented Protective Systems; Center for Chemical Process Safety, John Wiley & Sons 2007 ISBN 978-0-471-97940-1

[2] ANSI/ISA-84.91.01-2011. Identification and Mechanical Integrity of Instrumented Safety Functions in the Process Industry; International Society of Automation, Research Triangle Park, NC

[3] ANSI/ISA-84.00.01-2004 Part 1 (IEC 61511-1 Mod). Functional Safety: Safety Instrumented Systems for the Process Industry Sector - Part 1: Framework, Definitions, System, Hardware and Software Requirements

[4] Angela E. Summers, "Consider an instrumented system for overpressure protection"; Chemical Engineering Progress, November 2000

[5] ISA-84.00.01-2004 Part 3 (IEC 61511-1 Mod) Functional Safety: Safety Instrumented Systems for the Process Industry Sector - Part 3: Guidance for the determination of the required Safety Integrity Levels

[6] ANSI/ISA–18.2–2009. Management of Alarm Systems for the Process Industries

[7] Technical Report ISA-TR84.00.04-2005 Part 1 Guidelines for the Implementation of ANSI/ISA-84.00.01-2004 (IEC 61511 Mod)

[8] Layer of Protection Analysis; Center for Chemical Process Safety; American Institute of Chemical Engineers 2001 ISBN 0-8169-0811-7

Biography

Rajeev Limaye (Rajeev_Limaye@praxair.com) is Director of Control Systems & Instrumentation at Praxair, Inc. in their Global Hydrogen business unit in Houston, Texas. He has a masters degree in Chemical Engineering from IIT Bombay India and MBA from University of Houston. Rajeev has worked in process automation industry for over 25 years and holds a PE license in Texas. He is an advisor to the University of Houston Downtown for their degree program in Control Systems Engineering. Rajeev is a Certified Functional Safety Expert (CFSE) and a member of ISA84 standards committee.

Safety controls, alarms, and interlocks as IPLs

Angela E. Summers, Ph.D., P.E.
SIS-TECH Solutions
12621 Featherwood Dr. Suite 120, Houston, TX 77034
asummers@sis-tech.com

Prepared for Presentation at
American Institute of Chemical Engineers
2013 Spring Meeting
9th Global Congress on Process Safety
San Antonio, Texas
April 28 – May 1, 2013

UNPUBLISHED

Safety controls, alarms, and interlocks as IPLs

Angela E. Summers, Ph.D., P.E.
SIS-TECH Solutions
12621 Featherwood Dr. Suite 120, Houston, TX 77034
asummers@sis-tech.com

Keywords: safety controls, alarms, interlocks, SIS, BPCS, process control, layers of protection analysis

Abstract

Layers of Protection Analysis (LOPA) evaluates the sequence of events that first initiate and then propagate to a hazardous event. This semi-quantitative risk assessment technique can expose the role that automation plays in causing initiating events and in responding to the resulting abnormal operation. Automation that is specifically designed to achieve or maintain a safe state of a process in response to a hazardous event is now referred to as safety controls, alarms, and interlocks (SCAI).

Guidelines for Initiating Events and Independent Protection Layers addresses four basic types of SCAI: safety controls, safety alarms, safety interlocks, and safety instrumented systems (SIS). This paper discusses the design, operation, maintenance, and testing practices necessary for SCAI to be considered as independent protection layers (IPL). It also provides guidance on claiming multiple layers of protection in the basic process control system.

1. What are SCAI?

SCAI or *Safety Controls, Alarms, and Interlocks* are process safety safeguards implemented with instrumentation and controls, used to achieve or maintain a safe state for a process, and required to provide risk reduction with respect to a specific hazardous event (ISA 84.91.01-2012). They are the safety subset of instrumented protective systems (IPS), which are discussed in *Guidelines for Safe and Reliable Instrumented Protective Systems* (CCPS, 2007). An IPS is any instrumented system that addresses risk related to health and safety effects, environmental impacts, loss of property and business interruption costs. As shown in Figure 1, there are many terms that can be used to further classify SCAI.

Figure 1. Classification of SCAI (adapted from ISA 84.91.01-2012)

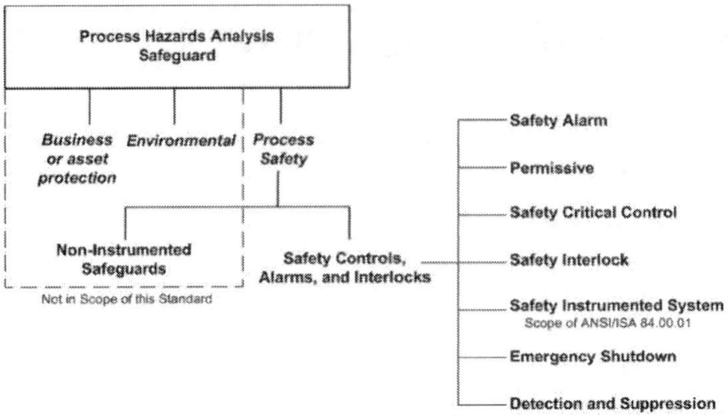

As with any IPL, SCAI should be independent of the initiating cause and any other IPL used for risk reduction for a particular hazardous event. To understand its capability in stopping the event, the SCAI function must be clearly identified, e.g., what does it detect, what is the setpoint, how does it affect the process, how fast does it need to act, and how does the process respond to its action. Typically, SCAI have sensors to detect the process condition, a logic solver to determine what action to take based on the process condition, and final elements to take action on the process (Figure 2). The decision and action can be manual (e.g., operator response to a safety alarm) or automatic (e.g., via continuous control, interlock, or SIS). SCAI also includes other interconnected equipment, such as human interfaces, wiring, process connections, communications, and utilities, which are necessary in order to achieve the desired functionality and to operate and maintain the SCAI.

Figure 2. Example elements required for SCAI successful operation

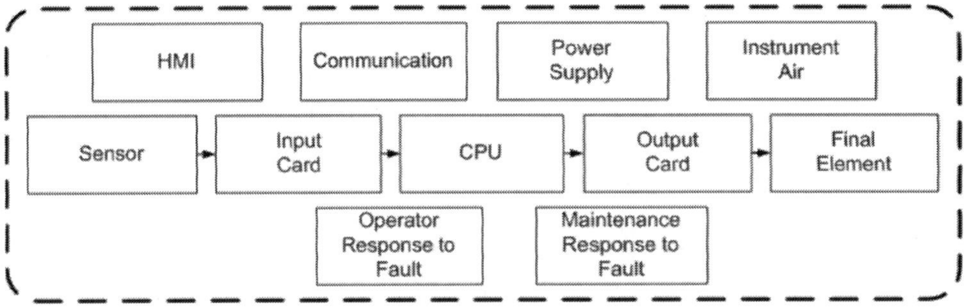

SCAI are complex systems requiring many different devices to operate successfully in order to stop the hazardous event propagation. The logic solver technology is typically implemented using electrical, electronic, or programmable electronic equipment. Older installations may use mechanical switches, which change state based on manipulating pneumatic (or hydraulic) flow and pressure. Field device technology ranges the gamut from simple switches to complex analyzers. While there is no limit on the designer's choice, the technology must provide the desired level of risk reduction in the operating environment and under the user's management system. As technology moves from the drawing board to sustained operation, the proof of integrity must also move from estimated performance to historical operation as evidenced through site mechanical integrity records. Historical demonstration of successful prior use is essential for assuring that equipment is fit for service.

The ability of SCAI to achieve the desired functionality and the claimed risk reduction is limited by the performance achieved by the installed equipment. Each SCAI device has distinct failure modes and has a failure rate that contributes to the overall SCAI performance. For this reason, equipment is selected based on its expected capability in the operating environment, but the SCAI is judged by the cumulative capability of the system. The actual level of risk reduction achieved should be substantiated by mechanical integrity and human reliability data. The records associated with any SCAI should confirm that the equipment operates as specified during all intended operating modes. Failure tracking and analysis is important to verify hazards and risk analysis assumptions and to support continuous improvement.

2. SCAI practices

According to *Guidelines for Initiating Events and Independent Protection Layers*, SCAI specifications should include:

- Functional requirements.
- Configuration, installation, and maintenance requirements to achieve and sustain the claimed performance.
- Failure modes, means used to detect these failure modes, and expected system and operator response to detected failure.
- Compensating measures required to continue safe operations with faulted SCAI
- Conditions required to safely bypass the SCAI (includes override or manual operation) and any compensating measures to be in place during bypass.
- Conditions required for safe reset of the SCAI.

SCAI equipment should be sufficiently robust to withstand the actual stresses of the operating environment and provide the required integrity and reliability. Some technologies used in control applications may not be acceptable for SCAI due to inadequate installation history, poor in-service reliability/integrity, or unpredictable/erratic behavior. For example, wireless technology as defined by ISA 100.11a is not generally considered an acceptable technology for executing SCAI, but may be used for monitoring, status or diagnostic communication (CCPS, 2013).

An equipment list should be maintained which identifies SCAI equipment by a unique designation, such as a tag number or equipment ID that is traceable to the mechanical integrity requirements necessary to maintain the required integrity and reliability throughout its life. Mechanical integrity includes a variety of activities, such as inspection, calibration, preventive maintenance, repair/replacement, and proof testing.

Inspections and proof tests are conducted as necessary to identify and correct equipment deficiencies. Visual inspection conducted during plant rounds may need to be conducted more frequently based upon process and environmental severity. Inspection and proof test are performed according to a written procedure that ensures that the required operation is demonstrated to the degree feasible under safe conditions. Typical information recorded includes as-found/as-left conditions, the tester, when tested, the procedure and equipment used, and calibration records.

It is inherently safer to implement SCAI equipment such that it takes the safe state action on failure rather than fails to an alarm (or dangerous) state. Examples of failures to consider are loss of signal, out of range values, loss of communications, power failure, instrument air failure, and loss of other utilities. If it is intended to continue operation of the process with out-of-service SCAI equipment, the risk should be assessed and compensating measures should be provided to address increased risk. Out-of-service periods should be tracked and minimized to the degree possible through equipment design and maintenance practices.

Documented procedures, training, and auditing ensure effective management of change control for the hardware and software. As with any IPL, SCAI are only bypassed (e.g., suppressed or setpoint changed) with administrative approval and after deployment of any necessary compensating measure. Administrative approval may involve formal MOC, bypass control procedures, or special operating procedures.

Information about SCAI should be included in operating procedures, since the operation of SCAI affects the process or requires that the operator take action. When programmable electronic systems are used, management of change and access security procedures are needed to ensure that changes to the embedded software and application program are reviewed and approved. Any change potentially affecting SCAI should be examined to determine the requirements for validating the functionality prior to the SCAI being placed in service. Access to engineering interfaces should also be controlled to reduce human error and prevent compromise of the system.

3. BPCS versus SIS

The type of equipment and the design and management practices used to implement SCAI limits achievable performance. SCAI can be implemented using basic process control system (BPCS) or safety instrumented system (SIS) equipment. To be counted as IPL, SCAI are designed and managed to achieve an order-of-magnitude risk reduction (PFD $\approx$ 0.1). Because of the typical design and management of BPCS equipment, SCAI implemented in a single BPCS is limited to a single order-of-magnitude risk reduction. Higher claims (e.g., PFD $\leq$ 0.01) can be supported by a SCAI designed and managed in compliance with IEC 61511.

SCAI claimed as IPL are limited to PFD = 0.1, unless they are designed and managed per IEC 61511. IEC 61511 describes the conditions necessary to allow two BPCS loops to be credited for the same hazardous event with an overall claim of PFD = 0.01. IEC 61511 identifies independence and separation of non-SIS (including control, diagnostics, and SCAI) and SIS as a critical requirement; otherwise the implementation must meet the full requirements of IEC 61511.

3.1 BPCS

3.1.1 Type

BPCSs are most commonly implemented in pneumatic control loops, programmable logic controllers (PLCs), distributed control systems (DCS), discrete control systems (e.g., on/off, relay), and single loop controllers (SLCs).

3.1.2 Practices

The design and management of the BPCS is usually focused on ensuring that it can reliably perform the control functions that maintain normal operation, such as PID (proportional-integral-derivative) and batch or sequential controls. The response to detected faults and momentary loss of communication is generally continued process operation using the last 'good' value. Failure of the BPCS in executing normal control functions is one of the leading causes of hazardous events. A BPCS is also capable of executing other functions such as alarming and process shutdown when special design and management practices are followed.

3.1.3 Performance capability

A BPCS's performance is limited by the hardware and software design. Most BPCS have little online redundancy in the input/output cards (I/O) or in the CPU and have limited diagnostic coverage of these components. When redundancy is available, it is generally a backup processor that does not operate until internal diagnostics (typically not themselves redundant) detect a problem. Redundant BPCS are generally comprised of hot standby or hot swappable processors, which significantly limits the potential performance claim.

With multiple devices required to operate correctly, individual devices within the SCAI usually need to achieve an installed performance an order of magnitude better than the required system performance. For example, if it is desired to claim that a BPCS loop achieves a PFD = 0.1, the BPCS controller would need to have a PFD $\approx$ 0.01. Industry data (ISA TR84.00.04 and PDS Data Handbook, 2010) does not support a PFD $\approx$ 0.01 claim for controllers that are not designed and managed in compliance with IEC 61511.

3.2 SIS

3.2.1 Type

SIS are most commonly implemented using discrete control systems (e.g., on/off relay) and programmable electronic systems.

3.2.2 Practices

The design and management of the SIS ensures that it can reliably take action to achieve or maintain a safe state of the process to stop propagation of the hazardous event. In the process industrial sector, the term SIS applies to instrumentation and controls designed and managed in accordance with international standard IEC 61511 (IEC, 2003).

Use of programmable electronic systems (PES) requires diagnostics on input and output cards to detect stuck-on or stuck-off conditions and on the main processor to detect a stuck or locked-up program execution. These diagnostics are necessary for the programmable electronic system to meet the minimum requirements for SIL 1 (PFD<0.1).

A PES relies on software - embedded and application – to execute the internal and user functions of the system. This software is patched, upgraded, and updated under a revision control system that ensures adequate testing and validation. Changes to embedded software require full system validation (ISA TR84.00.03), while changes to the application program must be tested based on their traceable impact to the system (IEC 6511).

Data communication and engineering interface to PES should be separated from other credited SCAI. Any interconnections that can compromise performance should be protected with firewalls and strict access security provisions.

Diagnostics are not required for hardwired logic solvers, such as relays or trip amplifiers, which have low failure rates and well-defined failure modes. Hardwired logic solvers are separate and diverse from other control systems and are not vulnerable to cyber-security risks or data highway corruption.

3.2.3 Performance capability

While it has many prescriptive requirements, IEC 61511 is fundamentally a performance-based standard that requires the establishment the safety integrity level (SIL). The target SIL is normally determined using some type of risk assessment. For example, a LOPA determines the PFD that the SIS needs to meet in order for the event risk to be reduced below the risk criteria. The required PFD establishes the target SIL. Table 1 shows the relationship between SIL and the PFD ranges.

Table 1. Safety Integrity Level Relationship to PFD and Risk Reduction (CCPS, 2013)

SIL Level	PFD is greater than:	But less than:	Risk Reduction
SIL 1	0.01	0.1	1 order of magnitude
SIL 2	0.001	0.01	2 orders of magnitude
SIL 3	0.0001	0.001	3 orders of magnitude

When claiming SIL 2 and 3, the standard has minimum redundancy requirements to protect against systematic errors. Redundant devices and signal paths using multiple sensors, controllers, or final elements to provide the same function may be necessary to ensure SIL 2 and are required to ensure no single points of failure for SIL 3.

4. The 4 SCAI and LOPA

SCAI are instrumented safeguards identified during the hazard identification and risk assessment that are necessary to achieve risk reduction related to a potential process safety event. As discussed earlier, SCAI can be referred to by many other names – the key to the identification is the function's purpose with regard to the event. This paper divides SCAI into four types of systems: control, alarms, interlocks, and SIS (CCPS, 2013).

IEC 61511 restricts the PFD claim that can be made for a BPCS loop to $\geq$ 0.1, while the PFD claim for SIS can be < 0.1. For the purposes of LOPA, both are given the rounded value of 0.1. Regardless of the PFD performance claim, SCAI must be under an appropriate mechanical integrity program that includes documentation, procedures, and administrative controls (ANSI/ISA 84.91.01).

4.1 Safety Controls

These normally operate to support process (or regulatory) control and during a process upset acts to stop the hazardous event progression. These controls are not typically designed or managed in accordance with IEC 61511, so they may also be referred to as BPCS IPL. The risk reduction claimed for a safety control (which does not conform to IEC 61511) must be below 10 (IEC 61511 Clause 9.4.2). In LOPA, a risk reduction factor of 10 or PFD = 0.1 is used.

4.2 Safety Alarms

These initiate an alarm that requires that the operator take action in accordance with a response procedure to stop the hazardous event propagation. The risk reduction claimed for a safety alarm (which does not conform to IEC 61511) must be below 10 (IEC 61511 Clause 9.4.2). The safety alarm includes not only the alarm, but also the interfaces and final control elements used by the operator to take the required action. Refer to ISA 18.2 and IEC 61511 for additional guidance related to the instrumentation and control design. Operator reliability in responding to the alarm should also be considered in making claims for safety alarms. Operator knowledge of required actions and the capability to take timely action should be validated using tests, drills, or simulated environments. When these conditions are met, a risk reduction factor of 10 or PFD = 0.1 is used in LOPA.

4.3 Safety Interlocks

These are instrumentation and controls that take automatic action to achieve or maintain a safe state of the process when a process variable reaches a defined limit. When an interlock is not designed and managed in accordance with IEC 61511, it may also be referred to as a BPCS IPL. The risk reduction claimed for a safety interlock (which does not conform to IEC 61511) must be below 10 (IEC 61511 Clause 9.4.2). With proper design and management, a risk reduction factor of 10 or PFD = 0.1 may be used. When the safety interlock is designed and managed per IEC 61511, it is an SIS.

4.4 SIS

Safety instrumented systems consist of instrumentation and controls designed and managed in accordance with IEC 61511 (IEC, 2003) that take action to achieve or maintain a safe state of the process in response to process demand. Unless the BPCS is an SIS, the SIS must be independent and separate from the BPCS to the extent that the safety integrity of the SIS is not compromised (IEC 61511 Clause 11.2.4). Safety instrumented systems are typically designed to operate in low demand mode but may operate in high demand/continuous mode.

The ISA84 committee has developed a series of complementary technical reports to provide guidance and practical examples related to various SIS topics and applications. Three of these technical reports, ISA-TR84.00.02, ISA-TR84.00.03, and ISA-TR84.00.04, provide a comprehensive overview of the SIS lifecycle and associated requirements.

5. LOPA - Approach A versus Approach B

In the initial LOPA guideline (2001, CCPS), two approaches were presented for considering the BPCS in LOPA. Approach A allowed the identification of the BPCS as providing an IE or an IPL, while Approach B provided the basis for additional analysis of the BPCS design and management system.

Whether choosing Approach A or B, special consideration should be given to situations where low-demand mode functions are implemented in a BPCS. It is inherently safer to implement low-demand mode functions in a separate and independent SIS that is specifically designed and managed for safety interlocks. Further, some application-specific practices require that safety interlocks be implemented in an independent SIS.

5.1 Approach A

The key criteria are (1) the initiating cause is not related to the BPCS and (2) the BPCS IPL is properly designed and managed to claim one order-of-magnitude risk reduction. This approach takes the position that a single BPCS loop failure invalidates all other functions that could be implemented in BPCS. It is used for LOPA because its rules are clear and conservative. It provides a high level of protection against common cause failures between the BPCS and SCAI.

5.2 Approach B

This approach potentially allows a maximum of two loops within BPCS to be identified for the same scenario. Approach B assumes that each BPCS is designed and managed such that it is capable of supporting the specified function and is sufficiently independent from the other such that the likelihood (probability) of simultaneous failure is low enough to support another order-of-magnitude risk reduction. Approach B requires that the analyst be knowledgeable in the BPCS design architecture, has adequate data available on the actual performance of the BPCS, and understands how to identify and account for common cause failures between the different automation layers.

Manufacturer claims regarding an individual device are not sufficient to justify sharing the device between two BPCS IPL or between BPCS IPL and BPCS IE for the same hazardous event.

5.2.1 Analytical Approach

LOPA is not the best tool for detailed evaluation of whether the BPCS architecture can support two loops. Analysis and test data should demonstrate that the two particular BPCS are designed and managed in a manner such that the two BPCS loops, in combination, can achieve the overall performance claim of ≈ 0.01/year. Consider use of advanced quantitative analysis techniques, such as fault tree analysis, to ensure that common cause (including systematic errors) is assessed fully. Since a BPCS failure could result in the simultaneous loss of control and safety protection, carefully examine the independence and separation of the control system and the SCAI. When strict independence is not maintained, the analysis can become quite complex necessitating higher expertise and knowledge of automation design.

If the risk reduction claimed for a BPCS is greater than 10, IEC 61511 Clause 9.3.3 requires that the BPCS be designed and managed as SIS.

At a minimum, the system analysis should include (CCPS, 2013):

- Assessment of potential common cause, common mode, and systematic failures between the BPCS IE and IPL or the two BPCS IPLs to determine that their impact is sufficiently low as compared to the performance claim.
- Written specification covering all functions within the system, e.g., instrument diagrams, P&IDs, loop diagrams, functional specifications.
- For the Generic Data approach to validation, the assessment of previous historical performance of the BPCS logic solver, input/output cards, sensors, final elements, human response, etc. (*NOTE: manufacturer information must be examined critically to ensure that it applies to situations similar to a particular installation*).
- For the Site-Specific Data approach to validation, evaluation of inspection, maintenance, and test data over a significant period to demonstrate that the system achieves the performance claim.
- Assessment of access security measures for hardware and software.
- Management of change and revision control for the hardware and software, including setpoints, fail configuration, and operator overrides.

5.2.2 Approach B – Example Configurations

Two independent and separate BPCS controllers can be implemented such that there is no communication (or interconnection) between the two BPCS loops (e.g., IEC 61511-2). This architecture is an inherently safer design, as it is physically separate as well as independent, so it provides a high degree of protection against data corruption and inadvertent software changes. This architecture can theoretically achieve the 0.01 PFD claim for the overall performance of the two BPCS loops if the two systems are designed and managed according to good engineering practices.

> *A maximum of two BPCS controllers (when designed and managed in accordance with good engineering practices) can be credited in LOPA for the same hazardous event as an initiating cause and IPL or as two IPL (CCPS, expected 2013).*

Figure 3. Illustration of two BPCS functions credited for the same scenario per IEC 61511 (Adapted from Figure 2, IEC 61511, 2003)

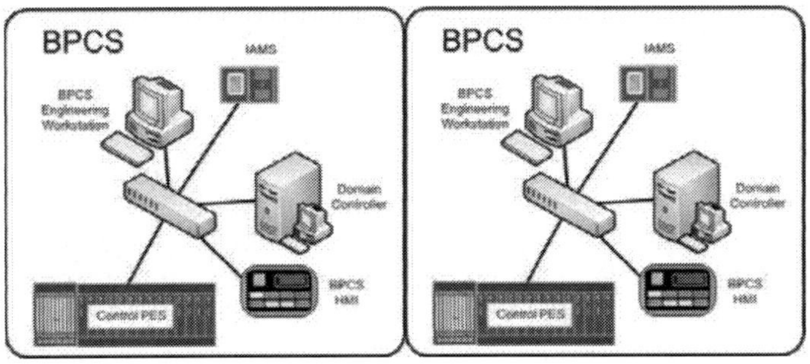

CCPS *IPS* (CCPS, 2007) recognized that the controllers could be functionally separate while safely sharing information. An example of this architecture (Figure 4) consists of separate controllers that communicate in a secure manner using firewalls and communication controls. When the controllers share data or interfaces, integrity of the two BPCS loops must be assured with rigorous management of change and access control. Ensuring cyber security and software/data integrity necessitate special design and management practices (dISA TR84.00.09 and ANSI/ISA 99). This architecture can also theoretically achieve the 0.01 PFD claim as long as the two systems are designed and managed according to good engineering practices.

Figure 4. Illustration of two BPCS functions credited for the same scenario per CCPS IPL/IE (Adapted from Figure 5.5, CCPS, expected 2013)

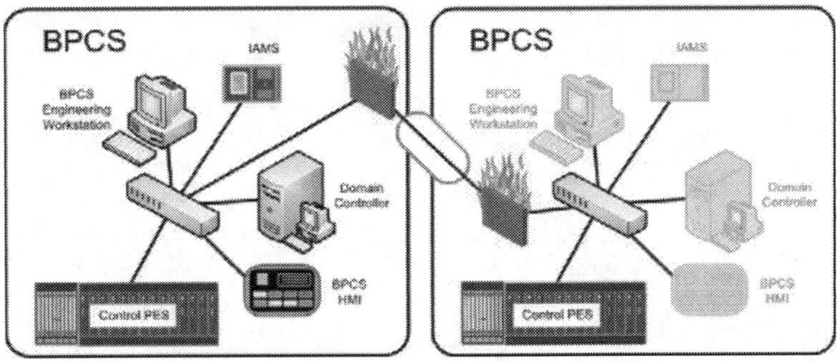

When both loops are implemented in the same BPCS (Figure 5), the shared equipment is generally insufficient for crediting the two BPCS loops for the same hazardous event. The CCPS LOPA book (CCPS, 2001) indicated that the BPCS CPU had at least two orders of magnitude better performance than the field devices. More recent field data demonstrates that the failure rate of a typical BPCS logic solver is greater than 0.01/year (PDS Data Handbook, 2010). In fact, industry and manufacturer data indicates a random hardware failure rate for typical BPCS equipment (such as DCS/PLCs) in the range of 10^{-5}/hr to 10^{-6}/hr (ISA TR84.00.04-2005 Appendix F.2.1), so the BPCS performance is equivalent to many other electrical, mechanical, or programmable electronic devices. Therefore, it is difficult to substantiate that a single BPCS qualifies for Approach B with a claimed failure rate of $\approx$ 0.01/year without advanced quantitative techniques.

When the two loops share equipment or support systems that can cause the dangerous failure of the integrated BPCSs, the resulting integrated system should be analyzed as a single BPCS. A BPCS architecture that uses a primary and hot backup controller is not sufficient for crediting as two independent controllers due to potential common cause failure. For example, a hot backup controller shares common equipment with the primary controller, such as the backplane, firmware, diagnostics, and transfer mechanisms. A hot backup controller improves operational reliability by reducing the likelihood of spurious process shutdown by a factor of perhaps 2 to 3, but it is not sufficient for claiming a dangerous failure rate $\approx$ 0.01/year.

Thus, designing a PE controller to achieve a PFD less than 0.01 requires advanced diagnostics, fault tolerant architectures, and rigorous software controls. IEC 61511 Clause 11.5 provides criteria and considerations for achieving this level of performance. Normal and customary BPCS practices will not meet these requirements.

Figure 5. Illustration indicating what is necessary to achieve ≈ 0.01/yr in a single controller

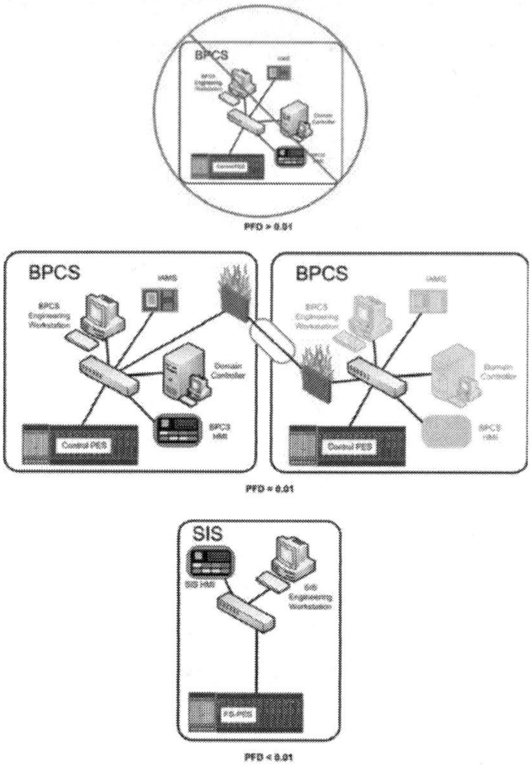

6. Summary

SCAI are an important subset of the process safety safeguards used to reduce the risk of hazardous events. The instrumentation and control systems need rigorous design and management practices to ensure that they are available when rare hazardous events begin to propagate. There are many words that can be used to identify, describe, define, or otherwise classify SCAI. As a rose is a rose, if it is an instrument or control, is process-safety related, and is required for risk reduction, it is SCAI.

When considering use of BPCS for SCAI, the first thing to remember is that risk reduction is not free. It requires effort. Getting an order of magnitude risk reduction from the BPCS is hard. Independence and separation require special design and management practices, focusing on eliminating single points of failure and human error. Claiming two orders of magnitude from two independent BPCS controllers must be supported with quantitative analysis (such as fault tree) that accounts for real world performance and human errors. Claiming two orders of magnitude from a single BPCS controller requires that the BPCS be designed and managed according to IEC 61511.

7. References

[1] CCPS (2007), Guidelines for Safe and Reliable Instrumented Protective Systems, Hoboken, NJ, John Wiley & Sons.

[2] CCPS (expected 2013), Guidelines for Initiating Events and Independent Protection Layers, Hoboken, NJ, John Wiley & Sons.

[3] ANSI/ISA 84.91.01 (2012), Identification and Mechanical Integrity of Safety Controls, Alarms, and Interlocks in the Process Industry. Research Triangle Park, NC, Print.

[4] ANSI/ISA 100.11a (2011), Wireless system for industrial automation: Process control and related applications, Research Triangle Park, NC, Print.

[5] IEC 61511 (2003), Functional safety - Safety instrumented systems for the process industry sector, Geneva, Switzerland, Print.

[6] ANSI/ISA 18.2 (2009), Management of Alarm Systems for the Process Industries, Research Triangle Park, NC, Print.

[7] ISA TR84.00.02 (2002), Safety Instrumented Functions (SIF) Safety Integrity Level (SIL) Evaluation Techniques, Research Triangle Park, NC, Print.

[8] ISA TR84.00.03 (2012), Guidance for Testing of Process Sector Safety Instrumented Functions (SIF) Implemented as or Within Safety Instrumented Systems (SIS), Research Triangle Park, NC, Print.

[9] ISA TR84.00.04 (2012), Guidelines for the Implementation of ANSI/ISA-84.00.01-2004 (IEC 61511 Mod), Research Triangle Park, NC, Print.

[10] CCPS (2001), Layer of Protection Analysis: Simplified Process Risk Assessment, Hoboken, NJ, John Wiley & Sons.

[11] ISA TR84.00.09 (expected 2014), Electrical/Electronic/Programmable Electronic Systems (E/E/PES) for Use in Process Safety Applications Security Protection Layers and Considerations Related to SIS, Draft 7.4, Research Triangle Park, NC, Print.

[12] ANSI/ISA 99.00.01 (2007), Security for Industrial Automation and Control Systems, Research Triangle Park, NC, Print.

[13] SINTEF (2010), Reliability Data for Safety Instrumented Systems Data Handbook, Trondheim, Norway, Gruppen.

Journey to Process Safety Excellence

Shanahan Mondal
Marathon Petroleum Company LP
Texas Refining Division
502 10th Street South
Texas City, TX 77590
smondal@marathonpetroleum.com

Brian McPherson
Marathon Petroleum Company LP
Texas Refining Division
bmmcpherson@marathonpetroleum.com

Prepared for Presentation at
American Institute of Chemical Engineers
2013 Spring Meeting
9th Global Congress on Process Safety
San Antonio, Texas
April 28 – May 1, 2013

UNPUBLISHED

Journey to Process Safety Excellence

Shanahan Mondal
Marathon Petroleum Company LP
Texas Refining Division
502 10th Street South
Texas City, TX 77590
smondal@marathonpetroleum.com

Brian McPherson
Marathon Petroleum Company LP
Texas Refining Division

Keywords: process safety leadership, process safety excellence

Abstract

This is a perfect time to call for action and focus on the reduction of process safety events. An organization that wants to achieve process safety excellence must look beyond compliance. This paper details how Marathon Petroleum Company's (MPC) Texas City refinery took on several initiatives to move the needle on process safety in the right direction.

1. Introduction

The refining and petrochemical industries have come a long way since the OSHA PSM standard was enacted in 1992. However, we are still seeing process safety events take the toll on the lives of our loved ones and property. There is still progress that we can make and therefore now is a good time for a call for action to further reduce process safety events. What are some ways we can proactively take steps towards achieving "process safety excellence"? Strong performing process safety programs are on the right path to achieving excellence.

This paper illustrates how a small Gulf Coast refinery embarked on the journey toward process safety excellence. Marathon Petroleum Company's Texas City refinery has a 76,000 bpd throughput and is one of seven refineries within the company. In order to be a sustainable business, the Texas City refinery must make efficient use of the people and resources. This story of the MPC Texas City refinery takes us several steps beyond process safety compliance. Folks throughout the refinery were able to embrace several new initiatives in process safety with that "can do" attitude Texans are known for. As a team this small plant has helped move the needle on process safety within our industry.

The main sections of this paper include:

- Everyone is a Process Safety Leader!
- Employee engagement
    - LOPC reduction efforts

  - Process Safety Bulletins
  - Process Safety Roundup
  - Celebrating our wins
- Industry outreach
  - API site assessment
  - AFPM Regional networks

2. Everyone is a Process Safety Leader!

2.1 Why is this important?

Many organizations have set aggressive goals to have ZERO process safety events. For these goals to develop momentum and come to fruition, they must reach the folks in the company that "have the boots on the ground" and are actively engaged in regular process safety activities. These activities include:

- Operating process units within safe operating limits
- Performing tasks to support the mechanical integrity of process equipment
- Identifying and mitigating hazards through process hazard analysis
- Auditing process safety programs for compliance with OSHA regulations
- Performing management of change (MOC)

An organization in which the leadership team empowers the entire workforce to be process safety leaders is one that is on the path to excellence.

2.2 Background

MPC partnered with Stars International to develop a "Leadership Skills for Extraordinary Safety Performance" course several years ago. Marathon employees and contractors participated in this "Safety 1" course and learned new ways to promote personal safety. One of the fundamental concepts taught in this class was that effective communication skills play a key role in achieving world class personal safety programs. The course focused on participants being able to identify which of the four personality quadrants they tended to define as their comfort zone: Controller, Thinker, Expressive, or Team Builder. The participants were then taught how to communicate effectively with others that were in different personality quadrants.

This course taught MPC employees and contractors various leadership skills to promote personal safety. Some of these include the following:

- Permission and pledge : "You have my permission to stop me if I am doing something unsafe and I pledge to give you the same feedback"
- Be in the moment : Safety outcomes are a function of choices we make in the moment
- Coaching others to be safe

The Safety 1 basic skills and refresher courses have been the foundation of Marathon's personal safety programs for the past five years. Marathon saw how this course was able to shift the personal safety culture in a positive direction.

2.3 What was done?

MPC partnered again with Stars International to build on this foundation but this time shifted the focus to process safety. The basic tenets of effective communication among different personalities still hold true to improve process safety. The key learning from the new course was that "Everyone is a Process Safety Leader! "

MPC Texas City hosted a pilot of this Process Safety Leadership course. A diverse group of employees were chosen as course participants. The refinery manager, department managers, refinery operators, lab personnel, training specialists, and support personnel were examples of those represented in the student demographics. The intent of inviting a diverse group of participants was so they could accept that process safety was everyone's responsibility.

The course highlighted a need to maintain a sense of vulnerability. Without a sense of vulnerability, complacency can set in. To illustrate how vulnerable we were, the course used a large scale Swiss cheese model to visually demonstrate how holes in the layers of protection can lead to catastrophic process safety events. Examples of real industry process safety events were used to make these personal to the course participants.

Once we had instilled a sense of vulnerability in the participants, they were given the task of figuring out how they play a role in blocking the holes in the Swiss cheese process safety model. We distilled down employee functions within the organization to 15 different categories. Each employee was given a placard with their category and on it there were various tasks they did that impacted the 14 elements of process safety. The "A-ha!" moment came in the course when participants realized what specifically they could do to prevent process safety events. A majority of the course participants found that they could block several holes in the process safety model through different tasks they were associated with. To make this even more personal, each participant was given a cardboard circle with a picture of their face on it. Through different interactive sessions during the class, they put their "hole-blocker" in the Swiss cheese model to cover up the holes. By the end of the class, the diverse group of participants had covered all the holes in the model and prevented catastrophic process safety events!

The take home message for course participants was that everyone can be a process safety leader by realizing that if they are not vigilant, holes can form in the Swiss cheese process safety model.

.

3. Employee Engagement

3.1 Loss of Primary Containment (LOPC) reduction initiative

3.1.1 Why is this important?

A site culture where small leaks are allowed to go unaddressed can lead to more catastrophic process safety events. A sense of vulnerability helps us realize that small leaks must be mitigated expeditiously in order to prevent larger incidents. An organization that focuses on a reduction in loss of primary containment (LOPC) events is on the right track to process safety excellence.

The reporting of LOPC events and subsequent incident investigations to address root causes is helpful in preventing process safety events. Facilities in our industry operate continuously 24 hrs a day 7 days a week. In order to capture the LOPC events that can occur at any given time, it is important to empower the entire shift workforce with the tools to identify and report LOPC events.

3.1.2 Background

Marathon Petroleum has adopted a suite of process safety metrics in accordance with API Recommended Practice 754 "Process Safety Performance Indicators for the Refining and Petrochemical Industries". This practice allows process safety incidents to be classified into different categories as shown in Heinrich's triangle below:

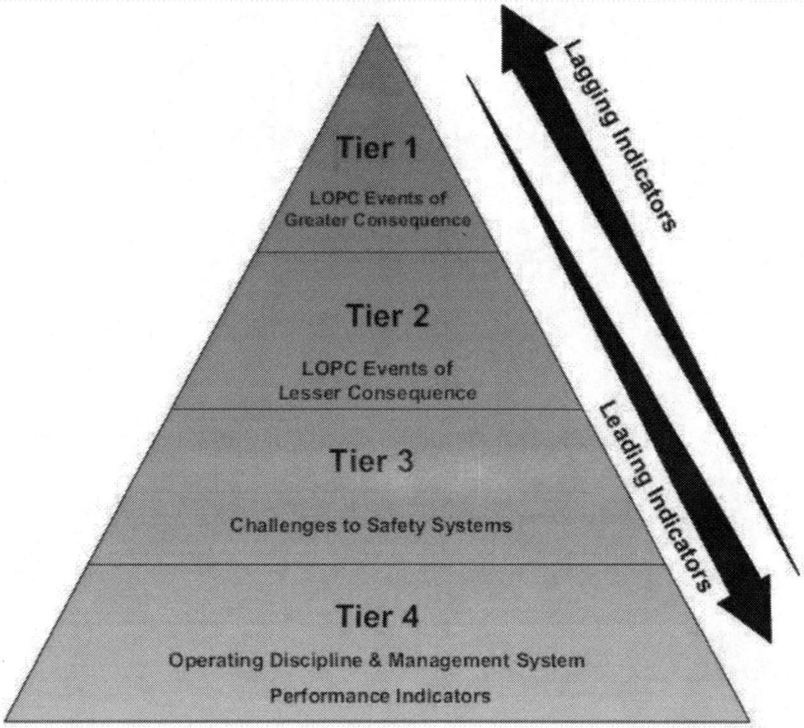

The PSE Tier 1 and Tier 2 events are lagging indicators of process safety while the Tier 3 and Tier 4 incidents are leading indicators. Small LOPC events typically are categorized in the Tier 3 bucket. Tier 3 events are leading indicators that could have led to more severe PSE Tier 1 and 2 events if conditions were different. If we keep highly hazardous materials in the pipes, there is less of a chance they will lead to the catastrophic process safety events.

MPC's personal safety culture is such that first aids and near misses are promptly reported. We have attributed the reporting of first aid injuries to preventing more severe personal injuries. Process safety is similar to personal safety in this sense. One can think of the LOPC Tier 3 incidents as precursors to Tier 1 and Tier 2 events. Focusing on keeping material within its proper designated process boundaries (eg. piping and vessels) can help prevent the catastrophic process safety events.

3.1.3 What was done?

MPC Texas City adopted a slogan "Let's Keep it in the Pipes!" to keep LOPC reduction in the forefront of the minds of employees. We developed a hardhat sticker contest for employees and contractors to participate in. They were to design a catchy logo to match the "Let's Keep it in the Pipes!" theme. We received several original submissions from all levels of the organization. Here is the logo Marathon Texas City ended up selecting:

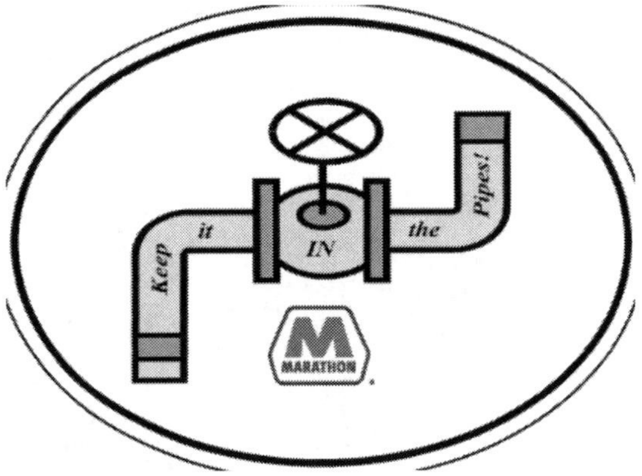

MPC Texas City employees and contractors see this logo and slogan on process safety communication and understand our focus is on reducing LOPC events. We have explained to our Marathon Texas City family that prompt reporting of small leaks and spills is the first way to prevent larger more severe LOPC events. We have seen the culture shift in the reporting of these small leaks and spills. Folks understand that by investigating the small leaks and spills we can help prevent the Tier 1 and Tier 2 process safety events from occurring.

3.2 Process Safety Bulletins

3.2.1 What does this mean?

One of the key aspects of a good Process Safety Management program is effective communication. The lessons learned from incident investigations can be turned into teachable moments with the workforce. Repeat events in our industry are a good justification for a process safety communication tool.

3.2.2 Background

MPC Texas City focused on communicating process safety Tier 3 incidents with the workforce. The intent was to develop a catchy medium through which we could grab the attention of the casual reader. A second challenge was finding a vehicle to deploy the process safety initiative through. Efforts were made to tie into our current Health, Environmental, Safety, & Security communication initiatives. A stand alone process safety initiative is often received as "flavor of the month" and may not make the desired impact on the workforce.

3.2.3 What was done?

In order to strengthen our communication of PSM, the decision was made to generate an "Event Sharing" tool at MPC Texas City. This would become known as the Process Safety Bulletin. Creating a monthly bulletin to highlight a short article on an element of PSM helped give a face to each element team. Additionally, the Process Safety Bulletin captured the refinery's PSE Tier 3 metrics to identify weaknesses which were likely contributing to loss of primary containment (LOPC) events. Finally, it conveyed a few process safety related incident investigations and gave the reader a visual picture of the incident, short description of what happened, and a take home "What can YOU do" message.

The Process Safety Bulletin was incorporated into the site's monthly Responsible Care Safety Meeting which allowed all personnel at the site to review within their workgroups. The impact of the bulletin was notable at Marathon Texas City. It was very well received by all; each month the bulletin was new, fresh, and exciting. As the development progressed and word got out, the bulletin became a company-wide PSM Best Practice and was shared with PSM groups throughout the company. This initiative has more than created a spark to our Process Safety Culture and has advanced our site further towards achieving process safety excellence.

3.3 Process Safety Roundup

3.3.1 What does this mean?

The journey toward process safety excellence is not just the journey of a single individual. All employees and contractors within the organization are on this journey together. Excellence can be achieved when folks are working together on process safety initiatives in unison – much like an engine firing on all cylinders. Just like any journey in life, it's important to pull over periodically and make sure you are headed in the right direction. Marathon Texas city's Process Safety Roundup activity provided such a venue for the reflection on process safety.

3.3.2 Background

MPC Texas City has implemented the concept of process safety element teams. An element team has an "Element Champion" which is often a department manager. The element champion leads a team of 4-5 folks that consist of subject matter experts, process safety engineers, and hourly workforce members. The purpose of these element teams included support of the following activities:

- Development of element specific goals

- Development of element specific leading and lagging indicators
- Identification of gaps between Marathon corporate policies and Texas City site policies
- Performing audits at the refinery to ensure we are doing what we say we will do in our PSM policies

As you can see these are teams with aggressive agendas that must work outside their traditional silos when it comes to process safety activities. Often a site's goals are shared across several element teams. Due to these shared goals, a forum through which PSM element teams could learn from one another was in high demand.

3.3.3 What was done?

For the past two years, MPC held an annual "Texas City PSM Round-Up." This meeting was held off-site in order to have the PSM element teams meet face to face without day to day distractions. In November 2012, approximately fifty element team members attended. The event was designed to facilitate communication styles, to discuss past accomplishments, and to establish future goals. The event also helped team members understand the goals of other PSM element teams and to gain a better perspective on MPC corporate and Texas City specific goals. As an extension of this effort, each team will be featured in the Texas City Process Safety Bulletin to make sure that the workforce understands who the teams are and what they are designed to achieve.

In 2012 we carved out a dedicated slot of time on the PSM Round-Up agenda to focus on one specific element. Mechanical Integrity (MI) was the spotlight last year. During this discussion we passed around a piece of piping that had failed due to corrosion in the presence of residual elements. This visual prop really engaged the audience and gave them a better understanding of the site MI programs.

3.4 Celebrating our Wins with our Family

3.4.1 What does this mean?

At MPC Texas City we have developed a sense of family among our Marathon and contractor employees. When we accomplished big goals, we have done them as a team. In order to have continued success in future initiatives, everyone in the organization should feel that they are valued for their contributions.

As mentioned earlier, occasional glances in the rearview mirror help us understand where we came from and give us purpose to head in the right direction. The same is true on the journey toward process safety excellence. The solution to the process safety puzzle is not just those two guys in the corner offices in the PSM department – it involves all employees and contractors. We all play a critical piece in the puzzle and as an organization we should thank them for their contributions.

3.4.2 What was done?

In the fall of 2012 MPC Texas City went one entire year with ZERO PSE Tier 1 events. This was a big success for the site and we wanted to celebrate it. It took all of the employees and contractors doing their process safety roles right each time to reach this milestone.

To acknowledge everyone's involvement in getting us to this milestone we did the following:

- Held a BBQ lunch for all MPC Texas City employees and contractors to say thanks
- Distributed the "Let's Keep It In the Pipes ! " stickers during the event
- Distributed a tri-fold pamphlet to further educate folks on the various tiers of process safety events
- At the end of the event, both employees and contractors signed commitment posters for the process safety elements that they felt related to their roles in the refinery.

4. Industry Outreach

4.1 Defining what "good" looks like

4.1.1 What does this mean?

Industry wide process safety statistics can be used to benchmark a site's process safety performance. A more meaningful tool is a thorough audit of a site's process safety management systems by a team with industry experience. In addition, a field audit to verify that the site is executing what is in the policies and procedures can help identify gaps in process safety. Both these activities can identify opportunities for improvement. When selecting an external process safety audit program, focus should be made to help ensure the external team has the right industry experience level.

4.1.2 Background

In 2012 the American Fuel & Petrochemical Manufacturers (AFPM) and the American Petroleum Institute (API) rolled out a suite of products to advance process safety in the industry. These included the following:

- Event sharing
- Process safety metrics
- Hazard Identification
- Regional Networks
- Training Certification
- Process Safety Site Assessment

The intent of the Process Safety Site Assessment (PSSA) was to have a team of assessors with significant industry experience assess the following areas of process safety:

- Process safety leadership
- Management of Change / Pre-Startup Safety Review
- Mechanical Integrity
- Safe Work practices
- Facility siting

- Process Hazard Analysis
- Operating Practices
- HF Alky RP 751

4.1.3 What was done?

Marathon Texas City volunteered in June 2012 to pilot the Process Safety Site Assessment. The motivation behind this was to see how we compared to how industry had defined what "good looks like". Industry through the API trade association had developed a set of protocol questions for each of the eight process elements being assessed. MPC Texas City compiled sets of answers on how the site would answer those questions. In addition, Marathon Texas City created "evidence binders" to support how they answered the protocol questions. These binders were very helpful during the assessment since the assessors were able to make more efficient use of their week long time at the facility.

The term "assessor" was used rather than the term "auditor" since the idea here was to instill a sense of learning during the assessment. During the week, Marathon subject matter experts were able to rub elbows with approximately 300 years of industry experience. We learned where our gaps were in our management systems and tactical execution of our policies and procedures. By working closely with the experienced assessment team, we received guidance on how these issues could be resolved in industry. The overall learning from this site assessment for Marathon Texas City was immense. Marathon Texas City now had a tool to identify what industry had defined as "best in class" and had an action plan in place to get to that level.

4.2 Sharing with Industry

4.2.1 What does this mean?

In order to move toward process safety excellence, an organization should learn from their peers in industry. Small companies with limited resources could seek out these networking opportunities. Through this networking, information can be shared regarding practices to prevent process safety events.

4.2.2 What was done?

MPC Texas City actively participated in the AFPM's Texas Gulf Coast Regional Network meetings. This meeting was an effective forum to share process safety topics among industry stakeholders. Texas City was able to share the following with their peer group:

- Value of participating in the Process Safety Site Assessment
- Marathon Petroleum Company's "Process Safety Advisory" as a means to communicate industry and company process safety events
- Interactive sharing of Texas City specific events

5.0 Conclusion

In order to move the needle on process safety in the right direction, an organization should continue to look for creative ways to go beyond traditional compliance activities. This paper has highlighted some of the activities that a small Gulf Coast refiner like Marathon Petroleum Company's Texas City has participated in to improve process safety. We value the sharing of what has worked for us and encourage our peers in industry to continue this exchange of information.

9TH GLOBAL CONGRESS ON PROCESS SAFETY

Management of Safety Critical Elements as a Base for Risk Management of Major Accident Hazards

Mariana Bahadian Bardy
Det Norske Veritas
Rua Sete de Setembro 111/12th floor
mariana.bardy@dnv.com

Flávio Luiz Barros Diniz
Det Norske Veritas
flavio.diniz@dnv.com

Paula Silveira
Det Norske Veritas
paula.silveira@dnv.com

Prepared for Presentation at
American Institute of Chemical Engineers
2013 Spring Meeting
9th Global Congress on Process Safety
San Antonio, Texas
April 28 – May 1, 2013

UNPUBLISHED

Management of Safety Critical Elements as a Base for Risk Management of Major Accident Hazards

Mariana Bahadian Bardy
Det Norske Veritas
Rua Sete de Setembro 111/12th floor
mariana.bardy@dnv.com

Flávio Luiz Barros Diniz
Det Norske Veritas
flavio.diniz@dnv.com
Paula Silveira
Det Norske Veritas
paula.silveira@dnv.com

Keywords: safety critical element, barrier, major accident hazard

Abstract

Considering the already established relevance of barriers to avoid Major Accidents, the objective of this paper is to present a methodology for management of Safety Critical Elements (SCE), from the identification of them, definition of relevant importance to each activity performed by the installation and establishing alternatives and contingencies for the failure or absence of the SCE. The proposed methodology, adapted from common use methodologies from Offshore Industry to Process Industries, is developed in 5 steps, being Step 1 the use of a Hazard Identification technique and indication of Major Accident Hazards. Following on Step 2 bowtie diagrams are developed for the MAH and SCE are identified. The SOOB – Summary of Operational Boundaries on Step 3 identifies the activities that may or may not proceed or caution is applied in the case the SCE is defeat and on Step 4 a Contingency Plan is develop to maintain operation for the cases indicated on the SOOB that operation may not proceed or proceed with caution. Finally, on Step 5, definition of prioritization of maintenance and inspection activities shall be in place for each SCE, including preventive maintenance routines, inspections routines and definition of spares, where applicable. This methodology can help on the identification of gaps and management of critical elements and consequently improving the performance of safety systems increasing their availability.

1. Introduction and background

Recent accidents have indicated the importance of safety barriers on management of major accidents, reducing its likelihood or minimizing consequence. Buncefield, Texas City and Macondo, just to indicate some, have stated in their accident investigation reports the failure of safety barriers or non-existence of adequate ones as potential causes for the major accident occurrence.

This paper presents a methodology for management of Safety Critical Elements (SCE), adapted from common use methodologies from Offshore Industry to Process Industries, from the identification of them, definition of relevant importance to each activity performed by the installation and establishing alternatives and contingencies for the failure or absence of the SCE.

Several reference have definitions of SCE and how they must be managed, as NORSOK[1], that indicated that Safety Critical Equipment is equipment that shall be in operation to ensure escape, evacuation and /or to prevent escalation.

According to HSE UK[2] "any structure, plant, equipment, system (including computer software) or component part whose failure could cause or contribute substantially to a major accident is safety critical, as is any which is intended to prevent or limit the effect of a major accident".

For this paper, SCE is defined as indicated by HSE UK, as being the Barriers that can avoid or mitigate Major Accident Hazards.

2. Description of Methodology

For the objective of systematic management of Safety Critical Elements, the methodology outlined in Figure 1 is proposed, covering the 5 steps described below.

Figure 1 – Methodology for SCE Management

2.1 Step 1 – Hazard Identification

The first step is to identify the accidental scenarios from the specific process under analysis. For that purpose, it is proposed to perform a Process Hazard Analysis (PHA) for identification of accidental scenarios and classification according to a Risk Matrix, defined by each company according to its risk management process. Figure 2 represents an example of spreadsheet to be applied for the PHA.

Process Hazard Analysis (PHA)											
System:			Hazard/Event Group:								
1.Hazard	2.Causes	3.Effects	4.Freq	5.Sev	6.Risk	7.Safeguards	8.Final Freq	9.Final Sev	10.Final Risk	11.Recommendations	12.#
											1
											2
											3

Figure 2 – Example of PHA Spreadsheet

The spreadsheet has 12 columns, and two classification of the risk for each scenario. Columns 4, 5 and 6 have the classification without considering existing safety barriers for the scenario. The barriers are listed on Column 7 and Columns 8, 9 and 10 indicate the classification of the risk, considering the existence of the barriers and these are operating or ready to operate when needed.

For the classification of severity, likelihood/frequency and risk, a risk matrix shall be used, representing the risk tolerability of the company. An example of risk matrix is on Figure 3, extract from ISO 17776:2000 [3].

Consequence					Increasing probability			
Severity rating	People	Assets	Environ-ment	Reputation	A	B	C	D
					Has occurred in E&P industry	Has occurred in operating company	Occurred several times a year in operating company	Occurred several times a year in location
0	Zero injury	Zero damage	Zero effect	Zero impact				
1	Slight injury	Slight damage	Slight effect	Slight impact	Manage for continued improvement			
2	Minor injury	Minor damage	Minor effect	Limited impact				
3	Major injury	Local damage	Local effect	Considerable impact				
4	Single fatality	Major damage	Major effect	Major national impact	Incorporate risk-reducing measures		Fail to meet screening criteria	
5	Multiple fatalities	Extensive damage	Massive effect	Major international impact				

Figure 3 – Example of Risk Matrix

Note that this matrix has 5 different severity ratings and analyze four different effects: people, assets, environment and reputation. A common approach to define Major Accidental Hazards (MAH) is to consider whose with highest consequence classification, as the ones classified with Severity Category 5, from the matrix indicated on Figure 3, that represents multiple fatalities as impact on people, extensive damage for environment, massive effect to assets and major international impact on reputation.

The main advantage to select the MAH to go to Bowtie, as described above, is that the barriers related to those events can be clearly identified and consequently be managed properly and in a focused way. On the other hand, when there is no distinction between MAH and other scenarios, with lower damage potential, the number of barriers to be managed increase reducing the focus on the major impact scenarios MAH.

Note that some safety barriers are normally identified on this PHA and shall be reviewed and detailed on the next steps.

2.1 Step 2 – Development of Bowties

The following step of this methodology is to develop bowtie diagrams for each of the MAH, or combination of MAH, if applicable, as exemplified in Figure 4. The BowTie methodology is designed to give a picture of the risks, to help people understand the relationship between the risks and organizational events and to identify where barriers in place can act, on the prevention or on the mitigation and consequently give a better overview if those are enough to mitigate the risks related to the MAH.

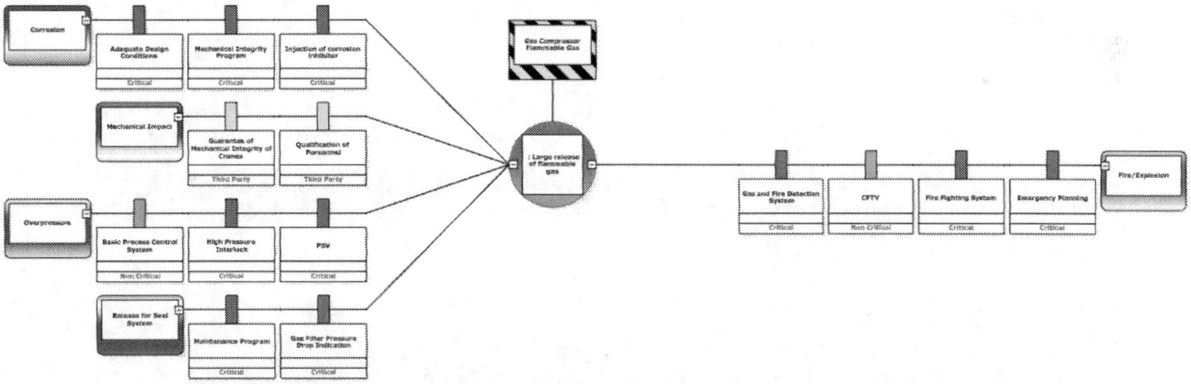

Figure 4 – Example of Bowtie

With the use of multidisciplinary team from the company, starting from a Top Event, localized in the center of the diagram, causes, preventive barriers, consequences and mitigating barriers are identified. Following, each barrier, preventive or mitigating, is classified as:

- Critical: essential barrier to avoid the causes or associated consequences.
- Non-critical: barrier that reduces likelihood or minimize consequence, but does not avoid the occurrence of the top event or associated effects.

- Third Party: barriers, critical or not, that are not under the company responsibility for management.

Also responsible person or function can be indicated on the bowtie for each barrier. The list of SCE are composed by those barriers classified as critical for each bowtie.

The Safety Critical Elements (SCE) can be an Equipment, System or Procedure. In the example presented in Figure 4, for Top Event *Large Release of Flammable Gas, from Compression System*, the following barriers were classified as Safety Critical Equipment or System:

- Safety interlocks
- PSVs
- Filter Pressure Drop Indication
- Injection of Corrosion Inhibitor
- Gas and Fire Detection System
- Fire Fighting System
- CFTV

The other critical barriers such as Mechanical Integrity Program and Emergency Planning are considered as Safety Critical Procedure.

All of those critical elements, equipment, systems and procedures, shall be managed but specially for the equipment and systems contingency procedures shall apply considering they are operating under degraded conditions or out of operation. As part of this scope a Summary of Operations Boundaries (SOOB) analysis is carried out as stated below.

2.3 Step 3 – Development of SOOB

The Step 3 of this methodology consists in developing the Summary of Operations Boundaries - SOOB analysis. This is based on a matrix which crosses main operations and activities with the Operational Risk Factors. Operational Risk Factors includes controls, identified in the BowTie analysis, under reduced effectiveness and risk factors such as severe weather/sea conditions. The matrix is completed row by row by reviewing all combinations.

The main objective is to examine if operations can be permitted or prohibited when certain controls have been defeated or running under reduced effectiveness and examine if operations can proceed in the case of occurrence of external factors that can potentially influence the risk of doing these operations e.g. severe weather conditions.

This will distinguish when a “stop work” is applied or if that shall be a “proceed with caution” condition, as indicated by IADC[4]. A “traffic light” system may be applied, indicating:

- Red: stop the work or do not proceed;
- Yellow: evaluate conditions, perform risk analysis or implement additional protection;
- Green: continuous normal operation.

Note that the activities will vary depending on the type of installation. Some examples are:

- loading or unloading of trucks or railcars;
- operation above normal conditions;
- increase of capacity;
- confined space entry;
- working at height.

An example of analysis is for Gas Detection System failure, it is allowed to proceed with works at heights and confined space entry; with caution, requiring additional evaluation normal operation and loading/unloading; and not permitted operation above normal condition, increase of capacity and hot work.

Operations vs. Operational Risk Factors	Normal Production	Operation Above Normal Conditions	Increase of Capacity	Loading/Unloading Truck	Loading/Unloading Railcar	Confined Space Entry	Hot Work	Working at Heights
• Safety interlocks	RA	X	X	RA	RA	P	RA	RA
• PSVs	RA	X	X	RA	RA	P	P	RA
• Filter Pressure Drop Indication	RA	RA	RA	NA	NA	NA	NA	NA
• Injection of Corrosion Inhibitor	RA	RA	RA	NA	NA	NA	NA	NA
• Gas and Fire Detection System	RA	X	X	RA	RA	P	X	P
• CFTV	P	RA	RA	RA	RA	P	P	P

P- Permitted
RA – Perform Risk Analysis
X- Do not Proceed

Figure 5 – Example of SOOB Matrix

2.4 Step 4 – Definition of Contingency Plan

The Step 4 of this methodology consists on the definition of Contingency Plan for each SCE. The immediate response action that will normally be applied are:

- to stop or limit operations to within the limits of remaining barriers; or

- identify and assess any temporary substituted safety system barrier that may be implemented to support continued operation.

The company shall establish and document contingency procedures and a system of approval and control of SCE to be used when those are under degraded conditions or out of operation.

The following items shall be considered:

- Implementation of alternative controls equivalent;
- Limitation and reduction of production;
- Isolation and stopping of equipment, systems, installations;
- Deadline for the temporary procedure to be allowed until corrective measures are taken.

A specific contingency plan is then developed for each SCE, using, for instance, the example indicated in Figure 6.

SCE	Permitted Acitivities	Activities with Restriction	Prohibited Activities
Gas Detection System	Confined Space Entry Working at heights	Normal Production Loading/Unloading	Hot work Operation above normal conditions Increase of capacity
Permitted Activities:	No limitation for the development or continuation of activity, event with loss of the SCE		
Activities with Restriction:	No limitation for the development or continuation of activity, event with loss of the SCE		
Prohibited Activities:	Not allowed to perform the activity and must be interrupted, even with		

Alternative Procedures for Activities with Resctriction		
Activity	Deadline	Responsible
Normal production to continuous with one extra Operation Supervisor per shift, with focus on Control Room supervision	One month	Operation Manager
If SCE not returned to full operation after first deadline, reduce production and safe stop production	-	Operation Manager
Performed loading/unloading activities with one extra fiel operator	One month	Operation Manager
SCE Resposible	Maintenance Manager	

Figure 6 – Example of Contingency Plan for SCE

2.4 Step 5 – Definition of Maintenance and Inspection Prioritization

The final step for implementation of this methodology of Management of SCE is to incorporate on maintenance and inspection routines and procedures prioritizations that will consider the findings of the analysis of the SCE. Some important points shall be considered:

- Guarantee that all SCE are classified as high priority for maintenance routines;
- Guarantee no delays for inspection routines for the elements associated with MAH and classified as SCE;
- Evaluate the need of spares of SCE, where applicable.

3. Conclusion

As initially indicated, this paper presents a 5-step methodology for management of SCEs, being those defined here as safety barriers that can avoid or mitigate Major Accident Hazards. The objective of each step as well as practical approach and examples are presented, adapting common use methodologies from Offshore Industry to Process Industries.

As extension of this work, considering all aspects for the presented, some improvements can be implemented. The inclusion of procedures as part of the analysis, after the identification of the critical procedures, with guarantee of correct training or certification of operators, is one of these points. One additional relevant aspect is to incorporate a 6th step on the above methodology of management of SCE, with the audit of process of management of the critical barriers.

Finally, it is importance to note that, this methodology was developed with the intention of giving support for companies to systemic manage Safety Critical Elements and comply with relevant regulation and best practices.

4. References

[1] Norsok S-001, Edition 4, February 2008, item 3.1.11. Norway. 2008

[2] Health and Safety Executive, A guide to the Offshore Installations (Safety Case) Regulations 2005, item 83. London. 3rd Edition. 2006.

[3] ISO 17776:2000, Petroleum and natural gas industries — Offshore production Installations —Guidelines on tools and techniques for hazard identification and risk assessment, Table A.1. Geneva. 2006.

[4] IADC, HSE Case Guidelines for Mobile Offshore Drilling Units, Issue 3.2.1, 2009.

Promoting Commitment to Process Safety

Russell A. Ogle,
Andrew R. Carpenter and Sean J. Dee
Exponent, Inc.
4580 Weaver Parkway, Warrenville, IL 60555
rogle@exponent.com, acarpenter@exponent.com, and sdee@exponent.com

Prepared for Presentation at
American Institute of Chemical Engineers
2013 Spring Meeting
9th Global Congress on Process Safety
San Antonio, Texas
April 28 – May 1, 2013

UNPUBLISHED

Promoting Commitment to Process Safety

Russell A. Ogle,
Andrew R. Carpenter and Sean J. Dee
Exponent, Inc.
4580 Weaver Parkway, Warrenville, IL 60555

Keywords: process safety management, safety culture, game theory

Abstract

Successful implementation of process safety management (PSM) typically requires the use of a combination of physical and procedural safeguards. Ideally, these safeguards must be maintained in a constant state of readiness throughout the life of the process facility. The management system provides the framework for maintaining and sustaining these safeguards, but the critical ingredient that ensures that the work gets done is the individual employee's commitment to PSM.

Game theory provides a useful framework for understanding how an organization can foster cooperation between management and its employees towards the common goal of safety. Using a specific game-theoretical model known as the repeated Prisoner's Dilemma, it is shown how cooperative behavior is the natural consequence of a balanced system of rewards and penalties. Here the term "balanced" means that both employees and their managers are subject to this system of rewards and penalties. By implementing a balanced framework for feedback, an organization can promote a commitment to process safety by fostering cooperation between both the employees and the managers.

1. Introduction

Successful implementation of process safety management (PSM) typically requires the use of a combination of physical and procedural safeguards. Ideally, these safeguards must be maintained in a constant state of readiness throughout the life of the process facility. The management system provides the framework for maintaining and sustaining these safeguards, but the critical ingredient that ensures that the work gets done is the individual employee's commitment to PSM. The social sciences, especially psychology and organizational behavior, can provide valuable insight towards understanding the management of employee commitment to safety. However, other fields of study can make useful contributions to our understanding. Game theory is one of these tools.

Game theory is the scientific study of interactive decision-making, and is especially useful in understanding the interplay of conflict and cooperation between different stakeholders. Consider the interaction between the individual employee and the management organization. The objective of both management and the individual employee is to secure a strong commitment to process safety. But there are many time and attention demands placed on the individual employee, and numerous financial pressures on management. There can be a strong temptation to cut corners and sacrifice operational discipline in favor of simplifying a task and getting it done faster or cheaper. In game theory, this situation is known as the Prisoner's Dilemma. The Prisoner's Dilemma is one of the simplest models of the conflicts that can arise between the goals of an individual and an organization.

Game theory provides the tools which can help an organization to understand how to foster cooperation between management and its employees to achieve the common goal of safety. Using a specific game-theoretical model known as a repeated Prisoner's Dilemma, it is shown how cooperative behavior is the natural consequence of a balanced system of rewards and penalties. Here the term "balanced" means that both employees and their managers are subject to this system of rewards and penalties. By implementing a balanced framework for feedback, an organization can promote a commitment to process safety by fostering cooperation between both the employees and the managers.

2. Seeking Commitment to Process Safety

The effective control of the risks associated with a chemical process requires more than just physical and procedural safeguards. It requires people who demonstrate a continuous effort to sustain these safeguards. In short, it requires commitment to process safety at all levels of the organization: employees, supervisors, and managers. The fundamental role of commitment to successful process safety management is discussed at length in the book, *Guidelines for Risk Based Process Safety*, by the Center for Chemical Process Safety (CCPS), a directorate of the American Institute of Chemical Engineers (AIChE) [1]. The CCPS risk based process safety paradigm is based on four pillars:

- Commit to process safety
- Understand hazards and risk
- Manage risk
- Learn from experience

The first component, the commitment of the organization to process safety, is perhaps the most fundamental. It is our thesis that a useful measure of this commitment is through the individual's operational discipline in executing their assigned process safety responsibilities.

Why is commitment to process safety important? Commitment is important because it is the individual's personal pledge to perform their assigned responsibilities. There are always going to be unanticipated or non-routine aspects to any work assignment. Individual commitment to process safety transcends simple compliance with a written work order or procedure. When confronted with the unexpected, it is the individual's commitment to process safety that spurs them to ask the important questions: 1) What can go wrong? 2) How bad is it? 3) How likely is it? 4) What can we do to manage the risk?

A true commitment to process safety introduces a difficult challenge for the employee and her supervisor. Commitment to process safety requires extra effort beyond simply getting the work assignment done. Commitment to process safety requires additional planning and thinking before the work is started. It requires a collaborative work approach with colleagues who are busy with their own work assignments. And, to be perfectly frank, it requires paperwork.

How is commitment achieved with employees, supervisors, and managers? The behavior-based safety literature tells us that what gets rewarded is what gets done [2, 3]. Individuals at all levels of the organization need to believe that they will be rewarded for their commitment to process safety, and furthermore, that they will be penalized if they fail in that commitment. This requires a sustained effort by upper management to observe, evaluate, and administer feedback—rewards or penalties—to the supervisors of the employees. The supervisors, in turn, must also observe, evaluate, and administer feedback. To successfully create and maintain a commitment to process safety from the front-line employees, the employees must see that this commitment will be rewarded, and that a failure in commitment will be penalized. Game theory is a tool that can inform management on how to develop a balanced system of rewards and penalties to achieve this commitment at all levels of the organization.

3. The Repeated Prisoner's Dilemma

Game theory applies mathematical modeling to understand the payoffs and costs associated with two or more individuals making interactive decisions [4]. By understanding the net cost benefit of an individual's actions given the available information regarding the potential decisions of the other players, one can develop strategies and predict behavior based on the mathematical results, and basic human instincts. In particular, the application of game theory has been useful in studying interactions between parties who are in conflict, but would benefit from employing a cooperative strategy. While the mathematics can become very complicated, valuable insights can be gained without heavy reliance on the more complex mathematical modeling components. The exercise of identifying players, considering each player's potential decisions, and determining the payoffs and costs associated with the various outcomes, can provide significant insight in solving problems involving conflict, and structuring optimized solutions.

When considering commitment to process safety, individual employees are left to optimize a multivariable problem. From an organizational level, multiple constraints are imposed on individual employees with respect to their job. First and foremost, employees are expected to perform their jobs tasks adequately, and when posed with the additional requirement of performing the job safely, an employee must sacrifice labor, attention, and thought that could otherwise be spent on their other responsibilities. It is then tempting for an employee to ignore, or shirk, their commitment to process safety. The benefit from shirking is twofold. First, a shirking employee does not need to invest the effort into the process safety management program. But additionally, if the other employees continue to put in the effort associated with maintaining a superior process safety program, the shirking employee also gains the additional benefits that all the employees realize with having a safe workplace. Since commitment to process safety is often an investment to prevent future loss associated with system failures, the benefits of a safe work environment can seem less tangible that an individual employee's day-to-day time and effort. However, if all employees shirk their responsibility, then the process safety management program becomes ineffective, and all the employees only reap the benefit of the additional time and effort that is not spent on their commitment to process safety. In the long run, this makes the employees more susceptible to accidents, and the high costs associated with not having an adequate safety management program. Management has often turned to incentive, reward, and penalty programs to reinforce individual commitment to process safety. The field of psychology is often used to develop these programs [5, 6]. This type of problem has also been extensively discussed in the game theory literature relating to economics and social sciences. Using game theory, management can better understand strategic programs to reaffirm and promote commitment to process safety.

This type of problem, where an individual can benefit by defecting from the beneficial collective action of a group but still reap the group's collective benefits is known in Game Theory as the free rider problem. From the perspective of this work, the free rider problem involves an individual (free rider), who reaps the benefit of a safe work environment (communal resource) without putting in any effort to generate that benefit. A free rider problem can be constructed as a two player game referred to as the prisoner's dilemma. On one axis, the individual employee has the opportunity to Commit to, or Defect from the process safety culture, while the rest of the collective group is represented as a single player on the second axis [7]. The payoffs for committing or defecting for the individual player are represented by the numbers in the two-by-two matrix below. There associated payoffs for the collective group represented by the second player have been omitted for simplicity.

Individual

	Enough employees Commit	**Not enough employees Commit**
Commit	Reward = 3	Sucker = 0
Defect	Temptation = 5	Penalty = 1

The four possible outcomes for the individual employee in this game construction are Reward, Temptation, Sucker, and Penalty. Reward corresponds to both the individual employee and enough other employees cooperating to maintain a superior safety management program. The Temptation provides the highest payout for the individual employee because in addition to reaping the safety benefits of, there are no costs (time, effort) to the defecting employee as others pick up the slack. But, if an insufficient number of other employees carry out their organizational safety responsibilities, then no one, including the individual employee, receives the benefits associated with good operational discipline. Instead, there is a smaller payout (penalty) associated with shirking that represents the employee not investing time or effort into organizational safety. Additionally, if an employee cooperates and fulfills their responsibilities but few others cooperate, then the individual receives the lowest payout and can be termed as a Sucker.

The payoff values shown in the matrix are for illustration only. The rankings are intentionally ordered such that: Temptation > Reward > Penalty > Sucker.[1] The evaluation of payoffs in terms of expected utility is an important subject, but it is quite involved. We refer the reader to decision analysis texts for a discussion of practical strategies for the evaluation of utilities [9]. For our

---

[1] For technical reasons, it is typically assumed that the payoffs are structured such that Reward > ½ (Sucker + Temptation). This relation ensures that continued cooperation results in a higher total payout than alternation of cooperation and shirking: which simply means that the dilemma cannot be beaten by players taking turns receiving the temptation payout. See Appendix B in the book by Axelrod [8].

purposes, the ranking of the payoffs is sufficient. In addition to the payout rankings, another key concept associated with the prisoner's dilemma is the appearance of a dominant strategy. Regardless of whether the collective group chooses to commit, or defect, the individual employee will also have a higher payout if they defect ($5 > 3$, $1 > 0$). Hence, the strategy of Defect strictly dominates the strategy of Commit.

Repeated interaction has a strong influence on the ability to coordinate the actions of multiple players. In the previously discussed matrix, one would predict the individual employee to defect. However, if there are repeated iterations of the prisoner's dilemma, it stands to reason that the defect strategy would soon dominate the entire population, resulting in the penalty payout. In this situation, one can use the promise of prolonged reward payouts in the future, to outweigh the incentive to defect in the present. This idea, of fostering cooperation and coordination to maximize the benefit for all employees is effective as long as the promise of future interaction is credible. If the game is played a finite number of times, the employee can benefit from shirking when the promise of future interactions has dwindled. If, however, the employee believes that the game will be played repeatedly without end, then the reward of continued cooperation minus the threat of future penalties outweighs the benefit of shirking. It makes more sense for the employee to cooperate.

When the reward for continued cooperation outweighs the incentives to defect, the coordination of action becomes self-enforcing. This is not only true mathematically, but it has also been demonstrated in laboratory and field work associated with the social sciences [8]. Furthermore, as discussed in Schneier's book, the importance tangible, meaningful rewards and penalties to foster coordination out of dominant strategies is critical [10]. Commitment to process safety is more than just a mission statement, or a policy, it requires active participation from players at all levels (employee and management) to reinforce the message, provide credibility, and rewards and penalties directly tied to performance, utilizing leading indicators or process safety.

4. Promoting Commitment through Cooperation

At facilities that handle highly hazardous chemicals, most employees have process safety responsibilities in their job. It is essential that those responsibilities are defined and measures of success are communicated and documented. Supervisors should then conduct observations, both scheduled and random, to verify that these responsibilities are being implemented. Studies have shown that the conduct of supervisory observations leads to higher levels of voluntary employee participation [11].

Process safety observations gathered by the supervisor should be classified into compliant and non-compliant categories. Employees should be rewarded for their compliance with their process safety responsibilities. Rewards should be consistent with the organizations value system and do not need to be expensive. Simple, frequent rewards like public recognition, a free lunch, or a gift certificate can contribute significantly to the employee's motivation for continued success [12, 13]. The key point is that regardless of the type of reward used, it must be delivered in a genuine manner with public recognition. Taken to its ultimate conclusion, employees who are leaders in demonstrating their commitment to process safety should receive special consideration for promotion over those who fail in this commitment. And serious consideration should be given to rewarding the employee's supervisor.

For those who fail in their commitment to process safety must feel the sting of penalties. The penalties should be consistent with the organizations values and should be delivered privately. Only in exceptional circumstances should a penalty be delivered publicly. The idea of the penalty is to encourage the employee to change their behavior. Penalties demonstrate the organization's commitment to process safety as an organizational value [14]. And just as with rewards, depending on the seriousness of the situation, consideration should be given to exacting a penalty on a supervisor with employees who fail to demonstrate the desired level of commitment to process safety.

5. Conclusions

Individual commitment is a cornerstone to a successful process safety management program. This commitment must be maintained at the employee, supervisor, and manager levels to ensure the safety and well-being of all those involved with hazardous processes. Additionally, this commitment requires continual reaffirmation and improvement. However, at times, the commitment to safety can be in direct conflict with financial constraints, deadlines, or production goals. This conflict can lead to individuals to shirk their commitment and make decisions based on factors that are not consistent with the safety management program. In these instances, the use of reward and penalties can reinforce the importance of process safety, and change behavior in impactful ways. Particularly, from a game theory perspective, it is possible to create scenarios where the incentives to ignore one's commitment are discounted to levels that make them insignificant in the day to day operations of a chemical process. By reinforcing positive safety habits, and correcting unsafe behavior, it is possible to improve plant wide process safety.

6. References

1. Center for Chemical Process Safety, Guidelines for Risk Based Process Safety. American Institute of Chemical Engineers, New York, 2007.
2. R.E. Medina, et al. "Behavioral Safety in a Refinery," *Professional Safety*, pp. 36-40, August 2009.
3. E.S. Geller. "Social Influence Principles," *Professional Safety*, pp. 25-31, October 2002.
4. J. Watson. Strategy: An Introduction to Game Theory, Second Edition. W.W. Norton & company, New York, p. 1, 2008.
5. E.S. Geller. The Psychology of Safety. CRC Press, Boca Raton Florida, 1998.
6. T.E. McSween. The Values-Based Safety Process. Wiley, New York 2003.
7. L. Fisher. Rock, Paper, Scissors: Game Theory in Everyday Life. Basic Books, New York, pp. 61-68, 2008.
8. R. Axelrod. The Evolution of Cooperation, Revised Edition. Basic Books, New York, 2006.

9. R.T. Clemen. Making Hard Decisions: An Introduction to Decision Analysis, Second Edition. Duxbury Press, Belmont California, Chapters 13 and 14, 1996.

10. B. Schneier. Liars and Outliers: Enabling the Trust that Society Needs to Thrive. John Wiley & Sons, New York, pp. 1-14, 2012.

11. S. Cook and T.E. McSween. "The Role of Supervisors in Behavioral Safety Observations," *Professional Safety*, pp. 33-36, October 2000.

12. T.R. Krause. "Motivating Employees for Safety Success," *Professional Safety*, pp.22-25, March 2000.

13. P.M. Gendinning. "Employee Safety Incentives: a Best Practices Survey of Human Resource Practitioners," *Professional Safety*, pp. 22-24, February 2001.

14. L.I. Perkinson. "Discipline in the Extremes," *Professional Safety*, pp.31-35, April 2005.

(86a) Explosion and Fire At Caribbean Petroleum Tank Terminal in Bayamòn, Puerto Rico

Vidisha Parasram and Maria Mazzocchi, Incident Screening, US Chemical Safety Board, Washington, DC

On October 23, 2009, normal tank filling operation became catastrophic when a 5,000,000 gallon atmospheric tank overflowed for approximately 30 minutes at the Caribbean Petroleum terminal in Bayamòn, Puerto Rico. The developing vapor cloud enveloped part of the tank farm and reached the waste water treatment area where it ignited. A flash fire and explosion ensued and the contents of the tank burned for 2.5 days. Neighboring communities were evacuated and oil contaminated the nearby wetlands. This incident revealed a number of safety system deficiencies at the facility and gaps in the regulatory requirements and industry standards addressing overfill prevention and emergency response. This incident also raises questions regarding the adequacy of safety management systems in place at tank terminals and the need for automated overfill prevention alarms including high, high high level alarm and independent alarms to detect and shut down flow into a tank during filling operations. In addition, due to the close proximity of the community to the Caribbean Petroleum tank terminal, this incident further demonstrates the need for a risk assessment process that addresses high hazard facilities. This presentation will provide an overview of the US Chemical Safety Board's Caribbean Petroleum investigation and discuss recommendations made by the CSB to prevent a similar incident.

(86b) Use Dynamic Simulation to Reduce Startup Flare Emissions for Ethylene Plants

Jian Zhang, Meiqian Wang, Jie Fu, Chuanyu Zhao, Ha Dinh, Shujing Zhang, Qiang Xu and Kuyen Li, Dan F. Smith Department of Chemical Engineering, Lamar University, Beaumont, TX

Ethylene plants play an important role in petrochemical industry. The products of an ethylene plant, such as ethylene, propylene, butadiene, benzene, toluene and other co-products, are all basic feedstocks in petrochemical industry. Ethylene production process is one of the biggest and most complicated processes in petrochemical industry. Startup of an ethylene plant is also very critical. A typical startup of an ethylene plant could flare 400,000 - 1,000,000 lbs of materials, and generate 200,000 lbs of CO and VOCs (volatile organic compounds). Reduction flare emissions will help reducing environmental impact and saving materials.

In this paper, rigorous steady-state and dynamic models of an ethylene plant was built. The models include CGC (charge gas compressor), chilling train, demethanizer, deethanizer, depropanizer, C2 splitter and C3 splitter sections. Both steady-state and dynamic models have many details so that they are more close to startup procedure. For example, performance curves were used in CGC section, so that the section can simulate scenarios on different feed and bypass flowrate. In chilling train section, information of equipment size and heat capacity is considered for simulating precooling process. For distillation columns, rigorous distillation model and tray sizing calculation are employed. Both steady-state and dynamic models are tuned and validated to make them close to real plant operations.

The validated dynamic model was used to simulate startup scenarios. A startup procedure has dozen of steps, and some steps have several flare reduction options. Through dynamic simulation, all major steps and options can be simulated and evaluated, and the flare emissions can be calculated. For example, CGC startup is the most critical step and has largest amount of flare vent, because charge gas has to be continuously produced through furnaces and sent to flare to keep inlet pressure of compressors. We simulated several scenario cases with charge gas of different flowrate, methane and nitrogen, and found the best scenario, which could largely reduce flare emissions of CGC section. Another case study is regarding the precooling of chilling train. A chilling train should not be chilled very fast, otherwise it could be damaged by thermal expansion. On the other hand, chilling train cannot work properly before chilled to certain temperature. Through dynamic simulation, several chilling scenarios were simulated. For each scenario, control strategies are employed to control the chilling rate within a certain range. Simulation results showed some scenarios could even cool down chilling train without any flare emissions.

Flare calculation showed that the total flare emissions could be greatly reduced through dynamic simulation. The final scenarios have been acknowledged by ethylene plants.

(86c) Development of Novel Reducing Agents for NOx Reduction by Sncr Method

Maoqi Feng, Div. of Chemistry and Chemical Engineering, Southwest Research Institute, San Antonio, TX and Reggie Zhan, Emissions and Vehicle Research Division, Southwest Research Institute, San Antonio, TX

Nitric oxides (NO_x), mainly including nitric oxide (NO) and nitrogen dioxide (NO_2), are of the most toxic pollutants formed during combustion and refining processes. Post-combustion treatment methods include selective catalytic reduction (SCR), selective noncatalytic reduction (SNCR), reburning, and the combination of these aftertreatment technologies. Among the post-combustion technologies, the SNCR technology is considered to be an effective approach to reduce NO_x, with reasonable capital investment and operation cost. To reduce NO_x in a lean environment, a reductant is needed. For SNCR, the most commonly used reductant is ammonia or urea water solution as an alternative to NH_3. One of the disadvantages of ammonia is the cost of the injection system. Different reducing agents for NOx removal by SNCR method will be evaluated in this paper, including carbon monoxide, hydrogen, char, hydrocarbons, and oxygenates, etc.

Rotating Zigzag Bed Industrial Application Research on Ammonia Stripping

Zhi-chao Xu, Xiao-hua Li, Guang-quan Wang, Jian-bing Ji

(College of Chemical Engineering and Materials Science, Zhejiang University of Technology,

Zhejiang Province Key Laboratory of Biofuel, Hangzhou, Zhejiang, China 310032)

Abstract: About 400 units of rotating zigzag bed (RZB) have been used in industrial distillation process. These successful applications fully demonstrated its advantages such as high gas-liquid mass transfer rate and small size. There will also be broad prospects of RZB applied in waste water treatment process. In this paper a preliminary attempt was made to study ammonia nitrogen removal from industrial waste water by RZB before emission. Firstly, stripping experiments were carried out to remove ammonia nitrogen from the waste water containing ammonia and ammonium sulfate in a small single-rotor RZB, obtaining ammonia solution with a proper concentration. Secondly, a unit of industrial-scale three-rotor RZB with outer diameter of 750mm was designed based on experimental data and used to treat ammonium sulfate solution with 2% ammonia nitrogen. The industrial results showed that the feed capacity was greater than 1500 kg/hr, ammonia concentration in the bottom was less than 30 mg/L, and aqua ammonia concentration in the top was greater than 25%.

With the improvement of human life, there is growing public concern about environment problems. Wastewater emission is one of the important factors contributing to environment deterioration. In the wastewater emissions, ammonia nitrogen concentration is a very important target which should be strictly controlled because excessive ammonia in the river will cause water eutrophication. The ammonia wastewater has a wide variety of sources, such as refining, fertilizer, inorganic chemicals, pesticide, ferroalloys, glass and food industries. Therefore, studies on removal ammonia from the wastewater and how to recover ammonia to reuse is of great significance. Rotating zigzag bed (abbreviated to RZB) has been successfully applied in about 400 cases of industrial distillation[1-5]. Due to its higher mass transfer rate it was expected to have a very wide application in wastewater treatment. Some traditional enterprises have demand for retrofit and optimization of the production process without occupied space increase because of limited land area. RZB can exactly meet this demand owing to its very small volume. In this paper lab-scale experiment of stripping ammonia from wastewater with RZB was conducted and ammonia solution with a certain concentration was obtained. Then, the experimental results were scaled up to design a unit of RZB successfully used in industrial production.

1.Lab RZB device and flowsheet

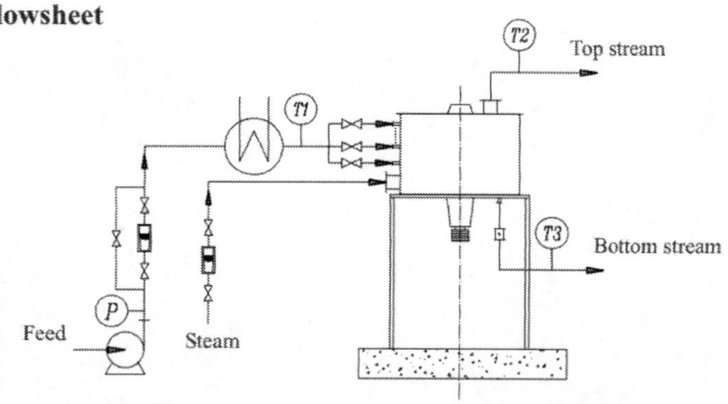

Fig. 1 Lab RZB device and flowsheet

Lab RZB device was set with multiple liquid feed inlets, which were used in variety of processes with different number of theoretical stages. For feeds with different concentration of ammonia, their optimal operation and equipment parameters can be determined by experiments.

In experimental flowsheet shown as Fig.1, the liquid feed was measured with a rotormeter and pumped into RZB device with multiple inlets after heating by a heat exchange. The steam with higher temperature enters into vapor inlet after flow measurement. The treated water was drawn from liquid outlet on the bottom of RZB. Vapor phase containing ammonia withdrawing from the top of RZB was condensed and absorbed in the next steps. T1-T3 were thermometers to measure and record the temperatures of several key positions.

2. Experiments and results analysis

In the experiments, the concentration of the prepared feed was close to that of industrial stuff, which was ammonium sulfate solution with about two percent ammonium. The feed position can be determined through the preliminary experiments. For this study, the flow ratio of feed to steam was varied and analyzed the ammonium compositions of bottom and top outlet stream as well as the power consumption of the motor. These data provided the basis for scale-up design of RZB.

Feed preparation: Ammonium sulfate solution with concentration of 15.5%, which pH was adjusted to about 13 by dripping NaOH solution, was prepared by adding ammonium sulfate powder into water. Ammonium sulfate solution has ammonium concentration of 3.3% and was chosen as the feed.

In the experiments, originally the top outlet vapor stream entered into an absorber but not a condenser. The system was maintained under vacuum condition by adjusting the relief valve at the end of absorber. The pressure of the system was 72kPa and the temperature of the top outlet vapor phase was about 80℃. The typical process and operation parameters in the experiments were shown in Table 1:

Table 1 Typical process and operation parameters

feed				steam		Bottom stream	Top stream
$(NH_4)_2SO_4$ %	Ammonium concentration %	pH	flowrate L/hr	flowrate kg/hr	pressure MPa	Ammonium concentration mg/L	Ammonium concentration(vapor) %
15.5	3.3	13	1000	120	0.35	<100	>48

If the top outlet vapor stream was connected to a condenser, solution with ammonium concentration higher than 20% can be obtained.

3. Industrial application of RZB in ammonium stripping

Fig. 2 Industrial equipment

After the results of industrial field tests were accepted by the customer, scale-up design was made for RZB. The section between the feed position and the top, corresponding to the rectifying section of distillation column, has a separation ability of 4.5 theoretical plates. This section can effectively prevent entrainment. On the other hand, if the actual ammonium concentration was lower than designed, partial reflux can be introduced to increase the ammonium concentration in the top outlet stream. The type of RZB is BZ750-3P, which means the largest one in the three rotors has a diameter of 750mm. Its appearance was shown in Figure 2.

The RZB was used in Taiwan, with industrial process flowsheet shown in Figure 3. The feed and alkali solution for adjusting the pH value, which flowrate were measured by flowmeter respectively, were mixed in a fixed proportion. The mixed stream was heated by the bottom outlet liquid and entered into the RZB. The steam was introduced by vapor inlet. The outlet vapor entered into absorption system which performs cooling and absorption steps. The absorption systems consist of four sections: pure water (absorbent), low concentrated ammonium recycle, high concentrated ammonium recycle and finished ammonium. The flow order is pure water section → low concentrated ammonia section → high concentrated ammonia section → finished ammonia section. The finished ammonia left from the bottom of the absorption system.

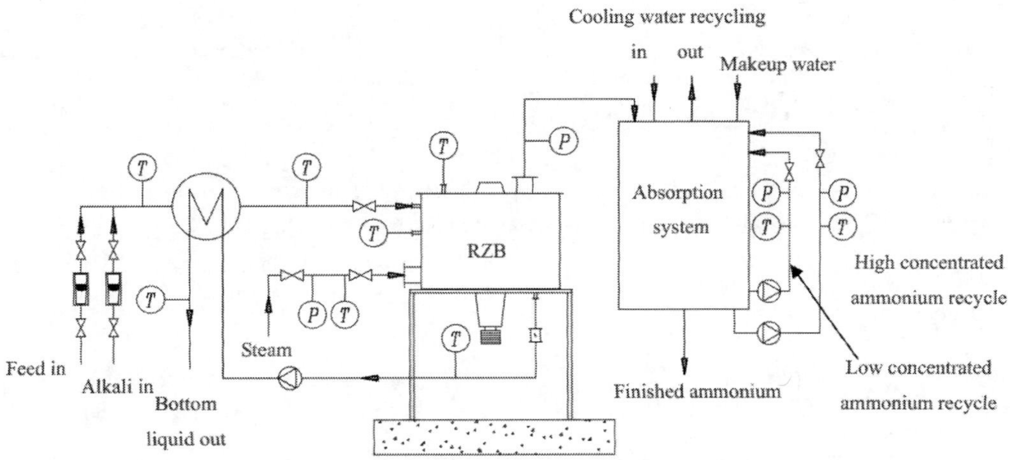

Fig. 3 flowsheet of industrial application of RZB
-Pump (P)-Manometer (T)-Thermometer

Actual operating parameters are as follows: the feed rate 1500kg/hr, ammonium concentration 18g/L, pH value 11.5-12, outlet vapor temperature about 80℃, no reflux, finished ammonia concentration > 25%, ammonia in outlet liquid<30mg/L, unit consumption of steam 90kg/ton feed, bottom temperature about 98℃. the temperature approach between inlet steam and heated feed was about 10℃. The customer was very satisfied with the results.

4. Conclusion

About 400 units of RZB used in a variety of industrial distillation processes demonstrated its higher separation ability. Its successful application in stripping ammonium from the wastewater was another good example.

RZB has a small volume with a height of lower than 1m. For traditional enterprises with small unused space in the factory, this kind of device can find its place here.

In this system, feed was heated by bottom outlet stream which sensible heat was recovered. So the system shows energy-saving effect.

5. Prospects

Modification of existing RZB device can realize liquid cut off between its rotors, which can extend its applications in recovery of ammonium from the wastewater. For RZB with more than three rotors, its top and bottom section perform absorption and stripping respectively, thus these two processes can be integrated in single equipment.

Reference

1 Ramshaw, C. (1983)., "Higee distillation—An example of process intensification, " Chem. Eng., February, pp. 13–14

2 Rao, D.P., Bhowal, A. and Goswami, P.S. (2004), "Process intensification inrotating packed beds(HIGEE): An appraisal," Ind. Eng. Chem. Res., pp. 1150-1162.

3 Li, X.P., Liu, Y.Z. and Li, Z.Q. (2008), "Continues distillation experiment with rotating packed bed," Chin. J. Chem. Eng., pp. 656-662.

4 Wang, G.Q., Xu, Z.C., Yu, Y.L.and Ji, J.B. (2008), "Performance of a rotating zigzag bed—A new HIGEE," Chem Eng Process, November, pp. 2131–2139.

5 Ji, J.B. , etc. (2004), "Equipment of zigzag high-gravity rotating beds," CN Pat., 01134321.4.

9TH GLOBAL CONGRESS ON PROCESS SAFETY

Operational Discipline: Panacea for Performance?

James A. Klein
North America Operations SHE, DuPont

Brian D. Rains
DuPont Sustainable Solutions, DuPont

Prepared for Presentation at
American Institute of Chemical Engineers
2013 Spring Meeting
9th Global Congress on Process Safety
San Antonio, Texas
April 28 – May 1, 2013

UNPUBLISHED

Operational Discipline: Panacea for Performance?

James A. Klein
North America Operations SHE, DuPont

Brian D. Rains
DuPont Sustainable Solutions, DuPont

Abstract

Over the last few years, many companies have had an increased focus on implementing or upgrading programs to improve Operational Discipline (OD) in pursuit of excellent PSM performance, leading to reduced process-related injury and incident rates. OD describes human behavior in following required systems, procedures, and practices to achieve reliable, high quality, and safe operations. DuPont defines OD as a deeply rooted dedication and commitment by each member of the organization to carry out each task, the right way, each time. But is a focus on OD enough? This paper will discuss the elements and precursors of a strong OD program, how leadership can help ensure successful improvement programs, and, ultimately, how a focus on OD and PSM performance must be integrated with operations and business processes throughout the life cycle of the facility.

Keywords: Operational Discipline, Safety Leadership, Process Safety

Introduction

DuPont has for many years defined Operational Discipline (OD) as the deeply rooted dedication and commitment by every member of an organization to carry out each task, the right way, each time [Klein 2005]. The corporate PSM standard requires that:

> *Sites shall implement and maintain programs and systems to achieve and maintain a high degree of OD for all PSM elements in a manner that supports business and operating objectives.*

Specific accountability for improving OD is assigned to line supervision, and sites are required to appoint OD Leadership Champions to help identify and support OD improvement activities. Ideally this person should meet the following criteria:

- A member of site line leadership (not SHE/PSM)
- Have a view of site operations and SHE, not just PSM
- Be able to impact site resources, priorities, and activities

The intent is that the Site OD Champion is a line leader who is able to provide significant leadership and influence site activities in support of improved OD. Successful OD efforts, of course, must involve all employees to achieve desired results.

Relationship of OD to PSM Performance

Once strong PSM systems have been implemented, the greatest opportunity to maintain and improve performance in avoidance of serious injuries and incidents often lies with a focus on improving OD. If people are not following system requirements, at all levels of the organization, then poor performance must result and the risk of serious injury and incidents necessarily increases. Safety programs are only as effective as the day-to-day commitment of everyone in an organization to rigorously follow system requirements correctly every time, changing safety systems from concept to reality in order to achieve desired performance. The failure of just one person in completing a job task correctly, just one time, can initiate or contribute to serious injuries and potentially catastrophic incidents. The design, implementation, and daily execution of effective PSM management systems are dependent on workers at all levels in the organization doing their job tasks correctly every time. Performance is ultimately a function of strong PSM systems and engaged workers rigorously following system requirements.

Are OD Programs Sufficient to Achieve Improved PSM Performance?

If PSM systems have been established for many years, such as for sites covered by the OSHA PSM regulation in the United States, is a new or renewed focus on OD sufficient? Most companies now have or can obtain the technical knowledge and capabilities to safely identify and manage process hazards and risks. As improved guidance and tools for implementing PSM have been developed, the focus on underlying causes of major incidents has shifted in many cases to understanding and improving organizational and cultural causes that relate to OD and other performance factors [Klein and Dharmavaram 2012].

What we've found, of course, is that strong PSM systems, culture, and OD excellence are closely related. Ensuring that many PSM systems are well-designed and implemented as planned is essential to achieving excellent OD. Excellent OD, in turn, is required to develop, implement, and maintain good PSM systems and to daily reflect the safety culture. The safety culture that underlies these efforts ultimately can support strong performance or may need to be improved to help address existing performance issues. This interdependence means simply that a focus on OD as a new or strengthened focus will probably not by itself be successful in achieving desired performance improvements. For example, good OD is certainly dependent on PSM systems and safety culture for development of comprehensive and useful operating procedures, training programs, and overall organizational capability, including fitness-for-duty programs. OD efforts that don't include a focus on these and other precursors as discussed in this paper are likely not going to achieve program goals.

<u>What's required for strong OD efforts?</u>

Strong Safety Culture

PSM/OD programs and performance will only be as effective as the underlying safety culture supports. If an organization has a weak safety culture, PSM efforts will probably be harmed by conflicting priorities, such as production needs or lowered cost. A strong safety culture, based on a commitment to SHE as a core value, and as evidenced by organizational policies, goals, metrics, and day-to-day decision-making, supports establishing strong PSM systems and achieving high levels of performance. Safety priorities are recognized, not as conflicting with other priorities, but rather as inherently necessary for completing any task the right way.

Focused Leadership

Management leadership and commitment which is visible throughout the organization is required for continued excellence in PSM. Beginning with the formation and fostering of a strong safety culture, continuing with policies that set high expectations for excellent safety performance, and implemented in day-to-day decisions that provide resources and establish high priority for safety-related activities, the actions of all levels of management must support and reinforce strong PSM programs and accountability.

Good OD will not happen on its own, but rather is something that requires continued management attention to implement appropriate systems, to monitor and evaluate actual performance, to demonstrate personal attention and dedication, to provide appropriate resources as needed, and to develop and support effective processes to facilitate employee understanding, involvement, and follow-through. The many benefits of OD in helping achieve world class manufacturing operations, reduce risk, and prevent injuries provide a basis for engaging leadership to support and maintain effective OD programs.

DuPont has begun implementing the Flame Model, which was developed to identify best practices for leadership processes in leading improvement of OD. Five key areas have been identified:

- **Future Direction** – An evergreen future-oriented vision process is in place that includes inter-dependent behavior that supports strong OD, has been well communicated and explained to engage site personnel, and is supported daily through leadership focus and behaviors.
- **Leadership Processes** – The leader is a role model for the site and actively works to build a strong leadership group committed to SHE and OD improvement. Coaching and other operations excellence practices are used to structure leadership activities as well as immerse site personnel in the desired culture.
- **Accountability** – Site and personal goals that include a focus on OD have been established with clear expectations on performance. People are challenged to contribute to improved OD, and commitments are periodically reviewed and reinforced. Recognition is provided as appropriate.
- **Managing Processes** – Metrics, audits, and other activities are defined to continually assess site progress and focus around OD improvement. Situational leadership practices are used to support site leaders and personnel as needed.
- **Engendering Trust** – Leaders focus on supporting a trustful work environment to engage and support site personnel through consistent leadership principles and behaviors. Programs are personalized in terms of impact to all employees, feedback is sought, and

inter-dependent behavior is encouraged. Site personnel know they are important to success and that their contribution is valued.

Effective Risk Management

Rigorous hazard identification, process hazard assessment, and risk evaluation and management are necessary for ensuring appropriate PSM systems are implemented to help achieve excellent PSM performance. If poor OD impacts risk management practices, then the necessary safeguards for risk management are not identified, implemented, or maintained, leading to performance problems. Risk management is essential in establishing the required foundations of effective PSM and OD programs. If this is not done well, a focus on OD will be like climbing a ladder placed on the wrong wall; it will be impossible to get to where you want to go.

Organizational Capability and Training

With increasing complexity associated with manufacturing processes, continuous addition of new PSM requirements, continued development of new RAGAGEPs and regulations, and potentially increased staffing turnover expected in the future, a focus on organizational capability and training is essential to achieving good OD. A well-defined training strategy should be developed and implemented, with refresher training provided at appropriate intervals. Effective management of change procedures associated with personnel turnover to ensure appropriate understanding of PSM systems and OD practices must be in place to maintain capability. In addition, fitness-for-duty programs based on alcohol, drug, and fatigue problems require continuous attention.

Well Designed Procedures

In many ways, the heart of a strong OD Culture is the Procedures. Within the procedures, the collective knowledge and know-how of an organization should be codified. Repeatable, excellent results are ideally delivered by the organization by robustly following the procedures. The discipline and organizational confidence to comply with procedural requirements comes foundational from the belief that the procedures are helpful, accurate, and ultimately, the best way to achieve the desired results.

Eleven steps for strengthening OD through High Quality Procedures have been developed. Details are summarized below:

1. A Procedures "Standard" that details requirements
2. Single Person Owners for every procedure
3. Well written, easily followed and visual
4. Consistent, comprehensive and user friendly content
5. Importance and Risk Classified
6. Routine Reviews by Users and Subject Matter Experts
7. Appropriate Verification
8. Change Management Process
9. Initial & Refresher Training plus Assessments

10. Revisions Controls while maintain Accessibility
11. Metrics and Audits

<u>What else is needed to achieve the desired PSM Performance?</u>

Clearly defined Roles and Responsibilities starting at the top

At DuPont, like most companies, Safety is a key performance measure. DuPont considers Safety a Core Value. Other companies express the relative importance of safety in other ways. All responsible companies are committed to keeping their employees safe. But Process Safety, as recent experiences have indicated, cannot be assessed with a high degree of accuracy with safety statistics only. Process Safety results will also be reflected in other key performance indicators including environmental performance, company reputation, asset productivity, business/supply chain continuity, and even employee morale. Because Process Safety performance is so critical, leadership responsibilities from the top of company must be delineated.

The attached describes the two key organizational functions. On the right side is Corporate Competency and how the DuPont Company is organized to ensure we are utilizing the best available knowledge and experience. The left side indicates the responsibilities of Line Management beginning with the CEO and the company's Board of Directors. Fundamentally, the Corporate Competency Team owns "What" the company program will be. Line Management is responsible for the "How, When, Where, and Who" of execution. In the DuPont PSM Standard, 14 pages of text are used to describe these duties for organizations within the Company, including Global, Regional and Site Operations, Corporate HSE and Engineering, Manufacturing Technology, Sourcing, Integrated Health and Business Leadership.

Said succinctly as written in the DuPont PSM Standard:

"Line management has the responsibility to implement this standard."

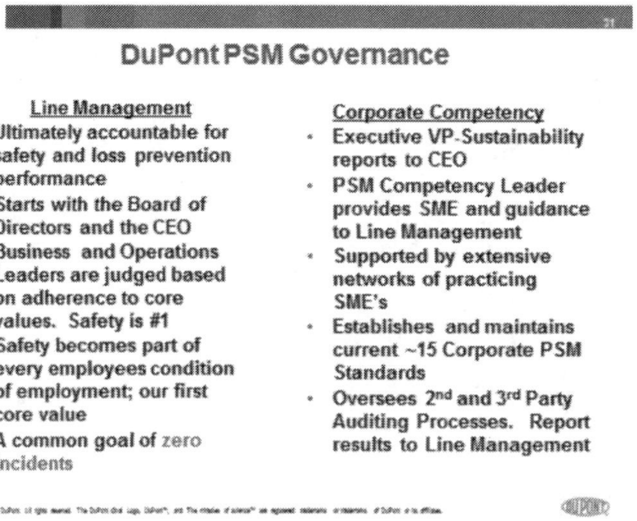

Operations & Business Integration

In recent years, the PSM technical requirements have been further incorporated into broader operations and production systems, including Operational Excellence (i.e. OEMS.) This degree of integration is critical as Process Safety Management must not be viewed by an organization as a Safety initiative, but rather how "we do business" within operations. The degree to which the PSM Managing System is adopted and integrated and to which all employees are involved in those elements that touch their day to day work will speak volumes about the quality of the overall PSM Managing System deployment and implementation. And will ultimately determine the organizations success in reducing the frequency and severity of Process Safety Incidents.

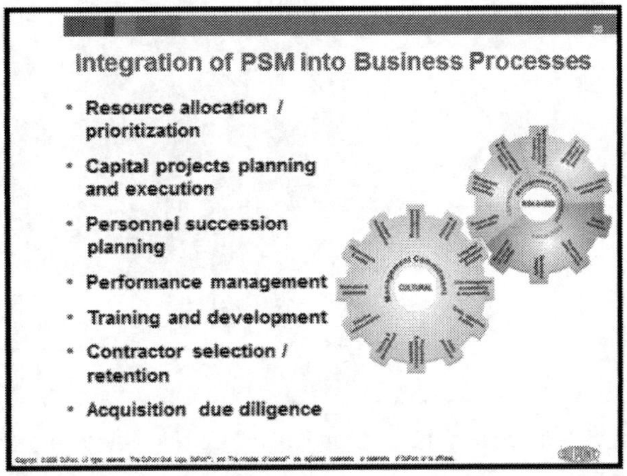

Strong Managing Processes

What brings together the Technical System Requirements with the Cultural requirements for successful implementation of Process Safety Management, including OD, is the PSM Managing System. Recommendation #2 from the Baker Panel Report into the BP-Texas City Incident described the recommended approach with the following words:

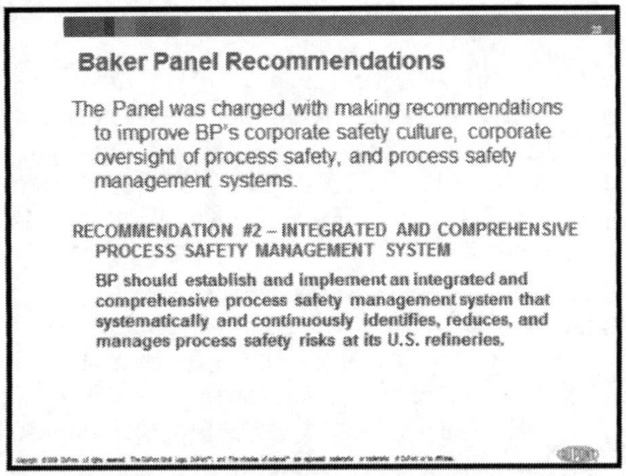

DuPont has graphically displayed the interdependent nature between leadership, culture and technical requirements of PSM Implementation for more than 20 years using the "PSM Wheel" also shown below.

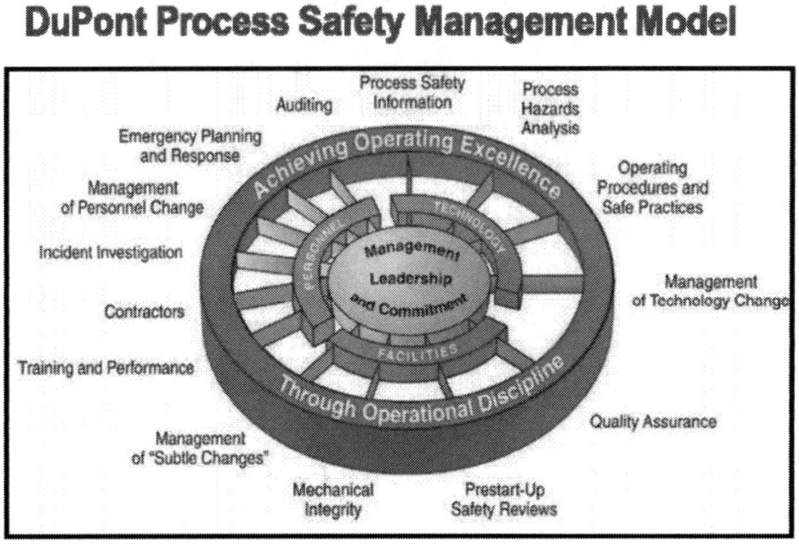

Management's Leadership and Commitment is in the center of the wheel where energy, drive and power and provided. From the center flows the resolve to make the wheel function as a smooth and single unit. Strength is delivered from the center by the 14 individual spokes or technical elements. Each spoke is important and weakness in any spoke will cause the wheel's "ride" to be uneven and sub-optimum. But where the "rubber meets the road" and where true risk mitigation occurs is in the outer rim that comes in contact with daily operations. Operational Excellence is achieved through rigorous adherence to the managing system requirements via a strong culture of OD. OD transforms an "on paper" managing system into an effective "in action" managing system.

PSM Standards incorporated into the Design/Build of new Facilities

An all too frequent conversation with companies around the world has been that the standard to which operating facilities are managing their risks (via a documented PSM System) is not the same standard that was used by the organizations that designed and built this same facility. The observation is usually expressed in the attitude of a complaint, as it is an enormous challenge to "catch up" the facility's process safety systems when those who delivered it to them "didn't do their job."

One area of particular weakness seems to be the management of Process Safety Information at the interface between the Project Teams and the Operations Teams. Too frequently there has not been an acceptable plan with clear ownership for such key project documents as Process Hazards Analyses, Quality Assurance inspections, Process Flow Diagrams, and Process & Instrument Diagrams. Once this information is lost, either literally or figuratively, it is extremely difficult (expensive and painful) to re-construct. Millions of dollars can easily be spent to re-create even a basic level of Process Safety Information.

"Beginning with the end in mind" is essential in the Design and Construction Phases of any capital project. Including as members of the capital project team the key, future operations, technical and maintenance leaders is a very effective way of ensuring operational continuity. It will also contribute to a common approach of PSM standards implementation in the design and construction of a new facility.

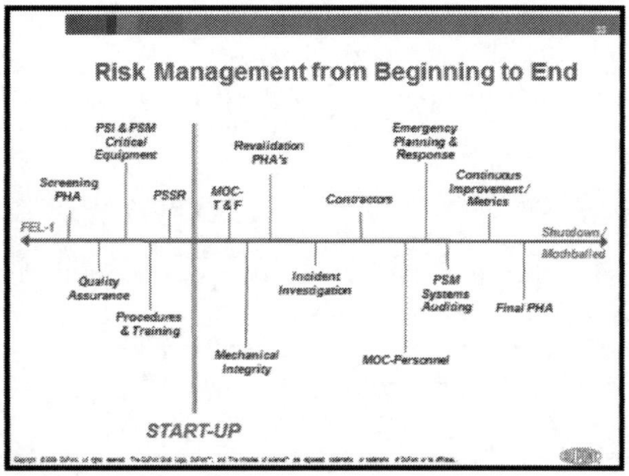

Active Employee Ownership and Involvement

Employee Participation is a required element of PSM Programs at USA sites that are covered by the PSM Regulation 29 CFR 1910.119. Key language in the regulation includes:

- *Employers shall develop a written plan of action regarding the implementation of the employee participation required by this paragraph.*
- *Employers shall consult with employees and their representatives on the conduct and development of process hazards analyses and on the development of the other elements of process safety management in this standard.*
- *Employers shall provide to employees and their representatives access to process hazard analyses and to all other information required to be developed under this standard.*

Compliance with these requirements is not optional but employee participation should not end here. True employee involvement and ownership of the PSM system should be a critical leadership objective and an essential part of the organization's safety and PSM culture.

From DuPont's, Process Safety Management Standard:

> *"Site leaders shall play a key role and be accountable for establishing and sustaining site safety culture, PSM expectations, accountabilities and objectives; resourcing program elements; providing reward and recognition; and measuring ongoing*

performance on a routine basis. Site management shall be responsible for demonstrating ongoing management leadership, commitment, and personal involvement in site PSM programs plus providing for and encouraging employee involvement."

Specific Employee Involvement requirements are detailed below:

Providing for and Encourage Employee Involvement
Effective PSM relies on both the individual and collective efforts of everyone involved in managing process safety. Recognizing that employees are uniquely knowledgeable about the process and are "key managers" of process safety (often the only "managers" immediately available), it is important that site management provide for and encourage a broad spectrum of employee involvement in the design, implementation, and ongoing operation of the site's PSM program within all elements.

Sites shall maintain programs to achieve employee involvement in process safety activities, including the following:

- *Membership on site PSM committees, subcommittees, or special project teams*
- *Participation in PHA teams*
- *Participation in incident investigation teams*
- *Participation in pre-start-up reviews*
- *Conducting equipment tests and inspections*
- *Writing and/or reviewing operating and maintenance procedures*
- *Participation in development of training procedures and programs*
- *Participation in process safety audits*
- *Participation in emergency response planning, drills, and critiques*
- *Participation in MOC-technology (MOC-T) and MOC-subtle facility (MOC-SF) change reviews and implementation*
- *Participation in preparations and implementation of MOC-Personnel changes*

Within each element of PSM and as part of the site PSM policy, sites shall solicit, on an ongoing basis, participation and input from employees regarding ideas for improving and strengthening site PSM programs and systems. Sites shall also establish, document, and communicate local mechanisms and accountabilities to collect, track, and respond to employee PSM concerns and suggestions.

Continuous Improvement driven by Metrics, Audits and ever-strengthening Standards

The goal of all Process Safety Management systems is the mitigation and control of operational risks such that significant process safety incidents are eliminated. Although great progress has been made over the decades that PSM has been practiced, all can agree that the desired "Goal of Zero" has not yet been consistently achieved and that more work is needed. Continuous Improvement of the robustness of the managing systems and rigor in their execution is needed.

Three tools that organizations have to drive PSM Continuous Improvement are metrics (lagging and leading), audits (first, second and third party) and ever strengthening PSM system requirements and standards.

Performance metrics have received considerable attention as we also seek to focus in on those vital few indicators that most accurately describe overall system health and which most consistently and accurately predict future process safety incident performance. Collaborative efforts sponsored by CCPS and API plus the Responsible Care Initiative have brought much new thinking to this field. DuPont has elected to collect corporately on a quarterly basis 10 fundamental metrics, as described below. Additional leading metrics are kept locally for periodic review by site and regional leaders. Finally, additional near-miss and other learning event metrics are recommended but remain optional for local leaders to consider in the journey via continuous improvement towards the Goal of Zero.

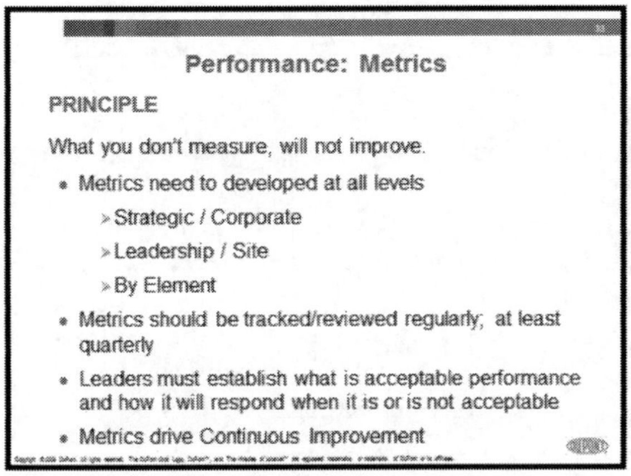

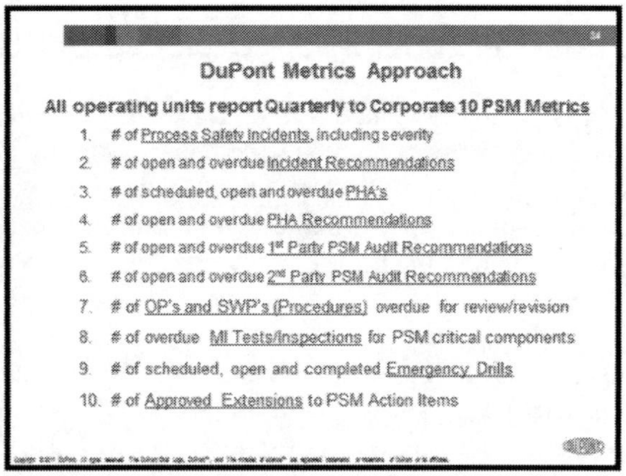

Proactive auditing of both the cultural and technical elements of PSM can provide leaders and the organization at-large valuable insight as to the relative strength of each. Insightful leaders understand the importance of seeking and obtaining unfiltered data that provides a clear picture of reality. Only then can leaders be confident that their decisions will achieve the desired outcomes.

First Party Auditing should be the most effective as long as the organization approaches the audit opportunity with brutal honesty and in the spirit of improvement at all costs. Second Party Auditing can provide valuable insight from outside the unit while keeping the learning's and auditor development within the company. Third Party Auditing should provide the most independent and potentially most valuable audit recommendations. Each should be incorporated into a comprehensive PSM Audit Program.

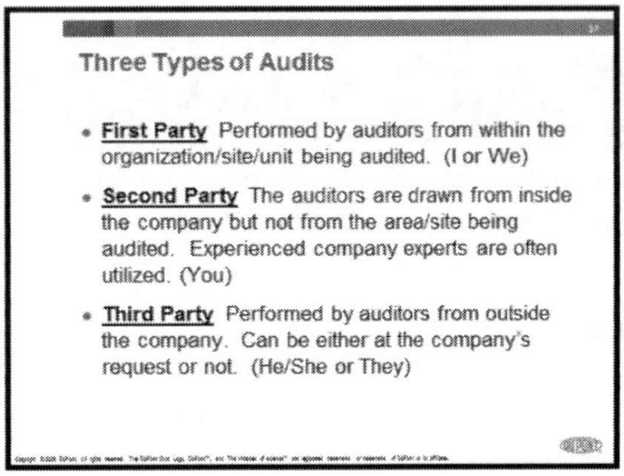

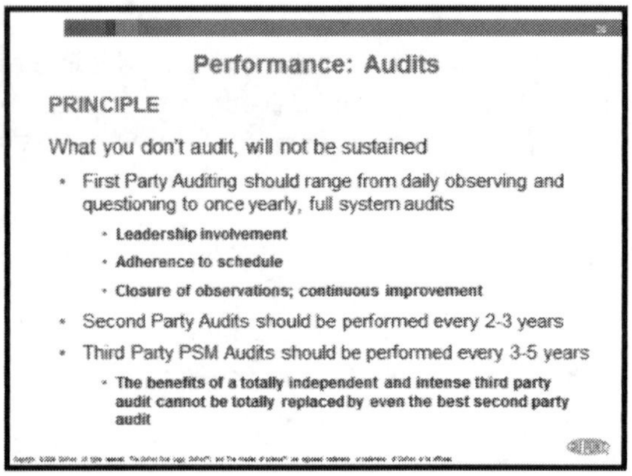

Many, if not most medium to large sized companies, especially in geographies that are not covered by prescriptive government regulations have developed comprehensive PSM Standards. The standard is intended to produce consistent implementation of company best practices and the highest level of performance. As mentioned earlier, PSM requirements are increasingly being integrated and incorporated into broader Operations Excellence or Asset Integrity Managing Systems.

One of challenges of establishing a PSM Standard is keeping it current as company and industry knowledge advances. Assuming that PSM Audits are based upon and conducted against this PSM Standard, PSM Audit questions must also be kept current. This requires effort and commitment.

The DuPont Company issued its first comprehensive Process Safety Management Standard in 1979. It was one of the approaches or systems to achieve PSM excellence that was considered when OSHA promulgated 29 CFR 1910.119. One might think that with such a rich history, the standard would be achieving a level of relative maturity and that changes would be minimal. In actual fact, the opposite has occurred. In the last 5 years, three revisions to the DuPont PSM Standard have been finalized and issued. And as detailed below, quite a number fairly significant changes have been introduced and incorporated into these three revisions.

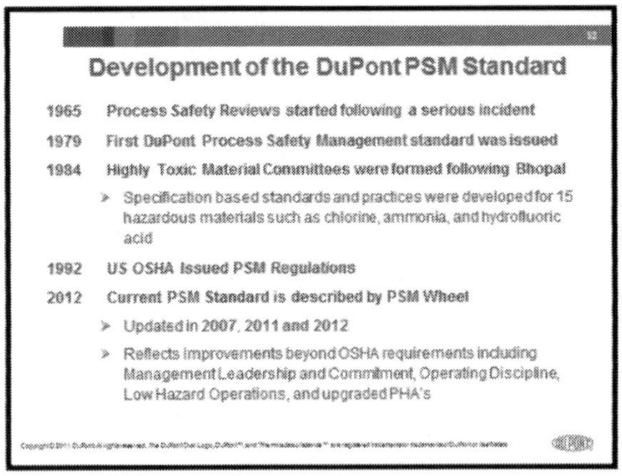

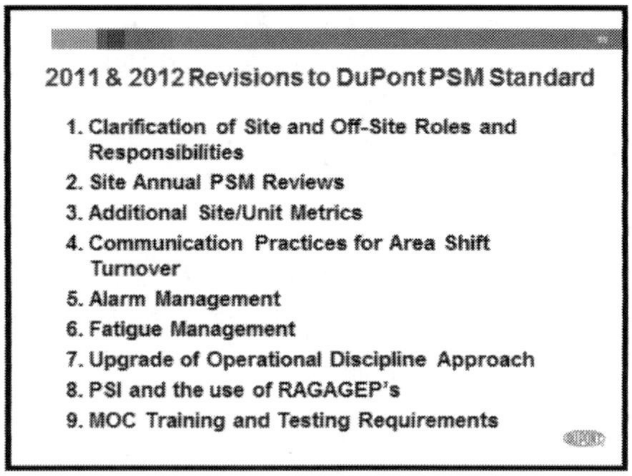

Continuous improvement in the standards can quickly elevate the entire company's PSM managing systems and its performance. Metrics, Audits and Standards are three of the most effective tools for driving and achieving PSM excellence via Continuous Improvement.

Conclusion

Process Safety Management has as its overall objective to mitigate operational risks such that process related incidents and injuries are prevented; DuPont's commitment is to zero. Organizations that embrace Operational Discipline are striving for excellence by having every member carrying out each task the right way, every time. As important as Operational Discipline is to achieving Process Safety Excellence, it should not be viewed as a panacea. It is an essential ingredient but in and of itself, is not sufficient.

What is needed to progress are robust PSM Systems, a strong Safety Culture and OD Excellence. These three legs of the PSM stool are highly interdependent and are mutually supportive. OD excellence needs focused leadership, effective risk management, organizational capability and well-designed procedures. A strong safety culture demands clearly defined roles, employee ownership/involvement and the integration of safety into operations and business processes. A robust PSM system requires an integrated and comprehensive approach, incorporation into the design and installation of new facilities, and continuous improvement as demonstrated by metrics, audits and ever-strengthening standards.

Knowing what to do is critical to preventing process safety incidents. Doing what we know saves lives.

References:

1 James A. Klein, Operational Discipline in the Workplace, Process Safety Progress, Vol. 24, No. 4, p. 228-235, December, 2005

2 James A. Klein and S. Dharmavaram, Improving the Performance of Established PSM Programs, Process Safety Progress, Vol. 31, No. 3, p. 261-65, Sept., 2012

(43m) Process Safety Information Management - It Is More Than You Probably Think

Mason Martin, Process Safety Management, Phillips 66, Borger, TX and **Patrick Nonhof**, Provenance Consulting, Borger, TX

Many businesses continue to struggle with managing the expectations of the Process Safety Management standard--even twenty years after the rule was first promulgated. In particular, the sheer enormity of Process Safety Information (PSI) continues to overwhelm many organizations. PSI is a foundational element of PSM upon which other elements such as Process Hazards Analysis are directly dependent. Relationships to the other elements may be less obvious, but PSI is ultimately the foundation upon which all risk-based decisions are based. PSI has to be based on the accuracy of core documents; tools must be created to identify and correct any discrepancies among the various systems that generate, store, and use PSI; and the same tools can be used to drive alignment outward into the other elements. By relying on these fundamentals, this paper details a system that can be put in place by management to improve a site's process safety culture through facilitating a better understanding of PSI and the inherent connections to other elements of the PSM standard.

Connecting the Dots – Managing the Little Things That Lead to Big Accidents

Dave Jones
Chevron Energy Technology Company
6001 Bollinger Canyon Road, San Ramon, CA 94583
dwjones@chevron.com

Anne O'Neal
Chevron Oronite
6001 Bollinger Canyon Road, San Ramon, CA 94583
anneoneal@chevron.com

Prepared for Presentation at
American Institute of Chemical Engineers
2013 Spring Meeting
9th Global Congress on Process Safety
San Antonio, Texas
April 28 – May 1, 2013

UNPUBLISHED

Connecting the Dots – Managing the Little Things That Lead to Big Accidents

Dave Jones
Chevron Energy Technology Company
6001 Bollinger Canyon Road, San Ramon, CA 94583
dwjones@chevron.com

Anne O'Neal
Chevron Oronite
6001 Bollinger Canyon Road, San Ramon, CA 94583
anneoneal@chevron.com

Keywords: Major accidents, details, dots, culture

Abstract

Major accidents often result from a failure to attend to seemingly little details. Examples might include an improperly designed injection point; a procedure that is mostly, but not completely followed; a small part with the wrong metallurgy; or other design, maintenance or operating discrepancies. In many cases, system flaws can exist for extended periods of time, detected or undetected, with little or no consequence. However, on occasion, and with the right initiating event, they can combine with other system flaws, allowing an unforeseen chain of events to propagate into a major accident.

This paper will review aggregated information from a range of process safety incidents, including "subtle" accidents resulting from a series of "little" failures. It will highlight the management system breakdowns that contributed to the incidents, and examine the complex interactions among people, process safety systems, and equipment as orchestrated by the overall organizational culture. The paper will reinforce the concept that an effective process safety culture is key to "connecting the dots" that, if unchecked, can result in a major accident.

1. Introduction

Following the Piper Alpha tragedy in the North Sea, in which 167 men lost their lives, Dr. Tony Barrell of the United Kingdom Health and Safety Executive said, "I wouldn't put it above or below other disasters. There is, actually, an awful sameness about these incidents."[1]

The "awful sameness" to which Dr. Barrell refers may reflect an observation that, with rare exception, all process safety accidents can be described within a common, recurring group of management system failures. For example, a failure to manage change was a key contributor to process safety events at Flixborough and Chernobyl, among others.[2] Similarly, the accidents at

Longford and Grangemouth were attributed to inadequate hazard identification/risk assessment.[2] Permit-to-work systems/safe work practice failures were identified as contributing to the Piper Alpha and Pasadena Texas (1989) fires and explosions.[2]

Clearly, major accidents have multiple management system contributors. The Bhopal tragedy involved, among other things, breakdowns in hazard identification/risk management, management of change, asset integrity, incident investigation, process safety competency, and emergency response.[2] Other events have different combinations of contributing factors.

However, when we evaluate a range of process safety incidents, we find that the management system failures leading to these events begin to repeat themselves into well defined categories. The CCPS *Guidelines for Risk Based Process Safety*[3] represent, in the authors' view, the current best compilation of these categories, often called elements (CCPS elements). The CCPS book discusses 20 elements altogether, ranging from process safety culture to management review and continuous improvement – and including key areas such as risk analysis, asset integrity and auditing. For a complete list and comprehensive treatment of these elements, see reference 3.

In this paper we will attempt to show that while the process safety management systems themselves are relatively simple, their actual application requires attention to innumerable "little things," any one of which has the potential to initiate or propagate a major accident. Identifying and managing these "little things" necessitates a combination of "big picture" thinking and rigorous attention to detail across all elements of process safety management and all relevant organizational functions.

2. "Little Things" That need to be Managed

In this section we will illustrate the dynamic nature and seemingly limitless numbers of details at work within each of the process safety management systems and highlight some of the complexities involved in managing them.

2.1 Accident Models

Many accident models have been developed. These include variations on the "Swiss Cheese" model, first proposed by James Reason[4] and the "Bow Tie" model emerging out of European and UK risk assessment applications. More recently, Nancy Leveson from the Massachusetts Institute of Technology has proposed the System-Theoretic Accident Model and Process (STAMP) [5] which attempts to include human error and culture into the accident equation. In the authors' view, all of these models have merit, yet none is perfect. For internal educational purposes Chevron uses a hybrid model as shown in Figure 1.

It is worth noting that each of the above models characterizes large-scale accidents as a coming together of multiple system failures. This theme is directly counter to the traditional concept of "double jeopardy" which holds that the combined failure of two independent systems is so unlikely as to not require special consideration. Industry accidents over the last 30 years have clearly demonstrated the fallacy of a "double jeopardy" approach to process safety.

2.2 Dynamic Nature of Process Safety

The "Swiss cheese" shown in Figure 1 represents basic prevention equipment, engineering controls, software, human actions, etc. These include facility-specific procedures, training, protective devices, inspection programs, and design integrity, among others and are the real-world manifestations of process safety management systems. They are the mechanisms that actually intercede to stop an accident from starting or propagating.

Figure 1 – Hybrid Accident Model

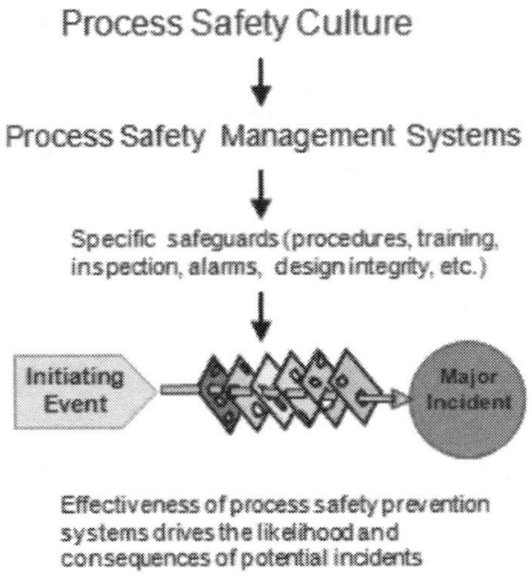

Driving the effectiveness of the "Swiss cheese" are the process safety management systems themselves - the CCPS Elements. Ultimately, the organizational culture as applied to process safety determines how well the management systems perform. In real time, each component of this and any other accident model is in constant flux. For example:

- The effectiveness of safeguards can change;
  - Safety systems can degrade, go untested, be compromised, or improve in functionality
  - Corrosion mechanisms can increase or decrease, or be introduced to previously unaffected areas
  - Procedures and training can vary in effectiveness, depending upon facility changes
- Process safety management systems change;
  - Critical controls such as management of change can get overwhelmed, allowing higher risk changes to "sneak" through
  - Personnel movements can lead to lost understanding of risks, or bring new, beneficial approaches
  - Process safety information can become quickly out of date
- Overriding process safety culture can degrade or strengthen with retirements and reassignments of key personnel, acquisitions or external pressures.

Given the constant change experienced by each of the above components in the accident chain, it is clear that process safety can never be viewed as a static condition. There is no milestone that can be achieved and final "victory" declared over major accidents. Effective process safety management approaches all aspects of the accident model with a sense of vulnerability, aware that one or multiple components may be compromised at any given time. Ongoing diligent monitoring and continuous improvement are the best defense.

2.3 Management Systems vs. Facility Complexity

The distillation of process safety into 20 key elements can mask the magnitude of the effort required to prevent major accidents. Given facility complexities, large equipment counts, variations in administrative controls, and human performance, the potential for flawed execution within each element is seemingly endless.

As an example, a process safety management system for Asset Integrity and Reliability requires that safety-critical shutdown systems be identified and tested at prescribed frequencies and repaired as necessary. Some of the factors that could influence how well this fundamental requirement is implemented for a specific shutdown system are suggested in Table 1.

Table 1 – Factors Potentially Compromising A Specific Shutdown System Integrity
• Failure to identify the system as safety critical • System identified but not included in testing program • System not designed to be testable on line • System designed to be mostly testable, except for final control element • System designed to be mostly testable, except for the process connection valves subject to plugging • Testing poses a potential to shut down or damage the facility • Procedure for testing is complex • Modifications to system design not included in testing procedures • Personnel not trained in testing the system • Shutdown system bypassed during startup and not reactivated • Shutdown system bypassed because of threat of spurious trips • Importance of shutdown system not understood • Software modification altering shutdown system set point • Internal control system failure • Testing poorly documented • Identified non-conformance not acted upon • ....ad infinitum

The list in Table 1, which is by no means exhaustive, helps illustrate the challenge in implementing just one aspect of one process safety management system for one piece of equipment. Nor does the above list include the human performance or cultural forces that can affect the viability of the system tests, such as operating distractions or lack of management emphasis and monitoring. More importantly, any one of the above shortcomings in execution could become part of a sequence of events ultimately leading to an accident.

If we began to construct similar facility-specific lists for all the CCPS Elements, applied to the entire inventory of equipment and activities, a picture would start to emerge of the enormous variety and complexity of the "little things" that can degrade process safety and ultimately contribute to a major accident. We don't have to look too far to find verification that it is these little things that present the challenge. In July 2000, the crash of the Air France Concorde jet was attributed to a small strip of titanium that had fallen off another airplane minutes earlier.[6] In 1980 the floating dormitory for North Sea oil workers *Alexander Keilland* sank, killing 123 people, when a cracked strut failed. A shipyard painter had painted over the crack during maintenance so it went undetected. [6] Similarly, one link in the accident chain that caused the Ocean Ranger drill ship to go down in 1982 off the coast of Newfoundland, killing 54 people on board, was the failure of small glass porthole window allowing seawater to enter the ballast control room.[6]

2.4 Picture the "Dots"

If we were to graph the relationships among the management systems and the "little things" ultimately precipitating an accident, they might look like Figure 2.

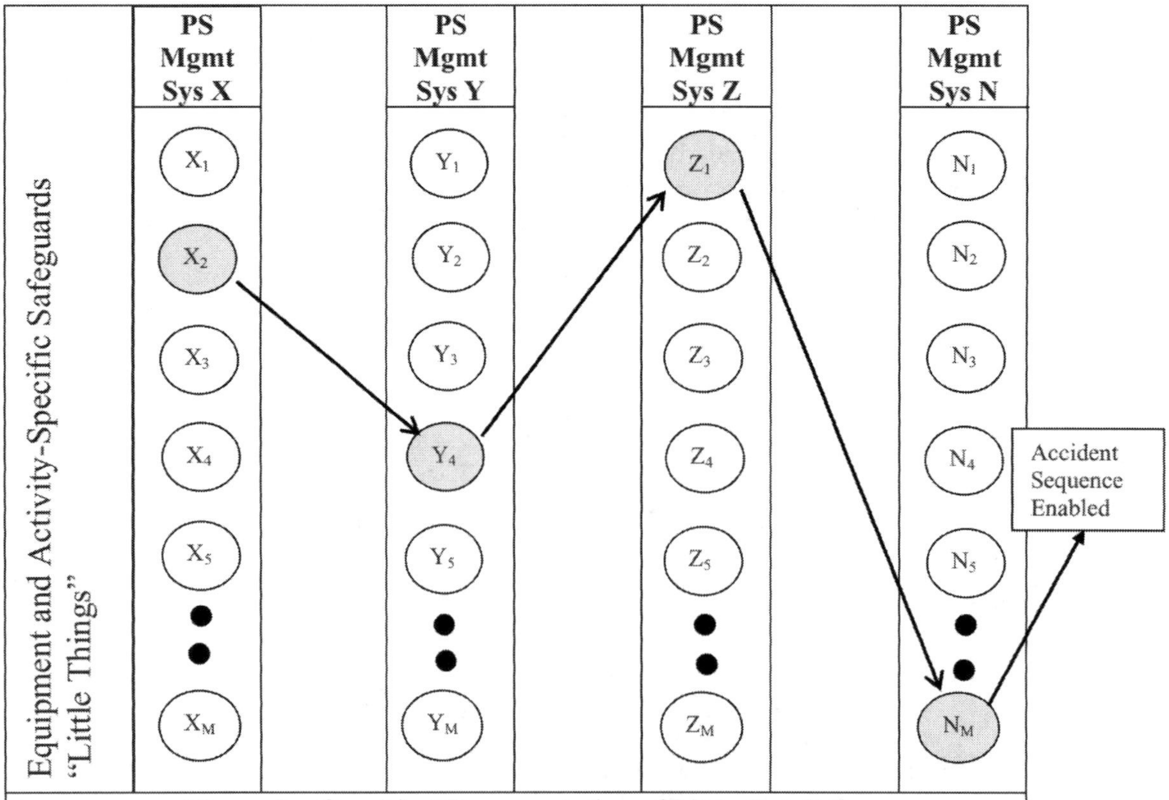

Figure 2 – Graphical Representation of Little Details"

3. Case Study[7]

This case study helps to illustrate the complex nature of the "little things" associated with just a couple of management systems – Operating Procedures, Training and Performance Assurance, and Asset Integrity and Reliability.

Figure 3 – LPG Treater Flow Diagram

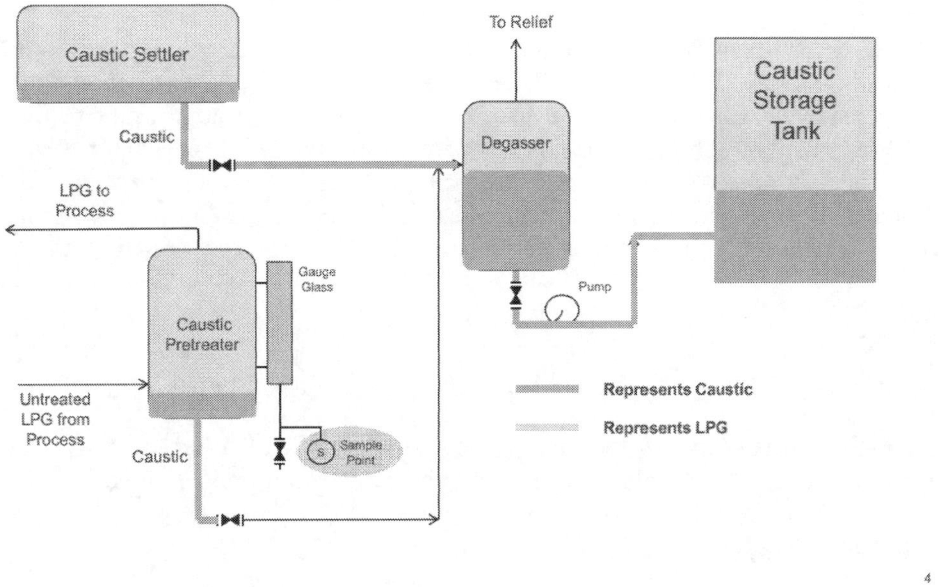

Figure 3 shows the flow diagram for a fairly standard process for treating liquefied petroleum gas (LPG) with caustic (sodium hydroxide) to remove certain impurities prior to further processing. The green represent caustic inventories, while the remaining volumes in the Caustic Pretreater are LPG. The Caustic Settler has an LPG inventory as well a vapor space. While performing a routine duty to drain caustic from the Caustic Pretreater, the draining operation was continued to the point where all the caustic material was removed and LPG was directed to the atmospheric Caustic Storage Tank resulting in a loss of containment at the tank. There were no injuries related to this incident.

3.1 Sequence of Events

The operator's routine duties included periodically transferring caustic from both the Pretreater and Settler to the Degasser and then to the Storage Tank. After sampling the Pretreater for caustic, the operator opened the manual valve on the outlet of the Pretreater to start the caustic transfer. The operator also started the Degasser pump to transfer caustic from the Degasser to the Storage Tank. The gauge glass on the Pretreater was dirty and difficult to read, but it was assumed that the caustic level was above the gauge glass. The actual level was most likely below the gauge glass but above the sample point.

The Pretreater level was then depleted by the operator, with LPG going to the Degasser and with flaring through the relief system increasing. The valve from the Pretreater was then closed. Caustic transfer was then started from the Settler, and the valve from the Pretreater was reopened, sending pure LPG to the Degasser from the Pretreater. The operator saw increased flaring as confirmation that procedures were being done properly, since two vessels were being fed to the Degasser, with increased vapor release from the entrained LPG in caustic. Some of the increased flaring was from the pure LPG being sent from the Pretreater.

With the Degasser pump continuing to run, the Degasser caustic level was depleted and LPG was transferred to the Storage Tank. The low level alarm on the Degasser was not functioning. As the level of caustic and LPG increased in the Storage Tank the operator stopped flow from the Settler and Pretreater. Attempts were then made to bring the level of caustic back up in the Degasser by opening the Pretreater valve, which sent more LPG to the Degasser since it remained depleted of caustic. Shortly afterwards, the Storage Tank high level alarm sounded, and material was released from the top of storage tank.

3.2 Analysis

The above incident description suggests that, as with many other accidents, there were "a lot of things going on" increasing the potential for incomplete diagnosis and overlooking of details. Focusing initially on just three of the management systems (CCPS Elements) involved we can see firsthand some of the "little things" that need to be managed to prevent incidents of this type.

3.2.1 Operating Procedures

The procedure for transferring caustic from the Pretreater did not specify whether the Degasser pump should be used simultaneously to the transfer. The purpose of such a restriction is to avoid inadvertently sending LPG to the Storage Tank, and also to allow sufficient retention time in the Degasser for disengagement of any entrained LPG from the caustic. Had either one of these process-related details been included in the operating procedures and training, it would have provided operating personnel with a broader view of the potential consequences related to this transfer. It might have helped them "connect the dots."

The investigation also concluded that the operator was monitoring too many items at once and did not notice that the caustic level in the Degasser had been depleted. Since the Degasser level alarm was dysfunctional, this would have meant that the operator would have to verify the level manually. This would be another "little thing" that might have compensated for the level alarm failure.

3.2.2 Training and Performance Assurance

The gauge glass on the Pretreater was difficult to read. Although included in field training the technique for blowing down and clearing the gauge glass was not fully understood by the operator. Improved training learning objectives might have provided the operator with a better understanding of the Pretreater level.

Risks and precautionary steps associated with LPG transfer during caustic transfer were not fully understood after classroom and field training. Improved training and learning objectives might have aided the operator in detecting potential LPG carryover to the Storage Tank.

3.2.3 Asset Integrity and Reliability

The Degasser low level alarm did not function correctly. It had been falsely indicating a low level condition for several months prior to the incident. A previous alarm replacement did not address the root cause of the malfunction.

3.3 Process Safety is all About Details

As the "Swiss Cheese" model suggests, had any one of the "Swiss cheese" layers been effective, this incident might not have progressed. It is also worth noting that the actual incident investigation went beyond the three process safety systems indicated above, in alignment with other areas identified in Figure 1.

Even the above limited review illustrates the many "little things" that contribute to an incident, and the many seemingly small opportunities for preventing it. Slightly modified operating procedures, better training on level gauges, clearer understanding of risks in transferring LPG, and others are all minor improvements on systems and activities already in place that might have stopped the accident from progressing It is this little bit extra that often means a difference between a non-event and an accident. For example, this can mean going one step further in training, adding one additional precaution in an operating procedure and, of course, ensuring that all alarms, even those perceived as less important, are functioning properly.

This incident also illustrates an important process safety opportunity. How do we know when one "little thing" can interact with another, progressing the accident sequence? How do we "connect the dots?" Put another way, how do we broaden our thinking so that when something abnormal is occurring in one area of an operation we are aware that other areas can potentially be impacted? For example, as the operator was having difficulty reading the Pretreater level gauge, the additional knowledge that the Degasser level alarm was not functional and that LPG could be sent to the Storage Tank might have driven different actions.

4. Connecting and Managing the "Little Things"

If we extend the lessons from the Case Study to a facility or even company-wide scale, several key focus areas are suggested that might help to prevent similar incidents.

4.1 Understand that Details Drive Accidents

Major accidents almost always come down to "little things" done, not done, or performed incompletely. This extends beyond the operating and maintenance roles into engineering and management. For example, a major fire occurred on an offshore platform because the compressor seal oil pumps had no back-up power supply – an engineering detail overlooked. On the management side, the Space Shuttle Challenger represents a clear example of where management ignored the recommendations of engineers and failed to fully appreciate the catastrophic consequences that could result from an O-ring failure if a launch were to occur in cold weather.[2]

4.2 Get the Big Picture

Process safety elements are often viewed in silos, rather than as holistically interactive management systems that depend upon each other to function properly. When one system has a breakdown assumptions may be made that other prevention devices will be working properly. Statements like, "the shutdown system will handle it" or "it will be picked up on operator rounds" reflect this tendency to rely on other mechanisms to stop the event. These other safeguards may or may not be working well enough to prevent the scenario from developing.

Before determining that a serious potential consequence is unlikely to occur after an initial failure, management, engineering and operating personnel should look across the spectrum of affected process safety systems, and verify that "backup" safeguards are functional and sufficiently effective. In some cases, simple questions might help broaden the identification process and identify seemingly small, but important breakdowns in the protection layers:

- What else can be affected? (Texas City)[2]
- How cold can we go? (Challenger Space Shuttle, Longford Gas plant)[2]
- Where is the feed going? (Three Mile Island) [2]
- Was the job finished and ready to return to service? (Piper Alpha) [2]
- In what order do I open these valves? (Feyzin) [2]
- What happens if we turn this up? (Chernobyl) [2]
- What does my Technical Authority say about this? (Flixborough) [2]

For the accidents indicated in the parentheses above, the answers to the preceding questions are obvious in hindsight. The challenge is to ask the right questions ahead of time and encourage personnel to explore consequences and safeguards beyond the obvious.

Understanding the big picture can be complicated by the natural segregation of functional activities. Performing separate but interrelated tasks in silos can make it difficult to fully comprehend how one effort might create "dots" that negatively combine with the "dots" of another effort. A review of the sinking of the *Herald of Free Enterprise* at the Zeebrugge, Belgium port might help demonstrate this point.

Reference 2 discusses the process safety system failures involved in the Zeebrugge accident. The ferry departed port with the bow doors open, and capsized about a mile offshore. 193 lives were lost. The "little things" contributing this tragedy included: (1) nobody checked to verify the doors were closed; (2) there were no clear procedures on who exactly was responsible for checking the doors; and (3) the Assistant Bosun, who had the task of closing the doors, was asleep in his cabin and did not hear the ship intercom call to attend to stations. Other contributing factors included basic design flaws, incompatibility with berth designs and overloading.[2]

Specific process safety management systems involved included hazard identification and risk analysis, operating procedures and incident investigation (there were five previous incidents of sailing with the doors open). In addition, the ferry designs were modeled after World War II beach landing craft, which were intended for a radically different purpose. The application of these designs to the safe transport of people should have been (but were not) subjected to a rigorous management of change procedure.[2]

Figure 4 – The Complex Pattern of the Zeebrugge Accident – Taken From Rasmussen[11]

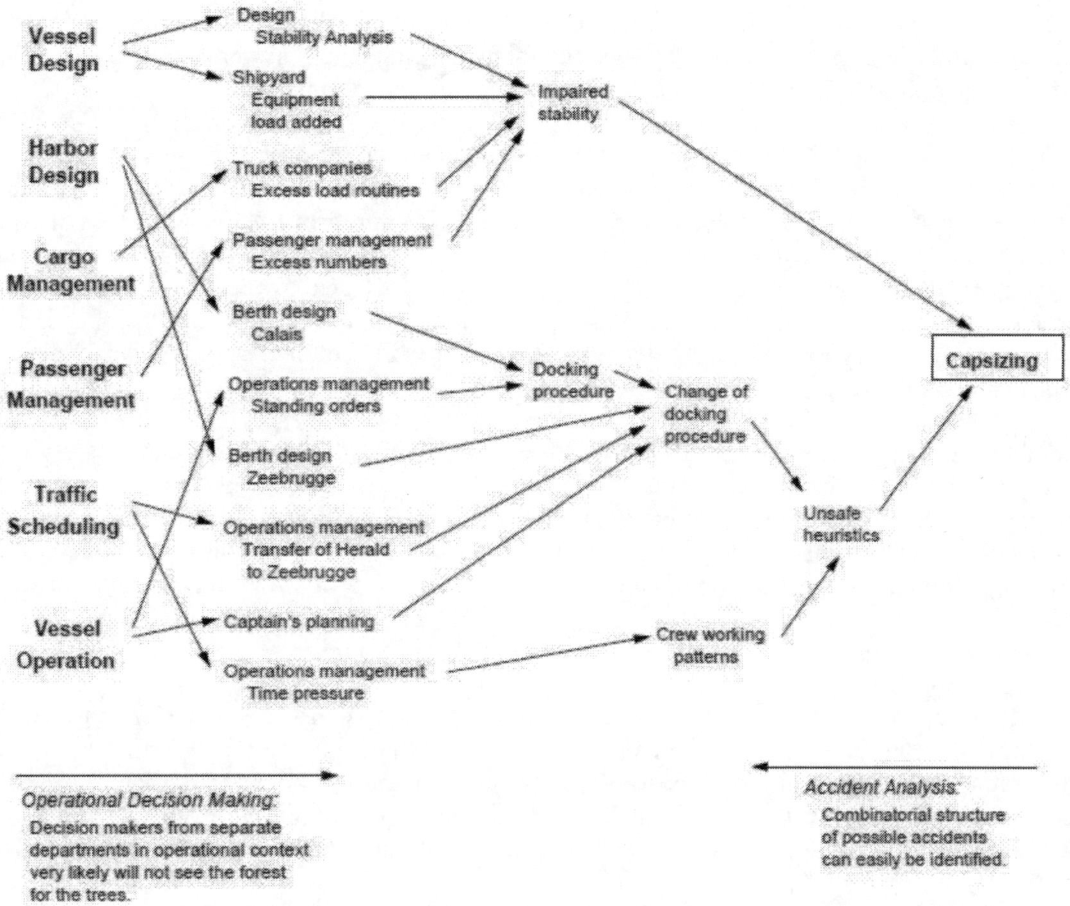

Figure 4 presents a causal tree for the Zeebrugge incident taken from Rasmussen.[11] He contends that major accidents are not a "coincidence of independent failures and human errors" but rather a coming together of different activities often working towards different objectives. The interactions among them can enhance the potential for accident.

As illustrated in Figure 4, each of the functions necessary for ferry operations (Vessel Design, Harbor Design, Cargo Management, Passenger Management, Traffic Scheduling and Vessel Operation) were conducted as relatively separate activities. Each also tried to optimize performance within their scope of work. This resulted in negative impacts to other functional areas. For example, optimizing cargo and passenger management meant adding additional weight to the ferry, directly compromising vessel stability.

Similarly, since the ferry was not designed for the berth at Zeebrugge, docking procedures had to be modified to allow passenger vehicle loading and unloading. These procedures negatively impacted requirements for traffic scheduling and standing orders for vessel operation.

Within the processing industries, engineering, operations, maintenance and other departments have the potential to operate in silos, each focused on achieving their own metrics. This brings attention to the importance of applying process safety management systems across all relevant parts of the organization to help identify and address interrelated impacts.

As a final example, a process plant upgrade included significant piping and equipment modifications, and new additional equipment. Parts of the piping were subject to a well-known corrosion mechanism. Piping and valves were specified appropriately to minimize corrosion. After installation and commissioning, one of the valves failed within 48 hours, requiring a widespread facility shutdown. The investigation revealed that while originally specified correctly, the Purchasing Department had substituted a different valve material, helping to meet their goal of lower cost and faster delivery. This valve did not undergo a positive material identification upon receipt or prior to installation.

The above example helps illustrate the breadth of application required for process safety to be effective across organizational boundaries. In this case, extending process safety to the purchasing department and developing appropriate procedures that ensure alignment with engineering specifications would have prevented the incident. It was also the "little thing" of not verifying materials of construction upon receipt of this valve that finally initiated the accident.

4.3 Perform Thorough Hazard Identification and Risk Assessments

Hazard and operability studies (HAZOPs) and other qualitative risk assessments represent the best formal opportunity for understanding the interactions among "little things." Fully "connecting the dots" requires the most experienced team members be made available. Also, the HAZOP team leader must be highly qualified and able to encourage the team to look beyond their normal, day-to-day experiences. Team qualification requirements can be difficult to sustain, study after study. However, it is essential risk assessments be done with the right personnel.

For a risk assessment to be effective, the team must be able to connect the dots in two distinct ways. Firstly, the team must have the ability to imagine a wide range of scenarios, including the unlikely. If the consequences are not fully developed and include low-frequency events, only a limited set of safeguards will be identified. These may not include the ones necessary to prevent or mitigate the full scenario.

Secondly, safeguards must be verified. This means confirming that "little things" are in place and working effectively: Do equipment or activity-specific procedures really exist and are they included in training? Is the alarm or shutdown tested? Is it testable? How often is it tested? What is done with the test results? Does the corrosion monitoring program include the piping system of concern? What are the inspection results? Is the relief valve sized for this case? When was the relief valve tested? Has the discharge piping been designed for the anticipated Joule-Thompson cooling?

All of above are indicative of the kinds of questions a qualified leader would guide the HAZOP team into exploring. Moreover, the team must have the expertise and documentation to be able to answer the questions.

4.4 Apply Discipline in Performing Basic Process Safety

The hunt for "little things" does not preclude the need to continuously manage all process safety systems. Alarms need to be tested, piping needs to be inspected, equipment needs to be maintained, changes need to be controlled, and engineering designs need to be appropriately executed consistent with relevant technical codes and standards, as examples. These overriding management systems, if effective, set the framework for finding the "little things" and allowing the interconnections among safeguards to be understood.

4.5 Stay Vigilant to Signals

Many major accidents were preceded by "dry runs." For example, a control valve slammed shut in a hot crude oil feed line, causing a pressure surge that damaged the feed pump seal. A year later, the same valve abruptly closed with the subsequent pressure surge causing a flange to open. Hot oil was released and a fire burned for several hours, requiring significant reconstruction. By thoroughly investigating these kinds of earlier incidents and taking corrective actions, the potential for later more serious events to recur is reduced. Reference 2 devotes an entire chapter to this subject, titled "Not Learning from Near Misses."

Since major accidents or even major near misses are infrequent occurrences, relying on them as indicators for corrective actions may be too late. When process safety management systems start to degrade, there are usually many small indicators. For example, important aspects of process safety information might be out of date, a few changes might be slipping through the system without proper management of change analysis, or there may be a handful of overdue inspections or corrective actions. Overall, the process safety systems appear to be functioning well, but with some gaps. Gaps can become "dots" in the accident chain. These leading indicators of system degradation represent the earliest opportunities to improve effectiveness. Special efforts should be made, through self auditing, to identify them.

Appendix A of the CCPS publication, *Recognizing Catastrophic Incident Warning Signs* (Reference 8) provides a useful checklist of "warning signs" as indicators that some degradation in process safety effectiveness is occurring. Of relevance, many of these checklist questions address "little details" ranging from "loose bolts" to "instruments bypassed."

Recently there has been renewed interest in the concept of "High Reliability Organizations (HROs)" Much of the early work on this subject emerged from research efforts on aircraft carrier and air traffic control operations.[9] Subsequent work attempted to identify key aspects of excellence exhibited by HROs. These include a "preoccupation with failure" and "reluctance to simplify interpretations" among others.[10] While a clear mapping between those organizations giving birth the HRO concept (highly repetitive yet variable activities) and complex processing facility operations has yet to be established, the "reluctance to simplify interpretations" is, in the authors' opinion, directly transferrable. By searching for a deeper meaning behind even the smallest of failures, organizations will be "connecting the dots" and evaluating those other components that, had circumstances been different, would have allowed that small failure to progress.

4.6 The Role of Culture

A wealth of literature is available reinforcing the principle that the overall organizational culture drives the effectiveness of the process safety management systems. Figure 1 conveys this concept. Leadership, starting with top management, sets the culture.

In Chevron, process safety is deployed within the context of our overall Operational Excellence (OE) efforts. As stated in Company literature:

> "Leadership is the single largest factor for success in OE. Leaders establish the vision and set objectives that challenge the organization to achieve world-class results. ....
>
> "Leaders visibly demonstrate their commitment through personal engagement with the workforce and by showing concern for the health and safety of every individual. They demonstrate the same commitment to protecting the environment and process safety risk mitigation."

The second sentence holds the key to success in process safety and OE. In order to affect the culture, leaders must do more than pronounce support. They must demonstrate it. Part of that demonstration is reinforcing the need for attention to detail and to the "little things" that lead to big accidents.

5. Conclusion

In this paper we've attempted to show that "the devil is in the details." Major accidents are often the result of small things that are easily overlooked or tolerated. Identifying the little details that can enable an accident to develop requires a "big picture" approach to understand the full breadth of the potential accident chain. It also requires a renewed rigor in such basics as qualitative risk assessments and incident investigation in order to ferret out those "little things." There is neither a Holy Grail nor a quick solution to process safety. Accident prevention comes down to implementing the CCPS elements consistently, thoroughly, and across all affected functions.

6 References

[1] Barrell, Tony, PhD, Former Head of UK HSE Offshore Safety Division (1998), as quoted in the BBC Video "*Spiral to Disaster*."

[2] CCPS, *Incidents that Define Process Safety,* BP; Center for Chemical Process Safety, New York, NY. (2008)

[3] CCPS *Guidelines for Risk Based Process Safety,* Center for Chemical Process Safety, New York, NY. (2007)

[4] Reason, James, "The Contribution of Latent Human Failures to the Breakdown of Complex Systems," *Philosophical Transactions of the Royal Society of London.* (1990)

[5] Leveson, Nancy, *Engineering a Safer World, Systems Thinking Applied to Safety,* Unpublished Draft. 2009)

[6] Chiles, James *Inviting Disaster, Lessons from the Edge of Technology* HarperCollins Publishers. (2002)

[7] *LPG Release from Caustic System,* API AFPM Operating Practices Symposium, Los Angeles, CA, Nov 13, 2012.

[8] CCPS, *Recognizing Catastrophic Incident Warning Signs,* Center for Chemical Process Safety, New York, NY. (2012)

[9] Roberts, K. H., *Some Characteristics of High-Reliability Organizations.* Organization Science, 1, 160-177. (1990)

[10] Weick, K. E., Sutcliffe, K. M., & Obstfeld, D., "Organizing for High Reliability: Processes of Collective Mindfulness." In B. M. Staw & L. L. Cummings (Eds.), *Research in Organizational Behavior* (Vol. 21, pp. 81-123). Greenwich, CT: JAI Press, Inc. (1999).

[11] Rasmussen, J., "Risk Management in a Dynamic Society: A Modeling Problem," *Safety Science,* Vol 27 No. 2/3 Pp 187-213 Elsevier Science Limited. (1997)

Practical Implementation of Safety Management Systems at Unregulated Upstream Oil & Gas Facilities

Kristin D. Norton, PE CFSE
Risk Management Professionals, Inc.
300 Goddard, Suite 200
Irvine, California 92691
Kristin. Norton@RMPCorp.com

Michael B. Saura
Colin R. Scholtz
Risk Management Professionals, Inc.
Michael.Saura@RMPCorp.com
Colin.Scholtz@RMPCorp.com

Prepared for Presentation at
American Institute of Chemical Engineers
2013 Spring Meeting
9th Global Congress on Process Safety
San Antonio, Texas
April 28 – May 1, 2013

UNPUBLISHED

Practical Implementation of Safety Management Systems at Unregulated Upstream Oil & Gas Facilities

Kristin D. Norton, PE CFSE
Risk Management Professionals, Inc.
300 Goddard, Suite 200
Irvine, California 92691
Kristin. Norton@RMPCorp.com

Michael B. Saura
Colin R. Scholtz
Risk Management Professionals, Inc.

Keywords: Process Safety Management, PSM, Upstream, Exempt

Abstract

Upstream oil and gas facilities are typically exempt from the OSHA Process Safety Management (PSM) standard; however, in light of recent regulatory scrutiny on the safety of upstream operations, many operating companies have extended their safety management programs to their upstream facilities creating challenges for implementation. Many of these facilities have been operating without significant incidents for many years without proper Process Safety Information, Process Hazard Analyses, Management of Change programs, etc. making the challenge of accepting and implementing these programs difficult to embrace. This paper is focused on providing insights and practical tips for the implementation of safety programs at upstream facilities to enhance the commitment to implementing process safety with limited resources.

1. Introduction

Numerous facilities process and store highly hazardous materials, both toxic and flammable, below the PSM regulatory threshold. While these facilities are not required to comply with the rigorous requirements of PSM, because these facilities process highly hazardous materials they still have a responsibility to identify hazards associated with the process and ensure the facilities are designed, operated, and maintained to minimize hazards. In order to meet this objective, many facilities have elected to implement PSM elements at the unregulated facilities.

For example, upstream oil and gas facilities with over ten thousand pounds of hydrocarbons meeting the definition of flammable liquids or gases (e.g., with a flash point below 100 F) are subject to the OSHA PSM standard, with several exemptions; however, many upstream oil and gas facilities do not meet this requirement and are therefore not covered under the standard. These upstream facilities are typically geographically dispersed and include facilities within each area for collecting oil from one or more oil well and separating produced oil, gas, and water.

Additional processing may be completed within the area or the products may be exported for further treatment and/or processing. This results in numerous nearly identical facilities throughout the geographical area that have the same functional design. Figure 1 below provides a simplified overview of upstream oil production.

Figure 1. Upstream Processing Overview

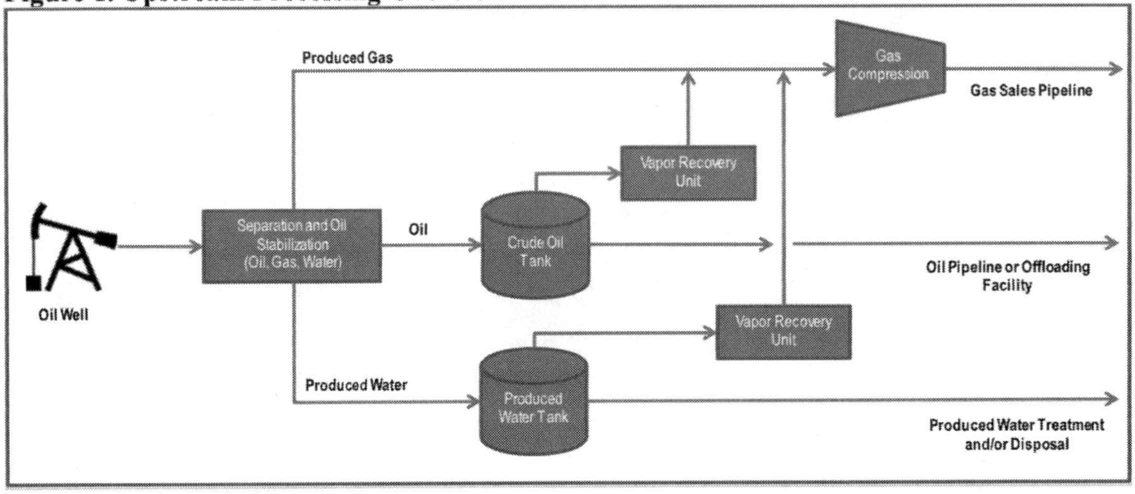

This paper gathers insights from working with companies with upstream operations of various sizes and complexity to provide guidance for organizing and managing the safety management system (both organization and recommendation tracking), as well as specific tips for the following PSM elements:

- Process Safety Information
- Process Hazard Analysis
- Operating Procedures

2. Safety Management System Management

2.1 Organization

PSM program responsibilities extend to various disciplines at the facility, including operations, maintenance, engineering, instrumentation, safety, etc. With numerous stakeholders managing their respective portion of the program it can become difficult to ensure the program is properly implemented and requirements can slip through the cracks. Designating a specific individual or group to be responsible for the overall management of the PSM program provides centralized focus; however, it is also important to have coordination between all disciplines to support the implementation of the program. Some of the most successful PSM programs have support from management at the top and are implemented from the bottom-up, with process safety being embraced as part of the daily duties of all personnel. For example, the Operations Department utilizes the Operating Procedures on a daily basis and as Operations identifies opportunities for improvement and implements changes, the documentation is then updated and coordinated with the PSM Manager. The coordination between the specific departments and the PSM Manager is key to ensuring effective implementation of the program. Thus, it is imperative to have a clear structure in place with defined responsibilities and systems in place to foster coordination among different department.

2.2 Recommendation Tracking

Recommendations are the result of various safety management system studies, including Process Hazard Analyses, Incident Investigations, and Compliance Audits. However, due to the volume of data to process from numerous facilities and studies, recommendation tracking and closure presents one of the most difficult tasks for companies with limited manpower resources. The mechanism to track recommendation closure is left to the discretion of the company and varies a wide spectrum - from simple spreadsheets to customized software solutions. While the specific tool utilized to track the data can streamline the overall management process, the most significant means to streamline data management is to develop and prioritize recommendations based upon risk (e.g., limiting the recommendations to those that will close a risk gap).

A simple way to reduce the number of PSM recommendations precipitating from the studies is to develop specific criteria to drive the team to develop a study recommendation – typically based upon risk and regulatory requirements (including codes and standards). As a result, all suggestions that the team develops that are not risk or requirement driven, but are "nice-to-have", can be tracked separately from the study recommendations, thus requiring less documentation when implemented. Having a separate mechanism to document the catch-all recommendations, such as a formalized "parking lot", allows the team to identify potential operational improvements without having these suggested improvements fall through the cracks and also fosters creative solutions. Additionally, another benefit of designating implementation criteria is to supply management with a tool to prioritize the implementation of recommendations based upon the amount of risk reduction achieved by the respective recommendation.

3. Process Safety Information

3.1 Piping & instrumentation Diagrams (P&IDs)

The successful implementation of PSM, even if not required, relies on accurate and complete Process Safety Information (PSI). An integral part of the PSI documentation is the P&IDs, which represents the process plant in a level such that it illustrates the process equipment, associated piping, basic process controls, and possibly safety instrumented systems. The P&IDs serve as a robust tool of understanding the facility. For legacy upstream facilities that may have switched owners several times, there is a significant gap in the availability of accurate P&IDs. Typically, either the P&IDs are very old and outdated or do not exist.

For facilities that do not have detailed P&IDs available it can be a significant undertaking to develop the P&IDs from scratch, requiring support from engineering and operations personnel via site walk-downs to work with the drafters to develop accurate P&IDs. This can be extremely time consuming and cost prohibitive for companies with limited engineering and operations support that are spread thin managing and operating numerous facilities. However, as mentioned previously, one characteristic of upstream facilities is that there are numerous facilities with the same functional design. While P&IDs are an efficient mechanism to document PSI and are considered best practices, there are other mechanisms a facility can utilize for documenting the technology of the process and the associated equipment. For simplified processes, such as those utilized for upstream operations, a process flow diagram can provide an overview of the technology of the process. Additionally, print-outs from the operations human machine interface (HMI) screen provide information regarding the normal pressures, temperatures, flow rates, shutdowns in place, etc. as well as the functional layout of the process. Figure 2 provides an example of a HMI screen shot for a Lease Automatic Custody Transfer (LACT) Unit that is utilized to accurately measure the amount of crude that is sold. These screen shots (if available) can be utilized as supporting PSI when conducting the Process Hazard Analysis, developing Operating Procedures, and implementing training programs.

Figure 2. HMI Screen Shot Example

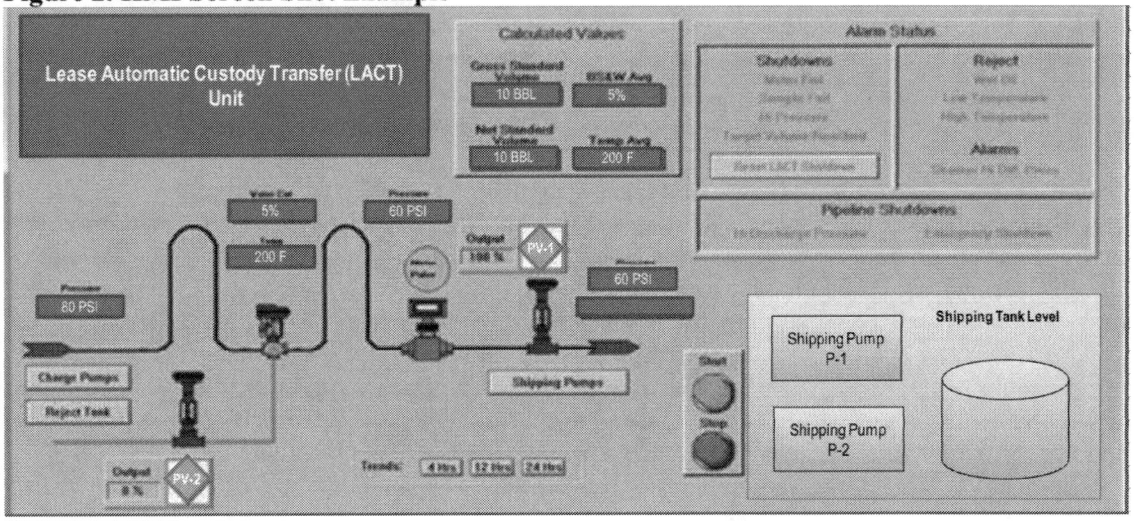

Additionally, for facilities with developed P&IDs, without a drafting department or document control program there is a potential for facilities to postpone or sometimes even neglect P&ID changes. One of the best mechanisms for a P&ID review is the Process Hazard Analysis. During the study, the team systematically reviews the P&IDs in detail and typically finds errors and improvements during the study (and in some cases specific systems are field verified). It is in the facility's best interest to ensure that the PHA sessions document these P&ID errors and changes on a "red-line" copy, typically assigned to one process engineer in the study. This copy can serve as the master red-lined copy until the changes can be implemented by a drafter. This provides a direct means of controlling the required changes to the P&IDs and the responsibility for implementation of changes lies with the assigned process engineer.

4. Process Hazard Analysis

The PHA is another element of the PSM program that requires extensive resources and commitment from the company. Additionally, since upstream facilities are typically dispersed geographically and may be under different operational management, there can be additional challenges ensuring consistency among operating groups. The following sections provide tips to streamline the PHA study process to ensure consistency of the results and make efficient use of available resources.

4.1 Characterizing Hazards of the Process

When managing numerous facilities it can be difficult to ensure consistency in risk ranking for the PHAs, since the ranking is based upon the collective risk perception of each specific team. In order to provide a mechanism to calibrate the risk ranking, management can provide supporting information for the hazards of the process and what level of risk this presents to personnel safety, health and the environment. For example, providing dispersion modeling for various H_2S concentrations, with associated release sizes (e.g., leak vs. rupture) and heights (e.g., process failures vs. relief valves to atmosphere) helps to illustrate the potential for personnel exposure and define whether the associated event can result in a fatality or injury. This can be applied at each facility PHA to assist in ensuring that the risk ranking is applied consistently.

4.2 Checklist Development

There are several PHA techniques that can be applied with various levels of detail, including What-If?, Checklists, and Hazard and Operability (HAZOP) studies. The HAZOP Study technique provides a comprehensive evaluation of the hazards of the process using a team-based, cause-by-cause methodology. While this provides a thorough evaluation, it also requires significant manpower resources when applied at numerous facilities. However, a typical characteristic of upstream operations is that while there are numerous widespread facilities, the facilities also tend to be patterned. By grouping similar facilities, such as tank batteries, compressor stations, vapor recovery units, gauge settings, automatic well test units, etc., we can evaluate each facility group systematically.

One effective approach to evaluate the facility groups is to conduct a baseline HAZOP Study for each facility type, then based upon the results of the study develop a checklist to evaluate the remaining facilities individually, while noting any differences in the evaluated facility from the baseline. These checklists are typically more detailed than the typical PHA checklists in that they are specific for each facility type, with a focus on mitigating specific consequences and safeguards.

4.2.1 Vapor Recovery Unit Checklist Development Example

The following section provides an example illustrating the checklist development for a Vapor Recovery Unit (VRU) Compressor (see Figure 3) based upon specific HAZOP scenarios. This example is not comprehensive, but provides an overview of the technique to transform HAZOP scenarios into checklist questions.

Figure 3. Block Flow Diagram Example –VRU Compressor

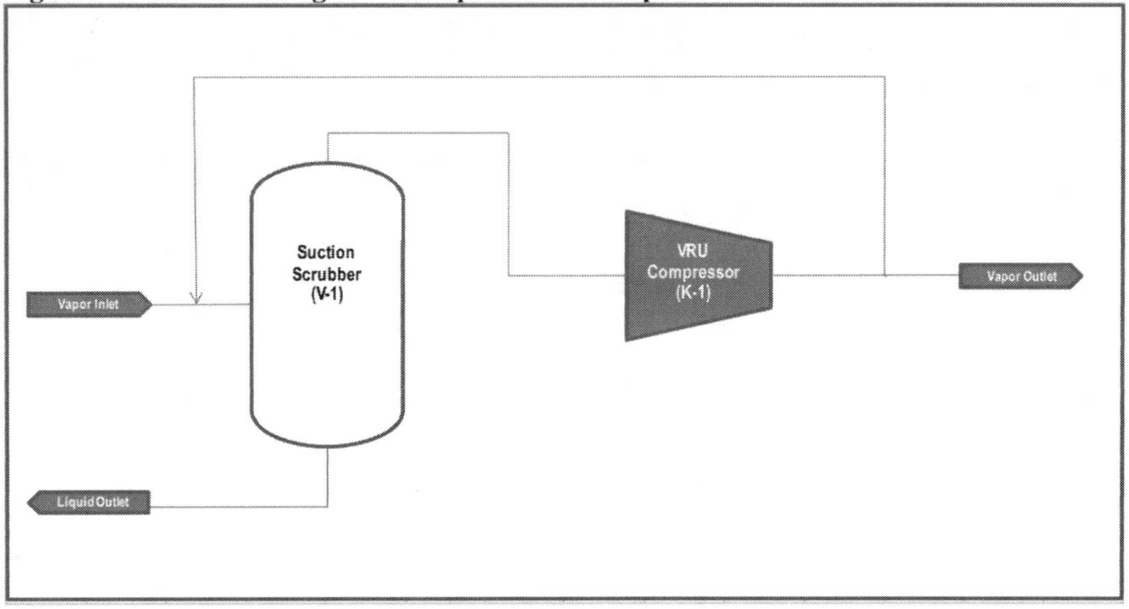

The following table provides an overview of several "No Flow" scenarios from the VRU Compressor baseline HAZOP Study:

Table 1. HAZOP Scenario Example – VRU Compressor: No Flow

HAZOP Scenarios					
Cause	Consequence	Safeguards	S	L	RR
1.1. Block valve on the suction of the VRU Compressor (K-1) inadvertently closed.	Potential loss of suction to the VRU Compressor resulting in increased temperature and bearing damage. This is an asset damage issue.	Operator Training and Procedures for valve operation.	4	4	L
		PALL-001 trips the VRU Compressor on low low suction pressure.			
		TAH-001 alarms on high VRU Compressor bearing temperature.			
		TAHH-001 trips the VRU Compressor on high high discharge temperature.			
1.2. Block valve on the discharge of the VRU Compressor (K-1) inadvertently closed.	Potential increased VRU Compressor discharge pressure resulting in overpressurization of the discharge piping. Potential loss of containment with fire/explosion and personnel exposure.	Operator Training and Procedures for valve operation.	1	4	M
		PV-001 opens on low VRU Compressor suction pressure to provide recycle.			
		PAHH-001 trips the VRU Compressor on high high discharge pressure.			
		PSV-001 on the discharge of the VRU Compressor is designed to relieve pressure for the blocked discharge case.			

In reviewing these examples, the following questions can be deduced from the consequences and safeguards:

Design Questions

- What are the operating suction and discharge pressures?
- What are the maximum suction and discharge design pressures?
- What are the operating suction and discharge temperatures?
- What are the maximum suction and discharge design temperatures?

Blocked Suction Questions

- Will a loss of suction flow result in increased temperature?
- If the temperature will increase, is there a potential for bearing damage?
- Is there any potential for loss of containment due to bearing damage?
- Are there alarms configured for high gas temperature? If so, will the Operator have sufficient time to respond?
- Are there alarms configured for high bearing temperature? If so, will the Operator have sufficient time to respond?
- Will the compressor trip on a loss of suction pressure?
- Is the suction piping or equipment designed for the maximum settle-out pressure for the compressor?

Blocked Discharge Questions

- Will a blocked discharge exceed the piping or equipment design pressure rating?
- If so, is overpressure protection (e.g., relief valve) provided to protect the piping and equipment?
- Will the compressor trip on high discharge pressure?
- Is there a minimum flow recycle to protect the compressor on a blocked discharge?
- Is the take-off for the minimum flow recycle upstream of the first block valve?
- Will a complete recycle of the flow result in increased temperature that could cause compressor damage?
- If so, are safeguards provided to mitigate increased gas temperature (e.g., cooler, high temperature alarm, high temperature trip)?

General Compressor Questions

- Area lock-out / tag-out procedures in place to minimize the potential for valve mis-operation?
- Is the compressor material selection appropriate for the expected vapor properties to minimize potential corrosion?
- For vibrating services, are pulsation dampeners included in the design?

- Is the discharge piping adequately supported for vibration?
- Is vibration monitoring provided for the compressor?
- Are emergency isolation valves provided in the design to safely isolate the compressor in an emergency?
- Are the emergency isolation valves (manual, automatic, remotely operated) appropriate for the specified service (location, fireproofing, fail position)?
- Are procedures in place for routine testing of the emergency isolation valves?
- Is fire water provided in the vicinity of the compressor?
- Can the fire water be safely operated in the event of a fire in the vicinity of the compressor?

After brainstorming potential questions for the scenarios, a checklist can be developed for these scenarios for the VRU Compressor. A sample abbreviated checklist format is provided on the following page:

Table 2. Checklist Example –VRU Compressor

VRU Compressor		
VRU Location:		
Checklist Team:		
Date of Completion:		
Design Parameters Design Pressure: Operating Pressure: Design Temperature: Operating Temperature:		
Checklist Questions	**Yes/No/NA**	**Comments / Recommendations**
Will a loss of suction flow result in increased temperature? • If the temperature will increase, is there a potential for bearing damage? • Is there any potential for loss of containment due to bearing damage?		
Are there alarms configured for high gas temperature? • If so, will the Operator have sufficient time to respond?		
Are there alarms configured for high bearing temperature? • If so, will the Operator have sufficient time to respond?		
Will the compressor trip on a loss of suction pressure?		
Is the suction piping or equipment designed for the maximum settle-out pressure for the compressor?		

VRU Compressor		
Will a blocked discharge exceed the piping or equipment design pressure rating? • If so, is overpressure protection (e.g., relief valve) provided to protect the piping and equipment?		
Will the compressor trip on high discharge pressure?		
Is there a minimum flow recycle to protect the compressor on a blocked discharge? • Is the take-off for the minimum flow recycle upstream of the first block valve?		
Will a complete recycle of the flow result in increased temperature that could cause compressor damage? • If so, are safeguards provided to mitigate increased gas temperature (e.g., cooler, high temperature alarm, high temperature trip)?		
Are lock-out / tag-out procedures in place to minimize the potential for valve mis-operation?		
Is the compressor material selection appropriate for the expected vapor properties to minimize potential corrosion?		
For vibrating services, are pulsation dampeners included in the design? • Is the discharge piping adequately supported for vibration? • Is vibration monitoring provided for the compressor?		
Are emergency isolation valves provided in the design to safely isolate the compressor in an emergency? • Are the emergency isolation valves (manual, automatic, remotely operated) appropriate for the specified service (location, fireproofing, fail position)? • Are procedures in place for routine testing of the emergency isolation valves?		
Is fire water provided in the vicinity of the compressor? • Can the fire water be safely operated in the event of a fire in the vicinity of the compressor?		

5. Operating Procedures

Though not required by PSM, it is considered an industry best practice to develop and implement rigorous Operating Procedures. A well-developed Operating Procedures program can provide significant benefits to both the facility and its operators, including minimizing risk, documenting safe operating limits, capturing the knowledge of experienced operators, and providing valuable training tools.

Upstream facilities deal with a number of hazards, so from a safety perspective, strong Operating Procedures are an effective management tool to minimize the risk of potentially disastrous incidents from human error. They are most commonly applied to frequent, low-risk tasks though they are often applied to infrequent but high-risk tasks as well. Outside of process controls, administrative controls such as Operating Procedures are a crucial element in minimizing the occurrence of hazardous events.

In addition to reducing the risk of process safety incidents, Operating Procedures also provide operational benefits. By the Center for Chemical Process Safety's (CCPS) "Guidelines for Writing Effective Operating and Maintenance Procedures" [2] definition, Operating Procedures are designed to keep a process within its operating limits. Operating Procedures capture these operating limits and ensure the process is operating within its design intent. This ensures consistency of operations, which is also highly beneficial to facilities with multiple operators operating the same process. [1]

Additionally, one of the most significant challenges posed to upstream facilities is knowledge transfer. Many facilities have been operating for years without robust Operating Procedures, relying primarily on the knowledge of experienced Operators to safely operate the facility. However, as operators retire or switch positions, the knowledge of that particular facility or system is no longer available to those left behind to operate the facility. Developing Operating Procedures from scratch is a significant time investment; however, in the long-run developing a knowledge transfer program with comprehensive Operating Procedures will not only help to support operations, but will also provide a valuable training tool. An effective way of implementing a knowledge transfer program is to develop detailed Operating Procedures utilizing the operator shadowing method (i.e., documenting actions for specific procedures based upon observing the actions of the experienced operator). When operating a facility becomes second nature, it is easy to overlook all of the implicit actions associated with a simple task. For example, when discussing starting a pump with an experienced operator is typically documented as a single task (e.g., Start Feed Pump #1). However, this could be expanded to include the specific valve line-up (including recycle), monitoring of specific pressure or flow indicators, expected suction and discharge pressures, troubleshooting advice based upon operating experience, etc. One of the best ways to extract this information is by coupling less experienced operators or engineers with the experienced operators for the shadowing. This method also provides a valuable training tool for the less experienced operators.

5.2 Keeping the Operating Procedures Current

The implementation of successful Operating Procedures is reliant on the accuracy of the procedure. A key method of ensuring accuracy is to conduct an annual review of the operating procedures by key stakeholders from management and operations. Additionally, Operating Procedures should be reviewed following the completion or revalidation of the PHA to ensure the consequences of deviations from the normal operating limits are documented accurately. The resultant updated consequences of deviation can be used in the O&M procedures. Both the annual review and the PHA ensure that the O&M Procedures are current, thus continually minimizing risk and ensuring that the process is kept within its operating limits.

6. Conclusion

Effectively applying PSM at upstream facilities presents numerous challenges due to the geographic extent of facilities and limited available manpower resources. However, as evidenced in this paper there are systems available to streamline the implementation which requires pre-planning and setting up the systems as part of the overall management structure, including the following:

- Defining a management system and clearly assigning PSM responsibilities.
- Developing and maintaining PSI, including P&IDs, and implementing document control policies.
- Standardizing the PHA approach, including characterizing hazards to consistently apply risk ranking, setting risk criteria to define and prioritize recommendations, and organizing facilities such that resources can be systematically applied to baseline HAZOP Studies and subsequent checklists.
- Implementing a knowledge transfer program to develop robust Operating Procedures.

7. References

[1] Scholtz, Colin R. and Maher, Steven T. "Tips for the Creation and Application of Effective Operating Procedures"

[2] "Guidelines for Writing Effective Operating and Maintenance Procedures", Center for Chemical Process Safety, American Institute of Chemical Engineers, 1996.

Overcoming PSM Challenges for Small Chemical Companies: Observations, Insights and Guidance from the Trenches

Philip M. Myers
Advantage Risk Solutions, Inc.
Sunbury, OH, USA
pmyers@arisksolution.com

Prepared for Presentation at
American Institute of Chemical Engineers
2013 Spring Meeting
9th Global Congress on Process Safety
San Antonio, Texas
April 28 – May 1, 2013

UNPUBLISHED

Overcoming PSM Challenges for Small Chemical Companies: Observations, Insights and Guidance from the Trenches

Philip M. Myers
Advantage Risk Solutions, Inc.
Sunbury, OH, USA
pmyers@arisksolution.com

Keywords: PSM, process safety, small company, limited resources, guidance, process safety management, process safety culture, chemical facilities

Abstract

Complete implementation of the OSHA Process Safety Management regulation can be a daunting task for small chemical companies. Small companies typically are much more resource limited in terms of finance and man-power. There may be no process safety professionals on staff, leaving the company without the associated current process safety expert knowledge, skills, abilities and attendant tools. So how can a smaller chemical company navigate the Process Safety Management (PSM) landscape and make steady progress in development and implementation of a practical, functional program that can be maintained and improved? From program development to the technical details of process safety information and mechanical integrity, management of ongoing changes, to the all important "buy-in" of personnel from the president to operators and technicians in creation of a process safety culture – how can it be achieved? This paper presents observations and insights from years of experience working "in the trenches" with small companies - regarding common pitfalls and limitations at small sites and guidance for successful methods to move forward and overcome obstacles encountered.

1. Introduction

Small companies are unique organizations, and there is a wide range of what constitutes a "small" company. Small chemical companies may have revenues in the millions of dollars, rather than the billions of dollars for many Fortune 500 companies. This also means that operating and capital budgets may be limited. Small companies are often privately held, may be family owned and managed, and/or have a single or few plant sites. Staffing is generally "lean" with the total number of employees often less than one hundred. Staff tend to "wear multiple hats," with responsibilities for more than one area that may be seemingly unrelated. It is not uncommon for a human resources manager to also be responsible for process safety at a site with a single process covered by the Occupational Safety and Health Administration (OSHA) Process Safety Management (PSM) standard [1] and/or the Environmental Protection Agency (EPA) Risk Management Program Rule (and Prevention Program - Level 3 provisions) [2]. Many small companies covered by the PSM regulation do not consider themselves to be "chemical companies," yet they use regulated chemicals above the threshold quantities in processes covered by the PSM standard in the conduct of business. While the PSM standard is not applicable to certain limited types of facilities and conditions, there is no general exemption or modification for small chemical companies. And while the EPA Prevention Program is tiered, it is based upon hazard and/or covered processes not on the size of the company. This leaves small chemical companies with the requirement to meet all aspects of the PSM standard. Observations and insights are provided herein, as well as guidance to small companies working to achieve PSM compliance and develop a positive process safety culture within the constraints of a small company environment.

2. Why is PSM important for small companies?

Small chemical companies, and other small companies that handle Highly Hazardous Chemicals (HHCs), may have significant hazards. Significant risks may be posed to employees and to the company without hazard identification, control, and risk management.

Small companies, that do not even consider themselves to be chemical companies, may not:

- Process toxic materials…but do have a large ammonia refrigeration unit, or use chlorine cylinders
- Produce explosive materials…but do use propane as a raw material to the process
- Produce corrosives…but use nitric acid (also toxic) in making a product
- Process at high temperatures…except for kettles, boilers, or heat exchangers
- Process at high pressures…but do have high pressure steam

and, yet, may be covered by the PSM standard due to the hazards and risks associated with those materials [3].

The Chemical Safety Board (CSB) is an independent United States (U.S.) federal agency charged with investigating industrial chemical incidents. From a review of CSB investigations, it is clear that many of those investigated are not Fortune 500 companies – most of the investigations are for accidents occurring at smaller companies and facilities.

Given small company constraints, management and staff may not be aware of the hazards or of regulatory requirements. Limited knowledge of the hazards, and in particular chemical reactivity, on the part of personnel at small companies has been a common theme and key factor in many of the CSB incident investigations.

At the onset, it was acknowledged that small chemical companies (and smaller companies generally) are unique. Process safety is not defined by the size of the company. There are small companies with excellent process safety programs, and much larger ones with programs that are lacking. In fact, much larger, but highly decentralized companies may effectively have problems similar to those at small companies.

The focus of this work is to highlight and provide insights regarding key factors and trends from observations while supporting small companies over a number of years - as well as apparent from CSB incident investigations. This is a solution-oriented approach to look at what makes small companies different, and to provide guidance regarding how to effectively improve process safety in a small company setting. This guidance may be equally applicable and valuable to small facilities within larger organizations as well.

3. The challenges

There are many challenges associated (rightly or wrongly) with small chemical companies and other small companies using Highly Hazardous Chemicals (HHCs). A number of these factors are apparent in specific consulting assignments, are readily recognized by staff at small companies, and are specifically identified in CSB investigations.

Resource limitations are well understood (at least superficially) – both in terms of finance and staffing. Smaller organizations with revenues in the millions of dollars have much more limited operating budgets as well as smaller and more focused capital budgets. Process safety activities need to be included in the budgeting process. This can be the first "hurdle" for a small company, to simply develop and include a process safety budget or line item. For example, process safety activities that will involve changes in equipment, non-destructive examination or testing, facility siting, or otherwise incur capital expenditure need to be communicated to management and included in the budget planning process. Smaller companies often do not have process safety professionals on staff, and operate as very "lean" organizations. PSM responsibilities are delegated from management to an environmental manager or engineer, engineering manager or engineer, human resources manager, or any of the many other possibilities.

Process safety knowledge as well as a complete understanding of the hazards and process safety risks is a difficulty for small companies. With limited staff that already have multiple responsibilities, and often with no process safety experience readily available onsite, adequately addressing process safety is certainly challenging. Even for those staff tasked with process safety and diligently attempting to do what is needed, it can be a matter of "you don't know…what you don't know."

Small chemical companies typically are very flexible and innovative organizations. This enables them to respond quickly to customer requests and needs. Many small chemical companies serve as suppliers or toll manufacturers for major chemical companies. Having to respond quickly, and given the potential for somewhat limited process safety knowledge, may lead small companies to be more prone to fall back on experience-based (i.e. "seat of the pants") solutions. Certainly, this can be a problem at larger organizations, but in those cases there are more people to ask questions, additional viewpoints, access to more professional process safety personnel and tools, and larger staffs and budgets to address the questions that arise. A person with process safety responsibility at a small company or site may feel he/she "has to do it all." At a small company, budget pressures may also drive someone assigned process safety to bypass the analyses and studies perceived to be "expensive" and, again, rely on experience or judgment. Experience certainly can play an important role, but it does not replace process safety expertise, tools, and analysis when they are needed.

Staff at small chemical companies may also have very limited budgets (if any) for training, conferences, and even reference materials. Many small companies do not have the resources to send persons with process safety responsibilities (as one of many) to specialized training courses or to national process safety events such as the American Institute of Chemical Engineers (AIChE) Global Congress on Process Safety. If financial resources are budgeted, it is typically for local industry association or trade group meetings. In addition, with a small staff, management finds it harder to justify sending someone away for several days of training, and providing the necessary "coverage" for that person's duties while they are away.

A significant problem that arises all too often is "normalization of deviation". "Normalization of deviation" is a state when a deviation (perceived or known by the company or not…and often it is not) continues for a long enough time period that it becomes the norm or expectation. Normalization of deviation can involve physical aspects of the plant equipment, site practices, procedures, use of methods or techniques, and others. This problem is sometimes exaggerated in small companies due to some of the factors already noted – small staffs where it is harder to obtain other views or opinions, limited process safety knowledge, and others. Small companies often have key personnel that have been with the company for years. However, if these deviations become in-grained in the key senior staff, they get passed along to new hires and the deviations are propagated. One example of "normalization of deviation" occurred at a small chemical company in the conduct of Process Hazard Analysis (PHA), with the technique misapplied so many times that it became "the way we do it." At another chemical company located in the United States (US), the first engineering and construction company hired was a Japanese firm. Throughout the plant flexible expansion joints could be seen - intended (in Japan) for earthquake design, but unnecessary at this plant location. The flexible expansion joints became unintentionally designed-in weak links in this plant, and were continued in later designs as the site norm.

Small chemical companies tend to use manual operations to a higher degree than larger counterparts (who are often customers). As small companies have to be responsive and flexible, sometimes it is not possible to hard-pipe or automate the process. Many more processes at small companies are batch, used to supply relatively small quantities when they are needed by the customer. In contrast, at major chemical facilities, a large proportion of the processes are continuous (or at least semi-continuous) as the processes are kept on-stream to meet larger, more regular product demands. At smaller companies, where processes may need to be routinely started up, shut down, and/or used to produce multiple products, more manual activities often result. All of these changes (often at a rapid pace) put pressure on the process safety program generally, and specifically on the Management-of-Change (MOC) process.

4. Overcoming challenges – a view and guidance from the "trenches"

4.1 Program leadership and the core "team" - first who…then what

The first and most important step a small company can take in developing or improving a process safety program is to focus on the individual to lead the process safety effort, and then on key person(s) for the core "team". This step will likely determine whether the process safety program will be successful, functional and compliant or an ineffective "drag" on the company. After leadership and the core "team" is established, the company can then focus on what needs to be accomplished. A defining characteristic of great companies presented in *Good to Great: Why*

Some Companies Make the Leap and Others Do Not by Jim Collins [4] is having the right program leadership – the "process safety leader." "Process safety leader" and core "team" are in quotations here because (1) at a small company process safety may be a functional role (not in title) with only a fraction of the responsible person's time spent on process safety, and (2) the "team" might be a "team" of one in some cases. The program is far more likely to be successful with a "champion" who is able to focus attention on the program and is committed to its successful implementation. After finding the right "process safety leader", next is identification of the core "team." This core team is key to a successful transformation. Jim Collins states that "People are not your most important asset. The right people are." [4] It is important to include persons that are highly respected and valued by the organization, and may include long-time technical managers, engineers, or other technical staff members. For small companies, this small core team will likely need support by outside expert resources to bring process safety knowledge, skills, abilities, and an industry-wide view and experience. The process safety consultant should be someone trusted (or in whom trust is built) and fully integrated as part of the core team from the beginning. This person can help guide the core team; take on specific, more difficult tasks requested; and/or serve as an expert advisor. It is critical that this core team works together well as a team. The team could consist of as few as two members (process safety leader and outside support) for a very small, single covered process site, or include two to three more for somewhat larger organizations with multiple covered processes.

Figure 1. Select the Right People First

4.2 Confront the brutal facts – initial assessment

Building a solid process safety program (or initiating a focused effort to improve one) begins with an accurate, honest, critical evaluation and appraisal of the "starting point" for the organization. The primary regulatory references are the PSM standard [1], OSHA compliance directive [5], and may also include (if applicable) the EPA Risk Management Program Rule [2], and EPA guidance for auditors [6] and inspectors [7]. Even in development of new process safety management programs, there likely is information from existing business operations that can be used. It is important to confront the brutal facts upfront – the whole, unvarnished truth. This can sometimes be a somewhat "uncomfortable" review process, with vested personnel, interests, long-standing procedures, and ways of doing the business. If a program is not working as intended, is inefficient, mistakenly "glosses over" serious risks, or if significant personnel issues exist, it can be tenuous to honestly assess the situation and develop a plan of action. Even within "good" programs or operations, there may be problematic areas – or simply complacency may have set in with certain personnel with important roles. To effectively move forward, the company needs to confront reality and document the "starting point" from the initial assessment – it will become the foundation for development and action. The initial review and assessment can be accomplished by leading with questions (not answers), engaging in open dialog, and conducting a solutions-oriented review geared toward the desired development or improvement. When a thorough formal review of the program or starting situation is conducted and the facts are gathered, the path forward often becomes apparent. A positive process safety culture is important at plant sites of any size. Smaller companies, however, are often focused on a limited process safety program that is compliant and works within the existing company structure and constraints. Most small companies are not interested in the "latest and greatest" and don't want to be on the "leading edge," rather, they want a program that is compliant, functional, and "good enough."

4.3 Decide "where to begin" – on what basis?

Once the upfront evaluation or appraisal of the organization's "starting point" is complete, the next step is to decide where or how to begin the work of developing or improving the program. For a new process safety program, the best place to start is with development of a written program that is documented in a PSM program manual. This is also a good starting point for any company without a written PSM program. It is important to have an overall written program in place. In developing the written program, many issues tend to get sorted out that would require resolution later anyway – when working on individual PSM elements. Development of an overall written PSM Program helps with assignment of responsibilities and to resolve other site issues. Through development of a written program, the core process safety team can begin to build the linkages between company or site management, the core team, and the program. It is important for the core process safety "team" and management to be "on the same page" before launching into development or improvement of the various individual technical elements and supporting site programs and activities.

Once a written PSM program (manual) is prepared, and the key issues have been sorted out with management, the team can then decide how to begin the detailed work for the program or individual elements. There are a number of ways that this can be approached, taking into account the results of the initial assessment of the company's "starting point." The first step is to determine if there are actions the site can take to exempt the process(es) from the OSHA PSM standard. This should be done even if the company will still develop a process safety program. In that case, the process safety program becomes driven by the site needs, without the regulatory requirements. For instance, if the maximum process inventory onsite or the chemical concentration can be reduced to below the regulatory thresholds, the OSHA PSM standard may no longer be applicable - though the OSHA (and/or EPA) General Duty Clause still applies.

It is important to prioritize work that will need to be done in development or improvement of a process safety program. For small companies, with limited staffing, this becomes even more important. There are a number of ways that that the tasks can be prioritized. Following are example approaches that can be used [8]:

- Begin with areas that have an existing basis and materials that only need to be improved or enhanced to include items specific to PSM
- Start with elements that can be implemented short-term and with little or no capital requirements
- Start with elements that present the highest risks (or losses) for the company
- Initiate work on elements with the longest anticipated time to completion (the most difficult problems to "tackle")
- "Divide and Conquer" – Implement several elements at the same time by assigning elements to different groups (e.g. engineering, environmental, research and development) and/or people

Figure 2. Choosing the right direction for a small company process safety program

However the core team decides to proceed, it is important to make sure that the program or element developed will work for a small company – given the type of facility, size, resources, management, and culture (unless the culture simply needs to be changed…a long term proposition). The program and its components need to be set up so that they can be routinely used as part of the mainstream business processes, and maintained on an ongoing basis once the initial development effort is over. If the program isn't developed in this manner, it may soon be worthless or effectively "shelved". At this point in the program development, it is important for the core process safety team to communicate with site management about resources required (financial and human), and to assist with provision of rough (conservative) estimates for budgeting and planning.

4.4 Plan for discrete, manageable, step-wise progress

As the core process safety team considers plans for further development of the process safety program, it is important to divide the work into discrete, achievable steps. A trusted process safety consultant can be invaluable in this process, as the company considers its "starting point," "where the program needs to go" (a vision of an optimal program for the company), and how the team can incrementally "get there". This involves a comparison to the regulatory requirements, as well as to current industry practices and norms, since the OSHA PSM standard is performance-based (as is the EPA Prevention Program – Level 3). There may be existing materials for certain elements of process safety that are "close" to what is needed for PSM, and simply need to be better or more specifically documented. Other elements may require development from little or nothing. The plan developed should:

- Begin with existing management systems (to the extent possible) that are used on a daily or regular basis. Small companies generally cannot afford to create another, entirely separate management system for process safety [9].
- "Begin with the end in mind." [10] Though the program will be developed in steps, it is good to think about the desired goal, and then build in "hooks" or "placeholders" as may be needed in the interim. For most small chemical companies, the goal is not to be "on the leading edge" of process safety. More often, the goal is to be in compliance, and to have a program that is functional and "good enough."
- Include business growth plans so that the PSM program isn't soon outstripped by the business activity.
- Clearly assign responsibilities and milestones – make sure that you know…and they know… (who is responsible, what is to be done, and when it is to be completed).
- Set reasonable targets – this work will take longer than the process safety leader or core team thinks – this is especially true for small companies where human resources often get diverted.
- Track work and completion progress and dates – what is measured and tracked gets done. Monthly status/progress charts can be used for safety/PSM meetings, and to communicate progress to management.
- Be adjusted/revised periodically – give the company time to make progress, but expect and understand that the plan will need to be revised and adjustments made as the work progresses. Some of these changes are likely to be significant if a new process safety management program is being developed.
- Include periodic checks and audits by others – the checks and "audits" can be formal or informal, as long as they are effective and keep the overall efforts and progress on track
- Include regular communication with company senior and executive management – communicate actions, what is being achieved, what is ahead/behind plan, and most importantly what will be needed from them – be "realistic" and don't understate resource needs, but also be credible…it's a balance.

4.5 People – the "wider" team - assign responsibilities and priorities

In the development or improvement of a process safety program or individual elements, personnel should be assigned specific responsibilities. This includes assignment of "owners" for the overall PSM program and for individual program elements. These individuals can be "point persons" for the development or improvement process, as they ultimately will be the primary persons responsible for that area or element. The "owners" can be supported by others with secondary responsibility for the element – potentially serving an important role both in development and in later function of the program. Still others may be expected to participate in the development of the program element through provision of inputs, though their overall level of involvement in development is lower.

At small companies, it is important to involve all technical staff and management to the degree possible – and "not let anyone off the hook." That includes senior and executive management. This is a chance for the core process safety team to communicate to management the important role management needs to fulfill (including visible commitment and leadership, provision of resources, authorizations and approvals), and that process safety is not "just" a compliance program. With a limited staff, it is helpful to identify how most company human resources can be assigned some process safety responsibilities that can be accomplished as part of day-to-day activities. An example is given in Table 1 (following the references) for assignment of process safety development and implementation responsibilities for the overall PSM program. The example is for a facility that is also covered by the EPA Risk Management Program Rule and Prevention Program requirements. [It should be noted that in the example table provided, the CEO is not involved in the day-to-day management or operations of the company. Rather, the CEO represents the company owners/shareholders, with a focus on overall company and financial performance.] It is recommended that a similar table be developed specifically for the mechanical integrity element due its importance to the PSM program and the level of effort typically required.

Assignment of responsibilities is very specific to the individual company and the knowledge, skills and abilities (KSA) of individual team members. Further, for smaller companies, a number of the roles listed in the example table may not exist (at any given organization or single location), or the roles may be functional, and without the title of "manager." While it is generally advisable to have one "owner' for the various program elements, there may be exceptions. One notable exception is for "visible support and actions" where management and leaders throughout the organization should consider themselves to have "primary" responsibility. Other cases may arise depending upon organizational preferences, personnel expertise, and/or a further functional split of responsibility – such as the engineering manager may have responsibility for mechanical integrity of new designs and engineering projects while the maintenance manager has responsibility for existing, in-plant equipment. The division of responsibilities does not necessarily appear to be logical at times, as with limited staff, sometimes it is simply necessary to review and approved each other's work. Assignment of responsibilities also depends upon reporting relationships, how "hands-on" an individual may be in a given area, and peculiarities of specific organizations. The table also highlights the need and expectation to get inputs to the program and individual elements from plant hourly personnel. In the division of process safety responsibilities, what makes sense for one organization may not work for another. Having the right people involved is important in all organizations – but is even more critical for smaller companies. The example roles and responsibilities matrix given (in Table 1) is provided to demonstrate the range of personnel that can be assigned various process safety responsibilities in a small chemical company.

4.6 Process safety knowledge and tools – identify areas of weakness and supplement

People at small companies are "motivated, hard-working, genuinely concerned people without the time, confidence and training" [3] for process safety. Often, they are missing the desired reference materials and tools as well. For a small company to develop an effective process safety program the "process safety leader" and core "team" should know and be focused on what is *required* and what is *needed* by the site. There is no "ignorance is bliss" application in process safety. And, even strictly in terms of compliance, a lack of knowledge or understanding is not a legal defense for non-compliance, but can potentially expose the company for gross negligence if an accident occurs. This is another reason that while PSM certainly is a compliance program, it is never "just" a compliance program. Personnel with process safety responsibilities at small companies want to "do the right things," but often have a more challenging situation given the difficulties associated (potentially) with insufficient process safety knowledge – unless local knowledge is supplemented with outside process safety expertise.

A difficulty for small companies comes in the area of process safety knowledge generally, as well as in a complete understanding of the hazards and process safety risks. Adequately addressing process safety is certainly challenging - with limited staff that already have multiple responsibilities, and often with no process safety experience readily available onsite. The often repeated "you don't know…what you don't know" is a real concern in process safety - even for personnel with process safety responsibilities and diligently attempting to do what is needed. Simply stated, staff may not know enough to be able to ask the right questions. Further, some process safety questions are difficult to address unless the person is immersed in process safety. For instance, how is a small company to know that the OSHA regulated concentration for formaldehyde is 37%? It was not published in the original list of HHCs, the list has not been updated with this information, and it only exists in an OSHA letter of interpretation. And, some process safety issues can "get into the weeds" even for process safety professionals, such as in defining upstream and downstream covered process limits (depending on the material, concentration, interconnections, co-location, etc.), inclusion or exclusion of various plant support systems and structures, understanding facility siting and human factors requirements, and so forth.

A more dangerous process safety knowledge issue occurs when "you don't know…what you think you know" – and have limited readily available resources to check into the issue or problem with others. For example, a chemical site had a large release of a toxic liquid, that covered a storage containment area, and nothing really "bad" happened. A level of "comfort" can develop that even if the worst event imagined goes wrong, it's still not a problem. Only…the release experienced was for a material with a boiling point near atmospheric temperatures, and the event happened on a cool spring day. What, then, happens on a hot summer day? The hazard zone or area affected could be significantly larger and have much different impacts. On that hot day, the plant "experience" with the previous release may be worth little. At a large company, when such process safety questions arise, the local site personnel may be able to access corporate Subject Matter Experts (SMEs) in process safety. At a small company, the "process safety leader" or team member may be left to find answers on their own…or with resources on the internet (and, hopefully, supplemented by a trusted outside advisor). A further complication can arise when checking with peers at other small, local chemical companies. While this is generally a helpful approach, the feedback should be carefully considered, as it sometimes may be misleading. Checking with and remaining comparable with peers doesn't help if they are also wrong. This situation arose for one small chemical company when comparing how a number of small local companies defined the limits of regulatory applicability. This situation involved bulk storage of a listed chemical (above the threshold quantity), and transfer of the material via process piping to downstream batch reactor processes using the material in small quantities (significantly below the threshold quantity in the reactor). In this case, the small chemical company had defined the limit of the covered process as the bulk storage and process piping – not including downstream batch reactors and equipment. Its local peers defined similar processes in the same way. The only problem was that they were all wrong. Working with the outside process safety expert, who also consulted (for further verification) with an individual at the regulatory agency who had primary responsibility for setting up the regulatory program, the batch reactor processes were then included in the PSM program. It is worth remembering that while benchmarking with or seeking the advice of other small chemical companies in the areas may be helpful, they also may "be in the same boat," with limited process safety knowledge.

The tools used to manage and implement the process safety program can also help set the stage for success…or frustration. Migration from paper-based systems to electronic ones – at least in part - is a positive, and generally advisable step for many small chemical companies. It is not uncommon, at least initially, for small company process safety programs to be paper-based systems. Transitioning to even simple, electronic (MS-Word®, MS-Excel®, MS-Access®) systems for a PSM program helps both in development and in the longer-term maintenance of the program. An electronic system helps makes it easier to manage and track various aspects and elements of the program, enables document control, provides easier access to information, prevents information from getting lost, and can be backed up with other business systems and data. For small companies with multiple covered processes use of a Computerized Maintenance Management System (CMMS) can become a worthwhile investment and a significant aid in handling all of the voluminous data, actions, and records associated with a mechanical integrity program. It does require effort to input information and build the database, but can yield significant benefits in the mid- to long-term. However, one word of caution…a number of small companies have found that trying to build a PSM program into an enterprise resource planning (ERP) system (designed for other business processes) can be a difficult and sometimes ineffective approach. And, it may be "overkill" for a small company. ERP systems often have maintenance functions, for example. However, use of a specialized CMMS – designed for and used in the process industries – may be a better choice. For companies with a single or few covered processes, simpler electronic systems may be sufficient.

4.7 Process equipment – limitations and finding a path forward

Small chemical companies often utilize used equipment that is in long-term storage or purchase used equipment on the open market, due to capital budget limitations, and with direct impacts on process safety. Often, the focus is short-term with the key driver being functionality at the lowest possible cost [11]. Compliance with codes and standards, facility siting, related process safety information, and other long-term issues that may require additional studies and incur additional costs may be "put off" to be dealt with at a later time.

Assembling a process with used equipment often results in a process with varying safe operating limits for each major equipment item, and equipment that may not have been designed for the currently intended service. Equipment may not be properly sized, including design for pressure control devices. A small company may change process parameters and the process design to enable use of the much cheaper used equipment. All of this leads to a process that may be designed to different codes and standards, with differing safe operating limits, different materials of construction, and a number of unknowns. Sometimes small companies will even forgo purchase of the vessel code documents (e.g. U-1 forms) that, if available, may be provided at an added expense in the equipment purchase price. Further, the used equipment may be physically modified to meet requirements of the project. In the case of physical modifications to equipment, if they are not made by a "code shop" the equipment is no longer a "code" vessel.

Documentation of the necessary information for process safety for used vessels purchased on the open market, in terms of the process safety information and information for the mechanical integrity program, can become a real challenge. This may then impact control of the process, management of change, as well as inspection, testing, and preventative maintenance for the equipment and entire process.

When purchasing used equipment, small companies should make every effort to obtain construction drawings, code information (e.g. U-1 forms), inspection records and certifications, and other information that will be needed for the process safety information and mechanical integrity program. This includes payment of additional fees in the equipment purchase price, if necessary, for vessel code documents and other information. It is well worth the small additional charge to have the equipment information, and will same time, money, and frustration later on.

If used equipment is incorporated in process design, the engineering team will need to carefully address issues regarding materials of construction and wall thickness; provision of properly sized pressure relief; and inspection, testing and preventative maintenance, for example. The OSHA PSM standard requires (in 29 CFR Part 1919.119(d)(3)(H)(iii)) that the company "…determine and document that the equipment is designed, maintained, inspected, tested, and operating in a safe manner." In addition, the safe operating limits for the process likely will be defined by the "weakest link" - the equipment or portion of the process with the lowest rating (e.g. pressure rating).

4.8 Build upon what already exists and develop what is needed

Small chemical companies have to be extremely effective in the use of human resources. It is vital that the process safety program is built upon information, methods, and systems actively used in the business, to the extent practicable. This should start with use and/or tie-in to an existing, functioning management system at the site. Small companies often use quality management systems or may have environmental management systems, for example, that may be used for process safety management as well.

Existing information that can be used as part of the process safety program (identified in the initial review and assessment) should be assembled for inclusion. Examples of materials that are often already in use at small companies and can be incorporated into the process safety program include:

- Material Safety Data Sheets (MSDSs)/Globally Harmonized System of Classification and Labeling Safety Data Sheets (SDSs)
- Procedures
- Batch tickets or protocols
- Safe work practices (e.g. hot work, line break)
- Maintenance program (in some form) and work order system

Plant sources of this information may include operations, research and development, the site safety program, environmental programs, operations, and others. The program development process can make use of this existing information so that not everything has to be built from the "ground-up." Some areas will need to be enhanced or improved, and others may require new development.

Key technical aspects or elements of process safety programs that are particularly important, and will need to be rigorously developed (and likely improved even if they already exist) include:

- Process Hazard Analysis – the primary means to identify the hazards.
- Management of Change – as it "touches" personnel throughout the business on a daily basis, is used to manage the risk associated with changes, and can be used to drive and build a positive process safety culture.
- Audit – formal/regulatory and more importantly informally as a means to evaluate progress and what is needed – feeds back into planning and may initiate "course correction."

4.9 Use available resources

There are a number of resources that small companies can use to develop process safety programs. The most important resource is a trusted process safety consultant. Working together with, and as an advisor to the "process safety leader" and core team, this individual should be ideally qualified and able to help point the core "team" to additional resources. Examples of available resources follow:

- Regulator websites (e.g. OSHA, EPA)
- Federal agencies (e.g. CSB)

- Reference materials, such as:
    - Center for Chemical Process Safety (CCPS) publications, including *Guidelines for Risk Based Process Safety* [12]
    - Standards from a number of organizations (e.g. American Petroleum Institute, National Fire Protection Association)
    - Journals (e.g. Process Safety Progress – provided with membership to the AIChE Safety and Health Division, Journal of Loss Prevention in the Process Industries)
    - Electronic and online services and libraries (e.g. Knovel® - provided as a benefit of AIChE membership)
- Networking sources:
    - Local networks (e.g. chemical companies)
    - Staff contacts at previous employers or vendors
    - Emergency services (e.g. fire department)
    - Professional contacts (e.g. trusted peers, mentors)
    - Linked-In
- Process safety websites, blogs, newsletters
- Electronic mailing lists and accident alerts/databases
- Professional associations (e.g. American Institute of Chemical Engineers – local and national level, American Chemical Society, International Society for Automation)

In many instances, quality information can be accessed for free on the internet. Due care should be taken, however, to ensure that information obtained is from a credible source – otherwise, the materials may be worth exactly what they cost…nothing. Most current standards are not available on the internet for free, as the standards organizations use income from sales to fund the organization's activities and further standards development.

4.10 Budget for process safety needs - operational and capital budgets

As early as possible in the process, the company planning and budgeting cycle should include process safety management resource requirements. Initially, at many small companies, process safety expenditures may be "ad-hoc" or as approved by management. It is important to develop a budget for process safety and include it within the normal business cycle. It is best to incorporate process safety into existing business systems and processes when possible. Inclusion of process safety as a budget item facilitates better planning and allows for steady progress. Two further recommended actions are to:

- Include process safety costs for specific customer-driven projects or changes in customer contracts – often the incremental cost to the customer is small and the impact in providing resources for a small company's process safety budget is quite large.
- Consider process safety projects or tasks as engineering projects that are planned, budgeted, and tracked.

4.11 Re-evaluate and adjust the process safety development or improvement plan

No matter how effectively and diligently the "process safety leader" and core "team" plan, changes to initial plans will be needed. Every small company is unique. The "starting point" for small companies can vary significantly, including the amount of information that can be used from existing business activities. And, the process and work likely *will take longer than initially planned.* This is especially true for small companies and teams – as personnel needed for other day-to-day tasks and projects in operations may become unavailable for process safety work planned. The "process safety leader" at a small company needs to be flexible and reasonable, while consciously working to hold staff accountable and ensure the process safety work progresses.

4.12 When an orderly process is not feasible – make progress when and where possible

All of the previous discussion assumes that a reasonable level of support is given by management, and that at least "pockets" of support can be found within the staff to develop a process safety program. If it appears that support is so limited that such a planned, orderly approach is not initially possible, a humorous alternative is presented by Berger in "The Process Safety Guerilla: Giving Your Company A Future, Even if it Doesn't Want One." [13] Berger explicitly states regarding terms used and the approach presented that "….their use is only in the metaphorical sense…not to advocate illegal activity of any kind." Berger's paper outlines how an employee concerned about process safety at a facility can (as a "process safety guerilla") accomplish positive change. The approach described by Berger [13] is to:

1. "Arouse and organize the people" – identifying supporters and involving people with similar concerns and interests.
2. "Achieve internal unification" – creating an identity and mission, and holding information discussions.
3. "Establish bases" – building a group of persons willing to promote process safety and networking with others.
4. "Equip forces" – gathering and obtaining materials and tools, and learning how to use them.
5. "Gain strength" – by becoming an authority and taking a stand as the opportunity arises.
6. "Gain control" – by converting senior or executive management into supporters, or by using a near-miss or an accident to draw attention to the importance of process safety.

Typically, support for process safety can be found somewhere within the company, so a planned, orderly path can be taken. However, at one point or another, the "process safety leader" at a small company may never-the-less feel like a "process safety guerilla."

5. Commitment and perseverance

Building, maintaining, and sustaining an effective process safety program within a small chemical company setting requires a high level of professional and personal commitment and sheer perseverance. At a small company there are many demands on each staff member, and most "wear more than one hat." Process safety can be compared to a heavy flywheel on an engine, using the analogy from Jim Collins [4]. It takes a lot of energy to get the flywheel going. Once it is going at the desired speed, it smoothes out the running of the engine, and continues the motion of cylinders.

Figure 3. The "Flywheel" – Through Steady Efforts Develop Momentum

The primary concept from the "flywheel" analogy that applies to process safety is to *build and maintain momentum*. Anyone that has been tasked with starting a new process safety program can express the frustration caused by a lack of momentum during initial stages of the program. It requires steady, hard work to bring the flywheel into constant motion and progress. The transformation of process safety programs works much like a flywheel – and, there is no secret or magical solution – it takes steady, hard work. There are many sound fundamentals that lead to a "good" process safety program, but no single answer. Process safety requires constant work, pushing in the same direction, more or less according to plan. To build, improve, maintain, and sustain a good process safety program – "work hard" and "work smart" are familiar and appropriate advice.

6. Summary

In spite of many challenges small chemical companies face the requirement, for a complete process safety management program and its full implementation, is the same as required for much larger organizations. Small companies have to be creative and highly effective in development of practical, functional programs that utilize existing resources and reasonably delegate responsibility for various program elements across the organization.

Key technical programs for process safety include Process Hazard Analysis, Management of Change, and Audit, as they are critical to identifying hazards, managing the ongoing changes affecting process safety and risk, and providing the means to identify gaps and areas needing further attention. The management of change program also can be effectively used to "drive" a change in process safety culture over time - with regular interactions of personnel throughout the organization and by focusing attention on changes, decisions and actions taken and the implications on process and company risk.

There are many challenges that small companies must overcome. It starts with good process safety leadership and development of a core team that is professionally and personally committed to process safety. The brutal facts regarding the company "starting point' should be identified early on, as well as identification of existing materials and processes that can be incorporated. This provides a basis for deciding "where to begin" with the work and enables planning of discrete, manageable steps in the process safety development or improvement process. At small companies, the process safety responsibilities generally have to be delegated across the small staff, with logical assignments made that recognize and incorporate other normal staff assignments and daily activities when possible. While there are a number of resources available to small companies, the best resource is likely to be a trusted process safety professional/consultant that becomes an integral part of, and advisor to, the core team. For small companies, one of the most difficult aspects of process safety to address is *insufficient process safety knowledge*. You don't know...what you don't know. This challenge can be overcome largely through judicious use of a process safety expert and additional resources he or she is able to provide. Throughout the entire process, it is important to integrate process safety into existing business systems and operations to the extent possible. And, this includes the company planning and budgeting cycle. Flexibility, commitment, and perseverance are all qualities needed in the "process safety leader" and core "team" for development, implementation, and maintenance of effective process safety programs at small companies – that when applied, can be used to creatively overcome the process safety challenges encountered.

7. References

[1] Occupational Safety and Health Administration (OSHA), 29 CFR Part 1910.119, Process Safety Management of Highly Hazardous Chemicals; Explosives and Blasting Agents; Final Rule, Washington, DC, February 24, 1992.

[2] Environmental Protection Agency (USEPA), 40 CFR Part 68, Accidental Release Prevention Requirements: Risk Management Programs Under Clean Air Act Section 112(r)(7); Final Rule, June 21, 1996.

[3] Howat, Colin S, "Simplified PSM for Smaller non-Regulated Sites: Development History, Program Overview & Recommendations," 8th Global Congress on Process Safety and 14th Process Plant Safety Symposium, Houston, Texas. April 2012.

[4] Collins, JC, *Good to Great: Why Some Companies Make the Leap...and Others Don't*, New York, NY: HarperCollins Publishers Inc., 2001.

[5] Occupational Safety and Health Administration (OSHA), Instruction CPL 02-02045, Process Safety Management of Highly Hazardous Chemicals - Compliance Guidelines and Enforcement Procedures, September 13, 1994.

[6] Environmental Protection Agency (USEPA), EPA 550-B99-008, Guidance for Auditing Risk Management Plans/Programs under Clean Air Act Section 112(r), August, 1999.

[7] Environmental Protection Agency (USEPA), EPA-550-K-11-001, Guidance for Conducting Risk Management Program Inspections under Clean Air Act Section 112(r), January, 2011.

[8] Goddard, Sheryl, "A Pre-Plan for Process Safety," 8th Global Congress on Process Safety and 14th Process Plant Safety Symposium, Houston, Texas. April 2012.

[9] Myers, Philip M, "Retool Your Process Safety Program for Efficiency and to Meet Multiple Management System Requirements," 4th Global Congress on Process Safety and 10th Process Plant Safety Symposium, New Orleans, Louisiana. April, 2008.

[10] Covey, Stephen R, *The 7 Habits of Highly Effective People: Powerful Lessons in Personal Change*, New York, NY: Simon & Schuster, 1989.

[11] Herber, John M, "Why Smaller Companies Struggle with Process Safety," 8th Global Congress on Process Safety and 14th Process Plant Safety Symposium, Houston, Texas. April 2012.

[12] Center for Chemical Process Safety, *Guidelines for Risk Based Process Safety*, Hoboken, NJ: John Wiley and Sons, 2007.

[13] Berger, Scott, "The Process Safety Guerilla: Giving Your Company a Future, Even if it Doesn't Want One," 5th Global Congress on Process Safety and 11th Process Plant Safety Symposium, Tampa, Florida. April, 2009.

Table 1. Overall PSM and Risk Management Program - Roles and Responsibilities Matrix

Roles/Positions	Vision, Goals, Objectives	Program Oversight & Direction	Provide Resources - People, Equipment, & Financial	Establish Organizational Roles, Responsibilities & Accountability	Visible Support & Actions	Overall Program & Reporting	Overall PSM Program & Reporting	Overall Risk Management Program & RMP & Reporting	Employee Participation	Process Safety Information (PSI)	Process Hazard Analysis (PHA)	Operating Procedures	Training	Contractors	Management of Change (MOC) & Pre-Startup Safety Review (PSSR) / Process Change Request and Review (PCRR)	Mechanical Integrity (MI)	Safe Work Practices (SWPs) - incl Hot Work	Incident Investigation	Emergency Management	Compliance Audits	Trade Secrets
CEO	A				*																
President	A	A	A/P	A	P	A	A	A	*									A	A	A	
Plant Manager/Site Leader	A	A	P	P	P	A	A	A	*				I		A	A		A	A	A	
Business Managers (Business Development)					*				*						I						S
Research & Development Manager	I				*	I	I	I	*	S	S	S	S		S	I		A		I	P
Controller/Purchasing Manager			S		*				*					S	I	A					
Human Resource Manager	I		S		*	I	I	I	*				S/A	I	S					I	
Engineering Manager	S	S	I	I	P	S	S	I	I	A	A	A	P	A/S	A	A/P	I	A/S	A	A	S
Environment, Health, Safety & Security Manager	P	P	I	S	P	P	A	P	A/S	I	I	I	A/S	I	I	I	A	P	P	P	A
PSM Manager/Coordinator	S	S	I	I	P	S	P	S	P	P	P	S	S	S	P	S	I	S	I	S	S
Maintenance Buyer			S		*				*	I				S	S	S				I	
Environmental Manager	I				*	I	I	S	*		I		S		I	I	I	I	S	S	I
Safety Manager	I				*				*	S	I	I	P	P	I	I	P	S	I	I	I
Process/Plant Engineer					*	I	I	I	*	S	S	P	I	I	S	S	I	I		I	S
Production Manager			S		P				*		S	S	I	I	S	I	I	I	I	I	
Systems Administrator					*				*			I			I					I	I
Instrumentation & Controls Engineer					*	I	I	I	*	S	I	I	I	I	S	S	I	I		I	
Maintenance Manager					P	I	I	I	*	I	I	I	I	S	S	P	S	I		I	
Chemist					*				*	S	I	I			I	I	I	I			I
Maintenance - I&C Technician(s)					*		I	I	*	I	I	I	I	I	I	I	I	I	I	I	I
Maintenance - Mechanical Technician(s)					*		I	I	*	I	I	I	I	I	I	I	I	I	I	I	I
Supervisors & Operators					*		I	I	*	I	I	I	I	I	I	I	I	I	I	I	I

A = Approve Activity or Awareness
I = Input and Participation Expected
P = Primary Responsibility
S = Secondary Responsibility or Support
* = Important Activity for All

Small Site MOC: Big Company Challenges without Big Company Resources

Rainer Hoff, Ph.D., P.Eng.
Gateway Consulting Group, Inc,
8610 Transit Road, Suite 100, East Amherst, NY, 14051
rhoff@gatewaygroup.com

Gordon C. Blake, P.Eng.
Gateway Consulting Group, Inc,
8610 Transit Road, Suite 100, East Amherst, NY, 14051
gblake@gatewaygroup.com

Prepared for Presentation at
American Institute of Chemical Engineers
2013 Spring Meeting
9th Global Congress on Process Safety
San Antonio, Texas
April 28 – May 1, 2013

UNPUBLISHED

Small Site MOC: Big Company Challenges without Big Company Resources

Rainer Hoff, Ph.D., P.Eng.
Gateway Consulting Group, Inc,
8610 Transit Road, Suite 100, East Amherst, NY, 14051
rhoff@gatewaygroup.com

Gordon C. Blake, P.Eng.
Gateway Consulting Group, Inc,
8610 Transit Road, Suite 100, East Amherst, NY, 14051
gblake@gatewaygroup.com

Keywords: Management of Change, MOC, business process improvement,

Abstract

Many have argued that the PSM regulations impact small sites particularly hard. While small sites must comply with the same regulations as much larger sites, small sites are hamstrung by a lack of resources. This is particularly true for Management of Change due to the inherent complexity of MOC.

This paper focuses on the challenges of improving the MOC business process at small sites. An MOC improvement initiative generally involves improving the MOC business process, and then assisting with some form of technological automation.

The authors interviewed representatives from small sites to obtain their first-hand lived experiences, in order to properly ground the study.

1. Introduction

Managing change effectively has always been important at chemical plants and refineries. Lack of a proper MOC program the prime cause in a number of well-known accidents, including the Flixborough explosion [1], and a contributing factor in other well-known accidents [2, 3]. Certainly small sites are not immune to accidents where improper MOC was a factor.

Since the promulgation of OSHA 1910.119 [4]—the Process Safety Management regulations—proper Management of Change is a regulatory requirement, for Covered Processes (as defined in the regulation).

The Process Safety Management regulations cover 14 elements, much more than just Management of Change. As a result, when the regulations were first promulgated in 1992,

substantial effort was needed to bring most plants into compliance. In this environment, with deadlines and potential penalties, Management of Change business processes were either hastily created, if they did not previously exist at the site, or quickly reworked and formalized from an existing business process.
Poorly or inadequately applied MOC processes have contributed to many incidents [2, 5] , and have been identified as issues in Refinery National Emphasis Program [6] and Chemical Plant National Emphasis Programs [7]. Based on the authors' experience, personnel who are required to use the MOC process, the "users", often have a negative view of the MOC process. User dissatisfaction may be due to a poorly-designed or poorly-implemented process.

Herber [8] notes that 85% of the Chemical Safety Board investigations occur at small companies, implying that small companies are disproportionally impacted by PSM issues, including MOC. Both Herber [8] and Goddard [9] elaborate on the reasons why small sites find PSM compliance so challenging.

Many small sites, whether triggered by an incident or simply a desire for enhanced process safety, attempt to improve their MOC systems. Improving MOC normally requires improving the MOC business process itself (the policies, procedures, forms, engineering aids, etc.). Efficiency improvements, and often better compliance, can be achieved by using electronic technology to implement MOC.

This paper summarizes the results of a survey of 64 sites who have attempted to improve their MOC processes. Some were independent small sites, while others belonged to a larger corporate entity.

This paper does not attempt to define MOC best practice, since that is well-covered by other authors [10-17]. This paper focuses on the effort, challenges and enablers that allow a small site to move on the path to best practice.

1.1 MOC Overview

A high-level model of MOC, referred to as a "lifecycle" model, is shown in Figure 1. The diagram depicts a permanent, non-emergency MOC process. A lifecycle is composed of "states": Initiation, Scoping, Change Design, etc. While the MOC is active, it could be considered to be "in" a state: e.g. the MOC is in the Approvals state.

Further details of lifecycle models can be found in the references [10].

Figure 1. Typical permanent, non-emergency MOC lifecycle.

2. Characterizing Small Sites

2.1 What is a Small Site?

There are various metrics which can be used to characterize the size of a site:

- number of units or processes
- number of employees
- physical size
- proximity of coworkers
- revenue

At a small site, there may only be a few or a dozen units or processes, unlike a large refinery, where there may be up to 100 units.

A small site often has fewer than 100 people on site (without regard to whether these are employees or contractors).

Also, a site may be physically small so that it is convenient to walk, say, from the administration building to the control room(s); in a large site, it may be more practical to drive.

At a small site, technical resources are usually in the same building or even the same room. In contrast, a large site is bigger and has more formal communications, often respecting organizational boundaries and hierarchies.

2.2 How Do Small Sites Behave?

The small size engenders a number of behaviors, which promote efficiency:

- informal communication
- rapid decision-making
- easier consensus-building

Communication tends to be less formal at a small site, since coworkers see each other on a routine basis.

Decision-making is rapid, because the people that need to be involved, are close at hand.

Consensus-building is often easier, since it's often based on personal relationships, which may be stronger in a smaller organization.

On the other hand, a small site has:

- fewer total resources
- fewer opportunities for specialization

With fewer individuals, a given person may have many roles and responsibilities, so it is more difficult to become an expert in one domain—staff at small sites are required to be generalists [9].

2.3 The Best/Worst of Both Worlds: Small Sites as Part of a Larger Corporate Entity

The prior discussion contains an implicit assumption, being that the small site is completely independent of any other site or corporate entity. There are comparatively few small sites that enjoy that level of independence. More common is for a small site to be part of a larger corporate entity. This provides opportunities for collaborating with other sites and leveraging corporate resources, but also has a concomitant obligation to conform to corporate policies, procedures and technology standards.

The position of a site as an independent entity versus part of a larger corporation has an important impact on site autonomy and how the site may proceed in any MOC improvement initiative.

2.4 Research Methodology

Small sites may not have the same opportunity to improve their MOC systems as large sites have, due to resource constraints. This paper reports the results of a qualitative research study with input from a variety of sites.

A total of 64 sites were surveyed for this research. However, the sample was not random. The sample was biased in favor of companies and sites that had expressed an interest in improving their MOC systems. Consequently, 0% of the survey participants believed that their MOC systems were best practice, and 100% believed that improvements were possible and desirable.

Small, independent sites were surveyed as well as small sites that belong to a larger corporate entity, the latter being more prevalent.

The results presented in the following chapters, are be ordered by the feedback that was most prevalent: that is, the most common issues are stated first, the second most common next, and so on. Since this is a non-randomized, qualitative study, presenting quantitative results (e.g. x% of respondents had characteristic Y), is not statistically valid. As is common for qualitative research, the results could be used to identify relevant questions for a future quantitative study.

3. MOC Business Process Improvement

MOC improvement initiatives are generally conducted in two phases:

1. MOC Business Process Improvement: This is where the MOC process itself (policies, procedures, guidelines, forms, engineering aids, etc.) are updated.
2. MOC Technology Improvement: This is where the MOC process is implemented (or updated) using some form of technology.

3.1 Independent Site MOC Business Process Improvement Challenges

Based on the survey results the most common challenges for small independent sites are:

1. Uncertainty over who "owns" the business process
2. Lack of user engagement at the proper time.
3. Lack of compliance with existing MOC process

There is often overlap between the Quality organization and the Process Safety organization at a site. Both the ISO9000 quality systems program [18] and OSHA1910.119 process safety management regulations [4] govern how procedures are modified, approved and released. This creates situations where both organizations (rare) claim ownership of the MOC process, or neither organization (more common) claims ownership of the MOC process. While there may be a perceived need for MOC improvements, no one organization feels that it has the mandate to advance this kind of initiative.

An MOC Improvement Team ("Team") may work diligently at improving the MOC process. This team will no doubt solicit input from interested parties at the site, in an attempt to sincerely address MOC requirements contributed by several disciplines. Often, the site response is lackluster, and meaningful feedback isn't received until after the Team has completed its work, has updated the MOC procedure, and is now syndicating the new process throughout the facility. Of course, at that point, it is too late to incorporate the feedback without redoing much of the Team's work.

Sometimes site personnel do not execute MOCs in accordance with their existing MOC procedures: checklists are ignored, forms are not completed, and signatures are overlooked. User reactions while being trained in the new MOC process may be that it is "a lot more complicated". In fact, the new process may be much simpler than the former process, but if the former process was ignored, then the mistaken perception is one of increased complexity.

3.2 Independent Site MOC Business Process Improvement Enablers

Based on the survey results the most common success enablers for small independent sites are:

1. MOC redesign can be funded from local (often engineering) budgets
2. Someone is willing to "champion" the initiative
3. A streamlined process, without unnecessary steps, leads to higher compliance

MOC process improvement initiatives are often done with a combination of local resources and outside expertise. For a small site MOC redesign, the cost of the outside resources is normally within the approval limits of local site management. Funds are often available in budgets for engineering studies.

Good leadership of an MOC improvement initiative is important. A project champion, like a good project manager, is likely to keep the initiative moving, and is able to overcome obstacles.

Needlessly complex processes discourage compliance. Even a well-intentioned person may find it difficult or impossible to comply with a confusing process or one with unnecessary steps. A simpler process, streamlined to the essentials should encourage better compliance with MOC requirements.

3.3 Corporate Site MOC Business Process Improvement Challenges

This covers the case where a small site desires or is required to participate in an MOC process improvement initiative in conjunction with other sites or in compliance with a corporate initiative. Based on the survey results the most common challenges for small sites that are part of a larger corporation are:

1. Lack of a regulatory mandate
2. Difficulty in arriving at a consensus on the path forward
3. Impacts of acquisitions and divestitures

Some companies have dozens of sites, but only some of them have any PSM-covered processes. Nonetheless, the corporation mandates that business processes will be consistent across the organization, and therefore the non-PSM sites must implement an MOC process. A non-PSM site has less impetus for deploying MOC than a regulated site, so the effort and cooperation are necessarily less at the non-PSM sites.

Many companies pride themselves on their adoption of consensus management. Consensus management normally implies that most or all parties must agree to a new initiative or approach. This becomes problematic when a consistent MOC process is sought to address multiple sites whose needs are inconsistent. Sites with simpler requirements require considerable coaxing to accept approaches that address the needs of other, more complex, sites.

The current business climate is very dynamic with acquisitions and divestitures of assets (e.g. entire plants or facilities) being quite common. Unfortunately these may not be rapid events, but multi-year processes as a site is made available for sale, groomed for the sale (by cutting all expenses beyond the bare essentials), becomes the subject of bidding and speculation, and is finally sold. Once a potential sale is announced, MOC improvement initiatives stop almost immediately, since the participants await direction from the new owners.

3.4 *Corporate Site MOC Business Process Improvement Enablers*

Based on the survey results the most common success enablers for small sites, which are part of a larger corporate entity, are:

1. Effective championing
2. Effective facilitation
3. Effective buy-in

As mentioned in §3.2, having a person willing to champion the MOC process improvement initiative is a big contributor to its overall success.

One of the difficulties of redesigning the MOC process, on a corporate or multi-site basis, is that none of the participants may be experts in business process design, and each of them may simply understand his/her local process and not fully appreciate the processes at other sites. A facilitator, experienced with business process redesign and having MOC experience, is a vital member of the MOC Redesign Team. Without effective facilitation, the Team may degenerate into cycles of the form: first "A" is seen as the solution, then "B", then "C", then back to "A", etc. The facilitator should avoid that.

An improved MOC process, with enthusiastic multi-site support, will experience better compliance than if such support were lacking.

4. MOC Technology Improvement

The second phase of MOC improvement initiatives is often to realize business process efficiencies by deploying an electronic MOC system.

Often, the MOC technology improvement is performed concurrently with the MOC business process improvement. That is a reasonable approach, since the available technology options put constraints on what is possible or practical during the redesign.

4.1 *Independent Site MOC Technology Improvement Challenges*

Based on the survey results the most common technology challenges for small independent sites are:

1. Budget limits
2. Positioning of MOC as a "tag-along" to another initiative
3. Lack of leadership

There are very real budgetary limits on what a small site can spend on a technology solution for what is perceived as an administrative process (in contrast with process control systems which are viewed differently). If the proposed technology fits into the site's spending limits, then the project can often proceed. If the project is more costly, then the justification process is so onerous, that site personnel either don't have the time or desire to pursue it.

There may be other technology upgrade initiatives taking place at the site, and the participants feel that they may be able to take advantage of economies of scale if MOC were upgraded at the same time. Often, there is a team with a mandate to implement better incident investigation and tracking; that team may be well-resourced. The person(s) responsible for MOC is invited to participate in the technology evaluation, again, with the hopes of obtaining a lower unit cost. The difficulty with this is that incident investigation (when measured by the number of tasks) is a much less complex process than MOC. So, the evaluation effort is actually inverted, whereby the simple process (incidents) receives a lot of attention, and the complex process (MOC) "tags along" and receives comparatively little attention. This will usually result in a sub-optimal MOC deployment, even if the technology selected is acceptable.

Similarly, there may be a team investigating electronic document management ("EDM") with the goal of managing process safety information and other important facility documents and drawings. Again, the person(s) responsible for MOC is invited to participate in the technology evaluation, again, with the hopes of obtaining a lower unit cost. This arrangement has a different problem. From a project management perspective, MOC is well-bounded. MOC lends itself to fixed-price, short-duration implementation because the MOC process requirements are similar from site-to-site—only the details differ. EDM, in contrast, is comparatively open-ended with many opportunities to expand the scope of the initiative by simply addressing more document types (PSI, environmental, maintenance, inspection, …). An EDM initiative is not typically fixed-price, and has a longer implementation duration than MOC. Combining MOC and EDM can be done. But the participants must understand that these different kinds of projects.

Lack of leadership is an oft-cited issue when implementing improved MOC technology at small sites. The difficulty lies in the fact that there are many people who want a better MOC solution—PSM Managers, MOC Coordinators, environmental representatives, the user community—but there may be no one who feels qualified to lead what is seen as an "IT project".

4.2 Independent Site MOC Technology Improvement Enablers

Based on the survey results the most common success enablers for small independent sites, undertaking an MOC technology improvement initiative, are:

1. Experienced partners
2. Effective championing
3. Supportive IT resource

Specifying and implementing an electronic MOC solution requires a number of specialized skills. It is unlikely that these skills exist at a small site, nor is it cost-effective to develop those skills. A technology partner with the requisite skills and MOC experience is an invaluable asset.

As stated previously, a project champion is a significant asset to an MOC technology improvement initiative. Technology changes often bring unexpected results, and the change the way people are accustomed to working. A champion is needed to reassure the doubters by explaining the merits of the "new" approach.

A small site electronic MOC implementation does not require much (measured in man-hours) support from IT. But it is usually difficult to accomplish with no IT resources. So a supportive IT person can significant streamline the implementation process.

4.3 Corporate Site MOC Technology Improvement Challenges

This covers the case where a small site desires or is required to participate in an MOC technology improvement initiative in conjunction with other sites or in compliance with a corporate initiative. The end-user survey had far more responses in this section than any of the others. Based on the survey results the most common challenges for small sites that are part of a larger corporation are:

1. An IT organization that wants to create a home-grown solution
2. An IT organization that wants to leverage existing vendor relationships, even when the existing vendor has little to offer
3. An IT organization that views MOC as an extension of the work order process
4. An IT organization that refuses to accept any solution a Team proposes, but offers no solution of its own

Many IT organizations have a mandate to provide technological solutions. IT may conduct "make versus buy" decisions, and decide that MOC is an application that can be accomplished in-house. Sometimes the decision is driven by the existence of programming resources that don't have any current assignments, so it appears that these resources are "free". The success rate of in-house MOC initiatives is poor. Oftentimes, the development team grossly underestimates the complexity of the MOC process and delivers solutions that are inadequate (by trimming the requirements to fit the budget) or consuming far more resources than planned (actual resource consumption of 10 times the planned resource consumption, is not uncommon). In-house development creates other forms of end-user dissatisfaction: as time goes on, the original developers move on to other projects, and the support for the in-house software wanes.

A common IT organization goal is to minimize the "technology footprint". In one sense, this can be restated as reducing the number of major application vendors that IT deals with. This is a reasonable goal, since each new vendor relationship has an associated cost. The problem arises when IT attempts to engage existing vendors that have no MOC solution or an MOC solution in name only (that is, something called "MOC", but lacking important functionality). The end-user community is quickly frustrated by this, because they see themselves teaching the vendor about MOC, when the expectation is that the vendor should already be proficient in this area, and, in fact, be able to share experiences and best practices with the end users.

Almost all sites have an electronic mechanism for generating work orders. Some even believe that MOC could be an offshoot of the work order system, illustrated at the top of Figure 2, and evidenced by checkboxes on the work orders entitled "MOC Needed ? ☑". The creates a situation where MOC is a reactive process. By the time the work order exists, the *final* solution has already been accepted and is ready for implementation. Even if an MOC is triggered at this point, it causes all the essential MOC tasks— risk ranking, task scoping, design proposal, impact analysis, and approvals—to be conducted in a hurried manner, after the fact. These are really the activities that minimize risk and ensure that the MOC has been safety designed, so these are not the tasks that should be rushed. In contrast, the MOC best practice begins with a *proposed* solution, and the MOC concept is developed and risk mitigated throughout its lifecycle. The latter approach, illustrated at the bottom of Figure 2, is the recommended approach. Work order systems are used to implement the MOC activities in the physical plant—so work order systems are important to the overall process. But a work order system on its own is too little, too late, to be a viable automation tool for MOC.

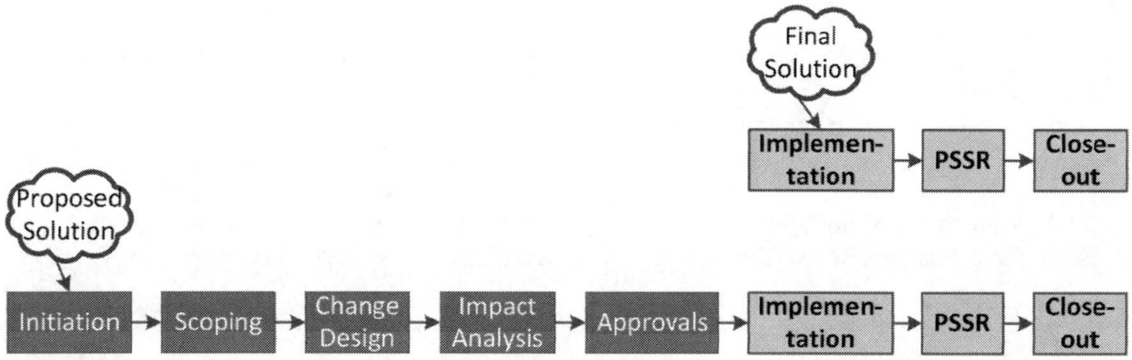

Figure 2. MOCs triggered by work orders (top) versus MOCs independently triggered by a need for change (bottom).

Finally, some IT organizations simply refuse to assist with an MOC improvement initiative. They are unwilling to accept the technology recommendations from the MOC Improvement Team, yet are unwilling to recommend an alternate solution. There may be legitimate reasons—manpower, budget, other priorities—but these reasons are often not exposed to the MOC Team.

4.4 Corporate Site MOC Business Process Improvement Enablers

Based on the survey results the most common success enablers for small sites, which are part of a larger corporate entity, are:

1. Supportive IT organization
2. Experienced partners who have experience with enterprise MOC rollouts
3. Executive commitment

Section 4.3 identified 4 barriers to success of an initiative to improve MOC technology, on a corporate basis. All of the barriers dealt with IT issues. The counterpart to that, and one mentioned by all the successful MOC Team, was the support of a capable IT organization. Much of the work is performed by third-party vendors, so the work load on IT is minimal, but the support of this organization must still be there.

Corporate MOC deployments are complex. Not only is the MOC process itself complex, but there are legitimate variations between business units (usually because the processes or hazards are different). An experienced partner will understand and highlight the trade-offs of various levels of standardization and site-specific deployment.

In previous sections, the importance of having someone champion the initiative was stressed. That is no less important for a corporate MOC technology initiative. But also, due to the cost of these larger initiatives and the impact on the organization, it is necessary to have executive support as well.

5. Conclusions

Independent smaller sites have some advantages over larger sites, since communication is easier and decision-making more rapid. They also have some disadvantage due to a lack of specialized resources and resources in general. Small sites that are part of a larger corporate entity may be able to leverage corporate resources, but may also be constrained by corporate policies.

An MOC improvement initiative generally has two phases: a business process improvement phase and a technology improvement phase. The two phases may be concurrent.

A survey of 64 companies identified a number of challenges experienced during MOC process improvement initiatives, primarily:

- uncertainty over who "owns" the MOC process
- lack of a regulatory mandate (at some sites).

The success enablers experienced during MOC process improvement initiatives were primarily:

- redesign efforts are comparatively inexpensive and can be funded from local engineering budgets
- effective championing of the effort

The challenges experienced during MOC technology improvement initiatives are, primarily:

- budget limits
- IT organizations who want to create their own MOC solutions, but may not be able to complete this project successfully

The success enablers experienced during MOC technology improvement initiatives are, primarily:

- experienced partners
- supportive IT organizations

6. References

[1] Parker, R. J., et al., 1975, "Formal Investigation into Accident on 1st June 1974 at the Nypro Factory in Flixborough," London.

[2] Baker III, J. A., Bowman, F. L., Erwin, G., Gorton, S., Hendershot, D. C., Leveson, N., Priest, S., Rosenthal, I., Tebo, P. V., Wiegmann, D. A., and Wilson, L. D., 2007, "The Report of the BP U.S. Refineries Independent Safety Review Panel," Report, Washington.

[3] Travers, I., 2010, "Learning the lessons from Buncefield," AIChE National Spring Meeting, AIChE, San Antonio, TX, p. 10pp.

[4] OSHA, 1992, "Process safety management of highly hazardous chemicals," 29CFR1910.119, OSHA, Washington.

[5] Board, U. S. C. S. a. H. I., 1998, "Investigation Report, Explosives Manufacturing Incident, Sierra Chemical Company," Chemical Safety Board, Washington, DC.

[6] OSHA, 2007, "Directive CPL 03-00-004 Petroleum Refinery Process Safety Management National Emphasis Program," OSHA, ed.Washington, p. 104.

[7] OSHA, 2010, "Directives -10-05 (CPL 02) - PSM Covered Chemical Facilities National Emphasis Program," OSHA, Washington.

[8] Herber, J. W., "Why Smaller Companies Struggle with Process Safety," Proc. 8th Global Congress on Process Safety, AIChE.

[9] Goddard, S., "A Pre-Plan for Process Safety Management," Proc. 8th Global Congress on Process Safety, AIChE.

[10] Hoff, R., 2007, "Lifecycles," MOC Best Practices, 1(2).

[11] Hoff, R., 2011, "Quantitative Models for Assessing MOC Performance," AIChE National Spring Meeting, 7th Global Congress on Process Safety, AIChE, Chicago, p. 15pp.

[12] Sanders, R. E., 2005, Chemical process safety : learning from case histories, Elsevier Butterworth Heinemann, Amsterdam ; Boston.

[13] Chosnek, J., and Clifton, R., 2008, "Improved Process Safety Management and Simple Metrics," Process Safety Progress, 27(4), pp. 284-288.

[14] Keren, N., West, H. H., and Mannan, M. S., 2002, "Benchmarking MOC Practices in the Process Industries," Process Safety Progress, 21(2), pp. 103-112.

[15] American Institute of Chemical Engineers. Center for Chemical Process Safety., 2008, Guidelines for the Management of Change for Process Safety, Wiley-Interscience, Hoboken, N.J.

[16] Chosneck, J., 2010, "Managing management of change," AIChE National Spring Meeting, AIChE, San Antonio, TX, p. 12pp.

[17] Garland, R. W., 2004, "Electronic Management of Change Process," Process Safety Progress, 23(4), pp. 244-251.

[18] International Organization for Standardization, 2000, "Quality management systems--Fundamentals and vocabulary," International Organization for Standardization, Switzerland, p. 36pp.

Ensuring Lower Life-Cycle Costs by Integrating Maintenance Execution and Reliability Management Work Processes

David A. Rosenthal, PE, CMRP
Reliability Delivery and Asset Management Manager
Jacobs Engineering Group
Houston, TX

Summary

The paper will detail how the Maintenance execution and Reliability work processes interact with each other in a manufacturing setting to improve asset condition and reduce their life-cycle costs. These life-cycle costs include the purchase price, maintenance, parts, utilities, etc. This paper will detail where both processes interface, what information is exchanged, the tasks involved, and the consequences of not leveraging this integration.

Background

Manufacturing competitiveness relies on first meeting and then exceeding productivity goals. Supply chain systems deliver customer demands based on assumptions of reliability for the manufacturing assets at each site. Business models that predict production, distribution and delivery capacity contain base maintenance assumptions for each facility including times for repair, preventative maintenance downtime, overhaul frequency, and turnaround duration.

Maintaining the condition of these assets to meet the above objectives is a strong function of the continuous improvement of equipment strategies aimed at reducing Mean Time between Failures (MTBF). These strategies contain maintenance tasks needed to maintain their intended function. Maintenance care for critical assets is generally based upon vendor recommendations, historical experience and addressing known equipment failure modes. When MTBF rates fall outside of excepted industry benchmarks, manufacturers must analyze failure history and the effectiveness of their equipment strategies.

Maintenance is in charge of only restoring function while Reliability designs care to ensure lower life-cycle costs. The continuous improvement of asset care depends on the exchange of information between these two work processes. One can draw the analogy of the application of Plan-Do-Check-Act cycle. Key "touch points" exist between processes to ensure correct asset care is provided to eliminate failure modes seen during operation. This paper identifies each of these interfaces and discusses the importance of managing the information that passes between these work processes that are managed by Maintenance and Reliability personnel. Poor execution of the above will lead to shorter MTBF times providing increased risks to supply chain assumptions which ultimately could affect the customer.

Maintenance Work Process

The Maintenance Execution work process (MEWP) has not changed much over the past several decades. A typical process is shown below. The mission of the MEWP is to safely restore function of assets that are about to or have failed. Failure in the present day context is when the asset no longer performs the function it was designed to meet. For instance, a pump designed for 100 GPM has failed when it only produces 50 GPM. In another example, a conveyer belt designed for 100 FPM has failed that is now only moving at 75 FPM.

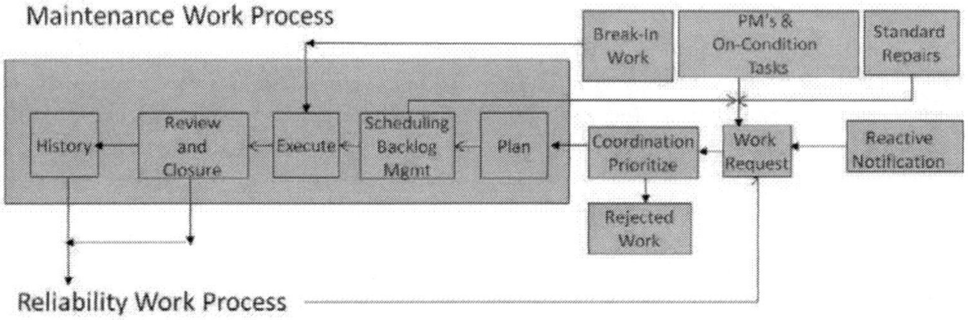

The MEWP receives many inputs from facility stakeholders. The obvious input is a work notification (work order) for a failed piece of equipment. Another could be a scheduled preventative maintenance task or another could be urgent work that must be addressed immediately. This process must then assess the priority of each type of work, plan what needs to be done (parts, resources, tools, procedures, permits) and then schedule resources to execute the work. However, that is not the end of its functionality.

After the work was performed, an analysis and history must be provided as an output for the record noted against the assigned component or asset. This history is vital for later analysis. It contains information of why it failed, what was done, and recommendations from the field to prevent future failures. This information is a key input for the Reliability work process which will be discussed. This information informs those who design and manage the care of the asset of the effectiveness of the current maintenance strategy.

Before we move on, there is a second interface as an input to the MEWP from the Reliability work process. Planned work such as preventative, predictive, time-based, and standard work prescribed by the Reliability work process drives another key interface. This tactical work is developed from the maintenance strategies for each critical piece of equipment. Failure to execute this work as prescribed increases the risk for failure.

The management of these two critical interfaces depends upon the asset care culture of the facility. Reactive organizations that are responding to excessive breakdowns do not allot time to their management. Proactive organizations drive recognition of the early signs of failure before function is lost. These organizations devote focus on continual improvement in asset care. Unfortunately reactive organizations are more plentiful than those that are proactive in their approach.

Reliability Work Process

The Reliability work process is shown below. It is shown tied to the Maintenance Execution work process. The Reliability work process is responsible for establishing asset maintenance, engineering, and operating strategies and continually improving them. The maintenance strategies are comprised of preventative, predictive, repair and life extension (overhaul) tasks. The mission of this combination of tasks is to enable the plant assets to perform at their desired function and thus meet the goals of the manufacturing organization.

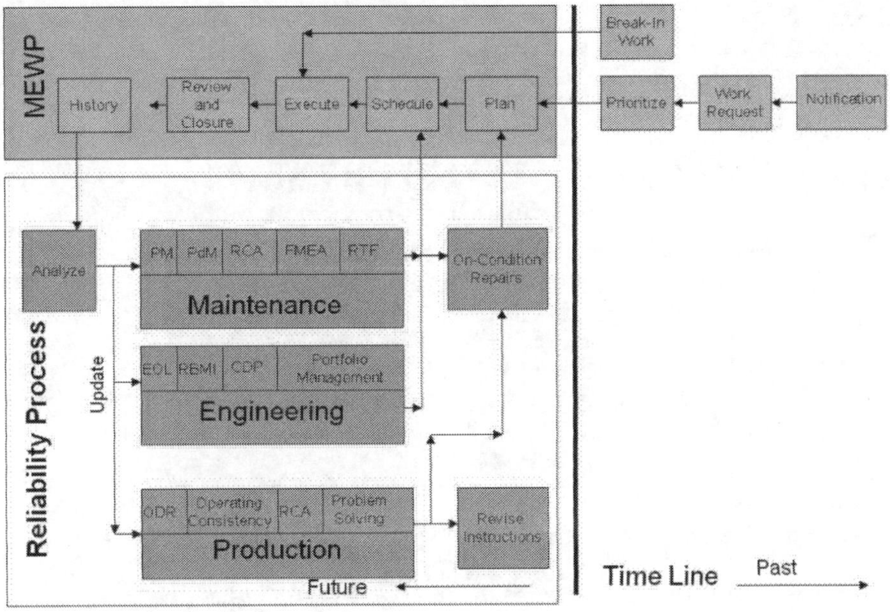

Note: PM – Preventative, PdM – Predictive, RCA – Root Cause Analysis, FMEA – Failure Modes and Effects Analysis, RTF – Run to Failure, EOL – End of life, RBMI – Risk Based Mechanical Integrity, CDP – Capital Deployment, ODR – Operator Driven Reliability

This process also includes Engineering and Production practices since the "triad" is responsible for the life-cycle practices associated with any asset. Although, the focus of this paper is the Maintenance interfaces, Reliability has equally important interfaces with both Engineering and Production. Especially, Production, since industry experience shows that operating practices have the greatest impact on maintaining the function of any asset. Operators have the most "exposure" to the assets and act as the "front line" to detect the earliest signs of failure. Their actions to adjust, calibrate, lubricate, monitor, and replace are essential to maintaining asset function. While, the Engineering practices have the next level of impact since it is estimated that 80% of the reliability of any asset is "built in" with the design.

Ensuring Lower Life-Cycle Costs by Integrating Maintenance Execution and Reliability Management Work Processes (Cont), David Rosenthal, PE, CMRP, Jacobs Engineering Group

Key Interfaces

There are two interfaces that will be described. They are the maintenance strategy – planning and scheduling interface and the work history - failure mode analysis interface. These two interfaces connect the Maintenance and Reliability work processes. These processes work in tandem to ensure the assets perform as per design. Their operation leads to the higher equipment availability, less downtime and a predictable supply chain performance for the business. Shown below are the interfaces and the typical components of the configuration "engines" for Production, Maintenance, and Engineering.

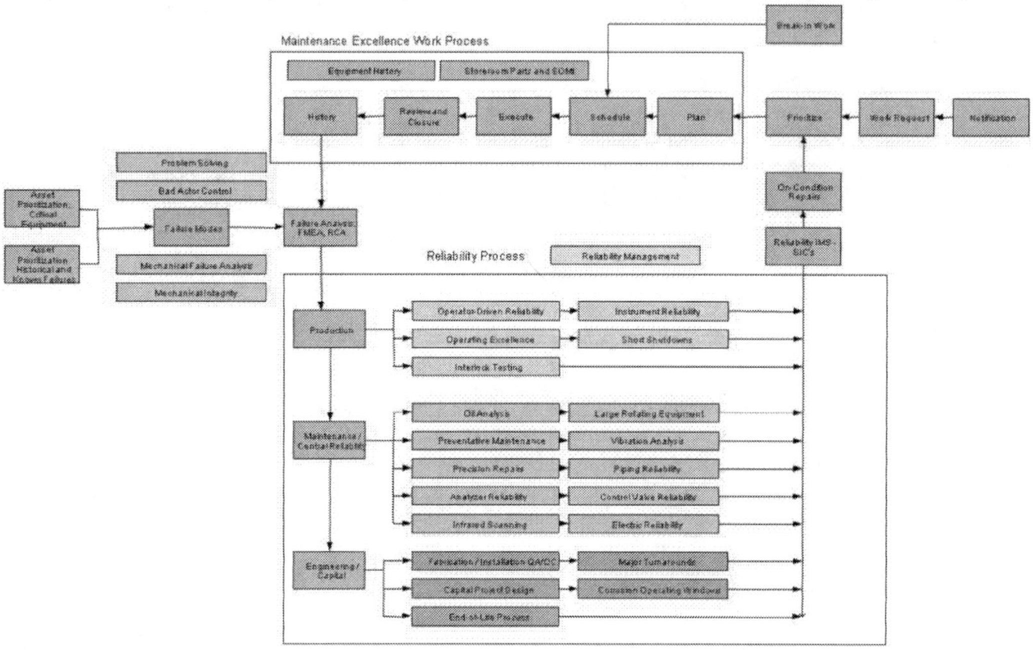

Maintenance strategy – Planning and scheduling interface

This interface between these processes ensures that prescribed care for first the critical assets and then the run-to-failure assets is executed. The planning and scheduling function receives this information through work orders, third-party work (Vibration, oil analysis), exceptions found during operator rounds, standardized work, etc. It is communicated electronically through the site's CMMS. Once it appears as a work order the planning function arranges the work for the maintenance schedule as planned work. Planned work is defined through several criteria including the presence of procedures, parts (both critical and non-critical), permits, tools, lifting and rigging plans (if needed).

If this interface is not working effectively, backlogged work grows, unplanned maintenance expands, maintenance costs rise, and the supply chain predictability is threatened. One major impediment seen at many sites is the prevalence of unplanned work on the schedule that "crowds out" the planned work that is needed to ensure high equipment availability. Moving towards performing more planned work is the subject of analysis, use of subject matter experts, and reliance on methods from previous case studies seen in the process manufacturing industry.

Work history - Failure mode analysis interface

The second interface allows for the continuous improvement in asset care. At the conclusion of most work orders, mechanics will write down what they found to repair the failure. Work history is vital for later analysis. It too though is a victim of a lack of understanding of how it is used and the information thought to be controlled only by mechanics. This work history must be recorded in the CMMS against not only the asset but against the specific component that has failed. The importance of having a complete asset hierarchy in the CMMS now comes into play. If work history is not recorded against the correct asset component or misplaced at a higher level in the hierarchy, an opportunity for later improving care is lost.

Work history includes not only what is recorded on work orders but what is recorded as "root cause" of the failure in the CMMS. Most CMMS have standard codes for failure type, cause, and remedy. These codes are often forgotten by those who "close out" work orders after their completion. This information truly aids those who are performing this later analysis.

For critical assets, work history should be analyzed to determine the existence of failure modes not previously used to design the current care plan. Generally, maintenance managers, maintenance engineers, or reliability engineers perform this task. Maintenance strategies for equipment (predictive, preventative) are designed from the both vendor supplied failure modes and those historically seen. Industry databases do exist which can provide a comprehensive failure mode view. The analysis involves reviewing equipment failure data, performing a failure modes and effects analysis (or based on experience) to determine changes in the maintenance care. For instance, a failure of a bearing on a rotating piece of equipment with no history might indicate the need for a revised bearing design, addition to the vibration route, or a note on an operator route to observe any abnormalities.

These changes to the asset care plan which are recorded in the site's CMMS will then be executed in the coming days, weeks, or months depending on frequencies. These changes represent a continual improvement in the care given to that asset.

Now that the two interfaces have been described, the implementation as per best industry practices will be detailed.

Implementation Scheme

The two interfaces described above are executed with a series of industry accepted practices. These practices must include assigned accountabilities and responsibilities for carrying them out. These are usually detailed in each site's maintenance execution and reliability work flows. These practices will now be described as observed over a multitude of process manufacturing industries.

Maintenance strategy – Planning and scheduling interface

Management of this interface generally is directed by the Maintenance Execution workflow and enabled through the site's Computerized Maintenance Management System (CMMS). Preventative maintenance work prompts are generated by the CMMS for the attention of the planner and scheduler. The majority of these tasks have standard work, parts, and procedures already attached to them. These components comprise what is termed a planned job. It is the job of the planner to order parts or reserve them if they are already in stock. Once this standard job is ready for scheduling, the scheduler determines where in the week's schedule this standard job will "fit" given current work and available resources.

Predictive work is generated both from the CMMS and from other information systems. Operator rounds are generally prompted through operations management systems. Fixed equipment inspections may be generated through the site's mechanical integrity management system. Third-party inspections for such tasks as vibration, oil analysis, thermography rounds are setup independent of the schedule unless they interfere with the operation of the facility. In that case, the inspections are placed in the maintenance schedule for the week. Some mechanic rounds are prompted through the CMMS for the planning scheduling function to manage.

All manufacturing sites struggle with scheduling all the work required to care for its assets. High levels of reactive (breakdown) work places the completion of these tasks in jeopardy since they cannot be accommodated in the schedule. The industry benchmark of greater than 80% weekly preventative maintenance work completion is a typical goal.

Work History - Failure mode analysis interface

This interface is managed both in the field and completed in the CMMS. It is the responsibility of first-line craft leaders to ensure that craft include the work history after work order completion. Especially for high priority breakdown work, this information is needed for analysis. Some sites attach a sheet to each work order with failure codes and reasons to assist field compliance. Utilizing troubleshooting tools like "5 Why's" can be helpful for field determination of the root causes of failures.

The next step involves the task of a planner, first-line craft leader, mechanic, or clerk to close out the work order. Accepted industry practice directs the mechanic to close-out work. However, this practice is dependent upon the staffing level of the organization and the labor culture. This involves ensuring all charges for parts and labor are assigned to the specific component that failed. This includes completing the failure codes and reasons that most CMMS offer in their functionality. Often these charges are not recorded against the right asset or at the right component in the asset hierarchy. Many times, information from the field is missing which leads to a missed opportunity to improve asset care to prevent the failure from re-occurring.

In most cases, a failure of an asset is from a known or forecasted failure mode for which the care tasks are designed. The reliability or maintenance engineering function of the site is responsible for determining if this is not the case and a new failure mode was observed. This detection can be done by reviewing regularly all high priority work orders completed. Some sites have a trigger system in place which is activated by the impact of the failure to either productivity or maintenance cost. Once the trigger is surpassed, the appropriate tool (Root cause analysis, failure modes and effects analysis) is applied. The purpose of these tools is to determine if the failure mode is new or previously observed.

Another technique is the use of a Pareto chart. This analysis used to identify "problematic assets" is generated from maintenance cost data and production loss time data. Assets are arranged from highest impact to the least against those two criteria. Generally, the work history from the top 3-5 assets is reviewed to determine the cause or causes of this higher impact.

Once the appropriate analysis tool is used, the appropriate change in the asset care plan is proposed. These changes can be change in the parts stocking, increased frequency of inspection, addition of preventative or predictive care tasks, or a design change. It is generally believed that two-thirds of these changes should be a movement toward more condition-based care moving away from preventative or time based care tasks. Only a third should be considered for design changes.

In closing, both the Maintenance Execution and Reliability work processes must be effectively linked to ensure manufacturing can reliability provide product to its customers. The failure to manage the two interfaces described above will put the supply chain at risk.

(108b) The Application of Lean Six Sigma within the Specialty Chemicals Industry

Darren Clark, Momentive Performance Materials, Waterford, NY

Lean Six Sigma (LSS) is a powerful and flexible methodology that has been utilized by a range of companies for delivering significant cost and growth productivity via simultaneous reduction of waste and defects from internal processes. LSS principles have been successfully applied to a full range of both batch and continuous processes in the chemical industry.

This presentation will provide an overview of the application of both the Lean and Six Sigma Methodologies within Momentive, a global manufacturer of specialty chemicals and materials. Examples will be shared on how this world class framework can promote safety, quality and operational effectiveness within the specialty chemicals industry.

(108c) Environmental Compliance Approaches in Manufacturing, and the Future of EPA Enforcement

Mary Ellen Ternes, McAfee& Taft, Oklahoma City, OK

The manufacturing industry sector is incredibly diverse, yet basic compliance approaches to achieve and maintain compliance are shared among industries. This presentation will review compliance assurance approaches, as well as methods of resolving issues of noncompliance and addressing potential enforcement actions arising from noncompliance. This presentation will also review the latest developments in EPA enforcement approaches and the future of EPA enforcement as conveyed by EPA Management.

ON THE CHALLENGE OF PRESENTING MEANINGFUL ENERGY DATA

Rein de Jonge, SABIC, Bergen op Zoom, The Netherlands
Deborah Downey, SABIC, Burkville, Alabama

Introduction

How do you present your site's energy data? Is it meaningful? Does it drive actions and progress? Finance, operations, business leadership may all have slightly different perspectives or methods to represent a plant's energy usage and performance. All of these perspectives and reporting methods may have value to their audiences. However, as this paper discusses, presenting meaningful energy data to drive improvement for a plant requires rigor, detail and analysis far beyond monthly utility bills.

Background

Energy has always been an important topic for the chemical industry – the chemical and petrochemical industries account for 30% of global industrial energy use, between 5-6 Exajoules/yr.[1] Publicly stated energy reduction targets are common today and have increased with the prevalence of Sustainability efforts and the advent of Corporate Social Responsibility (CSR) reporting. The world's top 10 Chemical companies[2] all published a CSR in 2011. CSR reporting for Fortune 500 companies grew from a meager 19% in 2010 to 53% in 2011[3]. This represents a staggering one year increase of nearly three-fold. Public interest in sustainability has traditionally been spurred on by significant and visible events, such as the first Earth Day in 1969, UN Earth summit Day in 1992, publication of the Kyoto Protocol and most recently, supplier sustainability demands from companies like Sony and Walmart[4].

Similarly, interest in energy efficiency is also on the rise; however, its importance can be tied to the more mundane topic of spikes in energy prices (as shown by the historical trend of Henry Hub Gulf Coast Natural Gas Spot Price[5] (Figure 1)). Energy intensive industries are both blessed and cursed with energy costs being a significant portion of controllable, operating costs. The "curse" is having to absorb fluctuating energy costs like NG spikes in the US to over $13/MMBTU just a few years ago. The "blessing" lies in the sheer magnitude / size of our utility bills. The gross dollar savings opportunity is simply too significant for management to ignore and affords leadership the opportunity to dedicate resources, time and efforts to a regimented approach to energy efficiency.

---

[1] Tracking Industrial Energy Efficiency and CO2 Emissions (2007), Industrial Energy Agency
[2] http://www.chemistryviews.org/details/ezine/2517771/Top_Ten_Chemical_Companies_in_2011.html
[3] Georgia Institute.com www.gainstitute.com
[4] Davies, John and the Editors of GreenBiz.com, "State of the Profession", January 2013
[5] US Energy Information Agency

Figure 1, Henry Hub Gulf Coast Natural Gas Spot Price

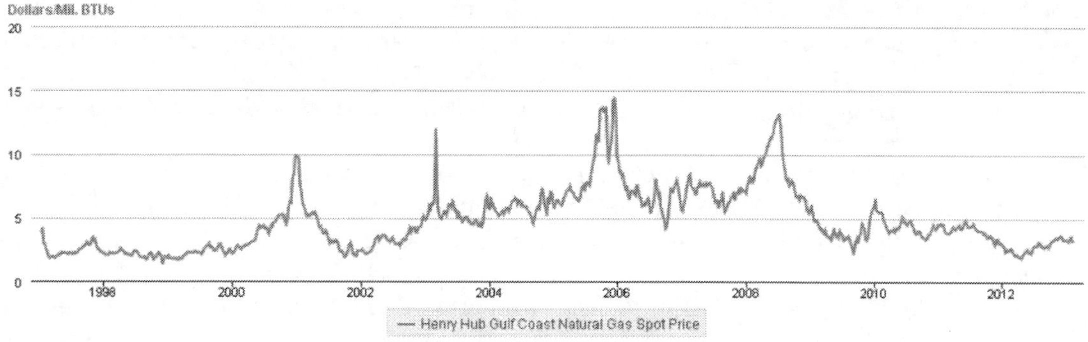

Fundamentally, chemical industry is interested in energy reduction and energy costs to become more profitable and increase competitiveness. There is a desire to improve energy usages, reduce utility consumption and conserve energy across our facilities. In order to make improvements, regimented techniques are not only helpful, but necessary to make lasting shifts in performance.

Process Improvement – Plan-Do-Check-Act

The Plan Do Check Act cycle, also called PDCA, Deming cycle or Shewhart Cycle can be found in nearly all process improvement designs and is clearly evident in today's Energy Management Improvement processes. As examples, Six Sigma DMAIC (Define, Measure, Analyze, Improve and Control), EPA Energy Star's Guidelines for Energy Management[6] and ISO50001 Energy Management improvement processes are all fundamentally based on variants of the PDCA cycle (Figure 2).

Figure 2, Shewhart Cycle, EPA Energy Star Guidelines for Energy Management

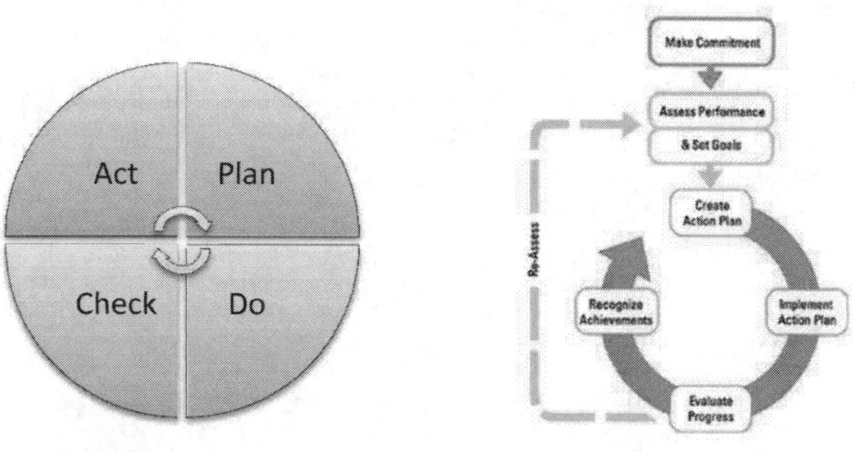

[6] **EPA Energy Star, energystar.gov**

In operations, the PDCA cycle fundamentally relies on having a valid metrics by which to measure progress. The below three elements of energy data are keys to maximizing the impact of energy efficiency improvement efforts. More accurately stated, these are the ways to keep delivering on your targets/goals:

1. Understanding the different levels of an organization and how data and metrics must be customized for each level
2. Driving a comprehensive approach -- many functions contribute to energy, relevant metrics must exist to provide feedback for each area's contribution
3. Driving a continuous process -- metrics must help maintain control, identify and eliminating leakage

The topic and title of this paper is critical for all three of these elements - Presenting meaningful Energy data to all levels of a corporation. The PDCA cycle may be revisited at different time intervals as appropriate for the different functions and/ or levels of business.

Different levels of an organization require different levels of data.

In driving a corporate initiative, goals and targets are often set for the business at a high level and aimed at external stakeholders (shareholders, customers and the general public). These targets are cascaded down to the lower level organization. Most companies are composed of many manufacturing sites, diverse in size, complexity, geography, etc. Making the targets meaningful at every level of the organization is critical to success.

In the case of Energy and Sustainability, the corporate, business unit and site energy KPI targets for reducing Energy Intensity are in units of GJ/ton. The use of GJ as an overall metric for energy makes sense as there is a need to aggregate all of the energy used in all forms (electricity, NG, gasoline, steam, etc) into a single metric. Even at the site level, the use of GJ/ton as a metric is of value because a site is often comprised of multiple units, making different products. However, at some point, the use of GJ is of less value – operators simply do not have levers that can directly affect "GJ". At lower levels in the organization, we transition from the use of generic GJ to more relevant, recognizable energy units of kwh, tons steam, etc).

Driving a Comprehensive Energy approach

This article is on the control of progress towards energy sustainability goals. We have learned that having a good Plan, Check and Act in place is crucial in our journey. As transparency between goals and execution rises, opportunities for improvement will reveal themselves. This however is insufficient to reach our goals. To reach our Sustainability goals, a structured approach to Energy Management is required. All functions (Figure 3) need to be engaged along with valid metrics to provide necessary feedback as to where the opportunity may lie.

Figure 3, SABIC Energy Value Chain

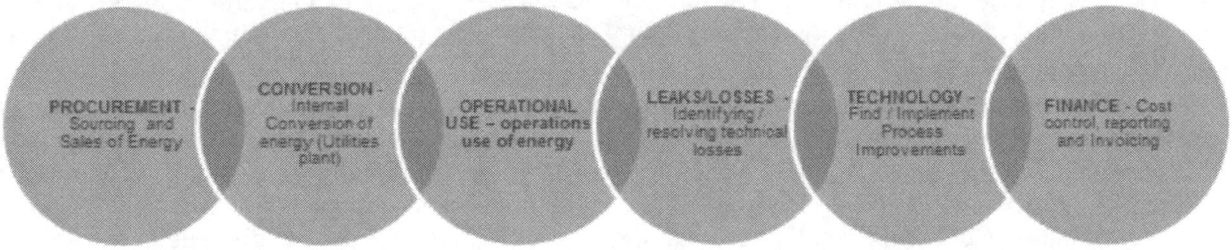

Continuous process, eliminating losses

Metrics must help maintain control, identify and eliminate the leakage of energy. So often, we focus on identifying and implementing new technology or improvement (CapEx) projects while taking an eye off the ball and allowing energy performance to slowly worsen over time. A basic, yet classic example is steam trap maintenance - steam traps have a life of approximately 10 yrs. Not having an ongoing steam trap management program will lead to deterioration in steam usage over time, often followed by a major steam trap repair capital expenditure. Other common examples are piping insulation and compressed air leaks. A continuous process will help mitigate these losses.

How to present meaningful data meaningfully?

Every company will, even if only for invoicing and accounting reasons, collect energy data. More complex sites will measure energy distribution to the various units within its boundaries. Typically this data is collected and disseminated to the plant managers and production engineers, compared to month over month, year over year, or quarter over quarter results. Within the Innovative Plastics Business Unit of SABIC (referenced as SABIC), we are no exception and until recently we would report absolute energy usage.

Figure 4, SABIC Site X, % of 2010 Energy Use

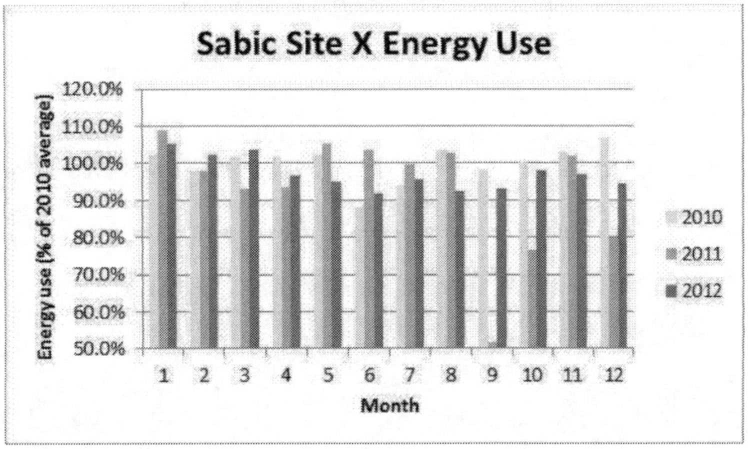

Try to assess the performance of the Site X asset from Figure 4. What conclusions would you make? Statistical tools are unable to distinguish between 2010 and 2012. We will demonstrate in this paper, how to turn this data into meaningful data.

The next evolution of measuring energy usage was to switch from absolute energy terms to energy per unit of product, intuitive since, from a business point of view, we normally look at conversion cost per unit of product. For Site X we have presented the Energy usage data in terms of Energy intensity in Figure 5. The keen observer will notice several anomalies.

1) A seasonal trend, during the mild temperatures of spring energy intensity is normally low.

2) 2011 was a year where a major turnaround was executed, with low volumes in Sept and Oct.

3) 2012 values appear to be consistently below those of 2010. Great! Statistical analysis confirms that there is indeed a significant drop in Energy Intensity.

Figure 5, Energy Intensity SABIC Site X, % of 2010 Average

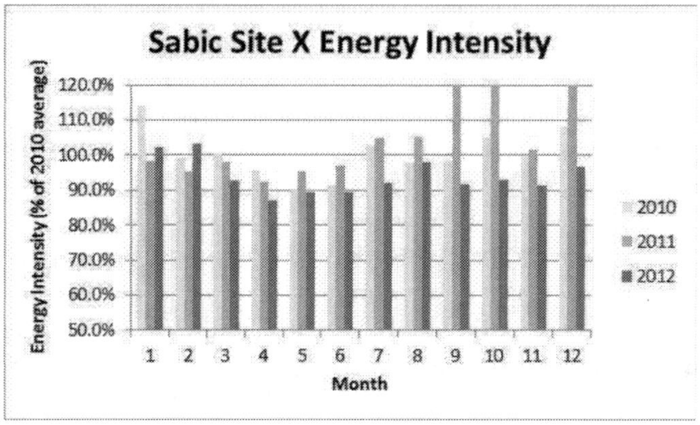

Volume Leverage

Our site energy leaders all intuitively understand that Energy Intensity has a volume component. The base energy use of a production unit is simply distributed over more units of product. Understanding the impact of volume on the energy use is a critical factor in advancing the game of the PDCA cycle in Energy Performance Management. Energy usage data requires normalization for volume and other factors not in control of the plant operators

Figure 6, Energy and Energy Intensity as a function of Volume

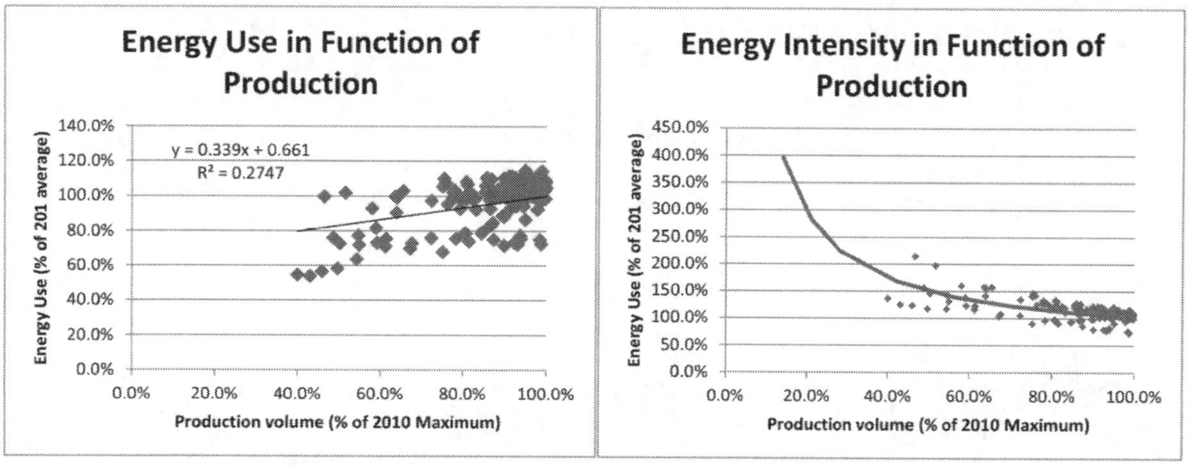

Energy Usage typically has a linear relationship with production volume. (Energy = A * Volume + B), Energy Intensity models should then have the form of (Energy Intensity = A + B / volume). Figure 6 shows the site energy use plotted against the sum of all plant's outputs. The resulting correlation is, with an R2 of 0.27, not impressive in quality. This is not surprising as we are trying to develop a "total Energy" model for a "complex" site, ten independent production facilities, and a utilities operation using CHP's to generate steam and electricity. This level of complexity is typical for a chemical site.

We have endeavored to enhance the Energy Usage model, by trying to build a linear model to the individual plants production volumes as well as external factors like temperature, imports and exports from the site. Finally, we've constructed a model that has the following form:

$$\text{Total Energy} = Base\ load + \sum_{i=0}^{n} Energy\ Factor_i Production_i - Temp_{Factor} * Ambient\ Temp$$

The result is of high quality, R2 = 0.87 and the 95% confidence interval is 1.2% of Total Energy for groups of predictions. We now have a power tool in our hands; we can accurately predict the energy consumption of the site. Understanding the usual now enables us to look for the unusual. We have observed variations in Site Energy Intensity that vary by 10% due to volume and temperature changes. These variations are in the same order of magnitude as our medium term improvement objectives (12.5% by 2015). Having a model describing the site energy intensity enables us to quantify these effects.

Looking for the unusual – Distance from Excellence

In the search for the unusual, we now look for trends in the difference between the models prediction and the actual Energy usage. For this we use the concept of 'Distance from Excellence' or as where excellence is defined as the model and the distance is the sum of residuals of the period under consideration, the sum of "actual – model" of all predictions in the period. The unit of measurement is in absolute Energy loss (GJ) and for convenience of our managers translated into currency by multiplying the Energy Loss with an average energy price. The distance to excellence principle can be adapted easily to other industries and a more generalized example can be found in the Energy Efficiency OEE methodology of McKinsey[i]

Figure 7, Distance from Excellence

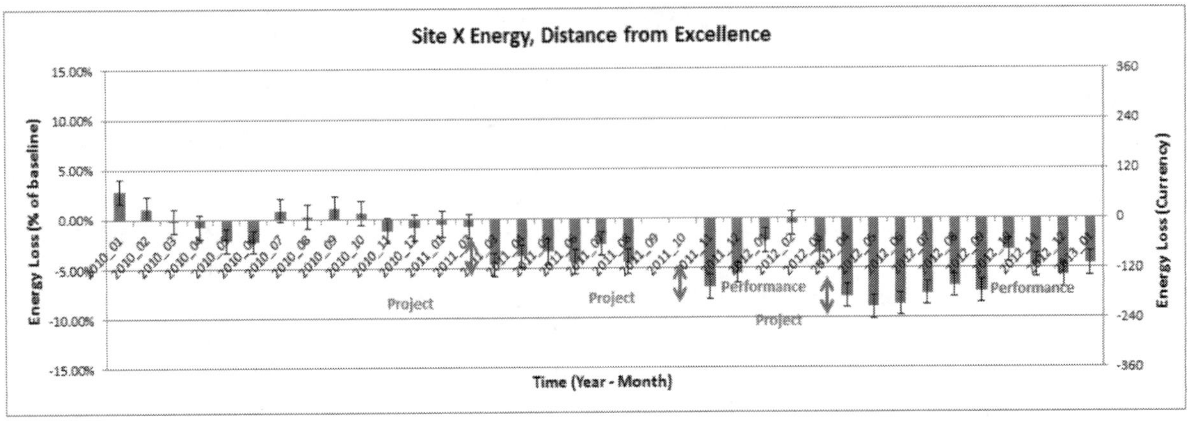

With this information in hand, we can have a robust conversation with our management. Figure 7 shows Distance from Excellence and, as we progress to ever more negative values, it means that we are improving versus the baseline. Since the graph is corrected for volume, temperature and other distorting factors, the trends shown in the graph clearly show improvements (downwards trends) and performance issues (Upwards trends). It is a solid foundation for robust conversations. Several key capex projects that were implemented are clearly visible as well as a number of significant performance and reliability issues.

Next step: moving from site to plant level

The method above is very useful for measuring progress towards the company's sustainability goals on various sites as it will give the corporate leaders insight in how each site is progressing, and where headwind towards the target is developing. Thus is obvious to try the same approach for each plant within a site. In our chemical operations, heat is the dominant force in energy usage. We therefore decided to move from "energy" to heat, the benefit of this is that we get better regressions if the energy sources are separated. For each plant we have again established models based on linear relationships.

With the exception of one plant, the modeling of the plant levels was straightforward and gave reasonable to accurate results with a basic linear model. The exception is a plant that runs two serial production lines, heavily integrated on energy, but not in production planning, using two different, exchangeable energy sources. Even though the developed model was limited in accuracy, we successfully deployed the methodology described in the previous pages and had encouraging results.

Combining "distance from excellence" graph of the site level view, with the plant levels gives an excellent understanding of which plant had a positive or negative contribution to the site's result. Site performance issues can be traced back to the originating plant. This gives the site management the opportunity to investigate with the respective plant manager and to check for issues in operations and projects.

Figure 8, Distance from excellence, Site View

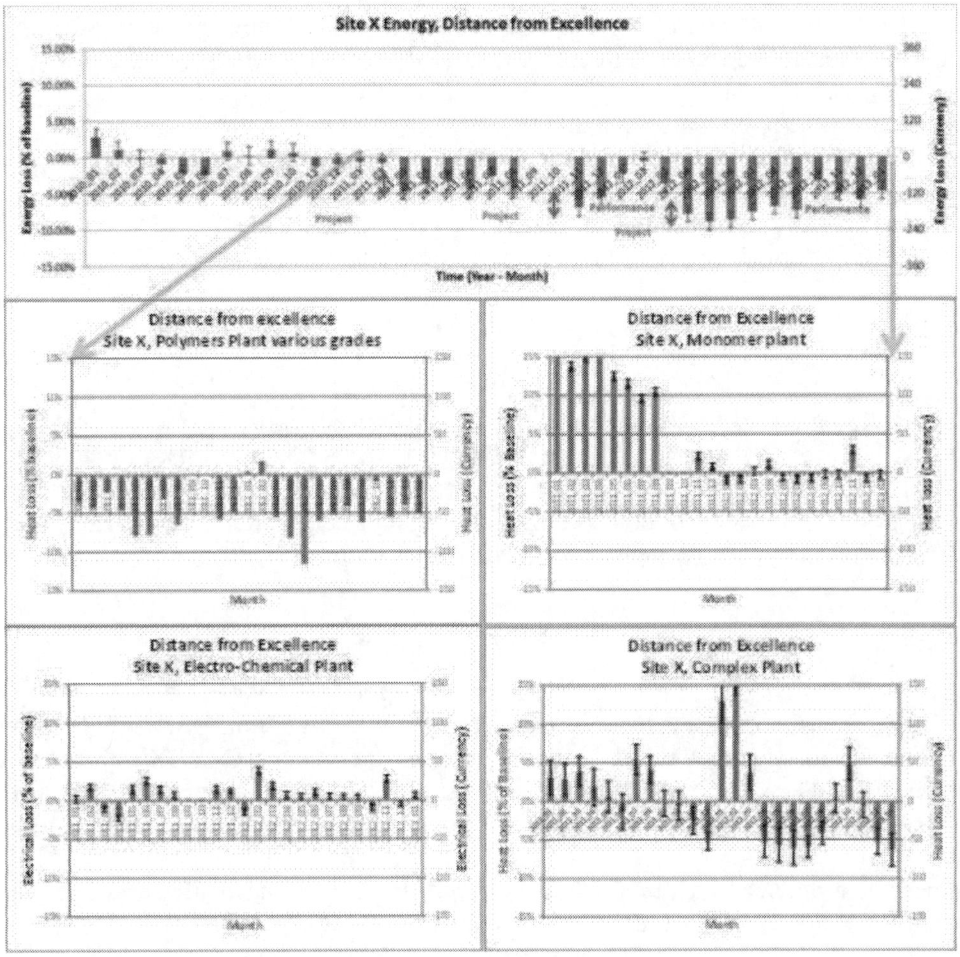

Figure 8 provides us with a good understanding of the site's performance with regards to primary energy usage performance and with key plant energy consumption performance. As a final step a performance indicator for energy conversion and distribution is required. The intuitive method is to do this in the "efficiency" form ie:

$$Energy\ Efficiency = \frac{\sum_i^n Energy\ Delivered\ to\ Customer}{\sum_i^n Energy\ Procured}$$

This certainly is not a bad way to evaluate your utilities operations. It will clearly show trends in performance. However, if the performance of the utilities operations is measured in loss, (and expressed in $ or Euro), the size of the issue and, perhaps more important, the opportunities immediately becomes apparent.

$$Energy\ Loss = \sum_{i}^{n} Energy\ Delivered\ to\ Customers - \sum_{i}^{n} Energy\ Procured * Cost$$

Note that for cost we do not use the "real" or allocated cost, but say for an 80% efficient boiler we calculate the cost of the steam delivered as 80% of the energy inserted as 20% is lost in space. Similar like the Site and plant graphs discussed earlier, we've constructed similar models and loss graphs for key utilities equipment on site, the Cogen units, boilers, air compressors, etc.

Using Distance from excellence in the operations is also something that we have tried. It works, but the perspective needs adjusting. Our operations departments tend to become complacent when the bar goes below zero. (I'm doing good, move on to the next challenge). A key learning is that when changes are made that improve performance, the baselines need to be adjusted, so that the average is reflected at zero.

One shortfall of this method is that if fails to predict the effect of implementing projects. A successful project will show up, however, unsuccessful projects or projects which lack in control plans fail to show up.

On Site X, the Energy Leader facilitates a weekly discussion with the production engineer on the Energy and Utilities consumption. The consolidated issues and actions list provides a good starting point for small improvement projects. We are also working with the operators to implement a review cycle on a shift cycle 8 hours basis, embedding reporting on Energy Anomalies in the Asset Operations work streams. As of today, this remains an elusive process to implement.

Conclusion

As Energy cost and Sustainability have become important business drivers, it is paramount to be able to have robust discussions on Energy consumption and transformation performance. This allows management to monitor the improvement from operations and CAPEX improvements towards the sustainability goals. We believe that the basis for this discussion should be a form of normalized metric. The method of "Distance from Excellence" as it was provided to us proves to be an answer to this conundrum.

Removing influences like production volume, grade and seasonal variations enables robust performance discussions between operations, management and engineering. How good did you operate, did projects deliver and if not, where are interventions required?? Management wishes to review the unusual and to reward excellence. This is only possible by clearly defining the usual, and bringing laser focus to the unusual.

References

[i] Plasschaert, Frank, Ken Somers, Gautam Swaroop, "Capturing the Lean Energy Opportunity in Chemical Manufacturing", McKinsey, Winter 2011

Project Management: The Good, the Bad, the Ugly

John H. Worstell
Fuel Project Engineer
ITSI Gilbane

Introduction

Whether a project is successful or not; i.e., meets schedule and budget, depends upon the onsite project manager or construction manager. This individual forms the focal point between owner, design engineer, and contractor. The onsite project manager translates engineering concepts into reality via the directed action taken by the contractor, thereby meeting the owner's goals. How well this individual performs this task depends upon communication and documentation, which are the foundation of every successful relationship. Communication and documentation with the owner focuses on budget and schedule; communication and documentation with the contractor focuses on translating drawings into results that meet specification. Understanding the relationships between owner, contractor, and design engineer is an understated concept in project management.

In this presentation, I discuss the communication and documentation tools required to develop and strengthen meaningful; i.e., functional, relationships between owner, contractor, and design engineer. I also discuss the proper techniques for reviewing construction budgets and schedules, thereby ensuring the owner value for the committed investment. My viewpoint throughout this presentation is that of a design engineer, construction manager and contractor on private and government sponsored projects.

The Players

The Owner funds the project based on a business or social need, depending whether the Owner is a commercial entity or a government entity. The Owner carries the burden for funding the project. The Owner empowers the onsite Project or Construction Manager[1] to ensure project quality and to protect the Owner from cost changing items; i.e., from design changes initiated by the Contractor at the project site.

The Design Engineer designs the project. All designs require approval by a registered Professional Engineer (PE). The Design Engineer may or may not be a PE; if not, then an independent PE will have to review and "stamp" the project design. Typically, the Design Engineer spends most of his/her[2] time prior to construction producing the Drawings and Specifications, which are used for bidding materials and construction. During construction, the Design Engineer will be "on call" if issues arise with regard to interpreting the design.

---

[1] Henceforth, I will refer to the Project Manger as the individual representing the Owner's investment and the Construction Manager as the individual who supervises construction. In some cases, one individual fills both roles.

[2] The current penchant for "he/she", "his/hers", and so forth is clumsy and repetitious. Its use suggests "his" alone is masculine, which it is not. "He" and "his" have traditionally been used in certain contexts when speaking and writing English; likewise, the same for "she" and "hers". Henceforth, I will use "he/his" and "she/hers" in the traditional English manner. (See M. Van Leunen, *Handbook for Scholars*, A. A. Knopf, Inc (1978).)

The Project Manager oversees and ensures project quality and manages the project budget. In some cases, the Project Manager will be onsite daily. In other instances, the Project Manager will oversee a Construction Manager identified by the Owner. The Project Manager represents the Owner; thus, the Project Manager manages the Owner's project funds, assesses cost changing proposals suggested by the Contractor, and serves as the communication conduit between Owner, Contractor, and Design Engineer, and Construction Manager, if one is utilized.

If a Construction Manager is utilized, he becomes the Project Manager's eyes and ear for project quality. The Construction Manager enforces Contractor adherence to the Drawings and Specifications issued for bid. The Construction Manager coordinates with third party inspectors to ensure that quality is professionally tested and documented. The Construction Manager also ensures that daily construction activities are documented. Of course, if a Construction Manager is not utilized, then the Project Manager performs these tasks.

The Contractor undertakes and performs the construction required by the project. The Contractor interacts daily with the Construction Manager and in most cases, weekly with the Project Manager, thus with the Owner. In an ideal world, the Contractor provides a quality product for the remuneration paid by the Owner. That quality product follows and fulfills the Design Engineer's specifications. But, the world is far from ideal; thus, the need for Project Managers, Construction Managers, and third party quality inspectors.

The third party inspector, henceforth "Tester", specializes in quality assurance, or control. The Tester can either be hired by the Contractor, Design Engineer, or Project Manager based on the written Specifications. The Tester ensures the quality of the Contractor's work. The Owner should think long and hard about allowing the Contractor to hire the Tester.

In large corporations, the Owner, Design Engineer, Project Manager and Construction Manager may all work for the same entity.

Project Phases

The project arises from a business or social need identified by the Owner. The Owner then incorporates the project in the annual budget and prioritizes it relative to other needs. Once the project funds are allocated, the Owner identifies a Project Manager who develops a Request for Proposal (RFP). The Project Manager then issues the RFP requesting companies to bid for Design Engineer. In large corporations, the RFP goes to the Design or Project Engineering group. Smaller companies and government agencies issue the RFP for bid. Typically, bids on the RFP are evaluated for best value rather than lowest price.

The Project Manager reviews the bid submissions and selects and contracts a Design Engineer, which may be an individual or a company. The Design Engineer then undertakes the design phase of the project. During this phase, the Design Engineer evaluates the Owner's needs and, working closely with the Project Manager, develops a solution that meets those needs.

This phase of the project may involve Other parties, who may be:

- Government agencies --- all projects require permitting and approvals;

- Property owners --- easements may be required from other property owners;
- Specialized engineers --- If the design is outside of the Design Engineer's expertise, then a contract is let to a specialized engineer to design that portion of the project;
- Lawyers --- a feature of modern life; they should negotiate easements with other property owners and be available for council if other issues that may arise.

The Design Engineer produces a final product consisting of Drawings and Specifications. The Drawings visually describe the project. Drawings are by trade: Civil, Architectural, Mechanical, Electrical, etc. The Specifications provide information about the materials and work quality required by the design. They are generally issued as a "book", segmented by trade. The Contractor must adhere to the Specifications.

The Project Manager joins the Drawings and Specifications with Contract Documents to form the Contract. The Contract also contains general and supplementary conditions to which the Contractor must adhere.

Following Owner approval of the design, the Project Manager advertises for bids to construct the project. This phase of the project, the advertisement phase, must be long enough to provide time for any and all bidding Contractors to review the contract and submit questions to the Design Engineer as well as provide enough time for bidding Contractors to obtain valid price quotes from material and equipment suppliers. If a question induces a design change, an Addendum to the Specifications must be issued to all interested Contractors. During this time, the Contractor assembles a bid and submits it during the Bid Opening.

Bid Opening is the period when all the Contractors submit their bids and learn the amounts bid on the project by each Contractor. During Bid Open, the Project Manager reviews each Contractor's bonding and insurance, and verifies that the Specifications and all Addendums have been met and acknowledged by the competing Contractors. If an issue arises with any of these items, then that Contractor's bid can be eliminated.

Bid Review generally lasts a week. In this phase; the Owner, Design Engineer and Project Manager review each Contractor's bid and evaluate the Contractor that offers the best value for the project. Picking the correct Contractor is crucial. The Design Engineer and Project Manager expend considerable energy making this choice. Note that most government funded projects are awarded to the lowest bidder.

Following the Bid Review, the Project Manager awards a contract. The Contractor will be given a time period to sign the Contract and return it to the Project Manager and Owner. The Project Manager must document justification of the contract award in case a disappointed Contractor submits a claim against the Owner for unjust selection. Thus, utilizing Lawyers during this phase is a must.

The Notice to Proceed period occurs when the Owner grants the awarded Contractor permission to begin the project. During this phase, the Owner may or may not assign a Construction Manager to the project.

Submittal/Shop Drawing Review occurs upon initiation of the project. During this phase of the project, the Contractor submits shop drawings for materials and equipment to the Project Manager. The Design Engineer must review all Contractor submittals and drawings. The Design Engineer return all submittals and drawings to the Contractor with one of the four criteria selected:

- Approved --- the submittal meets Drawings and Specifications and the product is approved for use;
- Approved as noted --- the submittal has minor issues that need to be changed. The Design Engineer will add comments to the submittal identifying the changes. It is the responsibility of the Project Manager to ensure that these changes are made when the materials or equipment are brought on site;
- Revise and resubmit --- the submittal has major issues that require change. If this is selected, the Design Engineer will typically make comments.
- Rejected – the submittal does not meet the Drawings and Specifications and another product must be submitted.

Work Commences, as a project phase, upon approval of all submittals and drawings. The Contractor mobilizes his crew onsite and has material and equipment delivered. The work commences and continues for the duration of the project. The Project Manager documents all budgeting activities while the Construction Manager, if one is appointed, oversees and documents construction activities. The Construction Manager informs the Project Manager of any issues arising with regard to the project schedule. The Project Manager should issue an open invitation to the Design Engineer to visit the project site and to inspect the integrity of the construction. The same invitation should be issued to any and all government agencies that issued permits for the project. The Project Manager or Construction Manager coordinates testing to maintain project quality.

The Project Manager and Contractor punch-list each portion of the project as it reaches completion. The punch-list arises from a walk through, detailed inspection by all concerned, with drawings "in hand". The result is an itemized list made by the Project Manager of the items that the Contractor needs to finish before the project can be commissioned.

Commissioning is the final part of construction or integration of the project into the Owner's existing facility. Upon approval by the Owner, the project moves to a point of Substantial Completion.

During Substantial Completion, the Owner agrees that the Contractor is ninety-five percent complete with respect to the contract. At this point, the Contractor will receive a fair portion of the retainage held throughout the project. Punch-list items and demobilization typically comprise the last five percent of the project.

Project Closeout is the final phase of the project. The Owner announces and advertises his satisfaction with the project and the Contractor. The Contractor then fully demobilizes his work force. Activity with regard to warranty work may occur during this period. The Contractor submits As-Built Drawings to the Owner, who then maintains them on file.

Typically, the Contract has a warranty period during which the Owner can request the Contractor to return to fix defective work. The Owner will rely on the end users or operators to communicate any such issues. In most cases, project warranties are effective for one year.

KEYS TO EFFECTIVE PROJECT MANAGEMENT

Understanding the Contract

A successful Project Manager knows more about the project than anyone else; that includes the Contract. The Contract documents the technical and budgetary aspects of the project. The technical portion of the Contract includes the Drawings and Specifications; the budgetary portion of the Contract includes the Contract Documents and General and Supplementary Conditions. To acquire technical competency with regard to the project, meet weekly with the Design Engineer. Have the Design Engineer thoroughly explain every feature and nuance of the design. Such knowledge helps communicate to the Design Engineer any issues that arise during construction. It accelerates the submittal review if the Project Manager can reject submittals that do not meet the Contract.

The Contract is a control tool to force the Contractor to perform quality work as well as stop work if there are major problems. The Project Manager needs to ensure that the Construction Manager understands the Contract as well, which can be accomplished through Contract review meetings. The better the Construction Manager understands the Contract, especially on the technical portion, the more likely he will identify an issue before it becomes a major problem. Expert knowledge of the Contract sets the tone of the project with respect to the Contractor.

Getting the Right Contractor

Choosing a reliable Contractor alleviates considerable problems for the Project Manager. Making this choice occurs during the Bid Review. Typically, the Project Manager investigates the credentials of the three lowest bidders and decides best value from their references. Of course, the reference check is only as good as the questions asked.

First, review the referenced projects and identify those most similar to your project. Second, interview the Owner, Project Manager and Construction Manager of those identified projects. Ask the below questions when investigating a Contractor's references.

- Did the Contractor meet schedule?
- Did the Contractor meet budget?
- If the Contractor went over-budget, why?
- Did the Contractor propose any change orders?
- How many change orders were there?
- Were the change orders warranted?
- Were the change orders priced fairly?
- How well did the Contractor understand the intent of the design?
- Were there any deficiencies noticed? And why?
- Did the Contractor assign a responsible person to site to ensure project quality?

Meeting schedule and meeting budget are the major indicators. If the Contractor does not meet schedule, it needs to be justified. Not meeting schedule indicates a mismanaged project and, most likely, the Contractor was not willing to add additional personnel to finish the project on time. Usually, liquidated damages are incurred for not meeting schedule, which covers the added time for the Project Manager and Construction Manager to oversee the construction past the finish date. If this is not properly calculated, the Owner loses money if the project extends beyond schedule.

There are many reasons for not meeting budget. First, the Contractor underbid the project to win the Contract. Second, the Contractor or the Project Manager or both mismanaged the project, thereby causing a budget overrun. Third, unnaturally inclement weather occurred during the Construction phase. Fourth, labor issues arise within the Contractor's workforce. Fifth, the Contractor issues and the Project Manager approves an inordinate number of change orders.

Change orders increase the cost the project. Change orders come in two types: those that are Owner requested; and, those that are Contractor requested. Change orders requested by the Owner typically indicates Owner

- satisfaction with the Contractor; therefore, the Owner increases the scope of the project;
- identification of a Design Engineer oversight.

Contractor requested change orders are usually formalized by value engineering on the part of the Contractor. A good Contractor will find cost saving changes as well as cost incurring changes. An inordinate number of Contractor initiated change orders indicates

- a poor design;
- an effort to gain additional remuneration on the project.

Additional questions to ask during Contractor investigation are:

- Did the personnel listed on the Contract appear and work at the site?
- How well did the Contractor work with the Construction Manager?
- How well did the Contractor work with the Project Manager?
- Did the Contractor effectively manage the construction?
- Was the Contractor fair?
- Any additional thoughts about the Contractor?

The last references that should be checked are with the bonding and insurance companies. Contractors typically dislike all contact with these entities and sometimes these entities will refuse to answer any questions without first consulting with the Contractor. There are typical questions that should be asked.

- Has there ever been a claim against the Contractor?
- Has the Contractor defaulted on a Bid?
- Has the Contractor defaulted on Performance?

If they answer yes to any of these questions, seek a different Contractor. A claim against the Contractor's insurance company indicates a dispute between a previous Owner and the Contractor or a claim from an ex-employee of the Contractor. Defaulting on the bid bond would indicate the Contractor signed the Contract but could not start the project. Defaulting on a performance bond indicates the Contractor did not fulfill the Contract. Both are signs the Contractor has funding issues.

Documentation

The Project Manager needs to document all activities so if the project enters litigation, evidence of events, meetings, discussions exists. Daily observation reports need to be filed. These reports should include the following.

- Date
- Project Name/Number
- Temperature
- Conditions
- Contractor Personnel and Trade
- Contractor Personnel Hours Worked
- Subcontractors Onsite
- Subcontractor Hours Worked
- Anyone Else that Visited the Site
- Equipment with Date Arrived and Date Departed
- Permits Required
- Description of Work
- Safety Issues
- Construction Issues
- Rework Items
- Testing Performed and Results
- Materials Delivered
- Signature of CM

Scheduled work cancelled due to inclement weather needs to be documented. Scheduled work delayed due to Contractor mismanagement requires documentation as well. These documents become evidence during litigation between the Owner and the Contractor. Thus, daily reporting is important.

Three critical items of the project require documenting by the Project Manager. These are the

- submittal log;
- RFI log;
- change order log.

The submittal log tracks the submittals the Contractor makes and notes how each submittal is returned to the Contractor: Approved, Approved as Noted, Revise and Resubmit, and Rejected. The submittals are key to the project because all materials and equipment need to be approved before they are installed. If a product is installed without a submittal approval, then work can be stopped until it is resolved, which causes future scheduling problems. Submittals are also important for long lead time items, such as customized items. The longer it takes the submittal to be approved, the longer it takes initiate manufacture of the item. Such a delay can also cause a future schedule issue. The typical submittal log contains

- Item Number
- Specification Section the Submittal Relates to
- Desciption of Submittal
- Required Submission Date
- Submitted Date
- Date Submitted to Engineer
- Date Returned from Engineer
- Date Returned to Contractor
- Submittal Approval/Rejection Status
- Additional Comments

It important during the submittal review to efficiently deliver the submittals to the Design Engineer and Contractor. Submittals should be held in binders or file drawers depending on the project size. The Construction Manager should always have a set in the field with him, as the Project Manager has in his office.

The second item that requires tracking is RFI's. RFI's are clarifications the Contractor will send to the Project Manager for potential design issues or product substitutions. These are usually items that require resolution by the Design Engineer. The log for this is quite similar to the submittal log and includes:

- RFI Number
- Specification or Drawing Sheet the RFI Relates to
- Description of RFI
- Submitted Date
- Date Submitted to Engineer
- Date Returned from Engineer
- Date Returned to Contractor
- Comments from the Engineer

In some cases, RFI's evolve into change orders if the design requires modification, which always increase project costs. If the RFI results in a change order, the Design Engineer typically reissues a Drawing or Specifications sheet for clarity. These sheets need to be added the Contract that the Project Manager uses. The old sheet should be clearly marked, so nobody is uses it.

The last item logged is changes to the project. Change proposals to the project are usually summarized in the Contract as to documentation requirements. Change orders consist of:

- Field Order – This is a non-cost incurring change. The form typically will document the change in writing with a supplementary Drawing if necessary. It is important that this change is agreed upon and signed by the Contractor so it will not return later in the project.
- Change Order Proposal (COP) – This is a proposal from the Contractor where a cost incurring change is proposed, which usually arises from a value engineering effort on their part or if any unforeseen field condition is encountered. It will usually include an estimate of the cost change and is submitted to the Project Manager for review by the Design Engineer.
- Work Change Directive (WCD) – This is a directive signed by the Owner where a cost incurring change is proposed. The Contractor will submit an estimate for the work that will be included in the document. A Work Change Directive either indicates that there was an oversight in the design or the Owner is satisfied with the work of the Contractor and added scope to the original Contract.

Once the latter two items are approved by all parties, the change will be documented by a change order, which adds to the project bid total, thereby increasing the overall cost of the project.

These items are not necessarily logged, but are retained in a secure filing system. Such documents may also result in Specification and Drawing sheet changes. Needless to say, it is important that all parties are using the same set of Specification and Drawing sheets.

Meetings

Developing a relationship between the Project Manager, Construction Manager, Design Engineer, Owner, Tester and Contractor is important for an efficient stream of communication during the project. Meetings, while time consuming, provide a relatively effective method for establishing an efficient communication network. Typical meetings for all parties include but are not limited to the following.

- Kick Off Meeting – This is a meeting where all the parties are introduced to each other and general questions are asked.
- Critical Construction Meeting – This is a meeting where an important event is about to happen requiring that all parties meet to ensure the work goes smoothly. Large concrete pours, punch-listings, commissioning, etc qualify for such a meeting.
- 50% Completion Meeting – This is a meeting where all parties assemble and organize the last half of the project.
- Project Closeout Meeting – This is a meeting where all parties meet and finalize the last few items that need to be completed to finish the project.

Agendas for these meetings vary based on the activity.

In addition to the above meetings, the Project Manager must organize a weekly meeting between himself, the Construction Manager, and the Contractor. These meetings typically last an hour and track schedule issues, submittals, change orders and any other issues with the project. A good format for such a meeting is:

- 4-Week Look Ahead Schedule – This item is supplied by the Contractor and shows the activities for the next four weeks. It is a way to coordinate inspections, testing and additional meetings.
- Safety Issues – This will be a review of any accidents from the prior week as well as potential safety issues from the items in the look ahead schedule
- Outstanding RFI's – This is a list of RFI's that still require resolution.
- Outstanding Submittals – This is a review of the submittal log with the Contractor.
- Change Order Status – This is a review of any future COP's and WCD's.
- Action Items – This is a list of activities that are typically Owner driven that the Contractor needs to expedite. Typical items include cleaning the jobsite, getting quotes for a WCD, etc.

Draft minutes from each meeting and distribute them to all attendees within the week. I prefer to just redline the agenda with any notes and add a section listing the attendees and their affiliations.

The Project Manager should schedule a monthly meeting with the Accounting Department to ensure that the Contractor is submitting invoices per schedule and procedure and that the Owner is paying per schedule and procedure.

Payment Application Management

The Contractor submits payment applications, a contractual invoice, monthly for payment. The Construction Manager and Project Manager ensure the numbers match what was accomplished the previous week. Getting the Contractor to segment costs is a major part of the payment application review. Such segmentation of costs provides a means for monitoring the productivity and efficiency of the Contractor. If the Contract is a unit bid contract, the project will be itemized by activities with a unit price associated with each. The Project Manager can easily monitor this type of Contract. If the Contract is lump sum, the Project Manager needs to have the Contractor segment the total project cost into activities that can be monitored easily. If a Contractor declines or resists this request, then the Project Manager must ask for monthly invoices describing the activities performed by the Contractor during the each month, which is painful to Contractors. Thus, the Contractor will more than likely prefer to segment the lump sum into job activities. If this is the case, the Project Manager must review the segmented costs and ensure their accuracy. To substantiate their accuracy, the Project Manager needs to understand the Contractor cash flow and material and equipment costs. In most cases, the Contractor will be in the negative for the first twenty percent of the project because the Contractor buys the materials and equipment but is unable to bill for these items until they arrive on site or are installed. If a lump sum is segmented and considerable money is mobilized, then early invoicing may be underway. The big money items generally enter the accounting process midway through a project.

Comparison of such activity with RS Means, a book that construction estimators typically use, will generally identify such activities.

CONCLUSION

Successful Project Management depends upon communication, relationship building, and careful scrutiny of cash flows.

(108f) Reliability Management Framework

Corey Evans, Shell, Deer Park, TX

The presentation will share an overview of the Reliability Management Framework (RMF) utilized by Shell Manufacturing sites to drive reliable operation. The RMF is intended to provide the framework of reliability elements that must be consistent and managed at each of Shell's Manufacturing sites. The RMF consists of the 16 most influential elements – many are core work processes, some are organizational behaviors, and a few are specific activities. Each of these elements is important on its own in driving site reliability, yet how these elements interface is also very important.

9TH GLOBAL CONGRESS ON PROCESS SAFETY

Novel Development of Asset Integrity Management for Refinery Equipment: Theory, Application and Effect of Setting Critical Operational TAN and Sulfur Level for Distillation Units

Dingrong Qu
SINOPEC Research Institute of Safety Engineering
No. 218, Yan'an 3rd Road, Qingdao, P.R. China, 266071
Qudr.qday@sinopec.com

Shanjun Mu
SINOPEC Research Institute of Safety Engineering
No. 218, Yan'an 3rd Road, Qingdao, P.R. China, 266071
mushanjun@vip.sina.com

Xiaohui Liu
SINOPEC Research Institute of Safety Engineering
No. 218, Yan'an 3rd Road, Qingdao, P.R. China, 266071
Liuxh.qday@sinopec.com

Xiu Jiang
SINOPEC Research Institute of Safety Engineering
No. 218, Yan'an 3rd Road, Qingdao, P.R. China, 266071
Jiangx.qday@sinopec.com

Yanyuan Li
SINOPEC Research Institute of Safety Engineering
No. 218, Yan'an 3rd Road, Qingdao, P.R. China, 266071
Liyy.qday@sinopec.com

Prepared for Presentation at
American Institute oF Chemical Engineers
2013 Spring Meeting
9th Global Congress on Pr°Cess Safety
San Antonio, Texas
April 28 – May 1, 2013

UNPUBLISHED

Novel Development of Asset Integrity Management for Refinery Equipment: Theory, Application and Effect of Setting Critical Operational TAN and Sulfur Level for Distillation Units

Dingrong Qu
SINOPEC Research Institute of Safety Engineering
No. 218, Yan'an 3rd Road, Qingdao, P.R. China, 266071
Qudr.qday@sinopec.com

Shanjun Mu, Xiaohui Liu, Xiu Jiang, Yanyuan Li
SINOPEC Research Institute of Safety Engineering

Keywords: naphthenic acid corrosion, sulfidation corrosion, crude distillation units, and mechanical integrity

Abstract

In today's market, refineries are continuously challenged by the low-grade crudes with increased concentration of sulfur and naphthenic acid. Operators need to know the limits of Total Acid Number (TAN) and sulfur content for their distillation units to avoid unexpected accelerated corrosion, which may lead to fires, explosions and toxic releases. New concepts and theory of critical TAN (Total Acid Number) and sulfur level will be discussed, and the procedure of such MI project is reviewed by using a successful case. Critical operational TAN and sulfur level are determined on the basis of comprehensive evaluation of high temperature corrosion risk for every important part of distillation units. A corrosion model, which included main factors such as TAN / sulfur content, materials of construction, and temperature, has been developed for this assessment. Historical corrosion inspection data is used to validate the accuracy of the model. Experience of setting TAN and sulfur limits for 11 refineries over last 10 years is summarized. Comments and recommendations to improve this technology were made. The benefit of such MI project has been widely acknowledged for the increased reliability of crude distillation units where it is implemented.

1. Introduction

A hierarchical asset integrity management (AIM) programme (Figure 1) is suggested and applied for a major refining and petrochemical company, which runs 30+ refineries with a total crude refining capacity approximately 1.6 billion Bbls/year. In this AIM programme, the operation system of AIM covers principle guidelines for many management aspects, such as organization and responsibility of AIM, resource and training, management of change, inspection, data management, material management, turnaround management, and preventative maintenance. The operation system is documented through standards and guidelines of different levels, and is embodied in AIM Software. AIM tools include some well-known tools for risk management and inspection planning, such as RBI (Risk-based Inspection), IOW (Integrity Operational Window) and SIL (Safety Integrity Level). Setting critical operational TAN and sulfur level is a recently developed AIM tool for distillation units. This AIM system also includes some necessary engineering techniques to ensure asset integrity.

Figure 1. Hierarchical structure of an asset integrity management progamme

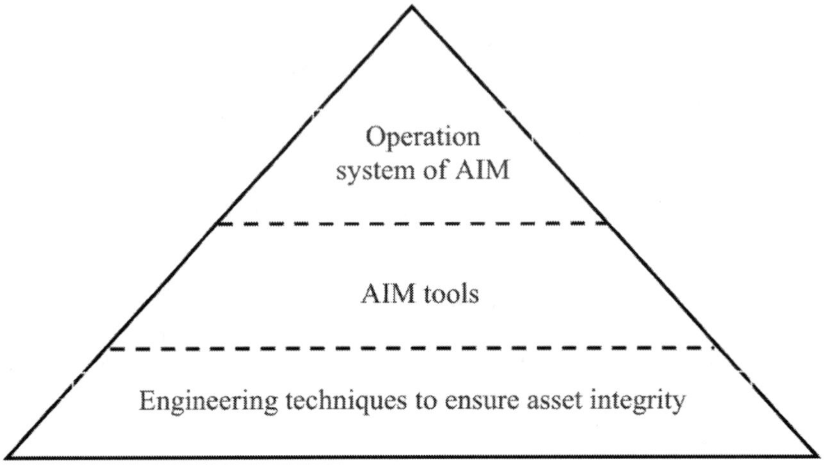

2. Challenge of Processing Corrosive Crudes

One of the worldwide trends in crude oil refining industry is purchasing and processing of corrosive crudes with increasingly higher TAN and sulfur content. Such "opportunity" crudes are intentionally purchased with great discounts. Since the cost of crude oil accounts for approximately 90% of the economics, the cost saved on crude purchases can potentially provide a huge economic advantage over industry competitors. On the other hand, inherently corrosive crude oil could negatively impact the refineries' profit margin by significantly increasing the corrosion rate, and thereby reduce the reliability and integrity of the equipment. Processing corrosive crudes and maintaining equipment integrity is a great challenge for refineries. Data shows that the average TAN and sulfur content of crudes for the company has increased markedly over the last decade. This change in crude qualities has resulted in equipment failures in both the high temperature equipment as well as the distillation overhead systems. As a result of these unpredicted failures, additional attention has been focused on these crudes that contain increased levels of naphthenic acid and sulfur.

Another great challenge for the company is the combination of uncertainty of crude resource and capability of distillation units bearing corrosive crude. Due to availability, refineries have to process any available crude oil. Normally, the crude slate is changed monthly. However, some refineries have had to adapt to crude slates that change weekly or even daily. As a result, processing crudes with various TAN and sulfur content could present a potential and unpredictable risk due to unpredicted corrosion. Moreover, most distillation units were designed and built decades ago for "benign" crudes with limited corrosiveness, such as organic acid and sulfur. During the life cycle of refining units, critical equipment and/or key pipes might be partially changed or upgraded. Therefore, components of distillation units have different remaining life and different capability of bearing corrosive crude. The company would like run its distillation units at a corrosion rate lower than the bearable corrosion rate, which is the maximum limit of corrosion rate in terms of safety and economic concerns. These facts make it of paramount importance for the company to know the bearable corrosiveness limit of its distillation units. And therefore the company can distribute crude slates appropriately among its refineries on the basis of their bearable corrosiveness limits.

3. Concept and Theory

The major problems induced by "opportunity" crudes are Naphthenic acid (NAC) and sulfidation corrosion (SC). Refinery equipment is attacked in the high temperature range of 425 °F (218 °C) up to about 750°F (400 °C). Over the last 60 years many efforts have been made to understand and predict NAC and SC[1-8]. Below are some rules of thumb and useful facts regarding NAC and SC:

- There are two different corrosion prediction tools for SC. The modified McConomy curves are used for SC in H2-free streams (which is the default type SC discussed in this paper) and the Couper-Gorman curves that are used for SC in H_2/H_2S-containing streams[9,10].
- SC in H_2-free streams is thought to occur at temperatures above 446 °F (230 °C). High temperature SC is typically a general corrosion. The protective iron sulfide scale that forms from SC will partially inhibit NAC[7].
- Limited plant experience shows that the modified McConomy curves generally overestimate SC rate. However, the modified McConomy curves could underestimate SC rate at some specific side cuts or locations.
- Fractions may vary significantly in their potential to cause SC due to the type of sulfur compounds. Some sulfur contents can be grouped as reactive or non-reactive, and therefore, their corrosivity could deviate from that predicted by the modified McConomy curves[11].
- Crudes with a TAN value higher than 0.5 mg KOH/g and crude fractions with TAN greater than 1.5 mg KOH/g are typically considered to be corrosive.
- NAC usually starts from 425 °F (218 °C) and severity increases with temperature up to about 750 °F (400 °C) according to API 571. However, some field evidence and laboratory tests show that the increment trend may end at 650 °F (343 °C) due to either thermo-decomposition of the acid or coking. Products of coking at high temperature may act as a physical barrier for NAC and SC, and decrease their rates.
- Naphthenic acids can also be grouped as reactives ("bad naphthenic acid") and inactives ("good naphthenic acid") by both their corrosivity and molecular weights. Some efforts are made to distinguish them by comparing their corrosivity and chemical structures[12]; however, engineering applications are not seen yet.
- Turbulent flow increases the NAC corrosion rate.
- Since SC and NAC can co-exist at the same temperature range, it is difficult and usually unnecessary to distinguish their respective contributions to the total corrosion rate. Occasionally, ''naphthenic acid corrosion index'' (NACI), a simple ratio of the corrosion rate to the weight of corrosion scale, could be used to compare the relative contributes of SC and NAC[13].

- API RP 581 provides a detailed database for estimating corrosion rates of concurrent NAC and SC of normal materials in refineries, including carbon steel, low alloy steel, austenitic stainless steel without Mo, and austenitic stainless steel with Mo.

However, precisely predicting NAC and SC rates based on common knowledge is still a challenging task. Compromise and assumption are needed to make an experiential corrosion predicting model simple and technically feasible:

- Sulfur compounds and naphthenic acids are similarly corrosive, supposing the source of crudes is disregarded. This assumption contradicts the knowledge of SC and NAC for specific crude; however, it is statistically true for most crude.
- Distribution of sulfur compounds and naphthenic acids over boiling points are the same, supposing the source of crudes is disregarded. Again, it is statistically true for most crude.

The corrosion model is used to calculate the bearable TAN and sulfur of distillation cut for every specific location. And then, the bearable TAN and sulfur of crude slate of every location can be calculated proportionally. Critical operational TAN (COTAN) and Sulfur Level (COSL) are then determined for the lowest bearable TAN and sulfur of crude slate of all locations in the distillation unit. The theory detail can be found in the reference[14].

4. Case Study

A small crude distillation with capacity 3 million Bbls/year suffers from heavy corrosion. The plant was originally made of carbon steel and partially upgraded recently. It was forced to shut down for inspection and reparation regularly even its major product paraffin wax was still profitable. Figure 2 through 5 shows that high temperature naphthenic acid corrosion and sulfidation corrosion extensively exist in high temperature part of unit, such as heat-exchanger, pipes, tower, and column internal.

Historical data shows that corrosives have been kept relatively constant over the monitoring period (Figure 6). The TAN of slate is around 1 mg KOH/g oil and sulfur content is about 0.25%. Concentrations of both acid and sulfur are predicted to increase slowly, but consistently in the future. Corrosives distribution is also measured for atmospheric and vacuum distillation as shown in Figure 7, which is used to proportionally calculate the bearable corrosives of distillation cut and those of crude slate for every specific location. Table 1 shows some selected corrosion model calculation results. Vacuum tower furnace piping is the weakest part of whole unit. It is recommended to upgrade the vacuum tower furnace piping with 316 stainless steel. Once the replacement is done, the COTAN and COSL of unit can be set as 1 mg KOH/g and 0.6%.

Figure 2. Corrosion pits on a heat-exchanger

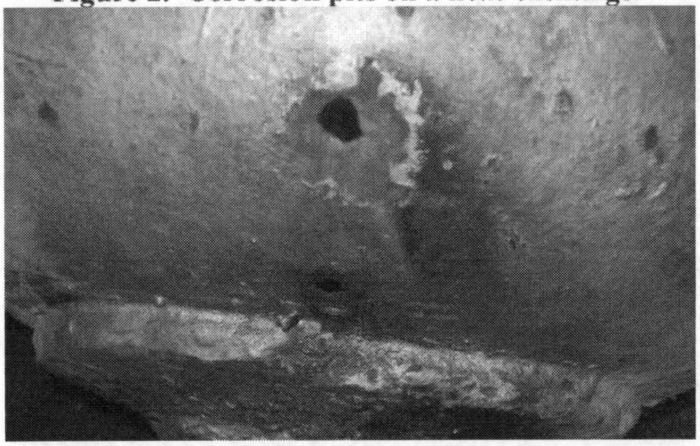

Figure 3. Corrosion of a tongue-like tray in an atmospheric distillation tower

Figure 4. Corrosion of a supporting beam in a vacuum distillation tower

Figure 5. Corrosion of a column tray in a vacuum tower with float valves missing

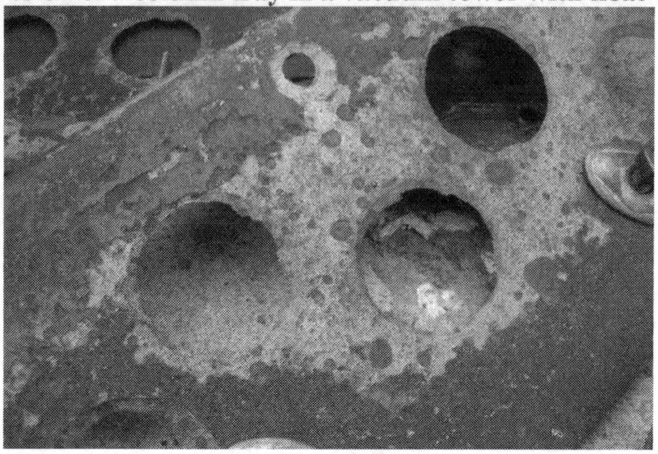

Figure 6. Historical date of acids and sulfur content for the crude slate

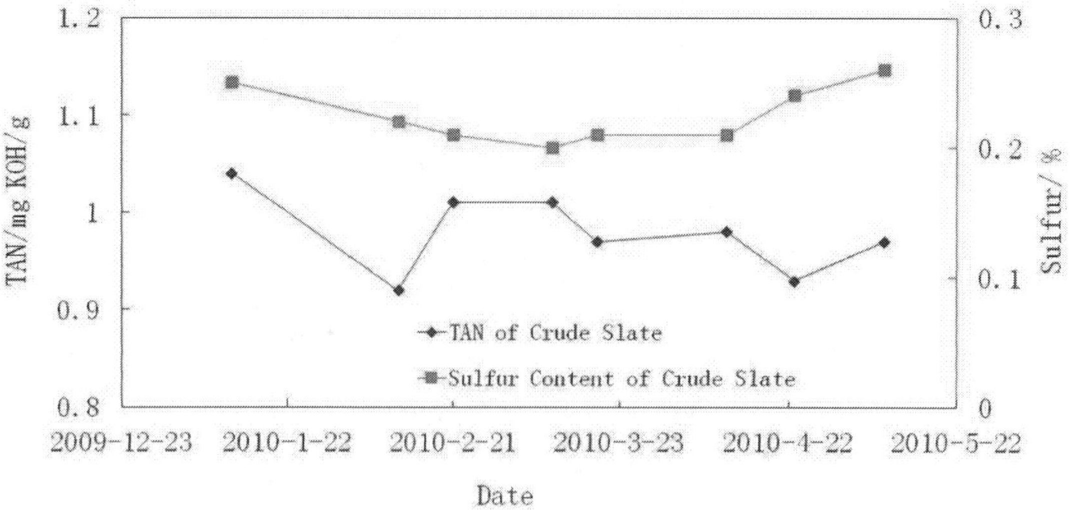

Figure 7. Corrosives distribution over atmospheric and vacuum distillation.

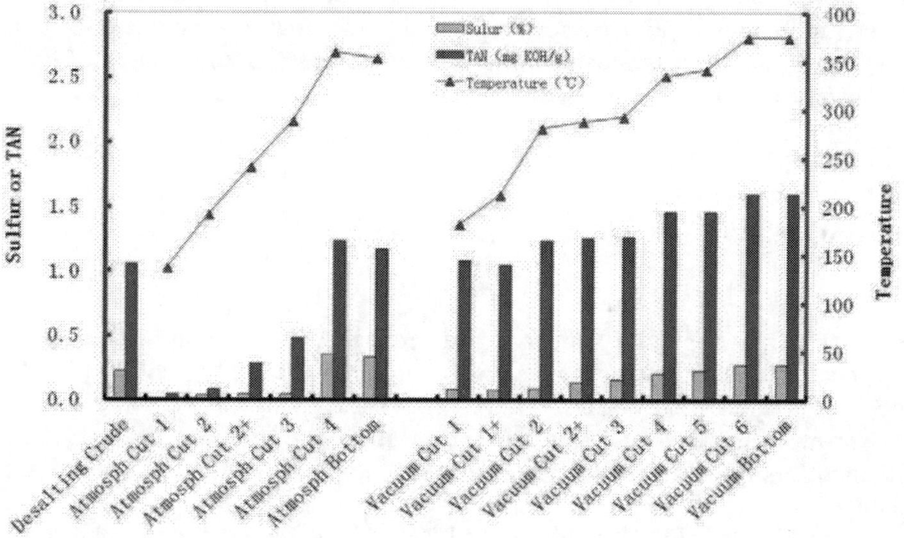

Table 1. Corrosion model calculation results.

Parts	Distillation Cut	Material	Bearable sulfur of crude (%)	Bearable TAN of crude (mg KOH/g)
Atmospheric tower furnace pipe	Desalting crude	1Cr5Mo	0.6	1.5
Vacuum tower furnace pipe	Atoms. bottom	1Cr5Mo	0.2	0.7
Pipe for Atoms. cut 3	Atoms. cut 3	Carbon steel	1	N/A
Pipe for Atoms. cut 4	Atoms. cut 3	304	N/A	N/A
Heat Exchanger E1009	Desalting oil/ Atoms. cut 2	Carbon steel	3	1
Heat Exchanger E1020B	Desalting oil/ Vacuum bottom	Carbon steel	0.6	2

N/A: not applicable, concentration of corrosives unlikely exceeds the bearable value according to the model.

5. Applications and Effects

The theory of COTAN and COSL originated years ago when the company was suffering a great challenge with "opportunity" crudes. The application of COTAN and COSL not only set sulfur and TAN limits for units, but also identified many units' "weak points". Some suggestions to execute necessary remedies were made for many units after applying the knowledge of COTAN and COSL, which had precluded many "would-be" accidents. The theory was gradually developed and the first application started from an old unit on 2007, and then gradually extended to others. So far, it has been implemented to 30+ units, and has contributed to the company's reputation of safety excellence.

6. Conclusion

A corrosion predict model plays a vital role in determining COTAN and COSL. Its accuracy is still controversial. Refinery experience shows that the corrosion rate predicted by the modified McConomy curves is normally conservative, and usually higher than the actual corrosion measured. However, more severe corrosion than the McConomy curves predicted is also observed. This diversity could be attributed to many factors, such as flow condition and source of crude slate.
Critical operational TAN (COTAN) and Sulfur Level (COSL) are still new concepts for refining industry. It is a successful attempt to apply current knowledge of NAC and SC on the basis of refineries experience.:

7. References

[1] W. A. Derungs, "Naphthenic Acid Corrosion – An Old Enemy of the Petroleum Industry", *Corrosion* 12, (1956): p.617.

[2] S Tebbal, "Critical Review on Naphthenic Acid Corrosion", *CORROSION/99*, paper no.380 (Houston, TX: NACE, 1999).

[3] K.R. Lewis, M.L. Daane, and R. Schelling, "Processing Corrosive Crude Oils", *CORROSION/99*, paper no.377 (Houston, TX: NACE, 1999).

[4] D. Johnson, G. McAteer, and H. Zuk, "The Safe Processing of High Naphthenic Acid Content Crude Oils Refinery Experience and Mitigations Studies", *CORROSION/2003*, paper no.03645 (Houston, TX: NACE, 2003).

[5] D. Qu, Y. G. Zheng, H. M. Jing, X. Jiang, and W. Ke, "Erosion-corrosion of Q235 and 5Cr1/2Mo Steels in Oil with Naphthenic Acid and/or Sulfur Compound at High Temperature", *Materials and Corrosion* 56, 8(2005):p.533.

[6] D. Qu, Y.G. Zheng, H.M. Jing, Z.M. Yao, and W. Ke, "High Temperature Naphthenic Acid Corrosion and Sulphidic Corrosion of Q235 and 5Cr1/2Mo Steels in Synthetic Refining Media", *Corrosion Science* 48, (2006): p.1960.

[7] G. Bota, D. Qu, S. Nesic, and H.A. Wolf, "Naphthenic Acid Corrosion of Mild Steel in the Presence of Sulfide Scales Formed in Crude Oil Fractions at High Temperature", *CORROSION/2010*, paper no.10353 (Houston, TX: NACE, 2010).

[8] H.D. Dettman, N. Li, and J. Luo, "Refinery Corrosion, Organic Acid Structure and Athabasca Bitumen", *CORROSION/2009*, paper no.09336 (Houston, TX: NACE, 2009).

[9] API RECOMMENDED PRACTICE 939-C (FIRST EDITION, MAY 2009), "Guidelines for Avoiding Sulfidation (Sulfidic) Corrosion Failures in Oil Refineries", (Washington, D.C: API).

[10] *NACE International Publication 34103* (2004), "Overview of Sulfidic Corrosion In Petroleum Refining", (Houston, TX:NACE, 2004).

[11] V. Kanukuntla, D. Qu, S. Nesic, and A. Wolf "Experimental Study of Concurrent Naphthenic Acid and Sulfidation Corrosion", *17th ICC* (2008), paper. 2764.

[12] D. Qu, Y. Zheng, X. Jiang and W. Ke, "Correlation between the Corrosivity of Naphthenic Acids and Their Chemical Structures", *Anti-Corrosion Methods and Materials* 54,4 (2007): p211.

[13] J.H. Lee Craig, "Naphthenic Acid Corrosion in the Refinery", *CORROSION/95*, paper no. 333 (Houston, TX: NACE, 1995).

[14] Dingrong Qu, Xiaohui Liu, Xiu Jiang, Zhenggui Lan, Guangbin Shan, "Setting Critical Operational TAN and Sulfur Level for Crude Distillation Units" *CORROSION/2011*, paper no. 11362 (Houston, TX: NACE, 2011).

9TH GLOBAL CONGRESS ON PROCESS SAFETY

Large Eddy Simulation – Based Mixing Tee Analysis and Comparison with In-Service Failure Data

Seetha Ramudu Kummari, Ph.D.
The Equity Engineering Group Inc., Shaker Heights, Ohio
srkummari@equityeng.com

Michael F.P. Bifano, Ph.D.
The Equity Engineering Group Inc., Shaker Heights, Ohio
mfpbifano@equityeng.com

David J. Dewees, P.E.
The Equity Engineering Group Inc., Shaker Heights, Ohio
djdewees@equityeng.com

Prepared for Presentation at
American Institute of Chemical Engineers
2013 Spring Meeting
9th Global Congress on Process Safety
San Antonio, Texas
April 28 – May 1, 2013

UNPUBLISHED

Large Eddy Simulation – Based Mixing Tee Analysis and Comparison with In-Service Failure Data

Seetha Ramudu Kummari, Ph.D.
Michael F.P. Bifano, Ph.D.
David J. Dewees, P.E.
The Equity Engineering Group Inc., Shaker Heights, Ohio
srkummari@equityeng.com

Keywords: Mixing tee, turbulent thermal mixing, Large Eddy Simulations (LES), Fatigue Crack Growth, Computational Fluid Dynamics (CFD)

Abstract

High frequency temperature fluctuations resulting from turbulent thermal mixing of hot and cold streams in a tee junction are known to cause surface cracks on inner surfaces of the tee junctions, a phenomenon called thermal stripping in the petrochemical industry. While temperature fluctuations due to turbulent thermal mixing have been linked to surface cracks in the mixing tees, to date, how such surface cracks grow and cause leakages has not been assessed in the literature. The current study evaluates, not only surface crack initiation due to turbulent thermal mixing but also the effects of operating fluctuations (global transients) on crack initiation and growth. An objective of the study is to develop a comprehensive methodology to address mixing tee failures utilizing Large Eddy Simulations (LES) and standard two parameter Computational Fluid Dynamics (CFD) turbulence models, finite element stress analysis, fatigue life estimation and fatigue crack growth approaches. The developed methodology is used to assess an industrial mixing tee that experienced cracks after 15 years of service. The results obtained from the assessment are benchmarked against in-service failure data. The study suggests that a combination of advanced analyses such as LES based CFD, FEA, fatigue life estimation, fatigue crack growth approaches are necessary to analyze the effect of turbulent thermal mixing and operating fluctuations for an evaluation of a mixing tee failure.

1. Introduction

Tee junctions are common in petrochemical industries and are used to mix two fluid streams at different temperatures, and often consisting of different compositions. The mixing process of hot and cold fluids in the tee junctions is chaotic (turbulent) in nature and can result in high cycle temperature fluctuations on the inner surfaces of the tee junction and/or downstream piping. Such high cycle temperature fluctuations can initiate surface cracks; a phenomenon well known as thermal stripping in the petrochemical industry. Though thermal stripping alone may not be sufficient to propagate initiated cracks, operating fluctuations such as substantial variations in flow rates, pressure, and temperature, and plant shutdown and startup could potentially propagate cracks. This could lead to leakage and structural integrity and process safety concerns. The high frequency temperature fluctuations resulting from turbulent thermal mixing cannot be sufficiently measured by modern thermocouples [1]. This makes it very difficult to identify and remediate locations where thermal stripping occurs. In the present study, advanced numerical methods such as Large Eddy Simulations (LES) are used to identifying mixing point locations prone to thermal stripping.

Standard Computational Fluid Dynamics (CFD) using two parameter turbulence models (Reynolds Averaged Navier Stokes Models, RANS) or unsteady RANS (URANS) cannot adequately predict the turbulent thermal mixing phenomenon [2], and advanced analysis, specifically LES, is required to obtain meaningful results [1, 3–7]. LES resolves detailed characteristics of the flow, such as the "large" eddies, and their propagation in time and hence the resulting temperature fluctuations. This paper presents a comprehensive analysis of a petrochemical industry mixing point. The analysis utilizes a combination of LES and RANS CFD turbulence models, finite element stress analysis, fatigue life estimations and fatigue crack growth approaches. The investigation is driven by extensive analysis of operational data, and is rigorously benchmarked against the observed in-service failure of a mixing tee. Specifically, the tee junction under consideration introduces cold hydrogen at variable flow rates into the primary hot hydrocarbon stream. After approximately 15 years of service, multiple cracks were identified downstream of the mixing point on the inside surface of the run pipe.

2. Methods

2.1. Severity of Mixing – Momentum Ratio Calculations

Severity of mixing depends upon the relative momentums of the two flows (branch and run pipe flows) and can be assessed by a parameter called, momentum ratio which is defined as:

$$MR = \frac{\rho_r U_r^2 (D_r \times D_b)}{\rho_b U_b^2 \times \pi \left(\frac{D_b}{2}\right)^2} \quad \text{(Eq. 1)}$$

Where:

ρ, U and D represent density, velocity and inner diameter of the pipes and the subscripts, r and b represent run pipe and branch pipe, respectively. In the mixing tee analysis literature [4, 5, 8, 9], flow regimes are classified based upon the value of the of the momentum ratio as follows:

Momentum Ratio	:	Flow Regime
< 0.35	:	Impinging jet
0.35 – 1.35	:	Turn jet
1.35 – 4.0	:	Re-attaching jet
> 4.0	:	Wall jet

A lower momentum ratio lead to a higher expectation of mixing, i.e., impinging jet flows result in heavier mixing compared to that of wall jets in which the branch fluid is forced against the upper portion of the run pipe. Though this type of flow regime classification can be used for comparing relative severities of mixing, the classification does not provide much insight into thermal stripping phenomenon and hence surface crack initiations. It is shown in the current study that even the wall type flow can result in severe local turbulent thermal mixing and lead to surface crack initiations. This suggests that the flow regime classification is not very predictive for typical petrochemical mixing points. Operating conditions that result in momentum ratios significantly higher than 4.0 still demonstrate enough branch fluid energy for random and turbulent penetrations into the run pipe fluid, followed by the forcing of these fluid structures back against the pipe wall. The result is alternating hot and cold shocks randomly distributed along the pipe that can be characterized in a statistical sense.

Momentum ratios are calculated using (Eq.1) for six months of actual operating history of the mixing tee under consideration. Based on the calculated momentum ratios, four momentum ratios are chosen to represent "heavy", "moderate", "less" and "very less" mixing conditions in the six months of operating conditions. Corresponding momentum ratios are as shown below. Note that classification of mixing conditions is relative to the operating history and applies only to the mixing tee under consideration.

Momentum Ratio	:	**Mixing Condition**
22	:	Heavy mixing
106	:	Moderate mixing
1038	:	Less mixing
7712	:	Very less mixing

2.2. Large Eddy Simulations – Simulation of Turbulent Thermal Mixing

A commercial CFD software, STAR-CCM+ version 7.02.008 is used for the CFD analysis of the mixing tee. Only one half of the mixing tee model is considered because of symmetries in the model geometry – the restriction this enforces on the turbulent flow is considered reasonable. The run pipe is extended by 3.0 times the diameter upstream and 9.0 times the diameter downstream of the tee to minimize the effect of the assumed inlet velocity distribution and backflow at the region of interest. Similarly, the branch pipe is extended by 5.0 times its diameter upstream of the tee junction. All the regions, in the model, except boundary layers, are discretized using polyhedral cells in STAR-CCM+. A focused mesh is used in the regions of interest and near the fluid/solid interface and where heavy mixing is expected. The model contains approximately half a million polyhedral cells. Figure 1 shows the discretized model of the mixing tee.

A Volume of Fluids (VOF) multiphase model with surface tension effects is chosen to represent the fluid interaction between the branch and run pipe streams. Velocity inlet boundary conditions are applied to both the run and branch pipe inlets and pressure outlet boundary conditions are defined at the outlet. The following power law (n = 7) velocity profiles are defined at the inlets to better represent fully developed inlet flows,

$$V(r) = \frac{(n+1)(2n+1)}{2n^2} V_{avg} \left(1 - \frac{r}{R}\right)^{\frac{1}{n}} \qquad \text{(Eq.2)}$$

$$\text{for } n = 7, \qquad V(r) = 1.2245\, V_{avg} \left(1 - \frac{r}{R}\right)^{\frac{1}{7}} \qquad \text{(Eq.3)}$$

Where:
r is radial coordinate of the fluid volume
R is radius of the pipe
V_{avg} is average velocity of the fluid in the pipe

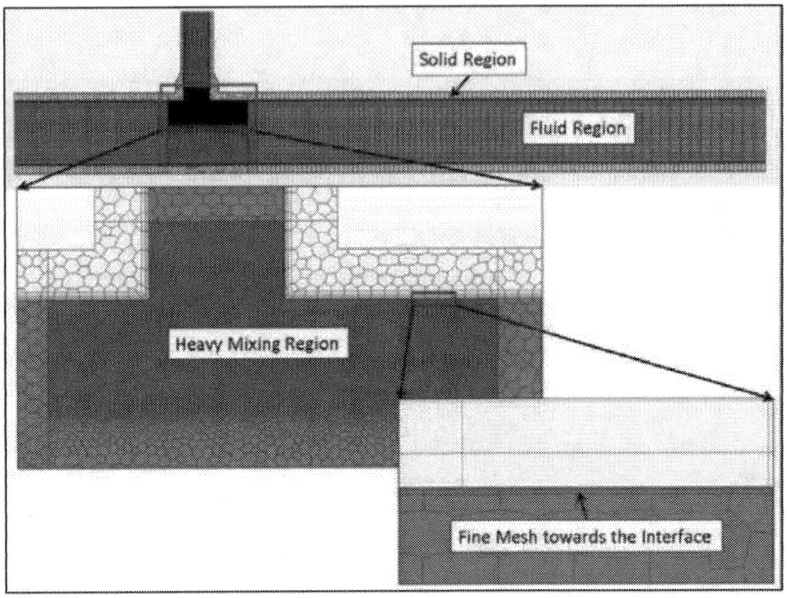

Figure 1: Different regions of the mixing tee CFD model are shown. The model is discretized using polyhedral cells in STAR-CCM+. Fine mesh is included in the regions of heavy mixing and near the fluid/solid interface.

The liquid hydrocarbon mixture in the mixing tee is modeled as an incompressible fluid, whereas the hydrogen gas is modeled as a compressible ideal gas. No slip conditions are defined at the fluid/solid interface. The effect of pipe insulation is simulated by defining an effective heat transfer coefficient of 0.568 $W/m^2.K$ and a sink temperature of 21°C on the outer surfaces of the pipes (Figure 2), which includes the combined resistance of the external insulation and free external convection in still ambient air. Initially, the CFD model is run with a conventional (RANS) k-ε turbulence model. To increase speed and stability in the iterative LES solver, the converged RANS solution is used as an initial condition. A time step of 5.0E-5 seconds is chosen for LES simulations. To remove all artifacts of the RANS solution, the LES simulation is run for 1.5 seconds prior to any data collection. Following the development of a stable LES solution, data is collected at the rate of 200 Hz. Fluid/solid interface temperatures are recorded at multiple locations along the length of the run pipe; at the center of the tee, weldolet, 6 inches away, 12 inches away and 18 inches away from the center of the tee (Figure 3). The simulation is run at the moderate mixing condition for about 2.0 seconds and then the flow is switched to the heavy mixing condition, and is run for an additional 2.0 seconds, in an attempt to capture any transient effects of unsteady operating conditions.

2.3. RANS Simulation – Simulation of Operating Fluctuations

Conventional RANS simulations are performed for the four mixing conditions; heavy, moderate, less and very less mixing conditions. The intent of simulating different mixing conditions is that the effect of bulk operating fluctuations is isolated. The CFD model and boundary conditions are identical to those described in the previous section except that only steady state RANS simulations are performed for each mixing condition.

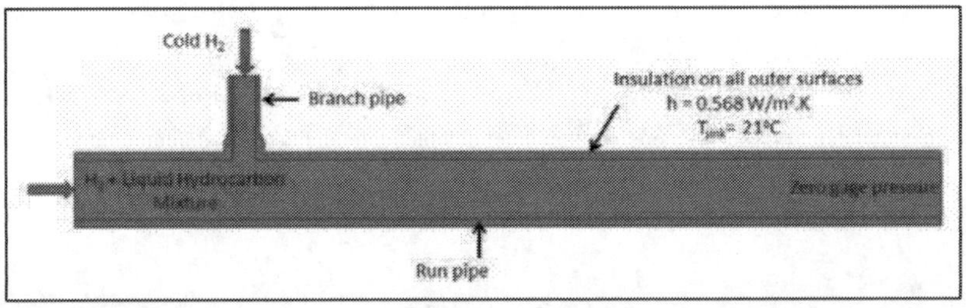

Figure 2: Velocity inlet and pressure outlet boundary conditions are applied. An effective heat transfer coefficient of 0.568 $W/m^2.K$ is applied to the outer walls of the pipe to account for the effect of insulation.

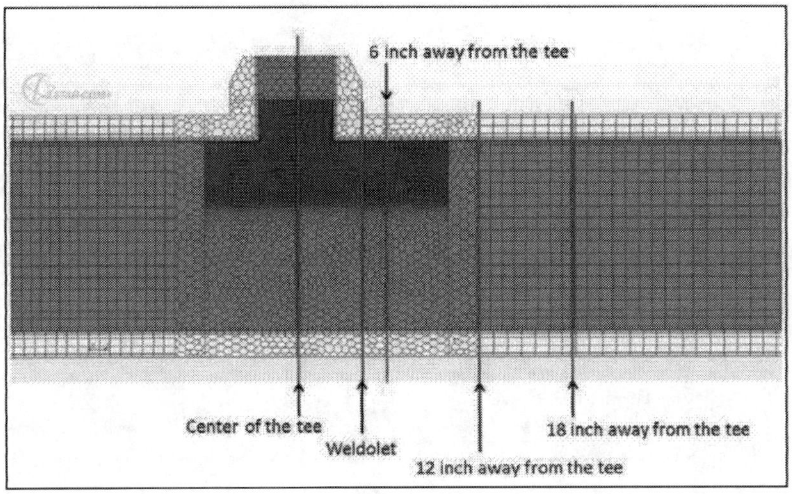

Figure 3: Five different planes are chosen downstream of the mixing tee for temperature data collection at fluid/solid interface.

2.4. Finite Element Stress Analysis

Stresses due to the temperature fluctuations are evaluated using a simplified axisymmetric finite element representation. The commercial finite element software, Abaqus Version 6.11 is used for the stress evaluations. An axisymmetric representation of the run pipe is discretized using quadratic coupled temperature–displacement axisymmetric elements (Abaqus element, CAX8RT). The finite element model consists of only one element along the length direction of the pipe, but contains 48 elements in the thickness direction with a fine mesh towards the inner diameter of the pipe. To evaluate stresses due to turbulent thermal mixing, the fluid/solid interface temperature fluctuations at various points in the LES CFD model are applied at the inner surface of the pipe as time dependent (transient) temperature boundary conditions. An effective heat transfer coefficient of 0.568 W/m^2.K with a sink temperature 21°C is defined on the outer surface of the pipe to simulate the effect of insulation and external convection. Adiabatic boundary conditions (no thermal gradients) are assumed on the transverse cut-planes of the pipe (Figure 4).

Additionally, the transient effect of switching mixing conditions (operating fluctuations) is simulated by changing the sink (fluid) temperature (steady state temperatures obtained from the RANS simulations) instantly at the inner surface of the pipe and specifying a nominal heat transfer coefficient of 1704 W/m^2.K. This value is calculated using standard forced convection pipe flow correlations and is typical for the relevant fluid properties and flow conditions applicable to the tee. The boundary conditions are shown in Figure 5.

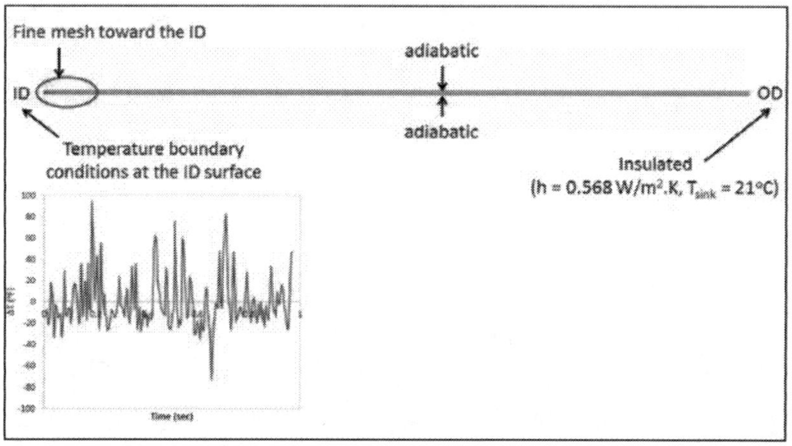

Figure 4: The boundary conditions on the axisymmetric finite element model to evaluate the stresses due to turbulent thermal mixing are shown. Temperature fluctuations from the LES are applied on the inner diameter as transient temperature boundary conditions.

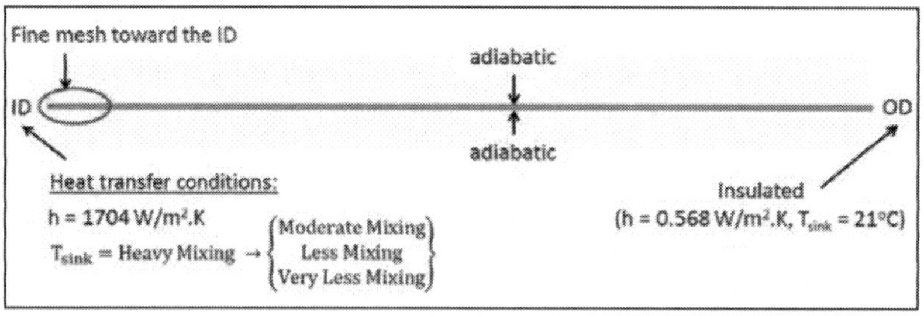

Figure 5: The boundary conditions on the axisymmetric finite element model to evaluate the stresses due to operating fluctuations are shown. Steady state temperatures obtained from the RANS simulations are applied as sink (fluid) temperatures.

2.5. Fatigue Life for Crack Initiation

Stress histories obtained from the axisymmetric finite element analysis are rain flow cycle counted to obtain the number of stress cycles that a particular location experiences. Then the number cycles to failure are obtained using the ASME design fatigue curve for stainless steel. The time to initiate a crack is determined by applying Miner's rule for all the cycles with stress amplitudes larger than the endurance limit of the pipe material (stainless steel, 13.6 ksi). A usage factor is calculated by dividing operating duration (15 years) by time to failure.

2.6. Fatigue Crack Growth Analysis

Following fatigue crack initiation, final crack depths are estimated with knowledge of the operating duration using API 579-1/ASME FFS-1 2007 Edition (API 579). In the present study, crack depths are estimated for 15 years of typical service. A semi-elliptical crack of size 0.04 inches in depth (a), 0.24 inches in length (2c) is assumed to be initiated at the time of crack initiation. The initial crack sizes are chosen based on engineering observations and fracture mechanics literature that indicate typical crack sizes between 0.5 mm (0.02 inches) and 3.0 mm (0.12 inches) at the time of crack initiation (i.e. at the end of a fatigue test). In this analysis a crack size of 1.0 mm (0.04 inches) is assumed to initiate at the time of fatigue crack initiation.

The resulting stress histories from the FEA model are rain flow cycle counted to determine the most common stress amplitude for the crack growth analysis. Stress intensities (K_{max} and K_{min}), corresponding to the maximum and minimum stresses and for the given crack size, are calculated according to Annex C of API 579. The crack growth rate is estimated using a Paris law model for stainless steel, given in Annex F of API 579. The Paris law used for the estimation of crack growth rate in both depth and length direction is given by (Eq.4) and (Eq.5). The final crack depth and length (a and 2c) are estimated by assuming that the crack growth is constant for 10 cycles. The validity of this assumption is verified by a sensitivity analysis.

$$\frac{da}{dN} = 8.61 \times 10^{-10} \left(\frac{E_{amb} \times \Delta K}{E_{at}}\right)^{3.0} \quad \text{for } \Delta K \geq \Delta K_{th} \qquad \text{(Eq.4)}$$

$$\frac{da}{dN} = 0 \quad \text{for } \Delta K < \Delta K_{th} \qquad \text{(Eq.5)}$$

Where:
$\frac{da}{dN}$ is crack growth rate per cycle
ΔK is stress intensity range ($K_{max} - K_{min}$)
ΔK_{th} is threshold stress intensity range below which $\frac{da}{dN}$ is assumed to be zero
E_{amb} is elastic modulus at ambient temperature
E_{at} is elastic modulus at assessment temperature

3. Results

3.1. Turbulent Thermal Mixing Results (LES Simulations)

Figure 6 show the distinction between the temperature distributions obtained using the RANS k-e turbulence model and the LES turbulence model. As can be seen from the figure, the conventional RANS turbulence model does not resolve eddies whereas LES models do. Temperature fluctuations of 50°F to 100°F at 5.0 Hz are common near the tee junction. Figure 7 shows temperature fluctuations due to turbulent thermal mixing at 27° (from the branch pipe), center of the tee and 30°, 6 inches away from the tee. Maximum temperature fluctuations around the circumference of the run pipe point to the locations of maximum thermal fatigue (Figure 8).

The results indicate that stresses corresponding to the simulated temperature fluctuations (Figure 9) may be as high ± 40 ksi (i.e. both in tension and compression). Figure 10 shows stress amplitudes verses mean stresses of the cycles obtained from the rain flow cycle counting algorithm. Fatigue analysis suggests that the time to initiate a crack varies from location to location. The allowable cycle limit can be readily determined as a function of the location.

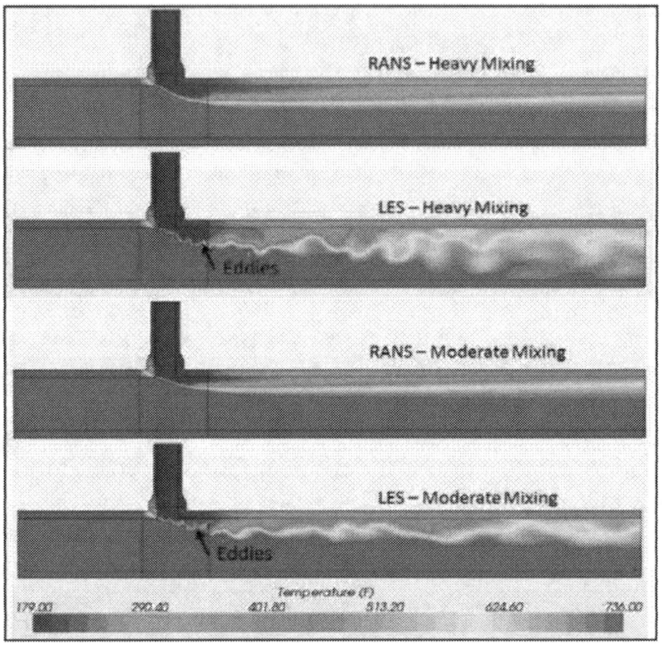

Figure 6: The distinction between RANS and LES simulations is illustrated for both heavy and moderate mixing conditions. LES resolve large eddies of the turbulent thermal mixing and hence the resulting high frequency temperature fluctuations.

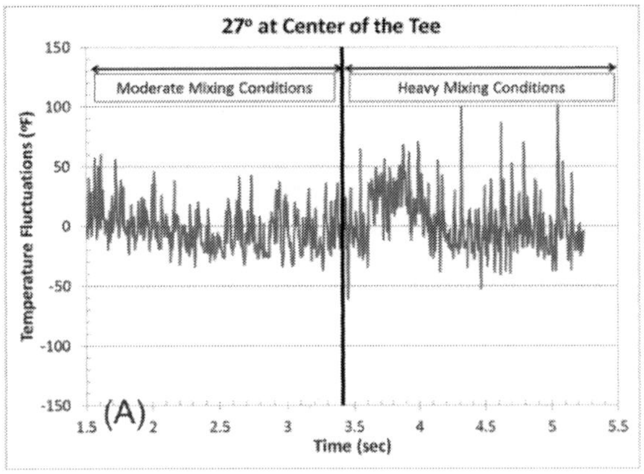

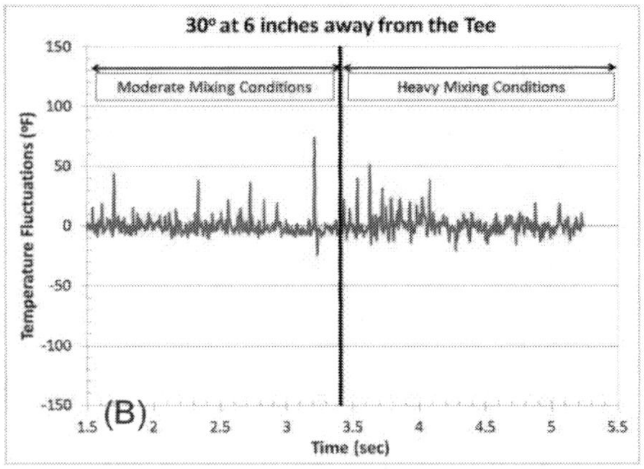

Figure 7: Temperature fluctuations due to turbulent thermal mixing at (A) 27°, center of the tee and (B) 30°, 6 inches away from the tee

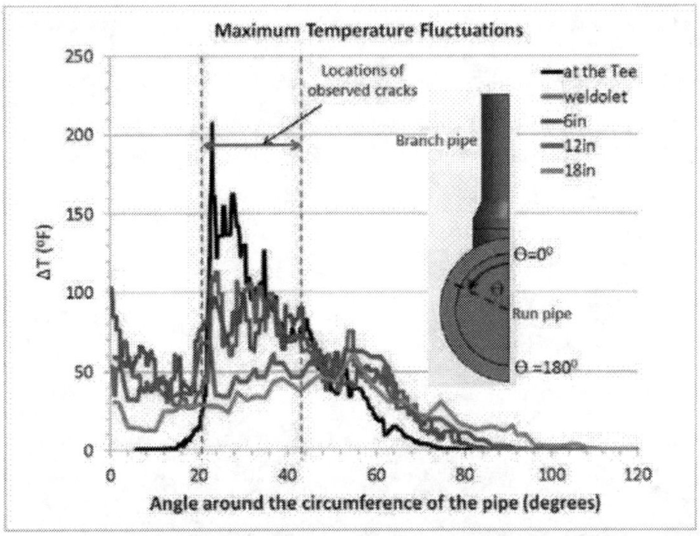

Figure 8: Maximum temperature fluctuations around the circumference of the pipe are plotted. Maximum temperature fluctuations location correlates well with the location of the observed cracking

The results of the crack growth analysis indicate that the stress intensity ranges (ΔK) for the maximum and minimum stress distributions caused by thermal fluctuations are below the threshold stress intensity range, ΔK_{th}. Therefore, while the temperature fluctuations are sufficient to initiate cracking, the thermal fluctuations due to thermal mixing do not penetrate far enough into the wall to create a sufficient through wall stress distribution capable of driving crack growth.

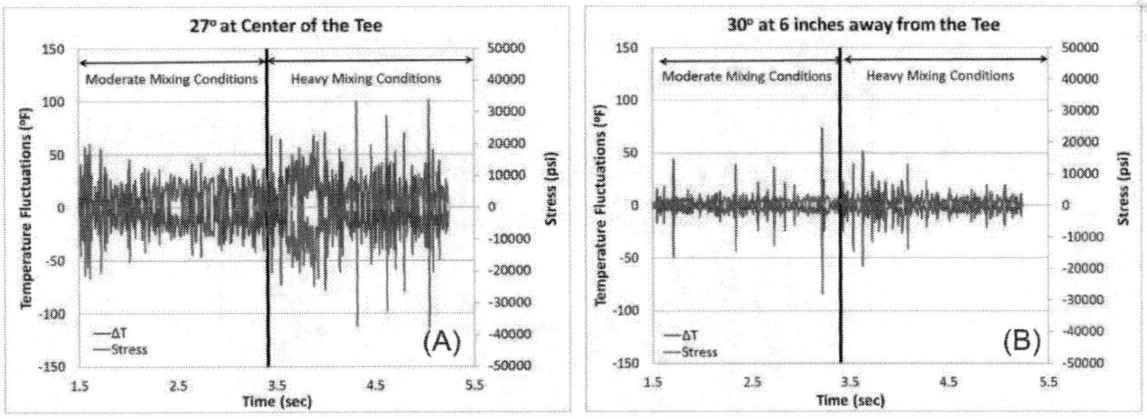

Figure 9: Stress fluctuations due to high frequency temperature fluctuations at (A) 27°, center of the tee and (B) 30°, 6 inches away from the tee

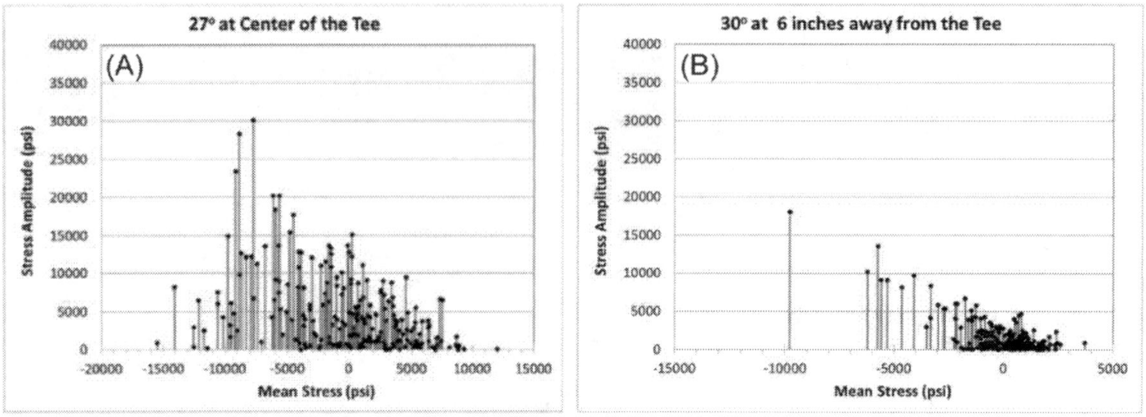

Figure 10: Stress amplitude verses mean stress from Rain Flow cycle counting algorithm for (A) 27°, center of the tee and (B) 30°, 6 inches away from the tee

3.2. Variations in Operating Conditions (RANS Simulations)

Figure 11 shows steady state temperature distributions for different mixing conditions simulated. Figure 12 illustrates the temperature variation that occurs around the pipe circumference when the mixing condition is changed from one steady value to another (operating fluctuations). The temperature fluctuations due to these operating fluctuations are as high as 450°F (Figure 12). The results indicate that the maximum temperature fluctuations occur at 6 inches away from the tee and near 40° from the centerline of the branch pipe. Remarkably, the results show that the locations of the maximum temperature fluctuations (30° to 40°, Figure 12) correspond to the locations of the observed cracks.

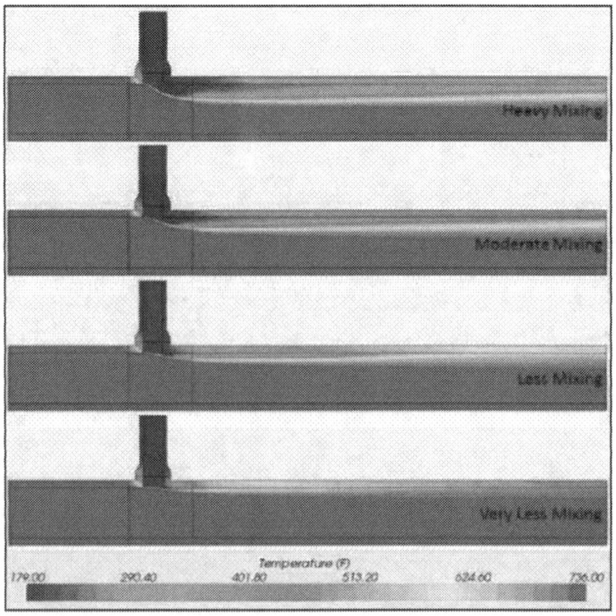

Figure 11: Steady state temperature distributions for different mixing conditions.

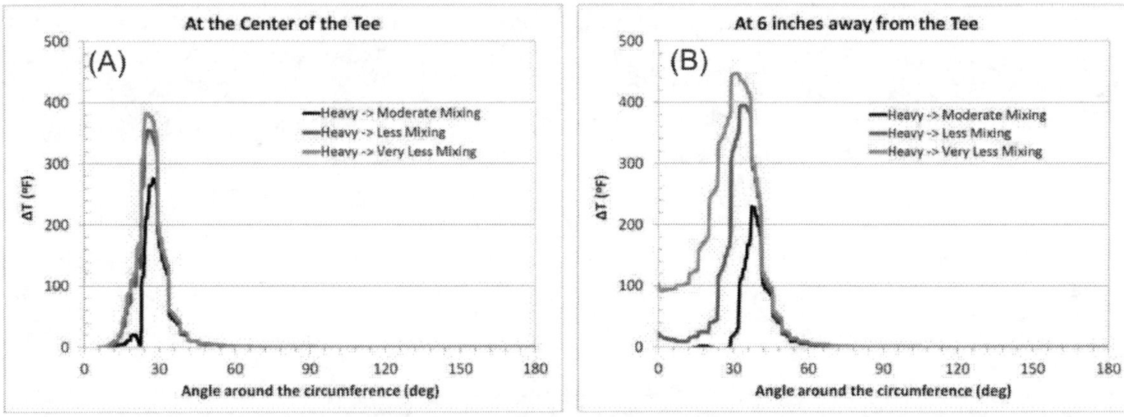

Figure 12: The effect of operating fluctuations is shown. The temperature fluctuations due to the operating fluctuations could be as high as 450°F and maximum temperature fluctuations occur between 30° and 40° from the branch pipe.

Stress analysis reveals that membrane plus bending stresses peak at a different time than that of the total stresses at the inner diameter of the pipe. The total stresses peak near 10 seconds after the change of mixing conditions whereas the magnitude of the membrane plus bending stresses peak near 22.5 seconds. The total stresses are used for fatigue initiation life estimations whereas through wall thickness stresses at which membrane plus bending stresses are maximum are used for crack propagation analyses (Figure 13). Fatigue life calculations for the operating fluctuations predict a usage factor of 0.25 suggesting that operating fluctuations exhaust ~25% of the fatigue life.

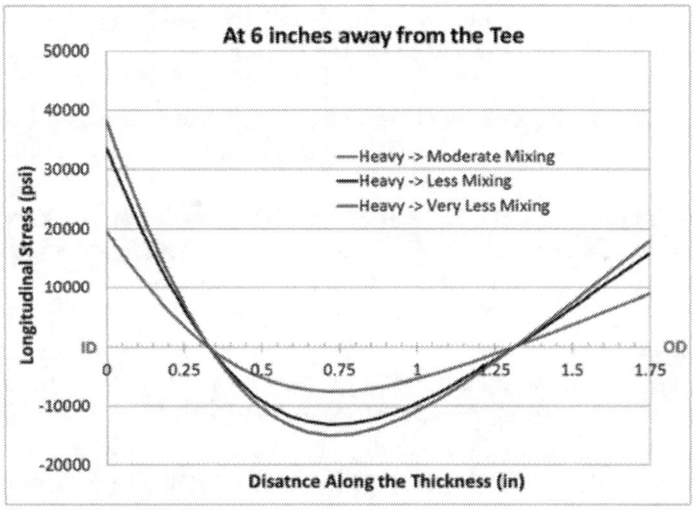

Figure 13: Through wall stress distributions due to operating fluctuations.

The results from the crack growth analysis are shown in Figure 14. The analysis predicts that an initial semi-elliptical crack of size, 0.04 inches in depth and 0.24 inches in length, grows by 0.0123 inches and 0.0113 inches in the depth and length directions, respectively, resulting in a final predicted crack size of 0.052 inches in depth and 0.251 inches in length. These results indicate that the operating fluctuations alone are unlikely to cause fatigue crack initiation but significantly contribute to fatigue crack growth. The stress fluctuations extend deep enough into the pipe walls to drive fatigue crack growth.

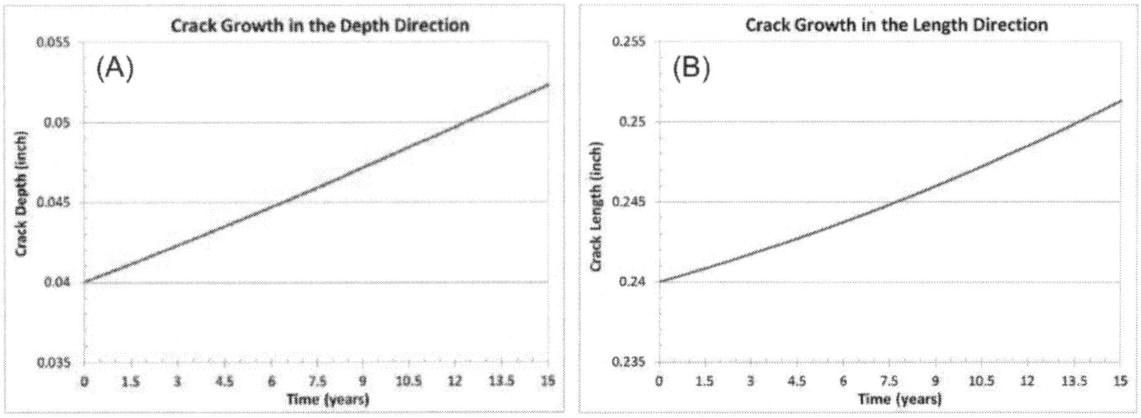

Figure 14: Fatigue crack growth in the operating duration of 15 years is shown. The simulated crack of initial size 0.04 inches in depth and 0.24 inches in length grows to a final size of 0.052 inches in depth and 0.251 inches in length.

4. Discussion

In conclusion, it can be stated that the comprehensive analysis including LES based CFD, FEA, fatigue crack initiation, and crack growth calculations, largely predict the cracking observed in the field. However it's important to understand the strengths and limitations of the methodologies used and how they can influence the results.

The results from the CFD analysis show clear distinctions between LES and conventional RANS turbulence models. The study shows that LES can be used to resolve eddies caused from the thermal mixing (Figure 6) and it is capable of obtaining the resulting thermal fluctuations at the fluid/solid interface (Figure 7). The current study also shows that the maximum temperature fluctuation locations correlate well with the locations of the actual observed cracks (Figure 8). This suggests an ability of the adopted methodology to locate the areas of possible thermal fatigue concerns.

Fatigue life estimations in this study are based on the ASME Section VIII, Div. 2 mean fatigue curve for industrial environments. There may be variations in the life estimation depending upon the fatigue curve used (for example, mean fatigue curve vs. -2σ fatigue curve). The results are also based on only two flow conditions (heavy and moderate mixing conditions). Practical time requirements and computational efforts limit LES analysis to only a few flow conditions, i.e., performing LES with actual transient inlet conditions or even a large number of inlet conditions is not a practically viable option.

The Paris law used for crack growth analysis is based on conservative estimates for crack growth rates. The final crack sizes may vary depending upon the constants in the crack growth model. The current analysis indicates that stress distributions due to turbulent thermal mixing do not penetrate deep enough into the metal to drive crack growth. This suggests that, though the high frequency thermal fluctuations are significantly capable of initiating cracking, the stress distributions are comprised of mainly peak stress at the surface of the metal and do not penetrate deep enough to propagate a crack front. Therefore variations in operating conditions such as pressure, temperature, and flow rates (as analyzed in the present study) are necessary to explain crack growth. The methodology adopted accurately predicts the most susceptible crack prone locations and a significant degree of fatigue crack growth. However, the method is not fully predictive since the projected crack depths are less than those found in service. Variations in operating temperatures and pressures likely account for these differences, and are logically the next element of complexity to add to the methodology adopted.

The major contribution of the present study is the finding that fluctuations in operating conditions, normally assumed to be steady in conventional LES studies, are found to be significant in driving a crack that is readily initiated due to turbulent thermal mixing. A multipart investigation that utilizes actual operating histories, LES CFD, conventional RANS CFD, finite element analysis, and fatigue crack initiation and growth models are necessary in identifying the vulnerability of industrial mixing points to fatigue damage.

5. References

1. M.H.C. Hannink, A.K. Kuczaj, F.J. Blom, J.M. Church, E.M.J. Komen. 2008, A coupled CFD-FEM strategy to predict thermal fatigue in mixing tees of nuclear reactors, EUROSAFE 2008, Paris.
2. C. Peniguel, 1998, Heat transfer simulation for industrial applications: needs, limitations, expectations, Int. J. Heat Fluid Flow, 19, 102-114.
3. J. Westin, P. Veber, L. Anderson, C. Mannetje, U. Andersson, J. Eriksson, M. Henriksson, F. Alavyoon, C. Andersson, 2008, High-cycle thermal fatigue in mixing tees. Large Eddy Simulations compared to a new validation experiment, Proc. 16th Int. Conf. On. Nuclear Engineering ICONE16-48731, Orlando, Florida, May 11-15.
4. L.-W. Hu and M. S. Kazimi, 2006, LES benchmark study of high cycle temperature fluctuations caused by thermal striping in a mixing tee, Int. J. Heat Fluid Flow, 27, 54.
5. M. Igarashi, N. Kimura, N. Tanaka and M. Kamide, 2003, LES analysis of fluid temperature fluctuations in a mixing tee pipe with the same diameters, Proc. 11th Int. Conf. On. Nuclear Engineering ICONE-11, Tokyo, Japan, April 20-23.
6. A.K. Kuczaj, E.M.J. Komen, M.S. Loginov, 2010, Large-Eddy Simulation study of turbulent mixing in a T-junction, Nuclear Engineering and Design, 240(9), pp. 2116-2122.
7. G. Grotzbach, M. Worner, 1999, Direct numerical and larege eddy simulations in nmuclear applications, Int. J. Heat Fluid Flow, 20, 222-240.
8. M. Igarashi, M. Tanaka, S. Kawashima, H. Kamide, 2002, Experimental study on fluid mixing for evaluation of thermal stripping in T-pipe junction, Proc. 10th Int. Conf. On. Nuclear Engineering ICONE10-22255 pp. 383-390, Arlington, Virginia, April 14-18.
9. S.M. Hosseini, K. Yuki, H. Hashizume, 2008, Classification of turbulent jets in a T-junction area with a 90-deg bend upstream, Int. J. Heat Fluid Flow, 51, 2444-2454.

9TH GLOBAL CONGRESS ON PROCESS SAFETY

Improving Reliability & Safety Performance of Solenoid Valves by Stroke Testing

Loren L. Stewart, BSME
exida
64 N. Main St., Sellersville, Pa 18960
lstewart@exida.com

Julia V. Bukowski, Ph.D.
Dept. of Electrical & Computer Engineering
Villanova University, Villanova, PA 19085
Julia.Bukowski@villanova.edu

William M. Goble, Ph.D., CFSE
exida
64 N. Main St., Sellersville, Pa 18960
wgoble@exida.com

Prepared for Presentation at
American Institute of Chemical Engineers
2013 Spring Meeting
9th Global Congress on Process Safety
San Antonio, Texas
April 28 – May 1, 2013

Improving Reliability & Safety Performance of Solenoid Valves by Stroke Testing

Loren Stewart, BSME
exida
64 N. Main St., Sellersville, Pa 18960
lstewart@exida.com

Julia V. Bukowski, Ph.D.
Villanova University

William M. Goble, Ph.D., CFSE
exida

Keywords: Stiction, static friction, binding, solenoid valve, emergency shutdown valve, valve stroke testing, proof testing, PFDavg, SIL rating

Abstract

Solenoid valves integrated into the design of emergency shutdown (ESD) valves used in industrial process systems, can tend to bind, i.e., to become stuck in one position, when not moved for long periods of time. This binding, also known as failure due to excessive stiction, has significant negative impacts on the valve's reliability and safety performance. It is a serious and costly problem normally addressed by expensive and time-consuming manual proof tests which typically require a process shutdown to perform testing. This paper describes an effective, alternative in-service testing protocol, known as valve stroke testing, which verifies whether or not the solenoid valve is stuck in position. It recommends a best practice procedure for implementing the valve stroke test. It provides a quantitative example of how valve stroke testing significantly improves safety performance when performed frequently (at intervals of one week or less) or even infrequently (at intervals of three to six months).

1. Introduction

In an industrial setting such as a chemical plant, oil refinery or a deep sea oil rig, there are safety instrumented systems in place which will promptly bring about a safe shutdown of a process if certain hazardous conditions are detected. Such safe shutdowns frequently involve isolating a process fluid flow, often accomplished by the use of an ESD valve.

Two of the ways an ESD valve can fail are it can function spuriously (also known as a spurious trip) or it can fail to function. If the ESD valve functions spuriously, it unnecessarily shuts down a properly working process in a safe manner; this type of failure is classified as a safe failure or fail safe. If the ESD valve fails to function, it loses the ability to shut down the process should a shutdown be required; this type of failure is categorized as a dangerous failure or "fail danger." If the process does not require shutdown while the ESD valve is in a state of dangerous failure then no harm is done provided the dangerous failure is discovered and repaired before a hazardous condition necessitating a shutdown occurs. If, however, the ESD valve is in a state of dangerous failure when a hazardous condition necessitating a shutdown occurs and the ESD valve does not perform its function when required, the worst case consequences of this "failure on demand" can be catastrophic including loss of life, significant destruction of or damage to equipment and property, and severe environmental damage.

Some ESD valve failures can be detected by various diagnostic procedures while the ESD valve is in service. If these detected failures are promptly repaired, safety performance is improved. Some ESD valve failures cannot be detected while the ESD valve is in service and consequently, safety standards, e.g. [1, 2] require periodic "proof testing" to verify that the ESD valve will operate properly if called upon to initiate a process shutdown. Such testing usually requires an interruption of process operations and in some instances may require physical removal of various ESD elements from the process for testing. Hence, proof testing is relatively time-consuming and expensive compared to in-service diagnostic testing.

ESD valves often include a solenoid valve of some type in the ESD valve design with the solenoid valve remaining in an energized state while the process operates normally and with the solenoid valve being de-energized, causing its seat to move to a different position, to initiate a safe process shutdown. Solenoid valves can fail to function on demand due to a number of different conditions called dangerous failure modes. In particular, based on a validated database of failure rates and failure modes for mechanical components [3, 4], one finds that a significant failure mode contributing to the dangerous failure rate of the solenoid valve is that of sticking or adhesion. This dangerous failure mode can lead to a failure to shut down when needed, i.e., a failure on demand.

Because the solenoid valve is one element of an ESD valve, it is often "at rest," i.e., stationary, during normal plant operation. The failure mode where it has become "stuck" in one position cannot be observed, i.e., the dangerous failure is undetectable in normal operation. Traditionally, this failure mode is discovered during proof testing and, for this particular ESD element, proof testing usually necessitates a process shutdown. As proof testing is a time-consuming and expensive proposition, the proof test interval is scheduled as infrequently as possible consistent with the reliability and safety objectives of the end-user.

Safety standards rate safety performance by safety integrity level (SIL) which is an order of magnitude rating based on the computed value of average probability of failure on demand (PFDavg) over the time interval $[0, T_P]$ as defined in (Eq. 1). T_P is the time interval between proof tests and PFD(t) is the time varying probability of failure on demand.

$$PFDavg(t) = \frac{1}{T_P}\int_0^{T_P} PFD(t)dt \qquad \text{(Eq.1)}$$

For a fixed value of T_P (Eq. 1) results in a particular value for PFDavg. Table 1 shows SIL levels assigned to different ranges of PFDavg.

Table 1. Relationship Between SIL Level and PFDavg.

SIL Level	PFDavg
1	$[10^{-2}, 10^{-1}]$
2	$[10^{-3}, 10^{-2}]$
3	$[10^{-4}, 10^{-3}]$
4	$[10^{-5}, 10^{-4}]$

But suppose we do not have a definite value of T_P in mind and want to set T_P as high as possible while still attaining a particular SIL rating. Then we are interested in PFDavg as a function of time as given in (Eq 2).

$$PFDavg(t) = \frac{1}{t}\int_0^t PFD(s)ds \qquad \text{(Eq. 2)}$$

Assuming that the rate of dangerous failures can be represented by the constant λ_D and that the device for which PFDavg(t) is being computed operates correctly when it is installed, then one major portion of PFDavg(t) can be reasonably approximated by

$$PFDavg(t) \cong \frac{1}{2}\lambda_D \times t \qquad 0 \leq t \leq T_P \qquad \text{(Eq.2a)}$$

Thus, PFDavg(t) increases approximately linearly as t increases up to T_P. The SIL rating that the device achieves can be altered by adjusting T_P. In general, mechanical devices do not undergo in-service diagnostic testing and it is not until the time of proof testing that dangerous failures are discovered and corrected. Just after the time of proof testing, assuming that the proof test was complete and perfect so that all dangerous failures were discovered and corrected, PFDavg(t) will return to zero and begin to increase linearly until the next proof test. Thus, over all t, (Eq. 2a) is properly written as

$$PFDavg(t) \cong \frac{1}{2}\lambda_D t - (j-1)\frac{1}{2}\lambda_D T_P \qquad (j-1)T_P < t \leq jT_P \quad j = 1, 2, 3\ldots \qquad \text{(Eq. 2b)}$$

Thus, the behavior of PFDavg(t) over repeated proof test cycles results in a saw-tooth plot as shown in Figure 1 where T_P has been arbitrarily set at 2 years. Note that the purple line in Figure 1 is the maximally allowed PFDavg to attain a SIL 3 rating. So with T_P set at 2 years, the example device would receive a SIL 2 rating. T_P would have to be decreased to approximately 1.25 years (the value of time when the PFDavg(t) plot and the horizontal purple line first intersect) in order to receive a SIL 3 rating.

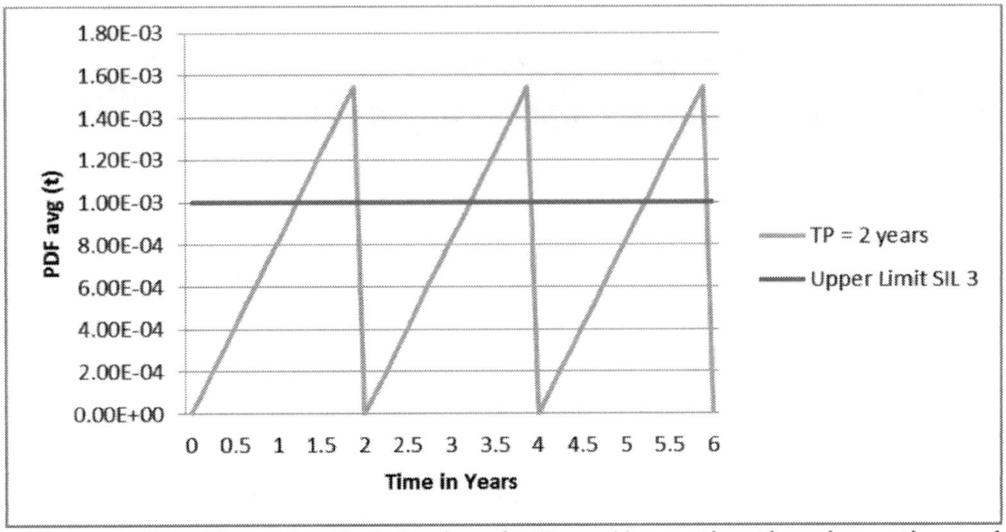

Figure 1. Plot illustrating the behavior of PFDavg(t) over time three inspection cycles when T_P equals 2 years.

Solenoid valve failures due to sticking or adhesion contribute significantly to reducing safety performance because they increase λ_D and result in reductions in T_P, i.e., they result in more frequent process downtime to facilitate proof testing, in order to meet safety objectives. Thus, finding less costly ways to address this adhesion failure mode has value.

The remainder of this paper

- Describes the basic operation of a solenoid valve to the extent required to understand the remainder of the paper
- Explains the concept of "stiction"
- Describes the effects of stiction on solenoid valve operation
- Presents evidence that stiction increases to some maximum level over relatively short periods of time
- Proposes a means to effectively reduce or eliminate the increased stiction through appropriately timed valve stroke tests
- Describes how to assess the impact of valve stroke testing on PFDavg(t)
- Details "best practices" guidelines for implementing solenoid valve stroke testing
- Provides a quantitative example illustrating improvement of reliability and safety performance when stroke testing is implemented

2. Solenoid Valve Basics

A solenoid coil generates an electromagnetic field which can be used to change the position of a metallic plunger. In a solenoid valve, this action can then be used either directly or indirectly to control process fluid flow. Figure 2 illustrates direct control of process fluid flow. In Figure 2a, when an electric current passes through the solenoid coil the resulting magnetic field attracts the plunger and holds it within the coil allowing the process fluid to flow from left to right. In Figure 2b, when the electric current is removed from the coil, the electromagnetic field vanishes and the plunger is no longer held within the coil but falls blocking the path of the process fluid flow.

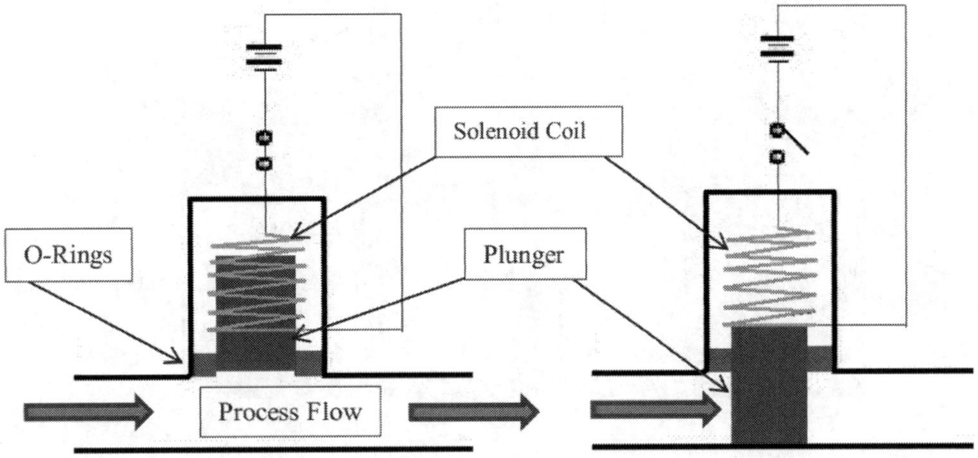

Figure 2a. Open solenoid valve | Figure 2b. Closed solenoid valve

Figure 2. Cross section of an open and closed soleniod valve, showing how plunger movement directly controls process fluid flow.

An example of using a solenoid coil for indirect *process* fluid control is a solenoid actuated pneumatic spool valve (see Figure 3). The solenoid valve typically controls compressed air which, in turn, positions another valve (not shown) which directly controls the process fluid. Thus the solenoid valve *indirectly* controls the process fluid.

Figure 3a shows the normal position of the valve for a particular application. In this coil design (which is different from that of Figure 2), when the solenoid coil is energized the magnetic field repels the plunger which is of a more complex design known as a spool. The spool, repelled by the magnetic field from the solenoid, is moved from left to right, changing the pathways allowing the flow of compressed air fluid. Consequently, when the solenoid coil is energized, the spool compresses the spring at the right end of the spool valve. In this spool position, ports x and y of the valve provide a pathway allowing compressed air to flow through the spool valve.

Figure 3b shows the emergency position of the valve in the event that an emergency shutdown of the process fluid is required. Now the solenoid coil is de-energized; the removal of the magnetic field means that the spool is no longer being repelled by the magnetic field. Without this magnetic force to push the spool to the right, the spring decompresses pushing the spool to the left, back inside the solenoid coil. In this spool position, port x is blocked which prevents compressed air from flowing through the spool valve to pressurize a valve actuator while ports y and z are open to provide a vent for pressure in the actuator.

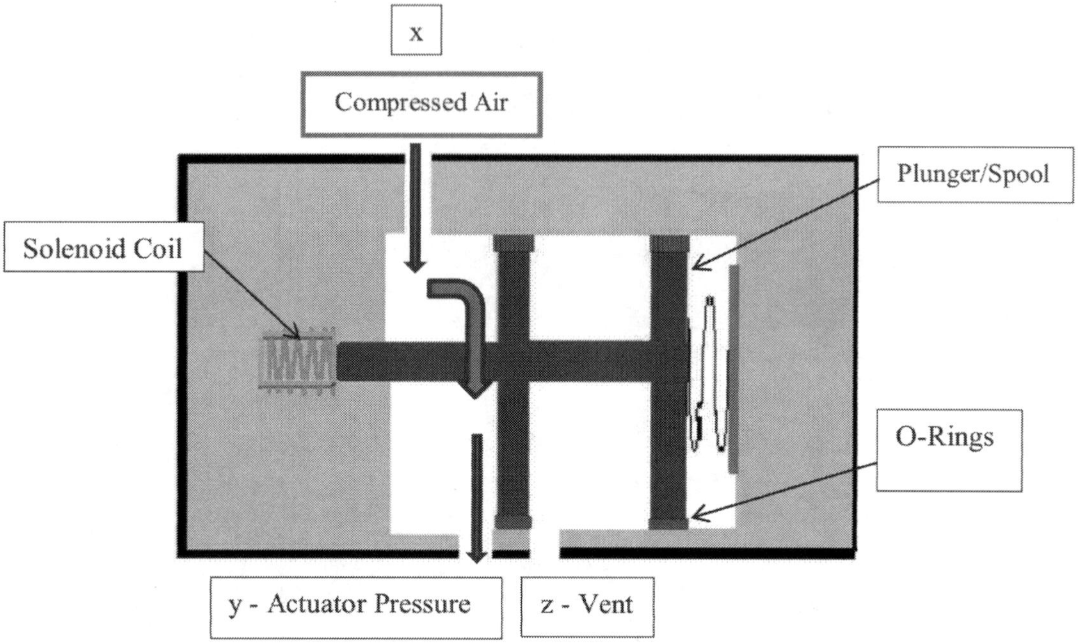

Figure 3a. Spool valve showing an open pathway from ports x to y for compressed air flow due to the soleniod being energized and compressing the spring.

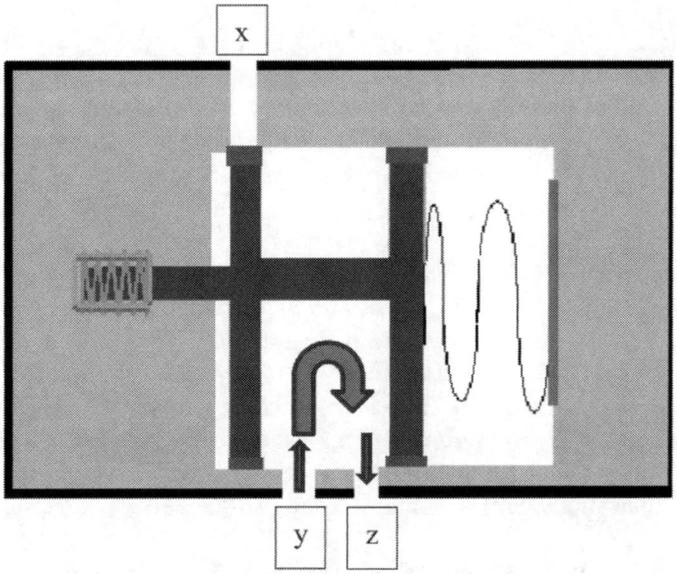

Figure 3b. Spool valve showing a closed pathway from ports x to y, blocking the flow of compressed air and venting air pressure by opening ports y to z. The soleniod is de-energized and the spring is not compressed.

Note the presence of O-ring seals (shown in blue) in both the simple solenoid valve (Figure 1) and the spool valve (Figure 2). They are in place to ensure that the fluid does not escape the valve chambers while also ensuring that no foreign elements can enter the fluid's environment. The O-rings surround both the plunger (Figure 1) and the spool (Figure 2) on the surfaces where the plunger or spool are in very close proximity (but not actually touching) the valve chamber walls.

While this introduction captures the main ideas needed to understand the remainder of this paper, the reader should be aware that actual valve designs are more complicated than presented here. Some use fluid pressures rather than springs to move the spool. Many other design differences exist. But all contain O-rings which are involved in a particular failure mode known to significantly affect solenoid valve reliability and safety performance. This failure mode is caused by excessive "stiction".

3. Stiction

3.1 What is Stiction?

Many studies and organizations have defined stiction, i.e., static friction, in different ways. A few examples of the various definitions are as follows:

- According to Olsson [5], stiction is defined as *"short for static friction as opposed to dynamic friction. It describes the friction force at rest. Static friction counteracts external forces below a certain level and thus keeps an object from moving."*

- According to Entech [6], *"stiction is a tendency to stick-slip due to high static friction. The phenomenon causes a limited resolution of the resulting control valve motion. ISA terminology has not settled on a suitable term yet. Stick-slip is the tendency of a control valve to stick while at rest and to suddenly slip after force has been applied."*

- According to Horch [7], *"The control valve is stuck in a certain position due to high static friction. The (integrating) controller then increases the set point to the valve until the static friction can be overcome. Then the valve breaks off and moves to a new position (slip phase) where it sticks again. The new position is usually on the other side of the desired set point such that the process starts in the opposite direction again."* This is an extreme case of stiction. On the contrary, once the valve overcomes stiction, it might travel smoothly for some time and then stick again when the velocity of the valve is close to zero.

- According to Ruel [8], the Instrument Society of America (ISA)(ANSI/ISA-S51.1-1979) described *"stiction as the resistance to the start of motion, usually measured as the difference between the driving values required to overcome static friction upscale and downscale."* The definition was first proposed in 1963 in American National Standard C85.1-1963.

- According to Ruel [8], *"stiction is a combination of the words stick and friction, created to emphasize the difference between static and dynamic friction. Stiction exists when the static (starting) friction exceeds the dynamic (moving) friction inside the valve. Stiction describes the valve's stem (or shaft) sticking when small changes are attempted."*

In the definitions above, all agree that stiction is the act of being "stuck" by static friction which prevents one surface from moving against another. Further, if the external force becomes greater than the static friction, the stiction between two surfaces will be overcome and the object will begin to move again [9].

Stiction between two surfaces can be the result of many factors including: corrosion, cold welding, i.e., metal on metal adhesion, breakdown of the lubrication boundary layer, change in the viscosity of the lubrication, build-up of deposits, contamination or foreign material on the contact surfaces, chemical reactions between the metal or lubrication, deterioration and breakdown of the valve's sealing components, etc.

This paper defines stiction as the resistance to the start of motion usually measured as the difference between the external force being applied in order to overcome the static friction and the force to maintain movement between the two contacting or working surfaces.

3.2 Stiction in the Context of a Solenoid Valve

In normal operation, the O-rings in the solenoid valve create a smooth transition while the plunger or spool is in motion. Because the O-rings are in direct contact with the valve chamber walls, in order for the plunger/spool to begin to move, it must first overcome the stiction between the O-rings and the chamber walls. Once in motion, the plunger/spool must overcome the sliding or dynamic friction between the O-rings and walls but, since the dynamic friction is usually significantly less than the nominal level of stiction, it is generally not of concern. By design the magnetic and spring forces in the solenoid valve are sufficient to overcome the *nominal* level of static friction and, depending on the design specifics of the solenoid valve, generally have excess force sufficient to overcome about 2 to 2.5 times nominal stiction. Once the stiction is overcome, the plunger/spool will continue to move. However, a common and troublesome problem can occur when the valve is at rest, i.e., stationary, for a length of time because the stiction between the O-ring and the valve chamber walls increases over time from its nominal level until it reaches some maximum value. It is possible that the increases in stiction may reach a level where the forces generated by the solenoid coil and/or spring (in the case of a spool valve) are no longer sufficient to overcome the increased stiction. Consequently, when the valve is called upon to close, it is unable to do so. However, if the excessive stiction is overcome, stiction reverts to its original nominal level though it does begin to increase again once the valve stops moving.

3.3 Evidence that Stiction Increases with the Length of Time the Contact Surfaces are Stationary

3.3.1 Expert Knowledge

Mechanical engineers and technicians who routinely work with solenoid valves are familiar with the experience of trying to "stroke", i.e., move through its range of motion, a valve that has been stationary for an extended time. They report that it is not uncommon for extra force to be required to move the valve after it has been stationary for a month or more. However, they are unable to say when within the month the increased stiction reaches a critical level that causes the valve to stick because it cannot generate sufficient force to overcome the increased stiction.

3.3.2 O-Ring Manufacturer's Findings

In the Parker O-ring Handbook [10] it is noted that the "coefficient of starting friction", i.e., stiction, increases when the O-ring has been stationary for between 1 week and 1 month after which time the stiction plateaus. One specific graph (Figure 5-7 in [10]) of "break out friction", i.e., stiction, vs. "delay between cycles" shows the stiction force plateauing at approximately 300 hours. "The theory has been proposed and generally accepted that the increase in friction on standing is caused by the rubber O-ring flowing into the micro-fine grooves or surface irregularities of the mating part"[10].

3.3.3 Experimental Study to Determine Equilibrium Rest Time

Zhao and Bhushan [11] conducted an experimental study on the effects of lubrication thickness, lubrication viscosity, and the length of rest time, i.e., the length of time the contact surfaces remain stationary, on the increases in stiction in magnetic thin-filmed disks. Based on their results they introduced the concept of equilibrium rest time, i.e., the time by which the maximum stiction is reached due to lack of motion. After conducting experiments with three different lubricants and holding the disks in a stationary position over extended periods of time, they found that a maximum stiction is reached by approximately 275 hours or about 11.6 days. The equilibrium rest time for the different lubricants are reached at slightly different points in time, but all are reached before about 275 hours [11]. Even though these experiments do not involve O-rings, it is interesting that a maximum level of stiction is reached and results regarding the time required to reach the stiction plateau are consistent with the information available in [10].

3.3.4 Reliability Declaration Per ISO 13849

In a declaration [12] regarding the reliability indicators and information for use of a specific solenoid actuated pneumatic valve with respect to the safety standard EN ISO 13849-1 [13], it is specifically stated that the "valve must be operated at least once per week or once per shift to insure the intended function." The data supporting this declaration was obtained by cycle testing during which the solenoid valve was never at rest for any significant time. This statement recognizes that the results of cycle testing are valid only if periodic valve movement is maintained. In the absence of such periodic movement, the failure rates derived from the results of cycle testing [14] cannot be considered valid when applied to components in applications where these components spend significant times at rest.

3.3.5 Section Summary

From this multi-source evidence we can conclude that, as a rule of thumb, O-ring stiction increases with valve inactivity and reaches a maximum value at approximately 275 hours. Further, stiction levels can be returned to nominal or near nominal levels if the increased stiction can be overcome by valve motion. This suggests that increased stiction can be dealt with in a more cost effective way than proof testing and that solenoid valve reliability and safety performance can be significantly improved.

4. Solenoid Valve Stroke Testing

In a stroke test, the solenoid coil is de-energized for a short period of time to see if the spring can fully decompress moving the spool in one direction through its full range of motion. Then when the coil is re-energized, the spool must again move, in the opposite direction, through its full range of motion. The reader may be familiar with the term "partial stroke testing" which refers to moving a valve through only a small portion of its range of motion to establish that it can be moved. In the case of solenoid valves, a partial stroke test is not possible. All stroke testing of solenoid valves will involve a full stroke.

If the solenoid valve is *directly* controlling process fluid, the stroke testing could lead to an unacceptable disruption of process fluid, so in the case of *direct* process fluid control, in-service solenoid valve stroke testing is impractical and without special redundant architectures, must be done when the process is shut down. However, if a solenoid valve is *indirectly* controlling process fluid, a full stroke of the solenoid valve (provided that the full stroke takes place over a short enough period of time) will generally not cause a significant disruption of the process fluid flow but will briefly alter the position of the actuator/valve assembly that the solenoid is controlling. This may be negligible or may result in a small perturbation to the process fluid flow that in most cases will be acceptable. At the completion of the valve stroke test, which is performed with the valve in service, either the solenoid valve has been verified to be operable or a dangerous failure due to stiction has been detected.

The success or failure of the solenoid valve stroke test can be determined indirectly by observing the actuator pressure and/or actuator valve stem movement because a full stroke test of the solenoid valve also performs a partial stroke test of the ESD. A small decrease in actuator pressure with corresponding actuator valve stem movement indicates a successful solenoid valve test *as well as* a successful ESD test. No decrease in pressure indicates a failure somewhere in the ESD but not necessarily of the solenoid valve itself. Some automatic stroke test devices measure the actuator pressure and actuator stem movement then automatically indicate pass/fail via a discrete output. Stroke testing the solenoid valve provides reliability and safety improvements not only to that device but also to the entire ESD as well.

4.1 Assessing the Reliability & Safety Improvements Due to Stroke Testing

This concept of solenoid valve stroke testing offers an opportunity to eliminate or at least address the failures of solenoid valves due to stiction as well as to address detectable non-stiction failures and, consequently, to improve solenoid valve reliability and safety performance. To aid in clearly understanding the beneficial reliability and safety effects of stroke testing, we note that the dangerous failures will be due either to the stiction failure mode or some non-stiction failure modes. The failures due to stiction are all detectable via valve stroke testing. The non-stiction failures are further divided into those failure modes which are detectable and those which are undetectable by valve stroke testing. We define the failure rates associated with these three cases as $\lambda_{Dstiction}$, $\lambda_{Dnon\text{-}stictionDetectable}$, and $\lambda_{Dnon\text{-}stictionUndetectable}$, respectively.

4.1.1 Case 1: No Valve Stroke Testing

In the case where no valve stroke testing is performed, the failure rates of all dangerous failure modes are included in λ_D, i.e.,

$$\lambda_D = \lambda_{Dstiction} + \lambda_{Dnon\text{-}stictionDetectable} + \lambda_{Dnon\text{-}stictionUndetectable} \qquad \text{(Eq. 3)}$$

and (Eq. 3) is substituted into (Eq. 2a) to compute PFDavg(t) at $t = T_P$. Provided all dangerous failures are discovered and repaired as a result of the manual proof test, the resulting plot of PFDavg(t) will be a saw-tooth plot where PFDavg(t) returns to 0 after each proof test.

4.1.2 Case 2: Infrequent Valve Stroke Testing

By "infrequent valve stroke testing" we mean testing that occurs less frequently than once per week. Typically infrequent testing occurs once every three to six months. As with Case 1, the failure rates of all dangerous failure modes are included in λ_D so that (Eq. 3) again applies. However, there are some noteworthy differences between Case 2 and Case 1.

In Case 2, we assume that any dangerous failures detected as a result of the valve stroke test are promptly and completely repaired. Indeed, if prompt repair (within 168 hours) is not made, the value in performing the stroke test is reduced. Thus, Case 2 has nearly the same effect as a manual proof test, albeit a slightly incomplete one. It will detect all but the failures associated with the failure rate $\lambda_{Dnon\text{-}stictionUndetectable}$. Thus, PFDavg(t) is governed by (Eq. 2a) expect that t now equals T_{VST}, the interval of the valve stroke test, and after each interval of length T_{VST} PFDavg(t) does not return to zero but is reduced to a small residual value (due to $\lambda_{Dnon\text{-}stictionUndetectable}$) which will not be eliminated until the time of a complete manual proof test. Thus, the plot of PFDavg(t) for this case will resemble a saw-tooth plot as is shown in the quantitative example in Section 6 but one whose peaks increase slightly each cycle until the time of manual proof test. [Note: for the example parameter values used in Section 6, this increase in peak values is too small to be seen in the plots.]. So, infrequent valve stroke testing with prompt and complete repair of detected failures will decrease PFDavg. As a result, it may be possible to increase T_P without negatively impacting the safety performance of the solenoid valve.

4.1.3 Case 3: Frequent Valve Stroke Testing

By "frequent valve stroke testing" we mean in-service testing that occurs at least once per week and includes complete and correct repairs of all detected failures within 168 hours. Since the stroke testing is performed before the increase in stiction reaches its maximum, there is a significant chance that the stroke test will succeed in moving the valve **and** disrupt the buildup of stiction returning it to its nominal level. Thus, while the valve stroke test is designed to detect stiction-related failures there is a **significant chance that frequent valve stroke testing will prevent the failures** due to this failure mode from occurring effectively setting $\lambda_{Dstiction}$ to 0. Therefore, for Case 3 only the failure rates associated with non-stiction failure modes are included in λ_D, i.e.,

$$\lambda_D = \lambda_{Dnon\text{-}stictionDetectable} + \lambda_{Dnon\text{-}stictionUndetectable} \qquad \text{(Eq. 4)}$$

and (Eq. 4) is substituted into (Eq. 2a) to compute PFDavg(t) at $t = T_{VST}$ which in this case is 168 hours or less. Thus frequent valve stroke testing clearly decreases PFDavg. As a result, it may be possible to increase T_P without negatively impacting the safety performance of the solenoid valve.

4.2 Setting the Frequency of the Valve Stroke Test

While valve stroke testing will have beneficial results on PFDavg even if performed infrequently, **ideally we would like not simply to detect failures due to stiction but also to actually prevent them**. So, is stroke testing once per week sufficiently frequent to insure that stiction build up will be disrupted before it exceeds the valve's available excess force? Unlike random failures truly represented by a constant failure rate (meaning that failures occur with equal probability at any time during the interval between diagnostic tests), failures due to stiction are unlikely to occur very early in the time interval between tests when stiction levels have not increased very much and more likely to occur once stiction levels have increased beyond the level that the excess force of the valve can overcome. Thus, while we know that, as a rule of thumb, the maximum level of stiction is reached at about 275 hours, we also need to consider how *fast* the stiction builds towards its maximum value in order to determine the most appropriate interval for valve stroke testing.

In [11], plots of $\log_{10}$ stiction force vs. $\log_{10}$ rest time (time of inactivity) are given for the three lubricants studied. These *log-log plots* show an approximately linear increase in stiction from a nominal stiction level until a plateau is reached at about 275 hours. This indicates that the stiction increases as a power function meaning the increase is proportional to t^n. Thus the increase in stiction force over time would be modeled as

$$\text{Increase in stiction force} = \alpha\, t^n \qquad 0 \leq t \leq t_{max} \approx 275 \text{ hrs} \qquad \text{(Eq.5)}$$

where α is a constant of proportionality. Now, for the purposes of this example let us assume that the maximum stiction is three times nominal stiction. This implies that the maximum *increase* in stiction force at 275 hours would be twice nominal stiction which when added to the initial nominal stiction would give a total stiction force of about three times nominal stiction. For the lubricants studied, $0.07 \leq n \leq 0.17$. Thus stiction can increase *very* rapidly to its plateau value. For example, if $n = 0.17$, then at one week (168 hours) the stiction would increase by almost 92% of its maximum possible increase making the actual stiction force equal to nominal + (0.92 * 2 * nominal) for a total stiction of 2.85 nominal value.

If these example values applied to the solenoid valve, would it have enough excess force to overcome this amount of stiction? Possibly, but maybe not. However, our problem is that the experiments described in [11] were not on solenoid valves and we do not know what values of *n* might apply to any given solenoid valve or even it the solenoid valve O-ring stiction rises as a power function.

4.3 Considering an Alternative to Valve Stroke Testing

In some end-user circles there persists the common belief that valve stroke testing risks a plant shutdown due to a false trip and that a better safety solution is to provide a much higher safety margin (greater excess force) to overcome increases in stiction. However, those who are well versed in solenoid design realize that a higher safety margin to overcome stiction actually results in a *greater* false trip rate. This occurs because a higher safety margin to overcome increases in stiction requires a bigger spring. A bigger spring requires a bigger coil which, in turn, requires more energy to power the coil. These increased energy requirements result in a higher likelihood that the coil burns out which results in a higher false trip rate. Therefore, if we seek to optimize the false trip rate while improving reliability and safety, then valve stroke testing provides a good solution.

5. Recommended "Best Practices" for Implementing Solenoid Valve Stroke Testing Based on Current Knowledge (March 2013)

Based on currently available information consistent over multiple sources, i.e., that stiction reaches a maximum value in about 275 hours, we recommend as best practice, the following approach to solenoid valve stroke testing:

- Begin solenoid valve stroke testing on a weekly basis recording whether each test verified valve movement or indicated the valve was stuck (in which case, initiate prompt repairs)
- After a few months of testing, if the number of times the valve is found stuck seems excessive (by engineering judgment) then halve the test interval (to 3-4 days)
- Repeat the above procedures with this new test interval until an optimum test interval is identified such that stroke testing generally verifies valve movement.

In some instances, the initial one week test interval may not be adequate. Because low power (less than 2 Watts) solenoid valves generally are designed with less excess force available, they may not develop adequate force to overcome increases in stiction that may build up even before the one week test interval. In these cases, a shorter initial stroke test interval may be warranted.

To avoid false trips, the partial stroke testing device which initiates the full stroke of the solenoid valve must be set up properly:

- the pressure drop in the actuator should not be allowed to go to zero during the stroke test. Many designs limit the pressure drop to a minimum value while still fully exercising the solenoid valve.
- the minimum actuator pressure should be established to overcome deadband yet provide minimum movement of the valve controlling the process fluid. Many valves allow a small movement without imposing a flow restriction. A small movement is all that is necessary.
- "quick exhaust valves" should not be used as these devices are non-linear and often lock in the pressure exhaust causing a false trip. Instead, linear pneumatic boosters should be used.

6. Assessing the Reliability & Safety Improvements Due to Stroke Testing

In order to present a realistic quantitative example of the reliability and safety performance improvements that can result from valve stroke testing, a failure mode effects and diagnostic analysis [15] was performed on a solenoid valve from manufacturer X, model Y. The predicted values for the various constant failure rates are given in Table 2.

Table 2. Values of failure rates from the solenoid valve (manufacturer X, model Y) used in computing PFDavg.

Parameter	Failures/10^9 hrs operation
$\lambda_{Dstiction}$	103.7
$\lambda_{Dnon\text{-}stictionDetectable}$	84.3
$\lambda_{Dnon\text{-}stictionUndetectable}$	1.9

Finally, consider the behavior of PFDavg(t) due to the solenoid valve for each of the three cases. For our example parameter values, PFDavg(t) was computed for Case 1 with proof testing performed every 2 years (T_P = 2 years), for Case 2 with valve stroke testing performed once every six months (T_{VST} = 6 months) and for Case 3 with valve stroke testing performed once per week (T_{VST} = 168 hours). The results are plotted in Figure 4. Figure 5 illustrates the plot of PFDavg(t) for Case 3 magnified to show its detail. Note that in Figure 4 the purple horizontal line represents the maximally allowed value of PFDavg for which the valve could receive a SIL 3 rating. Without valve stroke testing, the valve can only be rated at SIL 2. With valve stroke testing every six months or once a week the PFDavg(t) remains below the SIL 3 rating level without proof testing. Note that this does not mean that proof testing can be completely eliminated as other dangerous undetected failures can exist that need to be discovered and corrected through proof testing. It does mean that it may be possible to increase the proof test interval significantly while maintaining a SIL 3 rating.

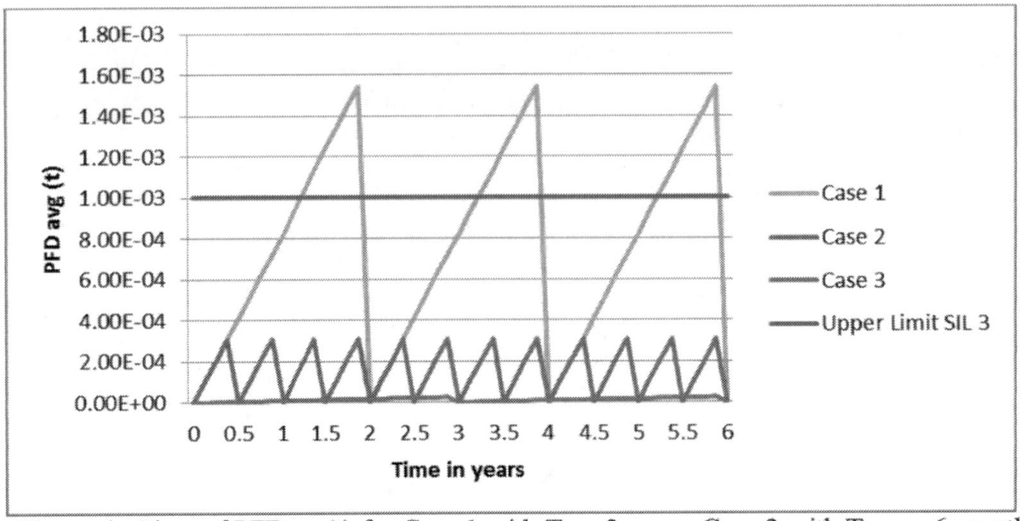

Figure 4. Plots of PFDavg(t) for Case 1 with $T_P = 3$ years, Case 2 with $T_{VST} = 6$ months and Case 3 with $T_{VST} = 1$ week.

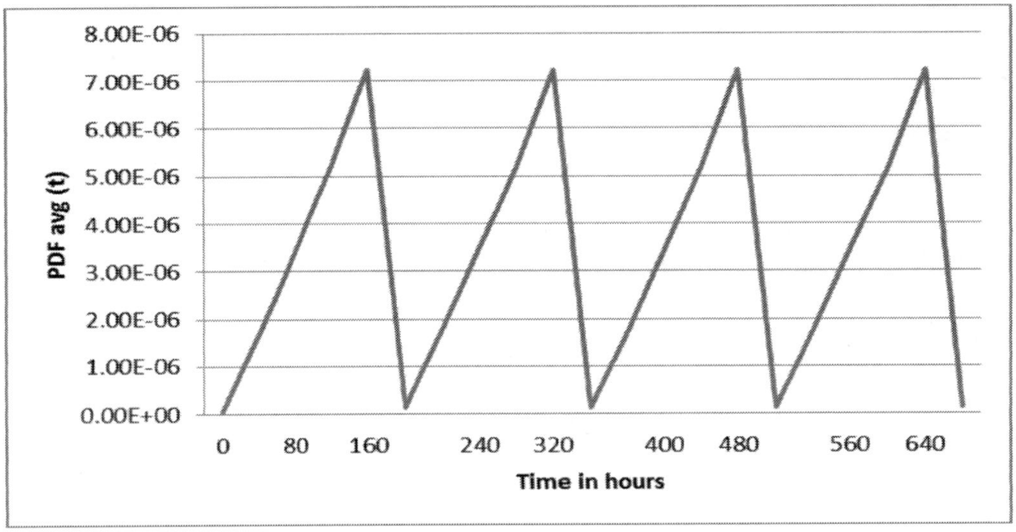

Figure 5. PDF_{avg} for the example solenoid valve as a function of time, in hours showing a weekly stroke test (every 168 hours).

7. Conclusions and Recommendations

Even infrequent solenoid valve stroke testing can have significant beneficial impacts with respect to improvements in reliability and safety performance. However, if a solenoid valve stroke test at a frequency of once per week (or more) is feasible within a given application, it should be implemented per the best practices strategy given above.

Clearly, the timing of valve stroke tests could be optimized if we had better models of how stiction increases with valve inactivity. Manufacturers may wish to undertake controlled tests and data collection on different valve designs within their product lines to better understand this stiction phenomenon. This would permit them to make recommendations regarding valve stroke testing that would improve reliability and safety performance and possibly increase the time intervals between required proof testing.

8. Acknowledgement

The authors would like to acknowledge the helpful discussions with and comments from Steven Close and Hal Thomas both of whom are with exida.

9. References

[1] IEC 61508 (1997), *Functional safety of electrical/electronic/programmable electronic safety-related systems,* Geneva: International Electrotechnical Commission.

[2] IEC 61511 (2003), *Functional safety – safety instrumented systems for the process industry,* Geneva: International Electrotechnical Commission.

[3] W. M. Goble and J. V. Bukowski, "Validation of a Mechanical Component Failure Database," *Proceedings of the 2009 Annual Reliability and Maintainability Symposium*, Fort Worth, TX, January 2009.

[4] *Electrical and Mechanical Component Reliability Handbook*, 3rd Ed, exida, Sellersville, PA, 2012.

[5] H. Olsson(1996), Control systems with friction, PhD thesis, Lund Institute of Technology, Sweden.

[6] EnTech (1998), *EnTech Control Valve Dynamic Specification (version 3.0).*

[7] A. Horch (2000), Condition Monitoring of Control Loops, PhD thesis, Royal Institute of Technology. Stockholm, Sweden.

[8] M. Ruel (2000), *Stiction:* The hidden menace, *Control Magazine.* http://www.expertune.com/articles/RuelNov2000/stiction.html.

[9] S. Choudhury, N. F. Thornhill, S. L. Shah, A Data-Driven Model for Valve Stiction, *ADCHEM 2003*, Jan. 11-14, 2004, Hong Kong.

[10] Parker Hannifin Corporation, *Parker O-ring Handbook: ORD 5700,* 2007.

[11] Z. Zhao and B. Bhushan, Effect of Lubrication Thickness and Viscosity and Rest Time on Long-Term Stiction in Magnetic Thin-Film Rigid Disks, *IEEE Transactions on Magnetics*, Vol 34. No 4, July 1998.

[12] Rexroth Bosch Group, *Declaration: Reliability indicators and informations for use with respect to the utilization of EN ISO 13849-1:2008-12*, 2009.

[13] ISO 13849-1 (2009), *Safety-related parts of control systems -- Part 1: General principles for design*, Geneva: International Standards Organization.

[14] TÜV Rheinland Report No. V372 2010 S1, TÜV Rheinland Energie and Umwelt GmbH, 2011

[15] W. M. Goble and A. C. Brombacher, "Using a Failure Modes, Effects and Diagnostic Analysis (FMEDA) to Measure Diagnostic Coverage in Programmable Elecronic Systems," *Reliability Engineering and System Safety*, Vol. 66, No. 2, November 1999.

10. Abbreviations and Terms

ESD – Emergency Shutdown

PFD – Probability of Failure on Demand

PFDavg – Average of the Probability of Failure on Demand on the time interval [0, t]

PFDavg(t) – Average of the Probability of Failure on Demand at time t

SIL – Safety Integrity Level

T_P – Time interval between proof tests

T_{VST} – Time interval between valve stroke tests

λ_D – Constant failure rate for dangerous failures

$\lambda_{Dstiction}$ – Constant failure rate for dangerous failures due only to stiction

$\lambda_{Dnon\text{-}stictionDetectable}$ - Constant failure rate for dangerous detectable failures due to causes other than stiction

$\lambda_{Dnon\text{-}stictionUndetectable}$ – Constant failure rate for dangerous undetectable failures due to causes other than stiction

IMPROVED BOILER SYSTEM OPERATION WITH REAL-TIME CHEMICAL CONTROL

Paper No. 295944

Deborah Bloom, Principal Consultant, NALCO, An Ecolab Company, Naperville, IL
David Puchan, Senior Industry Technical Consultant, NALCO, An Ecolab Company, Sugar Land, TX

ABSTRACT

The steam boiler system is a critical component of most manufacturing processes. Steam production reliability is often a key component in assuring product quality and overall production efficiency. Hourly steam load demands and varying system stresses make responsive water treatment of the boiler system difficult. A challenging production environment is made even more so by the volatile economic forces in today's world.

New technologies have been developed that provide real time monitor & control capabilities to help steam operations staff achieve more consistent, proactive boiler treatment by detecting system variability, determining the correct chemical or operational action, and delivering measurable environmental return on investment. These technologies are capable of communicating via web or DCS for monitoring and alarming existing conditions.

The new technologies will be described and case histories presented.

INTRODUCTION

Maintaining reliable and efficient steam operations in most manufacturing facilities requires good corrosion and scale control in the pre-boiler and boiler water system. Chemical treatments must be monitored and controlled within recommended ranges, an achievement that is often difficult with traditional methods. New state of the art systems provide real time monitor and control capabilities to proactively respond to system stresses and maintain tight control – increasing system reliability and asset protection, as well as typically providing chemical, water and energy savings.

As the U.S. marketplace has become more competitive, industry's environmental awareness has also increased. The competitiveness of the marketplace has resulted in a drive to extend equipment life (equipment such as a boiler system), reduce fuel and water costs, and optimize operational labor costs. The increased environmental awareness has resulted in major corporate initiatives to reduce greenhouse gas emissions, and fuel and water consumption.

SCALE AND CORROSION

There are many water-related challenges encountered when operating a boiler. Two of the most common problems are scale and corrosion.

Scale is the result of dissolved minerals, such as calcium and magnesium, exceeding their solubilities and forming deposits (1). Boiler heat transfer efficiency can be negatively impacted by scale. In addition scale can also create serious under-deposit corrosion problems. Scaling conditions, when not addressed in a timely fashion or when improperly dealt with, can result in ruptured boiler tubes or other boiler damage. This can cause boiler shutdown and lost or greatly reduced production.

For presentation at the AIChE 2013 Spring Meeting; San Antonio, TX April 28-May 2, 2013

Some of the more common ways to prevent scale from occurring in boilers include ion exchange (softening or demineralization) and reverse osmosis. Residual dissolved minerals are prevented from forming scale by the addition of chemical treatment.

Corrosion is a serious concern in the pre-boiler and boiler system for two reasons. First, it shortens the asset life of the boiler and pre-boiler system. Second, the corrosion by-products resulting from unchecked corrosion can deposit in high heat transfer areas of the boiler as scale. Oxygen corrosion is a specific and particularly devastating type of corrosion (Figure 1). It is a common problem that adversely affects the life of the key system components. Pitting attack from oxygen corrosion can cause premature metal failure.

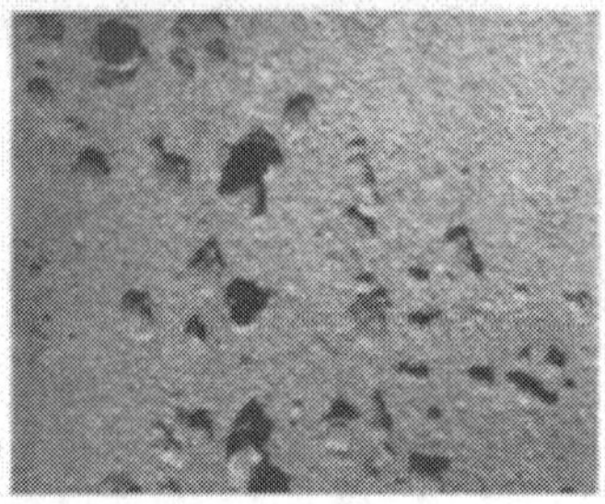

Figure 1. Oxygen attack, characterized by its classic pitting appearance

Iron oxide scale, just as the earlier described calcium and magnesium scales, reduces heat transfer efficiency. In uncontrolled scenarios, blown tubes (or other extensive boiler damage), boiler shutdowns and production losses can result.

Historically, the best practice for preventing oxygen corrosion in boiler and pre-boiler systems has been mechanical deaeration for primary oxygen removal, followed by chemical oxygen scavengers to remove remaining trace quantities.

For decades, the traditional practice employed to control many boiler water scale and oxygen control chemistries has been by sampling the blowdown. One primary reason for this was that the boiler concentrated the treatment chemicals sufficiently to enable them to be measured readily. Although modernization and improvement in test methods have enabled feedwater testing (e.g., cycling up is not required to run many of the tests), many operations hold onto the old practices of the 1930s through 1970s and test only in the blowdown. The major drawbacks resulting from this approach are related to the lag time between changes in feedwater conditions and those changes showing up in the blowdown.

In a blowdown based control scenario, the operator would sample the blowdown, perform an array of chemical tests, and adjust the feed rate of the scale inhibitor or the oxygen scavenger. This constant "add chemical - sample – test – adjust" control loop was the standard. This meant that if a test was missed, an opportunity for adjustment was also missed.

In addition to the challenge of the above testing loop, the physical capacity of the boiler often meant that a negative change in feedwater conditions might go completely undetected or not have a notable impact on the boiler blowdown composition for hours or even days. Feedwater conditions that were potentially adverse to system integrity (e.g., an oxygen excursion, a reduction in feedwater pH, or a badly contaminated condensate return stream) would only be seen after they had become history, if they were detected at all, because only the boiler water was being tested. What was missing was a proactive approach to protect the boiler as opposed to a reactive approach that adjusts after damage has occurred.

Ideally, one would want to measure and control boiler chemistry in the feedwater, because that would enable one to make adjustments to changing conditions in real-time, rather than chasing events or missing them entirely and suffering the outcome.

NEW CONTROL TECHNOLOGY

The new boiler feedwater treatment automation system is a combination of two different Nalco technologies. First is the patented solid state fluorometer, which is used to control boiler internal treatment (2, 3, 4). The second is the Nalco Corrosion Stress Monitor™ technology. This instrument works by measuring the oxidation reduction potential at the temperature and pressure of the boiler feedwater. The At Temperature Oxidation Reduction Potential, or AT ORP, detects changes in the corrosivity of the feedwater and can be used to control the oxygen scavenger feed. The AT ORP has a design rating up to 500°F and 2,800 psig. A uniquely designed controller controls both the fluorometer and AT ORP. Monitoring, On / Off Control, and PID (Proportional-Integral-Derivative) Control are all possible with this controller. These technologies are capable of communicating via web or DCS for monitoring and alarming existing conditions.

Scale Control

The use of fluorescent tracers for system diagnostics and dosage control of treatment chemicals is a technology that has been in practice for many years. This method involves adding an inert fluorescent molecule to a water treatment product in a known proportion to the active ingredients. The fluorescent molecules are inert and are unchanged by the conditions in the water being treated. They are also easily detectable in the water with an instrument called a fluorometer, and measurement of the tracer gives a quick and highly accurate value for the amount of treatment that has been added to the system. Typically, tracer measurements are used to control the feed of the scale inhibitor treatment.

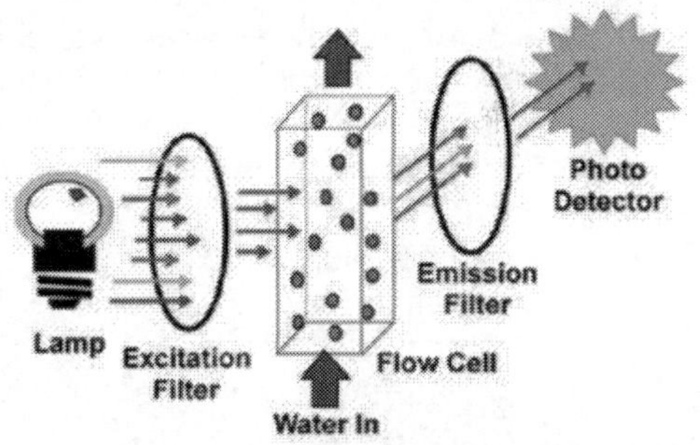

Figure 2. Principle of fluorescent measurement

Fluorometer

The fluorometer used with this technology can measure the presence of these fluorescent materials at the part per billion level of detection. Today's fluorometers are based on solid state electronic technology that makes these devices much simpler, more compact, less costly and more reliable, Figure 2.

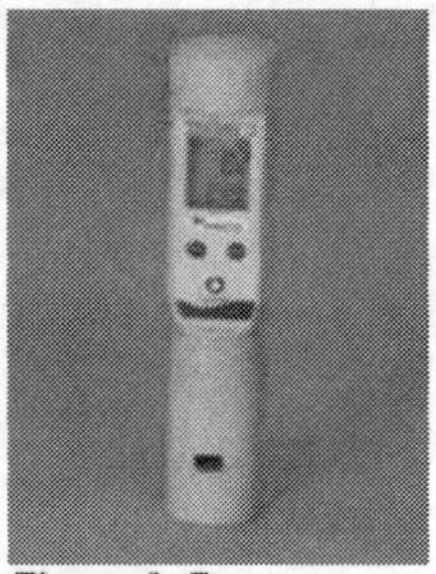

Figure 3. Pen fluorometer used with grab samples

The fluorometer enables us to monitor and control traced products either with a handheld measurement device using grab samples (Figure 3) or with on-line feed and control monitoring technology. In addition the fluorescent materials can be used by themselves as a system diagnostic tool for flow measurement, determination of boiler cycles of concentration, holding time index, boiler leak indication, or detection of boiler water carry over.

This unique tracer technology makes it possible to determine the mass balance of a boiler system with greater accuracy, Figure 4. It also makes it possible to accurately and automatically control and monitor system chemistry and operating conditions, respond to boiler system changes, and maintain optimum treatment levels of highly effective scale and deposit inhibitors to keep boilers clean. As an example, this technology has been used in conjunction with soluble iron (Fe^{+2}) monitoring in a boiler system to quantify flow-assisted corrosion rates.

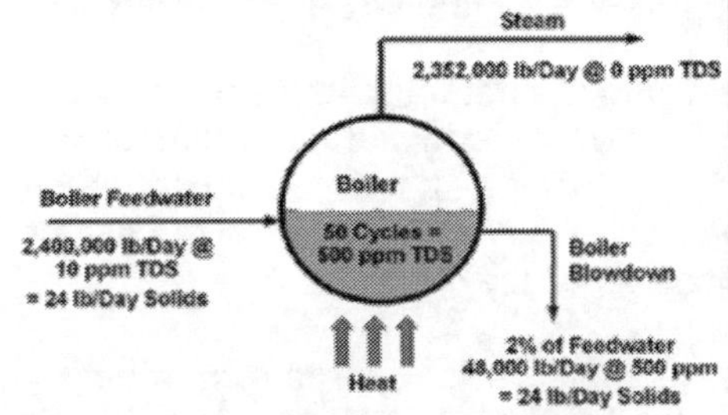

Figure 4. Boiler cycles of concentration

Corrosion Control

Corrosion of metals involves electrochemical reactions with charge transfer occurring at interfaces. The rate at which this transfer of charge occurs, or current, depends on the driving force, or voltage applied across the local system. Typically, as dissolved oxygen increases in a system, the ORP increases. When oxygen scavenger, or reductant, is added to the system, ORP decreases. The ORP reflects the composition of the system with respect to electroactive species. It is important to note the impact these species have on corrosion. While the ORP is not a direct measurement of corrosion potential, it is a useful indicator of the feedwater corrosivity (5).

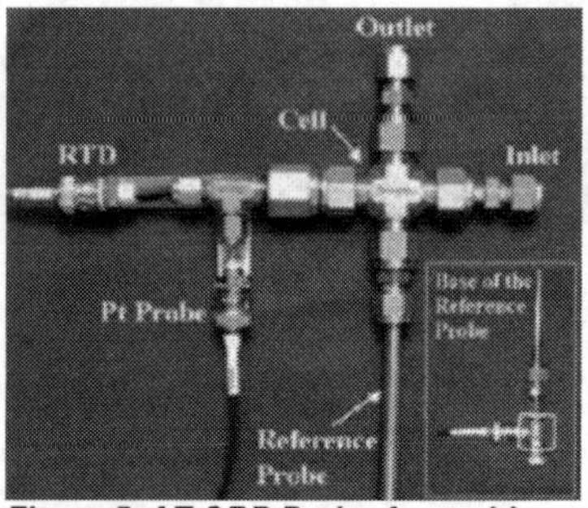

Figure 5. AT ORP Probe Assembly

At Temperature ORP

Figure 5 shows a commercial AT ORP probe. The probe assembly consists of a system water sample inlet and outlet, a connection for an integral RTD (resistance temperature detector) and platinum electrode (Pt Probe), and a separate external pressure-balanced reference electrode (Reference Probe). The reference electrode is filled with 0.1 N KCl in this case. All of the active portions of the three attached probes reside in close proximity within the cell itself.

The cell arrangement also consists of a high-pressure cell and high-pressure fittings that house the electrodes. The potential of the Pt Probe is measured against the external, pressure- balanced silver/silver chloride reference electrode (0.1 N KCl internal filling solution). The temperature of the water flowing through the cell is also recorded as a function of time.

There are a number of reasons why it is important to measure the feedwater ORP value at the boiler feedwater temperature and pressure:

- AT ORP is much more sensitive. It can see effects at high temperature that cannot be seen at room temperature.
- Corrosion rates vary with temperature and pressure. Corrosion is occurring at the operating temperature, so we need this correlation for proper corrosion control.
- AT ORP has a fast response time due to quick probe response coupled with sampling earlier in the stream.
- Scavengers/passivators work better at higher temperatures and pressures so control is more sensitive and realistic with AT ORP.

When comparing ORP readings, it is important to understand the differences between reference electrodes. Relative scales must also be standardized. Since both ORP and corrosion rates are very temperature dependent, AT ORP is a much more useful reading than Room Temperature ORP (RT ORP).

OPPORTUNITIES FOR ENERGY SAVINGS

Controlling boiler scale can have real energy savings benefits. There are many different sources of boiler scale – two of the most common are feedwater hardness and iron entering the boiler feedwater.

The new control technologies can have profound impact on minimizing the potential for scale, as well as mitigating its impact. Since the scale inhibitor is measured in the boiler feedwater and adjusted in real time, regardless of changes in steam load, underdosing and overdosing can be almost completely eliminated. Boiler feedwater corrosivity is likewise monitored and controlled in real-time, so feedwater corrosion excursions, and the resultant scale-forming iron can also be greatly reduced.

Other opportunities for energy savings spring from these technologies. It is important to remember that sulfite, one of the most common oxygen scavengers, contributes dissolved solids to the boiler feedwater. These solids increase boiler water conductivity. In a scenario where excessive levels of sulfite oxygen scavenger are being fed, improved control of oxygen scavenger feed can result in reduced boiler feedwater dissolved solids, higher cycles of concentration in the boiler, and reduced blowdown, all with associated water and fuel savings.

This technology also has the capability of monitoring and controlling feedwater pH, so energy and water savings from reduced feedwater dissolved solids, higher boiler cycles, and reduced blowdown can also be obtained by facilities that use caustic (sodium hydroxide) for controlling feedwater and boiler water alkalinity.

CASE HISTORY: GULF COAST REFINERY

A Gulf Coast refinery experienced several boiler tube failures on their 600 psig steam generators resulting in poor steam system reliability. The refinery management team wanted to improve the overall reliability and total system cost of its steam system. An Engineering Approach Analysis was done to understand system gaps and discover opportunities for improvement.

The 600 psig boiler system receives 100% sodium zeolite softened water. Operations ran samples to test for boiler feed water total hardness. The results are shown below in Figure 6.

Typical response to elevated boiler feed water hardness levels would include:

- Review of the sodium zeolite operation and corrective action where needed to bring the total hardness levels back into target
- Increased dosage of scale inhibitor
- Increased blowdown

Operations attempted to take corrective action with the tools and knowledge available.

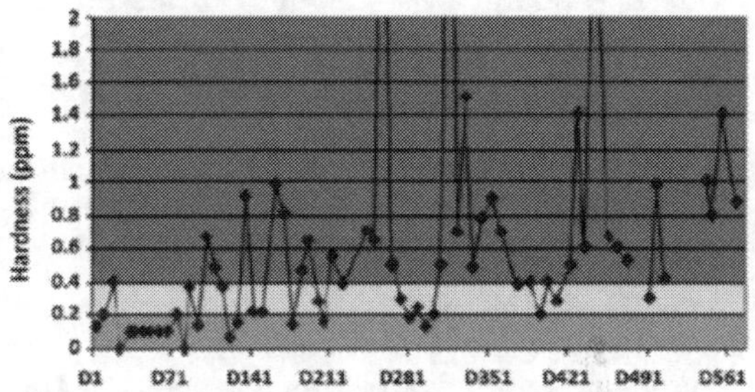

Figure 6. Boiler feed water hardness was in the target zone 45% of the time.

Boiler Cycles

Our analysis of the operational data found that boiler cycles were rather erratic and there was significant room for optimization as shown in Figure 7.

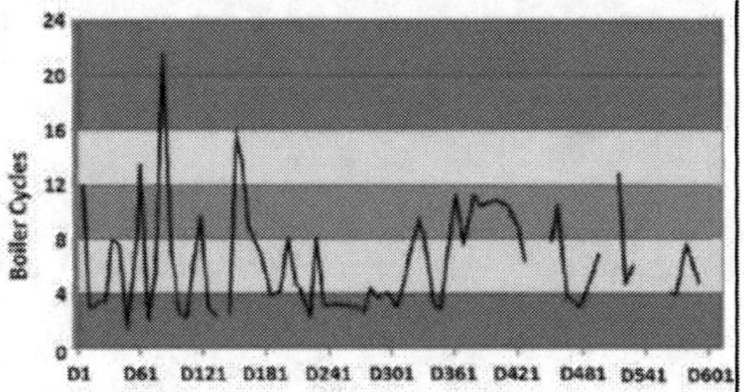

Figure 7. Boiler Cycles - In the Zone Analysis

Refinery Operations was taking daily measurements and attempting to make blowdown adjustments using a manual valve. Achieving good manual control on a highly variable system is very difficult.

Recommendations

After careful review of the customer's system, chemical treatment program, and daily operations using the Engineering Approach Analysis, several recommendations were made.

- Develop a plan to reduce hardness levels in boiler feedwater through additional operator training and implementation of an excursion reaction plan so that, when there is a hardness event, any negative impact can be minimized.
- Replace the current internal treatment product with the most advanced all polymer boiler treatment chemistry.
- Utilize boiler automation to continuously monitor and control the chemical treatment and other critical steam system parameters based on system stress and demand.

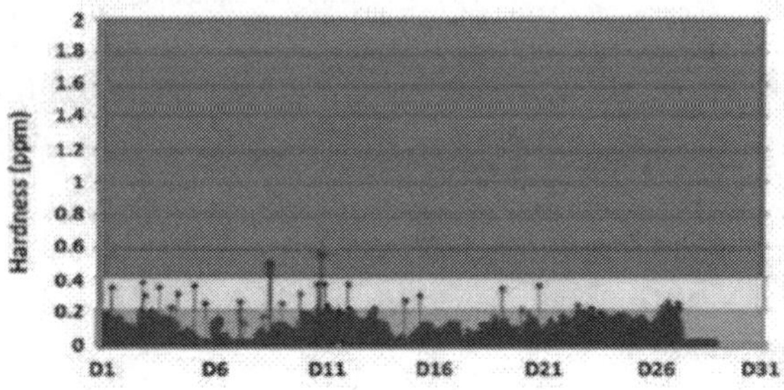

Figure 8. Boiler feed water hardness was in the target zone 89% of the time

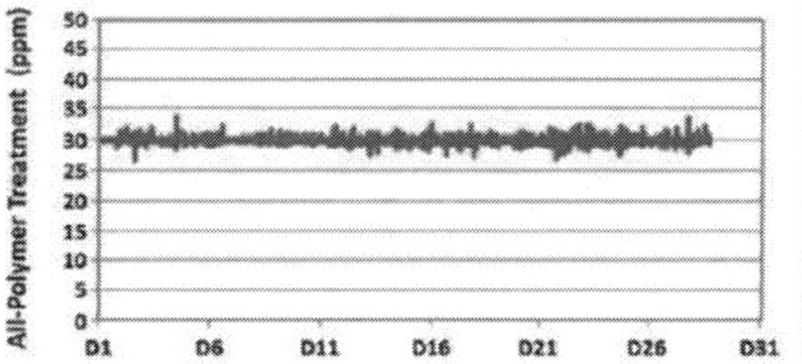

Figure 9. All-polymer internal treatment feedwater dosage

Boiler Feed Water Quality Results

The trend in Figure 8 shows the continuous boiler feed water total hardness measurement. Although there were occasional spikes above the best practice target, the off-spec measurements were quickly brought back into control. Plus, any time the hardness went out of target, the unit went into alarm mode and action was quickly taken to bring the hardness back into specification or control.

In the event of a significant hardness event, the controller would not only alarm, but would also automatically increase the all-polymer treatment dosage to account for the additional stress on the system. When the system stress was relieved, the product dosage automatically returned to its normal dosage.

Given the system design, an all-polymer/fluorescent tracer set point of 30 ppm was selected. The all-polymer dosage control was achieved by the automation system as shown in Figure 9.

The boiler automation controller is able to maintain a precise dosage even when there are system upsets. The trend shown in Figure 10 shows how the all-polymer set point of 30 ppm was maintained even with a sudden load increase on the #9 boiler due to another boiler going down unexpectedly. Although the boiler feed water flow rate increased by almost 30% in a matter of minutes, the new automation control system was able to maintain good control, assuring boiler reliability even under stressed conditions.

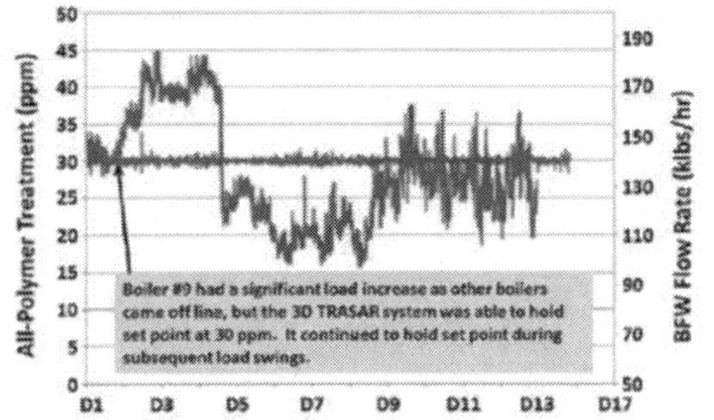

Figure 10. All-polymer dosage vs. feedwater flow

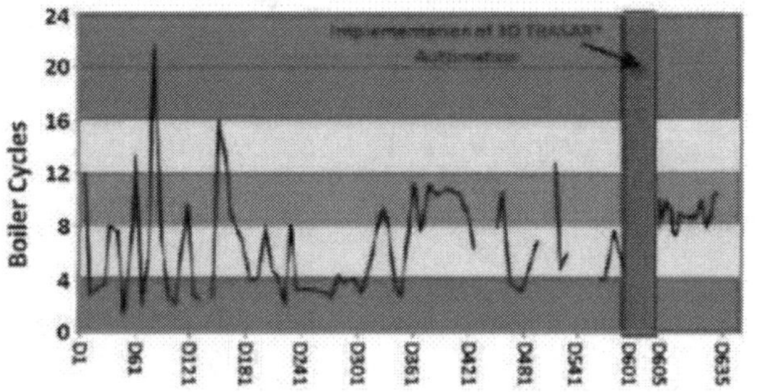

Figure 11. Boiler cycles after automation control

In fact, the boiler automation controller was able to maintain a six sigma control range of 29 to 31 ppm for the all-polymer internal treatment.

Table 1. Energy and Water Costs and Credits

		With Initial Conditions	Under New Conditions	Savings
Blowdown Energy Cost	$/year	669,614	334,807	334,807
Blowdown Sewer Cost	$/year	99,749	49,875	49,875
Makeup Water Cost	$/year	235,893	214,448	21,445
Sub Total (Costs)	**$/year**	**1,005,257**	**599,130**	**406,127**
Returned Condensate Fuel (Credit)	$/year	0	0	0
NET SAVINGS or (COSTS)	**$/year**			**406,127**
Calculated Cost of Steam	$/1000 lb.	9.83	9.39	0.44
NET CO_2 EMISSION SAVINGS (INCR)	Fuel is natural gas		Tons CO_2/yr.	3,236

Boiler Cycles Results

Prior to implementation of the new automation control technology, plant operations had to manually control boiler cycles. As a result, the cycles ranged anywhere from 3 to 22 cycles. This variability created stress on the boiler system and resulted in poor reliability as well as wasted water and energy.

The new technology can be used to optimize and automatically control boiler blowdown. With boiler cycles control, the boiler cycles were automatically maintained between 8 and 10 cycles, Figure 11.

This boiler cycle control will result in significant annualized financial and sustainability savings of over $400K as shown in Table 1. In summary, implementing the new automation technology resulted in tremendous improvement in operational control leading to improved boiler reliability, lower total cost of operations, and a significant eROI™ value.

CASE HISTORY: EAST COAST CHEMICAL PLANT

After a turnaround, the deaerator supplying water to the 900 psig Utility boilers was unable to maintain dissolved oxygen levels within normal ranges (<7 ppb). In fact, dissolved oxygen readings were consistently in the 200 – 300 ppb range as shown in Figure 12.

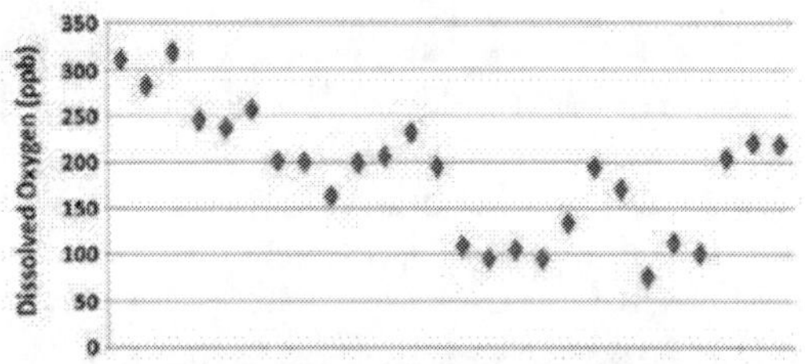

Figure 12. Manual Boiler Feedwater Dissolved Oxygen Test Results beyond normal ranges of less than 7 ppb.

The chemical plant determined that major modifications were needed to the deaerator. The challenge was to maintain boiler operation with high feedwater dissolved oxygen concentrations until the deaerator repairs could be completed. Since boiler feedwater was used for steam attemperation, the use of sulfite oxygen scavenger was not an option. The chemical plant was using Nalco ELIMIN-OX® oxygen scavenger, which is both an oxygen scavenger and metal passivator.

All of the Utility boilers had an economizer, which is the area that is most susceptible to oxygen corrosion in a boiler system. One of the economizers was two years old; the other economizers were 20 years old. All economizers had outlet sample points that could be utilized to monitor iron-related corrosion in the economizers. The challenge was to utilize the passivating properties of the ELIMIN-OX oxygen scavenger and control the feed to a level that minimized the pre-boiler corrosion, especially in the economizers.

Solution

Initially, only daily grab samples were tested for dissolved oxygen, ELIMIN-OX oxygen scavenger residual, and total iron before and after the economizer. An on-line dissolved oxygen analyzer was installed on the boiler feed water to continuously monitor the actual dissolved oxygen levels. The results of the manual and on line monitoring are shown in Figure 13.

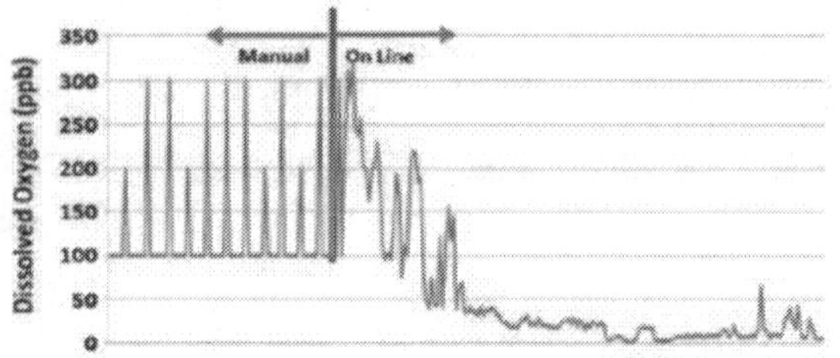

Figure 13. Dissolved Oxygen measurement was steady with on line monitoring.

To determine the amount of potential corrosion taking place with the high dissolved oxygen concentrations, total iron was monitored on the outlet of the economizers. As shown in Figure 14, economizer outlet iron levels remained extremely low. On-site total iron values were cross-checked with inductively coupled plasma (ICP) analysis. ELIMIN-OX feed rate was maximized and controlled to minimize the economizer outlet iron values.

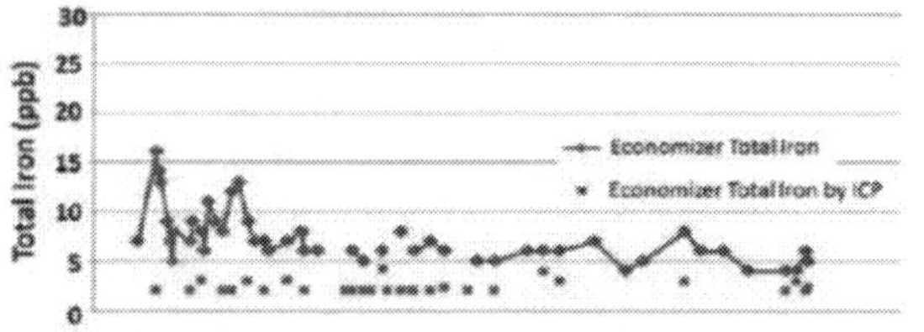

Figure 14. Economizer outlet total iron values on-site and by ICP.

The next step was to install the AT ORP to determine the overall pre-boiler corrosion potential using at temperature ORP. As shown in Figure 15, the AT ORP is far more responsive than the dissolved oxygen analyzer in detecting corrosion stress.

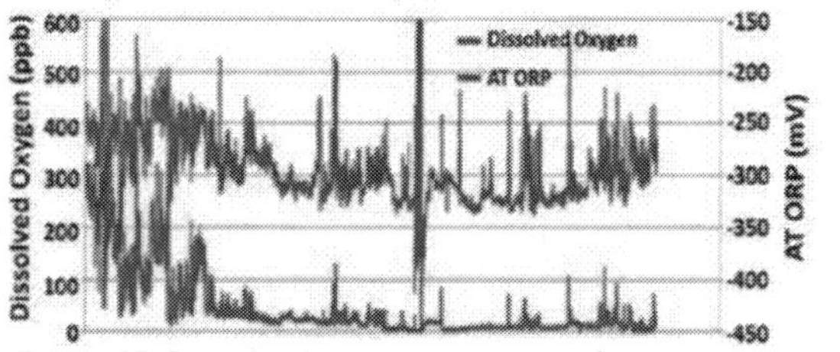

Figure 15. Dissolved oxygen versus AT ORP readings.

The AT ORP was used to optimize the ELIMIN-OX feed rate. As shown in Figure 16, ELIMIN-OX feed was reduced from 20 ppm to 13 ppm (35% reduction) with no measurable increase in corrosion potential as shown by the AT ORP readings. These results were also correlated to dissolved oxygen, as shown in Figure 17.

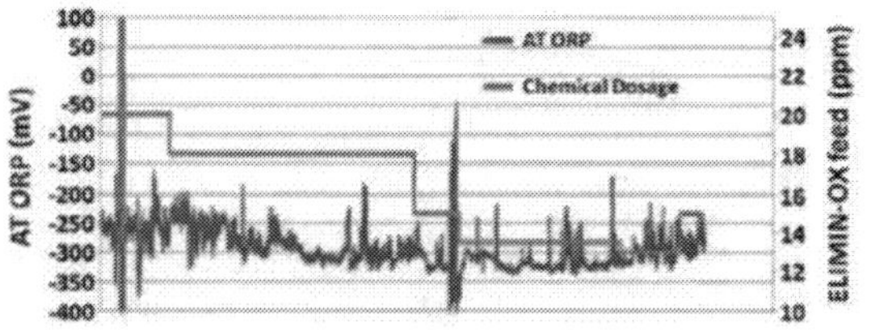

Figure 16. Chemical Feed rate versus AT ORP readings, showing the ability to reduce ELIMIN-OX feed while maintaining corrosion potential.

The final monitoring step was to correlate the economizer outlet total iron reading with each step-change in ELIMIN-OX feed and AT ORP reading to validate that no economizer corrosion was occurring. Figure 18 shows that, even with the reduction in the ELIMIN-OX feed rate, economizer outlet total iron values remained low.

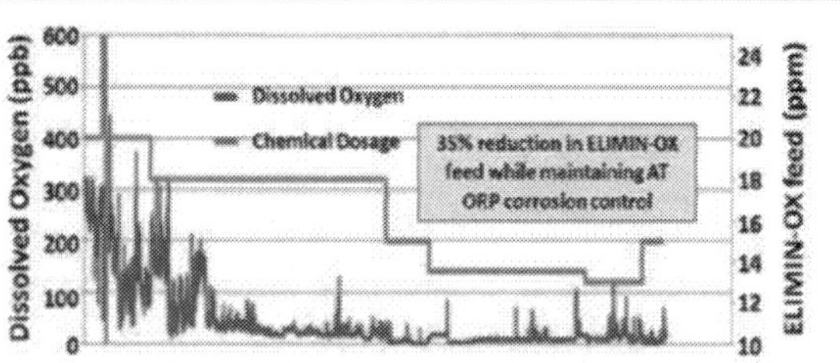

Figure 17. Chemical feed rate versus dissolved oxygen.

Results

All data showed that by utilizing the passivating properties of ELIMIN-OX technology and maintaining control with the AT ORP, the plant was able to operate their boilers confident that the long term reliability was not compromised. The plant easily made their goal of safely operating the boilers until the deaerator could be repaired.

However, for final verification, a tube sample from the newest economizer was extracted and sent to an external lab for metallographic analysis. The analytical report stated: "Internal corrosion is not extensive, and some areas are completely unaffected. The deepest pits were less than 9 mils deep (<0.009"). Overall tube thickness meets or exceeds 0.165" nominal."

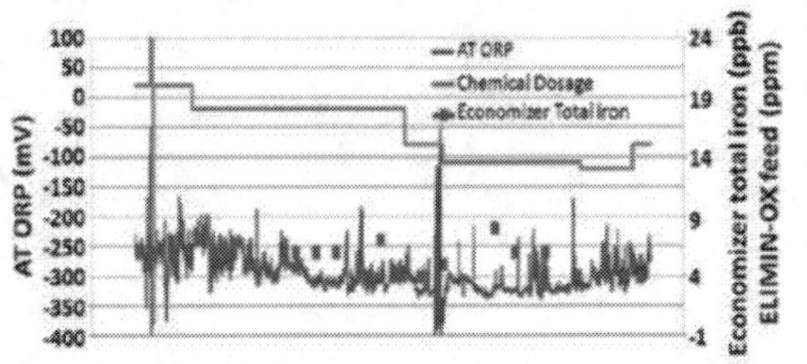

Figure 18. Dissolved oxygen, AT ORP, and chemical dosage.

The conclusion from the metallographic analysis was that the economizer tube that had been in service for two years showed no significant internal corrosion, thereby confirming the success of mitigating the potential corrosion effects of an improperly operating deaerator.

The eROI values achieved were:

- Asset Protection and Reliability – Allowed the refinery to operate for more than a year with high dissolved oxygen levels without damage to downstream equipment until a replacement deaerator could be procured. By minimizing oxygen corrosion to the economizer, the economizer life was extended.
- Chemical Savings – Allowed optimization of the oxygen scavenger feed while maintaining asset protection (35% reduction in chemical feed).
- Energy Saving – By minimizing incoming iron values, optimum boiler efficiency was maintained.

CONCLUSIONS

Even the best RT ORP probes are not as sensitive as the AT ORP probes. With this new boiler feedwater treatment automation, chemical is now fed based on a true at-temperature corrosion demand. Production events can occur quickly, and this new automation has shown that it can respond to the events quickly and with the appropriate response. The solid state fluorometer and AT ORP are both valuable instruments that can protect the boiler and feedwater systems from scaling and corrosion events. In addition, this technology can be used as a troubleshooting tool to help diagnose issues and correct them.

Given the recent economy, it is becoming even more important for all boiler house managers, engineers, and operators to take a fresh look at ways to reduce operating costs, protect asset life, and improve productivity. This technology has clearly shown that it can help find issues that previous technologies cannot see, and allows users to protect their pre-boiler and boiler systems.

REFERENCES:

1. Nalco Water Handbook, 2nd Edition, 1988, McGraw Hill Publishing
2. United States Patent #6,255,118 "Method for Using an All Solid-State Fluorometer in Industrial Water Applications"
3. United States Patent #5,736,405 "Monitoring Boiler Internal Treatment with Fluorescent-tagged Polymers"
4. "Monitoring and Control in Boilers With All Solid-State Fluorometers", Martin Godfrey, Corrosion / 2002, Paper No. 02012, NACE International Conference, Houston, TX (2002).
5. "A New Approach to Feedwater Corrosion Control: A Detailed Nalco AT ORP Study at a Paper Mill", Peter Hicks, David Grattan, and Paul Desch, Nalco ADV-1105.

SOLID-LIQUID FILTRATION TRENDS FOR PARTICLE FINES REMOVAL IN MANUFACTURING ENVIRONMENTS

Barry A. Perlmutter
BHS-Filtration Inc.
Charlotte, North Carolina 28270

Biography

Barry A. Perlmutter is President and Managing Director of BHS-Filtration Inc., a subsidiary of BHS-Sonthofen GmbH. Barry has over 30 years of technical and marketing experience in the field of solid-liquid separation, filtration, centrifugation and drying. He has published and lectured worldwide and has been responsible for introducing and creating growth for many European companies into the marketplace. He has a BS in Chemistry from Albany State, NY, MS from the School of Engineering at Washington University, St. Louis and an MBA from the University of Illinois.

ABSTRACT

In the manufacturing of chemicals, petrochemicals, pharmaceuticals, bioenergy etc., after the initial mixing-reactions, there is always the necessity to remove the mother liquor from the process slurry. In some cases, the liquid is the valuable component, in some cases it is the solids and in other cases it is both compounds. The nature of the requirements will determine the type of equipment needed for the separation process.

Generally, after the initial mixing - reactions, there will be bulk separation to remove the larger product solids or unwanted solids. This is the easy step. As processes have become more sophisticated and quality requirements increased, there is a trend to be able to remove the residual particle fines from the slurries. These fines are very small, typically in the 1 – 5 micron range and smaller and are very low in concentration down to ppm levels.

This paper provides a brief overview of crude separation equipment and then continues with a detailed view of technologies that can remove – recover small particle fines. The general term for this is "clarification" technologies and this will cover alternatives to traditional bag and cartridge filtration systems.

The paper concludes with a review of the installations in chemical, refinery, bioenergy and pharmaceutical plants to provide engineers with new and creative ways of thinking to evaluate process filtration problems and solutions.

SOLID / LIQUID SEPARATION FOR PROCESS SYSTEMS

Filtration, pressure or vacuum, is the art of finding a filter media which allows the liquid to pass through while retaining the solids. The driving force may be gravity, vacuum, or pressure.

The ultimate solid phase is always drier than that from gravitational systems and a great deal of processing can often take place in the actual filtration stage, especially with regard to the conversion or exchange of the residual cake moisture.

The driving force, causing the liquid to exit while leaving the solids behind, can be divided into four groups:

1. Gravitational, which is really little more than draining, but can be very useful to reduce large quantities to more manageable proportions. They can be batch operated or, more usual, continuous.
2. Vacuum, this is in many ways mechanically the simplest driving force available. Vacuum filters can be batch operated but are normally continuous. In general, the solid (cake) thickness can be controlled within close limits; there are few limits to materials of construction and some of vacuum filters offer the best solids washing possibilities.
3. Pressure, with or without compression, which obviously involves mechanical constraints. All pressure filters are batch-operated units with one exception, the exception being the Rotary Pressure Filter. Cake washing can be excellent and the final cakes are usually as dry as can be expected without heat input.
4. Centrifugal, which can sometimes offer a compromise between vacuum and pressure filtration. Centrifugal filters can be continuous in operation or they can operate in an automated continuous batch-mode. The nature and behavior of the solids play a great part in the success or failure of centrifugal filters. Cake washing can be good depending on the type and the behaviour of the solids.

TYPES OF FILTRATION SYSTEMS

With the above in criteria in mind for chemical and pharmaceutical production operations, there are several basic choices for a process engineer. These are segmented by particle size removal as well as whether or not the filter will discharge the cake batch-wise or continuous.

BATCH FILTERS

Autopress: This horizontal pressure filter is tubular in construction and operates as a contained filter press. Circular plates with welded metal or synthetic media are contained in a pressurized housing. This allows for pressure filtration, cakes washing and vacuum or pressure drying. After the cycle is completed, the housing is moved and automatic cake discharge is via scraper knives that move between the plates

Bag filters: A bag with a connection for a high pressure inlet held captive in a metal cage. Filtrate runs through the bag and the solids stay inside.

Basket strainer: Similar to the bag filter, except that the strainer may be the filter medium in which case the solids have to be dug out, or the basket may be lined with a bag, for easy cake solids disposal.

Basket centrifuge: A metal basket rotating at centrifugal speeds and a cloth bag is fitted inside the basket. The drive mechanism runs the basket up to high speeds and feed is poured into the basket. Filtrate runs through the basket and is collected in an outer mantle. When sufficient solids have collected inside the bag, the machine is stopped and the solids removed or it is slowed down to allow a mechanical scraper to remove the bulk of the solids (a "heel" stays behind). In a horizontal version, the Peeler Centrifuge, the bulk of the cake can be removed without slowing the machine.

Cartridge filters: Pressure vessels inside which are fitted a number of filter cartridges. Almost without exception the flow is from the outside inwards. Cartridges are available in cleanable or (much more common) disposable form. A very wide variety of materials, textures, pore size and physical sizes are available.

Filter press: The original filter presses were of the "plate and frame" construction. Modern presses use almost exclusively "recessed plates". A very wide variety of sizes, configurations, plate supports and degrees of automation exist.

Leaf filter: Leaf filters consist of a pressure vessel inside which a number of filter leafs are vertically mounted. The leaves are normally metal and fitted with a synthetic media. For removal of the solids different arrangements exist, ranging from opening the bottom of the vessel to mechanically spinning or vibrating the cake into a chute.

Membrane presses: The membrane press is basically a filter press but instead of having drainage grooves in the plates, the plates are fitted with an elastomer sheet (which has the drainage grooves) and which sheet can be inflated. By inflating the sheet at the end of the filter cycle any residual moisture will be expelled and the cake itself will be squeezed, usually resulting in better cake moisture figures.

Nutsche filter: Circular or rectangular filters with a drainage bottom onto which a filter medium is fastened. If the drainage section is connected to a vacuum source the filters are often open top. If they are closed at the top, they can be pressurized and thus benefit from a higher driving force.

Tube press: These may be vertical or horizontal and consist of a metal tube and inflatable membrane. Inside this membrane is mounted a central filter core consisting of a drainage tube with a filter medium. The feed is introduced into the tube under pressure, filtrate exits from the central filter core. After filtration and / or washing the membrane is inflated to squeeze the cake solids to extremely low residual moisture values. To discharge the cake, the membrane is relaxed, the bottom section of the tube opened and the tube slightly or totally withdrawn allowing the cake solids to drop off or to be scraped off.

CONTINUOUS FILTERS

Disc Filter: These consist of a flat disc mounted on a hollow shaft. The disc is made of metal and has on either side an open cloth support structure which connects with the hollow shaft which carries the filtrate. A filter cloth is fastened to the disc and the hollow shaft connects to a vacuum source. The disc, usually 30 % submerged, rotates slowly in a feed trough where it picks up the solids, which can later be scraped off just prior to re-entry into the feed trough.

Rotary Pressure Filter: The Rotary Pressure Filter is a continuously operating unit for pressure filtration, cake washing and drying of slurries up to 50% solids. The filter has a rotating drum inside a housing. The drum has cells within which are cell inserts fitted with the filter media. Filtration is conducted via pressure of up to 6 bar. Positive displacement washing or counter-current washing follows filtration. Multiple washing steps as well as solvent exchanges, steaming and extraction are accomplished. Finally, the cake is dried by blowing hot or ambient gas through the cake. The Filter has a uniquely designed discharge system, which provides for atmospheric discharge from pressure filtration. After automatic cake discharge, the filter cloth is washed; the clean filter cloth the re-enters the feeding / filtration zone thereby continuing the process. All solvent and gas streams can be recovered separately and reused in the process to minimize their consumption.

Pusher Centrifuge Horizontally mounted filtering centrifuge with a filter cage made of metal bars the smallest opening of which is about 100 micron. Inside the cage is a "pusher" arrangement, basically a sturdy ring which lies very close to the filter medium. The ring is connected to a central shaft which rotates at the same speed as the filter cage but reciprocates (approx 1 – 3 cm) causing the collected solids to be pushed out of the machine. Multi stage machines exist. In these the solids are pushed onto the second or third stage which is each time of a larger diameter. The higher linear velocity as well as the tumbling action can be beneficial for dewatering and / or cake washing, although it may cause the crystals to fracture.

Rotary Vacuum Drum Filter: The original rotary drum filter consisted of a cylindrical drum, fitted with a drainage grid on the outside over which a filter cloth was stretched. The drum had perforations and was connected to a vacuum source. The drum itself was mounted on a horizontal shaft and submerged for about 30% in a feed trough. The vacuum caused the liquor phase to be sucked through the cloth, leaving the solids behind as a filter cake. The cake was scraped off just before re-entering the feed trough.

Vacuum Belt Filter: The two main types are carrier belt and tray (fixed or indexing). These filters have a continuous belt as the filter medium which is subjected to a vacuum source. Feed is introduced at one end, where it is normally allowed to settle under gravity for a few seconds, causing the coarser fraction to form its own pre-coat. The suspension is then subjected to full vacuum. Since the cake thickness can be controlled by adjusting the speed of travel or indexing time, the optimum cake thickness to obtain the required residual moisture and / or cake washing efficiency is achieved. The horizontal configuration makes this filter one of the most efficient cake washing filters.

CLARIFICATION OF SLURRIES & RECOVERY OF SOLIDS

In the previous section, batch and continuous filtration systems were discussed. As processes have become more sophisticated and quality requirements increased, there is a trend to be able to remove the residual particle fines. These fines are very small, typically, smaller than 1 – 5 micron range and are very low in concentration down to ppm levels.

Candle Filters and Pressure Plate Filters are installed for clarification and recovery applications from liquids with low solids content. The candle filters are vertical candles while the pressure plate filters are horizontal plates. The cake structure as well as the process parameters determines the optimum thin-cake technology.

Description and Operation of the Candle Filter

Candle Filters provide for thin-cake pressure filtration, cake washing, drying, reslurry and automatic discharge as well as heel filtration in an enclosed, pressure vessel. Units are available from 0.17 m^2 up to 200 m^2 of filter area per vessel.

Filter Candles & Media

The filter candles, as shown in Figure 1, consist of three components: single-piece dip pipe for filtrates and gas, perforated core with outer support tie rods and filter sock. The filtrate pipe is the full length of the candle and ensures high liquid flow as well as maximum distribution of the gas during cake discharge. The candle can be a synthetic, stainless steel or higher alloys. The outer support tie rods provide for an annular space between the media and the core for a low pressure drop operation and efficient gas expansion of the filter media sock for cake discharge. The filter media is synthetic with a clean removal efficiency to less than 1 micron (um). As the cake builds up, removal efficiencies improve to less than 0.5 um.

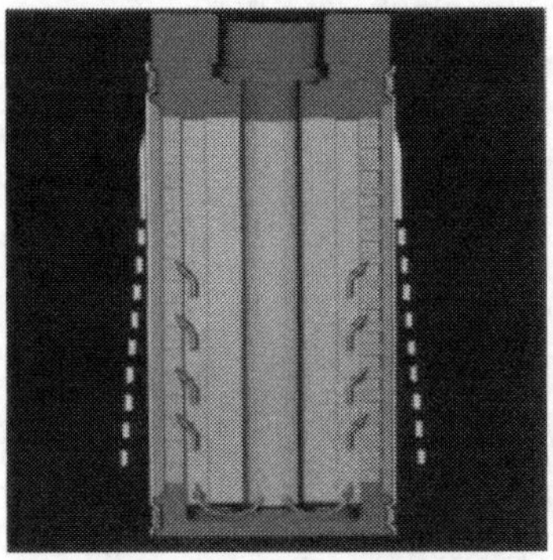

Figure 1: BHS Candle Showing Gas Flow to Expand the Filter Media Sock for Cake Discharge

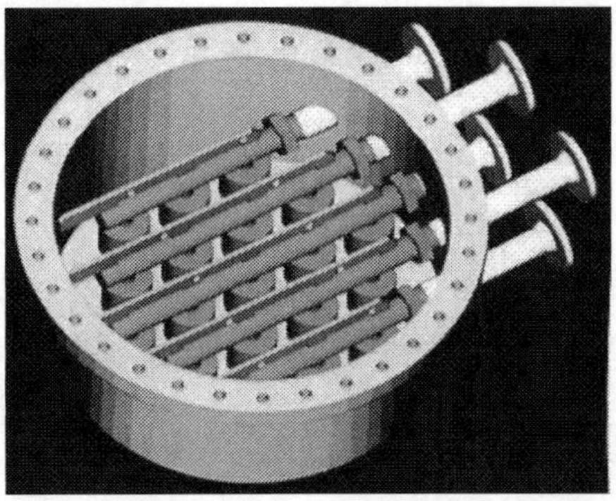

Figure 2: BHS Candle Filter

Filter Vessel & Candle Registers

The candle filter vessel is constructed of stainless steel or higher alloys. Within the vessel are horizontal manifolds called candle registers. Each candle is connected to a register with a positive seal to prevent bypass. Each register may contain from 1-20 candles depending upon the filter size. The registers convey the liquid filtrate in the forward direction as well as the pressure gas in the reverse direction for filter media sock expansion. Each register is controlled with automated valves to ensure optimum flow in both directions. Figure 2 illustrates the candle filter vessel.

Automatic Process Cycles

Filling: The slurry feed enters the bottom of the filter vessel.

Filtration: The slurry is either pumped or pressurized from the reactor into the vessel. Cake will deposit on the outside of the candle; the separated filtrate will flow through the filtrate pipe and the registers. This process continues until one of the following conditions is achieved: maximum pressure drop, maximum cake thickness, minimum flow or time.

Washing: Displacement washing or recirculation washing.

Drying: Blowing gas, steam or "shock" drying.

Heel (Falling-Film) Filtration: The liquid remaining in the vessel cone after filtration or washing is completely filtered.

Cake Discharge: Gas flows sequentially through each of the candle registers, down each of the filtrate pipes and then is distributed by the perforated core. The filter media sock gently expands by the gas flow and pressure allowing for cake discharge, as shown in Figure 1. Alternatively, the cake can be discharged as a slurry.

Description and Operation of the Pressure Plate Filter

The pressure plate filter has similar operating characteristics to the candle filter. The filter design is shown in Figure 3.

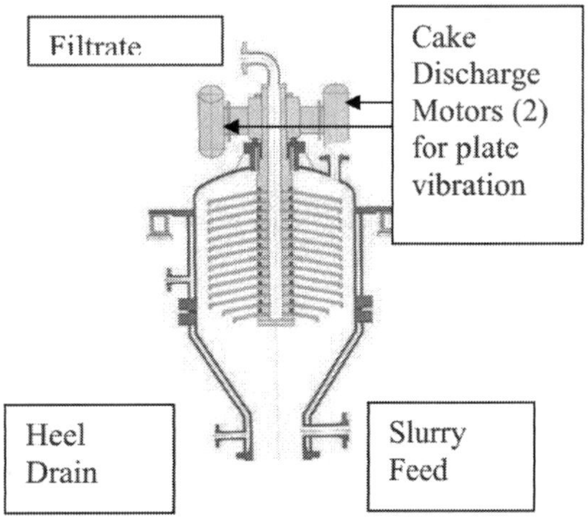

Figure 3: BHS Pressure Plate Filter

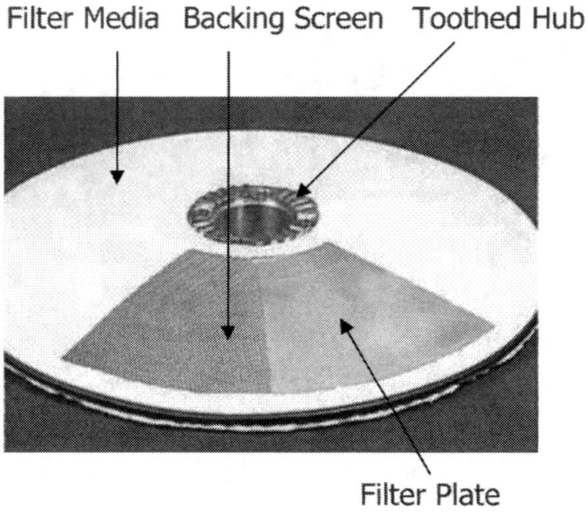

Figure 4: Filter Plate

Automatic Process Cycles

Filling: The slurry feed enters the bottom of the filter vessel.
Filtration: The slurry is pumped under pressure into the vessel or via gas pressure through the reactor. Cake will deposit on the top of the plates. The separated filtrate will flow through the plates to the center main filtrate outlet. This process continues until one of the following conditions is achieved: maximum pressure drop, maximum cake thickness, minimum flow or time.
Washing: Displacement washing or recirculation washing.
Drying: Blowing gas, steam or "shock" drying.
Heel Filtration: The liquid remaining in the vessel cone after filtration or washing is completely filtered.
Cake Discharge: As shown in Figure 5, the motors on the top of the filter operate at different frequencies and the plates gently vibrate for cake discharge. The plates vibrate in the vertical and horizontal planes and the solids are conveyed in an elliptical pattern to the outside of the vessel. Gas assist helps in the discharge process. There are no rotating plates, gears or bushings and mechanical seals are not required.

Description and Operation of the Pressure Plate Filter (Continued)

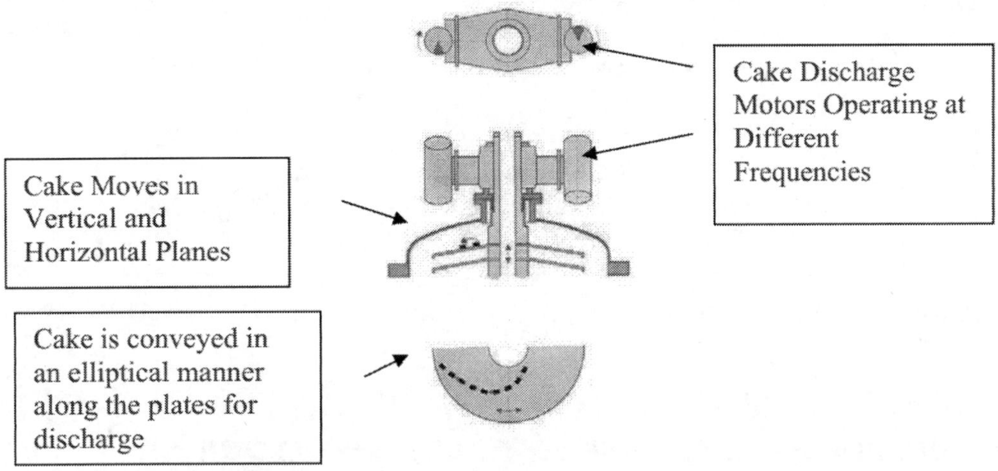

Figure 5: Cake Discharge in the Pressure Plate Filter

SELECTION OF CANDLE VERSUS PRESSURE PLATE FILTER TECHNOLOGIES: CAKE STRUCTURE AND PROCESS PARAMETERS

The major difference between the two technologies depends on the cake structure that is formed. Some cakes are better handled in the horizontal and some in the vertical.

Cake Thickness and Filtration: The candle filter is limited to cake structures that can be formed to about 5 – 20 mm. The pressure plate filter can handle cakes up to 75 mm. Both units can conduct filtration up to 150 psig.

Filter Media: The candle filter uses only synthetic media with a clean removal efficiency from 1 –3 micron range and finer down to 0.5 microns. The pressure plate filter can also use metal media. For the pressure plate filter, the clean micron range removal efficiency is also 1 – 3 microns and finer.

Cake Washing: If the process requires washing of the solids, then generally the pressure plate filter is a better alternative. If washing is not as critical, then the candle filter may be the optimum technology for clarification and recovery.

Heel Filtration: The remaining liquid in the vessel (liquid heel) after filtration or washing can be removed from the candle filter or pressure plate filter by circulation, heel filter in the cone of the vessel, or additional heel filter plates in the pressure plate filter.

Cake Drying: The candle filter can produce cakes with approximately 10% moisture. This moisture level depends upon the specific cake but the moisture lower limit is that moisture just above the cake cracking point. The pressure plate filter can produce bone dry cakes.

Cake Discharge: Both designs can easily discharge most cakes equally with no residual heel.

Clean-In-Place (CIP)/ Steam-In-Place (SIP): Both units conduct CIP / SIP operations in identical manners by filling and circulating cleaning fluids while blowing gas in the reverse direction to the filtration direction, which creates a turbulent mixture or a quasi-ultrasonic cleaning effect. The pressure plate filter further enhances this operation with plate vibration.

APPLICATION 1:
REPLACING A MANUAL PLATE FILTER AND BAG FILTER COMBINATION

This specialty chemicals manufacturer produces various resins that require filtration. Current production includes a neutralization step which yields metal salts. These salts are filtered out with a manual plate filter followed by a bag filter for polishing. Two solvent washes follow the filtration step to recover as much resin as possible. After washing, the filters are steamed and opened. The solids are disposed manually for each batch and the filter paper is replaced. The goals are to eliminate exposure to heptane, reduce the maintenance and operation on the two filters and to recover a dry, as possible, catalyst. Current production is 3000 gallons in 4 – 5 hours.

Results & Conclusions

The filtration flux rate from the BHS laboratory tests ranged between 10-30 L/m^2min at approximately 20 psi feed pressure. The filter cloth for the sock is polyester with an air permeability of 1.0 cfm/ft^2.

The tests showed that one BHS candle filter with 10 m^2 of filter area can complete the cycle in a time of 4.3 hours and replace the manual plate filter and bag filter.

The cycle time is as follows:

Cycle Times			
	Filling	5	min
	Filtration	10	min
	Wash	4	min
	Drain	10	min
	Dry	5	min
	Vent	2	min
	Discharge	5	min
	Reserve	9	min
	TOTAL	**50**	**min**

Figure 6: Dry Cake After Discharge

APPLICATION 2:
REPLACING A FILTER PRESS FOR A HOT SLURRY FILTRATION

This specialty chemicals manufacturer produces various grades of polyols from ethylene glycols that require filtration of small particles in the 1 – 2 um range. Current production includes a manually-operated filter press at 200 – 250 degrees F. The goals are to eliminate operator exposure, reduce the maintenance and operating costs for the filter press and to increase the polyols yield. Current batch sizes are 45000 pounds of slurry.

Results & Conclusions

BHS conducted the lab tests in the BHS laboratory in Charlotte, North Carolina. The slurry temperature was between 200 – 250 degrees F at a pressure of 90 psig. The filter cloth for the sock was successful using PTFE or PEEK material.

The tests showed that one BHS candle filter with 27 m^2 of filter area can complete the required batch is three (3) cycles, each of 6-hours. The cycle time is as follows:

Cycle Times			
	Filling	12	min
	Precoat	8	min
	Filtration	60	min
	Drain	15	min
	Dry	3	min
	Vent	2	min
	Discharge	15	min
	Reserve	5	min
	TOTAL	**120**	**min**

Figure 7: Jacketed Candle Filter with Register-Filtrate Piping & Junction Box

APPLICATION THREE:
REPLACING CENTRIFUGE & CARTRIDGE FOR TIN HYDROXIDE SLURRY

This process was previously done with a centrifuge which did not provide much separation at all and the fines were causing issues in the downstream operation at their sister plant.

Process Observations:

Precoating with Hyflo (5 lbs) and water – filtrate goes to sump
Precoating based on volume and then switch over to feed slurry (0.4 – 0.5% body feed)
Filtration - Pressure built to around 20 psi in about 30 minutes, batch size is approximately 3000 lbs. (10-15 gpm)
Washing with water- 40 psi until filtrate looks clean with 12 um PEEK filter sock
Draining- initially through candles first (25 psi) and then the heel drain
Drying – Blowing of gas for 5 minutes at 20 psi
Cake discharge – dry cake with a cake thickness of 15-20 mm

Results:

The BHS candle filter with 10 m^2 of filter area replaced the centrifuge and cartridge polishing filter. The results are below.

Cycle Times		
Filling	5	min
Precoat	10	min
Filtration	45	min
Wash	5	min
Drain	10	min
Dry	5	min
Discharge	5	min
Reserve	5	min
TOTAL	**90**	**min**

APPLICATION FOUR:
REPLACING A FILTER PRESS USING PAPER MEDIA

This specialty lube oil process, for final clarification, is conducted in a filter press using a paper media at 170° C, and using carbon, lime or celite as a body feed. There are five (5) grades of lube oils and cleaning between each campaign is required. The material is dried with nitrogen and the solids are removed and burned in the power plant as fuel. The process has the manual problems associated with a filter press plus the safety concerns of being burned with hot oil. The average particle size distribution (PSD) is 0.7 um.

Results:

The pilot testing demonstrated that the oil slurries could be filtered to meet the customer's specifications, that the resultant cake could be dried to the customer's desired moisture content, and that the cake could be automatically discharged satisfactorily from the candles. The purchased system was a 10 m^2 jacketed BHS candle filter using PEEK filter socks for the current five (5) products with expansion possibilities.

APPLICATION FIVE:

DUAL STAGE CANDLE AND PRESSURE PLATE FILTERS FOR FINES REMOVAL AT REFINERIES, GAS PLANTS AND GREY WATER AT COAL GASIFICATION PLANTS

The fines generated in grey water as well as by cat crackers and cokers are generally less than 1 um in size and very low concentration. For these large flows, upward to 800 gpm, simple bag or cartridge filters or filter presses cannot be used. For these applications, BHS has developed a dual-stage process using candle filters for concentrating and then followed by a pressure plate filter for the final filtration, cake washing and drying. The benefits include increased reliability, much lower consumption (25% less) of compressed air/gas for drying and cake wash water. A typical example and photo is below.

Process & Results:

Total Flow:	94 gpm
Design Solids Load:	28 lb/hr (600 ppm)

- Two (2) Candle Filters, each with 20 m^2 of filter area, for concentration
- One (1) Pressure Plate Filter, 4 m^2 of filter area, for final filtration, cake washing and drying

Figure 8: Complete Skid Package with Candle Filters, Plate Filter, Tanks, Pumps and PLC Controls

SUMMARY

There are many choices for the initial filtration steps in a process. The take-away is that one filtration system may not achieve those quality requirements. Engineers must evaluate all outcomes to make an informed and successful decision. A secondary filtration system for fines removal, while adding capital expense to the project may result in an overall more reliable and optimum process solution. While not discussed in this article, laboratory and pilot testing is critical for a successful decision and project.

Kneader Technology for the Direct Devolatilization of Temperature Sensitive Elastomers

Synthetic elastomers have been produced for over 50 years. Advances in catalyst systems and polymer formulations have been somewhat diminished by the continued use of the same processing technology. In particular, the use of coagulation, steam stripping, mechanical dewatering, and convective drying for the devolatilization of temperature sensitive elastomeric solutions can be replaced with direct devolatilization using kneader technology. A two-step, direct devolatilization process has demonstrated energy savings and advantages in environmental emissions and process/product flexibility when compared to the conventional steam stripping process.

Meeting: Spring AIChE National Meeting & 9th Global Congress on Process Safety

San Antonio, TX – April 28th through May 2nd, 2013

Topical 3: Chemicals Manufacturing for the 21st Century

Title: ***"Separation Technologies for the New Biochemical Applications"***

Authors: Glenn Shiveler, Sulzer Chemtech USA Inc. – Tulsa, OK

Dr. Antonio Garcia, Sulzer Chemtech USA Inc. – Houston, TX

Abstract:

Advances in the next generation of biofuels will include development of commercial process technologies for the global transportation markets. R&D has led to new biofuels from renewable feed stocks: grains, bagasse, switch grass, sugar cane, beets, turnips and wood pulp. Fermentation of these feed stocks produces alcohols along with contaminates that include acetates, aldehydes and acids. Separation of the valuable biofuel components from these contaminants is made challenging due to azeotropes, separation of the chemicals from the fermentation biomass, potential use of extraction solvents, efficient heat integration and dehydration of the biofuel product.

Additional challenges are: process economic feasibility and fast-track commercialization process.

For many years Sulzer has been active in providing mass transfer equipment in many existing 1st generation biofuels production facilities. Currently, Sulzer is supporting the development of 2nd generation biofuels with separation process technologies, process equipment, skid mounted systems for demonstration scale and basic engineering and key equipment for commercial plants. The new biochemicals plants involve various design considerations that are described in this paper.

(132e) In-Process Particle and Droplet Characterization for Process Design and Optimization in Manufacturing Operations

Anjan P. Pandey, METTLER TOLEDO, Columbia, MD

Particles and droplets are pervasive in the chemical, consumer products, and bio-pharmaceutical industries. Particles grow, agglomerate, break, dissolve, and change shape - all of which can have significant impact on process performance and product quality. Droplet distribution can impact emulsion stability, polymerization kinetics, and oil-water separation rate. In order to ensure repeatable product yield, purity, and bulk density at production scale, it is critical to understand and control particle/droplet distribution. However, manufacturing conditions (high pressure, high temperature etc.) often make it extremely difficult in obtaining samples for offline measurements. In addition, challenges of obtaining representative sample and extensive sample dilution could lead to erroneous information about particle/droplet distribution.

With FBRM® and PVM® probes the researcher can track the entire dynamic particle and droplet system in production scale at standard operating temperatures and pressures, without sampling or sample dilution. For over 20 years, scientists and engineers in chemical/pharma/oil/consumer industries have routinely used FBRM® and PVM® technologies for tracking particle/droplet distribution in real time to ensure less batch failures, batch to batch consistency, and greater production throughput. This presentation will focus on real case studies applying FBRM® and PVM® technologies at production scales to:

- Ensure batch-to-batch consistency during crystallization
- Target a desired crystal endpoint distribution
- Optimizing filtration rates and product throughput
- Optimize droplet distribution to control oil-water separation

(132f) Saving Agitator Power Costs by Using the "Off" Switch (and other techniques)

Gregory Benz, Benz Technology International, Inc., Clarksville, OH

Though agitator power consumption may not be one of the primary energy costs in most processing plants, it is still a contributor which may be minimized by judicious equipment selection as well as operating procedures, such as shutting the agitator off when agitation is not needed, or using variable speed drive when less agitation is needed. The right approach depends on the kind of agitation problem. In this article, we will focus on 3 major problem classifications: liquid motion, solids suspension and gas dispersion.

(135a) Hybrid Processing for CO_2 Capture – A Model Based Approach for Process Synthesis

Anna-Katharina Kunze[1], Philip Lutze[1] and Andrzej Górak[2], (1)Department of Biochemical and Chemical Engineering, Laboratory of Fluid Separations, TU Dortmund University, D-44227 Dortmund, Germany, (2)Department of Biochemical and Chemical Engineering, Laboratory of Fluid Separations, TU Dortmund University, Dortmund, Germany

Hybrid processing for CO_2 capture – A model based approach for process synthesis

A. Kunze, P. Lutze, A. Górak

TU Dortmund University, Laboratory of Fluid Separations

Emil-Figge-Strasse 70, D-44227 Dortmund, Germany

Phone: +49 (0) 231 755-3034, E-Mail: anna-katharina.kunze@bci.tu-dortmund.de

In general, separation of gaseous components is an important process step in chemical processes. But process synthesis for this application is a challenge, because there is a huge variety of possible unit operations to fulfill the same separation task. A restriction to a small number of possible unit operations for gas separations is not that easy, because there is no as dominant unit operation as for example distillation for liquid separations. This leads on the one hand to the question, how to decide which unit operation should be used in each individual separation task and operates best at different process conditions. On the other hand, due to the large number of possible unit operations, hybrid processes might be an opportunity here to gain process intensification. Hybrid processes are defined as combinations of at least two separation techniques in one operation by using the synergy between them [1].

Gas separation processes are often developed based on knowledge [2]. This might end in fixed decision pathways. A model based approach connected to a superstructure analysis will be presented to obtain a higher flexibility and a more general procedure in process synthesis for gas separations. In contrast to purely knowledge based approaches, this method offers the opportunity to take more process configurations into account and evaluate them. A scheme of this multi-step method is shown in Figure 1.

Figure 1: Method scheme for model based process synthesis

Therefore, the first step after defining a separation task is to use a technology database to summarize suitable unit operations. This database contains information about various separation technologies for gaseous mixtures including further reactions and necessary recycle runs. Not only industrial mature technologies can be taken into account, but also promising new technologies like membrane contactors. For each technology the process windows, structural and logical constraints are implemented to restrict the number of process options. The next step is a superstructure analysis based on short cut models of each separation step. If necessary mixed integer non-linear programming (MINLP) has to be used to gain a number of reasonable process options. Those options will be ranked following the objective function, for example energy consumption and the most promising process options are set. In the last step, rigorous modeling can be used to proof the results of the short cut analysis and define the process configurations in a more detailed way.

To figure out a method to find advantageous process configurations, CO_2 capture is used as a case study as a first step. CO_2 can for example be separated with different absorption methods, like washing with aqueous monoethanolamine or piperazine promoted potassium carbonate as well as different absorption technologies such as conventional columns and membrane contactors. But also membrane and adsorption techniques might be suitable to separate CO_2 from flue gas stream. Another possible application of the developed method at a later stage could be the adjustment of biogas upgrading plants to the process conditions of each individual place of action. Therefore, the developed method has to be generally applicable and easily adaptable to other, more complex separation tasks, which include for example more components.

Literature

[1] A. Stankiewicz et al., Re-Engineering the Chemical Processing Plant. Industrial & Engineering Chemistry Research. New York: Marcel Dekker, Inc., 2003

[2] S.D. Barnicki et al., "Separation system synthesis: a knowledge-based approach. 2. Gas/vapor mixtures". Industrial & Engineering Chemistry Research 1992, 31 (7), 1679–1694

(135b) Covalent Organic Polymers (COPs) for Exceptional CO2 Adsorption

Mert Atilhan, Department of Chemical Engineering, Qatar University, Doha, Qatar and **Cafer T. Yavuz**, Graduate School of EEWS, KAIST, Daejeon, South Korea

Efficient CO2 scrubbing without a significant energy penalty remains an outstanding challenge for the fossil fuel-burning industry where aqueous amine solutions are still widely used. One of the main challenges still remains to develop more efficient separation methods that can remove efficiently high fractions of acid gases from natural gas (or any other source) at pre and post combustion streams. For this purpose, for different sources of CO_2 different scenarios are being investigated in order to deal with the removal of CO_2 from various gas streams at various different pressure and temperature conditions. Thus, effective CO_2 capture is essential to investigate the further options to re-use the captured CO_2. Porous materials have long been evaluated for next generation CO_2 adsorbents. Porous polymers, robust and inexpensive, show promise as feasible materials for the capture of CO_2 from warm exhaust fumes. We report the syntheses of porous covalent organic polymers (COPs) with CO_2 adsorption capacities of up to 5616 mg/g (a world record - measured at high pressures, i.e. 200 bar) and industrially relevant temperatures (as warm as 65 ^{o}C). COPs are stable in boiling water for at least one week and near infinite CO_2/H_2 selectivity is observed. Theoretical calculations refer to an amorphous extended framework as density is likely the main reason for exceptional CO_2 capacities. COPs 1-2 feature basic nitrogen sites that show chemospecific affinity towards acidic gases such as CO_2. COP-3 has reasonably high surface area (418 m^2/g), effective for low pressure operations. Post combustion carbon capture from fossil fuel power plants demands pressures of up to 6 bar and a minimum temperature of 40 ^{o}C. By tuning their architecture, we show that COPs reach to 3 mmol CO_2/g sorbent at 6 bar and 45 ^{o}C. High and low pressure capacities make these porous polymer structures viable alternatives to amine scrubbers.

(135c) The Syntheses of CO2 Adsorbent Microporous Materials in Amine Solutions

Fahd Mohamed, Mohammed Shurair, Baligh Dakdakay and Mert Atilhan, Department of Chemical Engineering, Qatar University, Doha, Qatar

This work focuses on the design, production, characterization and testing of materials that potentially have a high capability of capturing carbon dioxide from natural gas precombustion (high pressure adsorption) feed and post combustion (low pressure adsorption) effluent gases. It is anticipated that the proposed work will result in high surface area solid materials that can be utilized for the optimized CO2 capture. Aqueous amine solutions have long been used to remove CO2 and H2S (natural gas sweetening) from industrial gas streams. Alkanolamines such as monoethanolamine (MEA) and N-methyldiethanolamine (MDEA) are some of solvents commonly used for CO2 removal. On the other hand Prussian Blue (PB) analogues have been investigated widely in gas uptake field for their promising gas uptake behaviors. It is known that PB complexes are prepared mostly in aqueous solutions. In this project we will prepare PB complexes in alkanolamine solutions to incorporate these amine molecules into the pores of the aforementioned microporous materials. The amine molecules that occupy the micropores will increase the CO2 adsorption of these materials due to the well-known interaction between CO2 and -NH2 groups. Convenient and conventional amine molecules such as MEA, DEA, and MDEA will be used in this project. Obtained complexes will be characterized with powder X-ray Diffractometer (XRD), Infrared Spectroscopy (FTIR), Thermogravimetric Analysis (TGA) and BET isotherms as full characterization scheme. CO2 adsorption of these clusters will also be investigated using Rubotherm® magnetic suspension balance.

*** Formatting Issue - Best Available Paper/Graphic ***

(135d) Development of Stable Multiphase Ni-Mgx-Aly-O Catalysts for CO2 Reforming of Methane

Jun Zhang Sr., Low Carbon Energy Conversion Center, Shanghai Advanced Research Institute, Shanghai, China

Development of stable multiphase Ni-Mg_xAl_yO catalysts for CO_2 reforming of methane

J. Zhang[a], Q. J. Chen[a], B. R. Pan[a], W. B. Kong [a], Q. W. Jin[a], T.J. Zhao*[a], Y. H. Sun*[a,b]

a: Low Carbon Energy Conversion Center, Shanghai Advanced Research Institute, Chinese Academy of Sciences, Shanghai 201203, China

b: State Key Laboratory of Coal Conversion, Institute of Coal Chemistry, Chinese Academy of Sciences, Taiyuan 030001, Shanxi, China

---

Abstraction:

CO_2 reforming of methane to produce synthesis gas is becoming more research attractive due to the increasing interest in reduction of carbon dioxide emissions and more efficient utilization of natural gas. From an industrial standpoint, Ni-based catalysts have drawn significant research attention for their high activity and wide availability. However, metal sintering, oxidation, and especially significant carbon formation result in the severe deactivation of the Ni catalysts. Some catalysts such as NiO/MgO solid solution, Ni/Al_2O_3, $NiAl_2O_4$ spinel, etc. were reported to have excellent performance for CO_2 reforming of methane. However, those kinds of catalysts still have their own disadvantages, for example, hydrophobia of NiO/MgO, low active site dispersion for Ni/ Al_2O_3, difficult reduction for $NiAl_2O_4$ spinel, etc.

Fig .1 XRD for Ni-Mg_xAl_yO catalysts

Fig .2 Stability of Ni-Mg_xAl_yO for CO_2 reforming of methane

In this work, a high stable multiphase Ni-Mg_xAl_yO catalyst for CO_2 reforming of methane was developed by HT precursors. For comparison, NiO/MgO solid solution, Ni/Al_2O_3, and $NiAl_2O_4$ spinel catalysts were also prepared and tested. The Catalysts were characterized using ICP-MS, BET, XRD, H2-TPR, TEM, and TG. Catalysts screening for the carbon dioxide reforming of methane showed that Ni-Mg_xAl_yO catalysts with

***** Formatting Issue - Best Available Paper/Graphic *****

multiphase (periclase-spinel) obtained by hydrotalcite-type (HT-type) precursors exhibited superior performance to other Ni–Mg–Al–O catalysts with different phases. This special multiphase catalyst exhibits very high and stable activity over 1000 h without deactivation at a GHSV of 79, 000 ml/(hg) at 850 °C. Both commercial Ni/Al_2O_3 catalyst and the NiO/MgO solid solution with single periclase phase deactivated quickly after the ignition of reaction, and the $NiAl_2O_4$ spinel show less activity for their low reductively. BET, XRD and H2-TPR results indicated that the activity of the catalysts were strongly depended on the structures, the sizes of the nanoparticles, surface area and active site dispersion. In fact, those characteristics are strongly depended on the types of the phases in the catalyst, and the interaction between them. It was believed that the synergy between periclase and spinel could also significantly improve catalyst performance and reduce carbon formation. A high metal dispersion or small ensemble size can be obtained by NiO/MgO solid solution. The excellent stability of metal was enhanced by spinel phase exist in the NiO/MgO solid solution, which further promoted the metal dispersion and resistance of metal sintering. The synergistic effects between crystal phases and phase cooperation, the phenomena responsible for the variation of activity in the methane reforming as a function of the degree of spinel substitution in periclase were deeply discussed.

Keywords : Natural gas; CO_2 Methane reforming; Ni-Mg_xAl_yO multiphase catalysts; phase cooperation

(138a) Synthesis of 2-Acetoxy Propanoic Acid Via Reactive Distillation an Intermediate in Synthesis of Biobased Acrylic Acid From Lactic Acid

Aspi K. Kolah, Lars Peereboom, Carl T. Lira and Dennis J. Miller, Chemical Engineering and Materials Science, Michigan State University, East Lansing, MI

The selective formation of 2-acetoxypropanoic acid (2-APA) by reaction of lactic acid with acetic acid has been carried out in continuous reactive distillation. The reaction takes place in a six-meter tall, 50 mm ID Pyrex column operated at atmospheric pressure and is catalyzed by Amberlyst-15 cation exchange resin supported in Katapak-SP11 structured packing in the middle four meters of the column. Lactic acid solutions in water were used as feedstock in the experimental study which was reacted with glacial acetic acid in the reactive distillation column. High conversions of lactic acid and its oligomers were achieved along with high selectivity to monomeric 2-APA. Experimental studies are complemented by process analysis using AspenPlus simulation software. By implementing a kinetic model and phase equilibrium data generated in our laboratories, we are able to predict the observed pilot-scale column behavior and simulate commercial-scale facilities.

The formation of 2-APA represents the first step of a process to produce acrylic acid from lactic acid. In the second step of the process, 2-APA is cracked under high temperature to form acrylic acid and liberate acetic acid that can be recycled.

Acrylic acid, its esters, and acrylate polymers are critical chemical building blocks used in a wide gamut of applications including superabsorbents, hard plastics, paints, detergents, and adhesives. The global acrylic acid market is estimated at 3.5 million tons and $8 billion annually, with a growth of 4% per year. Acrylic acid synthesized from biobased lactic acid is a drop-in replacement for the petroleum-based product; lactic acid is a mature biobased feed stock with a proven track record of high yield (>95%) from glucose and low cost.

(138b) Revolutionizing the Production of Transportation Fuels and Chemicals

David St. Angelo, Engineering, Joule Unlimited, Bedford, MA

Many technological alternatives to petroleum-based fuels have been explored, and with good reason. Limited energy supplies, pricing volatility and environmental concerns have converged to create one of the greatest challenges of our time. Joule® has an answer that surpasses biomass-derived fuels, channeling the sun's energy into a new, industry-changing category of direct solar fuels. The company's ***Helioculture***™ platform incorporates proprietary, engineered photosynthetic microorganisms to directly produce infrastructure-ready diesel, ethanol and multiple chemicals with no dependence on biomass feedstocks, agricultural land or fresh water. In parallel, Joule has developed a novel ***SolarConverter®*** system to enable the direct, continuous process with productivities greater than biomass-dependent methods, which require numerous energy-intensive steps and downstream processing to achieve an end product. In contrast, using sunlight, non-potable water and waste CO2 from industrial emitters or pipelines, Joule ultimately targets productivities of up to 15,000 gallons of diesel and 25,000 gallons of ethanol per acre annually, at stable costs as low as $50/bble and $1.28/gallon respectively (without subsidies).

(138c) Production and Upgrading Of Renewable Bio-Crude From Large Scale Algae Ponds

Neil Osterwalder, Dan Sajkowski and Ben Saydah, Sapphire Energy Inc., San Diego, CA

Renewable fuels from algae are moving closer to commercialization as Sapphire Energy showed with their Integrated Algae Bio-Refinery (IABR). The IABR in New Mexico was commissioned in 2012 and is producing algae derived bio crude oil on a continuous basis using Sapphire's proprietary conversion and extraction process. This algae bio-crude contains a wide range of hydrocarbons, fatty acids, aromatics, and nitrogen compounds and other heteroatoms. This chemical profile suggests that algal bio-crude can be economically processed by blending with existing refinery streams and co-processing the combined streams as opposed to the more costly development of algal-specific refineries. Different options of co-feeding the algae crude to a refinery can produce a full suite of refinery products from petrochemicals to gasoline, diesel and other refinery products. This study describes the products that can be produced from algae bio-crude and its implementation into existing refining infrastructure.

*** Formatting Issue - Best Available Paper/Graphic ***

(138d) Membrane-Assisted Downstream Processing for Biobutanol Purification

Sebastian Heitmann, Daniel Welz, Philip Lutze and Andrzej Górak, Department of Biochemical and Chemical Engineering, Laboratory of Fluid Separations, TU Dortmund University, Dortmund, Germany

One of the main challenges for chemical industry is the change from fossil resources to a sustainable production of chemicals. The change to biobased raw materials and biotechnological processes often leads to product inhibitions of microorganisms and diluted system, resulting in an expensive downstream processing. One bulk chemical that can be produced in a sustainable way is biobutanol. One method of producing biobutanol utilizes the so-called acetone–***n***-butanol–ethanol (ABE) fermentation that can be carried out on basis of e.g. cellulosic and lignocellulosic biomass or agricultural wastes. However, the productivity of this fermentation is still limited, because ***n***-butanol is toxic towards the microorganisms used in ABE fermentation. That is why the concentration of n-butanol in the fermentation broth is often limited to values lower than 2 wt.-percent [1]. For n-butanol recovery mainly distillation processes are used which - as result of the low concentrations - have a high energy demand. About 80 % of the total energy demand of the downstream process is required for the separation of acetone, n-butanol and ethanol from the fermentation broth. Alternatively to distillation the fermentation can be coupled with a pervaporation unit, thus n-butanol can be separated from the broth continuously to prevent the n-butanol concentration from reaching the toxicity limit. Because of the mild operating conditions, pervaporation enables an energy-efficient n-butanol separation and facilitates the recycling of the broth with lower n-butanol content back to the fermenter.

Therefore, the aim of this work was to evaluate if it is more efficient to separate n-butanol by means of pervaporation or distillation or a combination of both. For this purpose extensive experimental studies were performed to characterize the permeation behavior of a commercial available poly(dimethylsiloxane) (PDMS) membrane (Sulzer Pervap^{TM} 4060). These studies provide a wide data basis for the downstream process analysis and are used to identify possible operating windows for the application of a pervaporation. An experimental comparison between the pervaporation of a binary mixture of n-butanol and water, the pervaporation of a synthetic medium and the pervaporation of real fermentation broth was carried out. The synthetic medium contained acetone, n-butanol and ethanol in mass ratios of 3:6:1 together with acetic and butyric acid, which are the solvent precursors. Additionally process parameters like feed composition, pervaporation temperature and permeate pressure were varied. The detailed experimental procedure is described elsewhere [2]. Next to the PDMS membrane, composite membrane materials containing ionic liquids are promising for separation of n-butanol from an aqueous solution. These membrane materials show advantageous permeation properties compared to conventional polymer membranes [2]. Therefore different supported ionic liquid membranes (SILMs) have been tested to evaluate their potential in n-butanol pervaporation.

Based on the experimental results a correlation has been developed describing partial fluxes of acetone, n-butanol, ethanol and water. Fluxes of acetic and butyric acid were neglected. With the help of this correlation modular-based process studies have been carried out using Aspen Custom Modeler^{TM} to find an optimal process configuration for n-butanol separation from the fermentation broth. It was found that pervaporation as stand-alone process as well as distillation is comparatively expensive. Because n-butanol should be almost completely separated from the broth, the concentration at the outlet of the membrane module becomes small, leading to a large required membrane area and a small permeate concentration of n-butanol.

In contrast to the application of distillation and pervaporation as stand-alone processes, a combination of pervaporation and distillation was found to be best suited for efficient n-butanol recovery. A possible membrane-assisted process is shown in Fig. 1. Pervaporation is used for continuous n-butanol removal in order to extend the fermentation cycles. Because the mass fraction of n-butanol $w_{B,Rec}$ in the recycle stream is only slightly smaller compared to the mass fraction of n-butanol in the fermentation $w_{B,Ferm}$, the required membrane area is comparatively small and a higher permeate concentration can be obtained. The complete removal of n-butanol from the broth is then carried out in distillation columns. In this case, the extension of the fermentation periods and the reduction of the stream to be processed in the energy-intensive distillation lead to a reduced energy demand and lowered downstream costs for n-butanol. The permeate stream and the solvent phase from the distillation can be processed further in a conventional distillation sequence. As existing biobutanol production plants mainly rely on distillation sequences, retrofit of a pervaporation thus enables an increase in the biobutanol production efficiency, while existing paid-off plants can still be used.

Fig. 1: Solvent removal from fermentation broth by distilliation (a), pervaporation (b) and a membrane-assisted process (c).

Acknowledgement:

The research leading to these results has received funding from the European Union Seventh Framework Programme (FP7/2007-2013) under grant agreement n° 241718 EuroBioRef. We acknowledge Sulzer Chemtech AG and the Laboratory of Biochemical Engineering, TU Dortmund University for providing PDMS membranes and fermentation broth.

[1] T.C. Ezeji, N. Qureshi, H.P. Blaschek, Bioproduction of butanol from biomass: from genes to bioreactors, Curr. Opin. Biotechnol. 3 (2007) 220-227.

[2] S. Heitmann, J. Krings, P. Kreis, A. Lennert, W.R. Pitner, A. Górak, M.M. Schulte, Recovery of n-butanol using ionic liquid-based pervaporation membranes, Sep. Purif. Technol. 97 (2012), 108-114.

Cool Planet Energy Systems and the Production of Carbon Negative Fuels

Mike Rocke, Vice President Marketing
Cool Planet Energy Systems, Inc.
mr@coolplanet.com

Abstract: Cool Planet Energy Systems is a 3-year old start-up producing gasoline blendstock from renewable non-food biomass feedstock. We have devised an economical biomass-to-liquid thermo-chemical conversion process that simultaneously produces liquid fuels and sequesterable biochar useful as a soil amendment. Fuel produced from this process is termed negative-carbon fuel since more carbon dioxide is removed from the atmosphere than released from combustion. The process for producing this fuel relies in part on a unique thermochemical decomposition of biomass that ramps the biomass temperature while the biomass is being subjected to pressure shocks. Co-produced biochar is converted into a soil amendment. We have built a 200,000 gallon per year pilot plant using this process and have demonstrated in laboratory studies that biochar generated using this process can exhibit excellent soil amendment properties. Fuel produced from this pilot plant has been fleet tested in a blend formulation in several standard configuration automobiles. The process is widely tunable in the produced biochar/fuel ratio. We also describe our plans to mass produce trailer-sized modular refinery units using this technology. We are backed by some of the world's largest energy companies and top Silicon Valley venture capitalists. We will discuss the company's trajectory from infancy to present status, and its fit in the bio-economy.

Overview

Cool Planet Energy Systems is a CA based start-up that has developed ground breaking, disruptive technology that economically converts non-food biomass into a sustainable, high-octane gasoline blend stock. Cool Planet's technology is based on twelve awarded, US patents. Hemi cellulosic or lignocellulosic biomass is processed through a mechanical biomass fractionating system that uses pressure and heat to create a series of useful volatile components that then go through proprietary catalytic systems to make high octane fuel. The process also captures the residual carbon to create a byproduct, called biochar, which is an inert, high surface area carbon that enhances soil quality through improved water and fertilizer retention and thereby increasing crop production on marginal or arid farmland. This carbon negative process removes atmospheric carbon to reverse the effects of fossil fuel consumption resulting in up to 150% carbon footprint reduction for every gallon of fuel used.

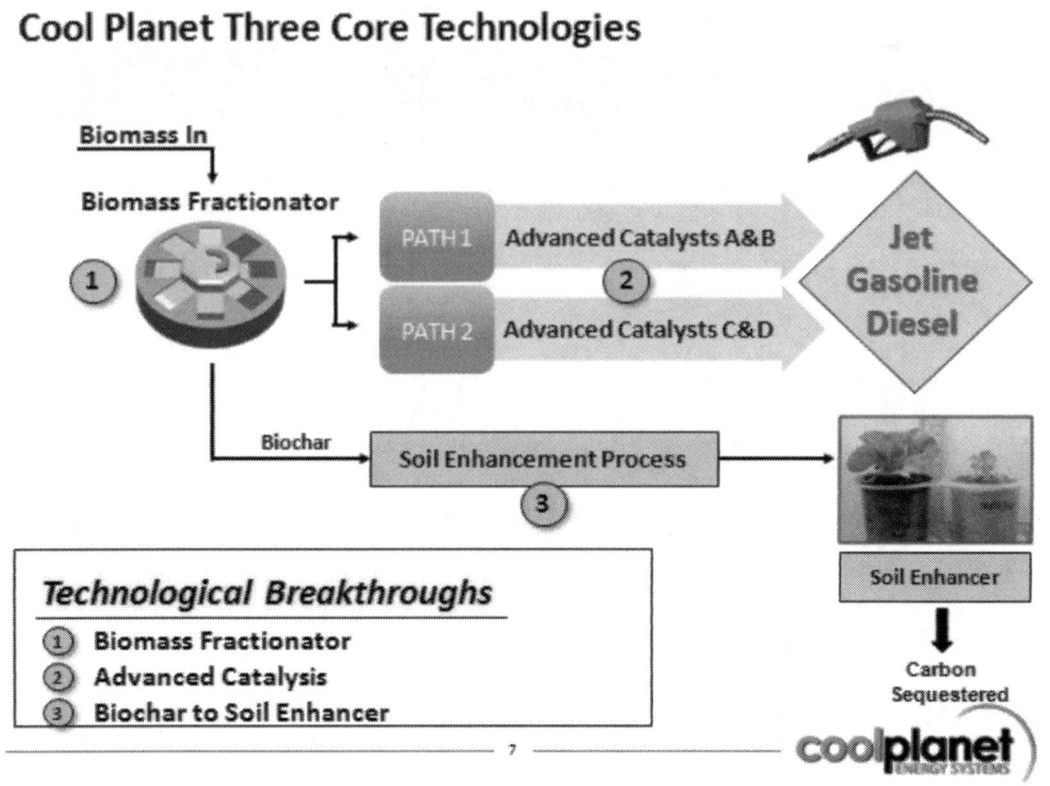

Technology of the Biomass fractionator

Ground biomass is transported by a commercial, mechanical transport system to a load hopper that pre-stages the biomass for the fractionator. Methods of introducing biomass into a processing station can include introduction via conveyor belts, hoppers, and/or pulverizers. The biomass is fed to the top of a sliding gate valve above a load station, and a precise amount of biomass is placed in a moving tray that holds the biomass for the next stage of processing. The biomass moves to the 1st heating and processing stage where the function is to drive off entrapped water, and to perform 1st step of de-hydration of the biomass while mechanically compacting it for more effective heat transfer at later stages of processing. The processing stages rotate or use a shuttle drive from step to step where the biomass is progressively heated and mechanically compacted from an initial stage temperature of 150 degrees C. As each of the following stages heat and compress the biomass, it will outgas commercially viable bio-intermediary compounds that are then taken off to catalytic reaction chambers. At each successive processing stage, the biomass is heated to a higher temperature profile and the material is compacted to a thinner layer with increasing level of elemental carbon in the remaining biomass. The processing stations thus produce volatile and non-volatile components when heated. Because the material can be heated within a relatively narrow temperature range, the non-volatile component (in the form of partially formed char) has additional volatile components embedded therein, which are extractable and processed in subsequent processing stations. The output of these processing stations is connected to an array of catalysts as described in the next section.
The final step or stage does not compact but controls the final biochar product output quantity and quality.

Advanced Catalytic Reaction Arrays

The Cool Planet system has a programmable number of processing stations, connected to an array of catalysts; a means for subjecting biomass within these stations to programmable temperatures and increments to produce a volatile and non-volatile component; a means for incrementing the non-volatile component temperature to further produce additional volatile and non-volatile components; and a means for subjecting the volatile components generated in each station through a series of catalysts. The catalysts involved are selected from the group of dehydration catalysts, aromatization catalysts, and gas-upgrading catalysts to produce different types of renewable fuel. The chemical compounds from biomass decomposition are routed to these distinct catalyst chains to produce products containing renewable, pure HC fuels and other value added products. Biomass is typically comprised of a wide array of compounds classified within the categories of cellulose, hemicelluloses, lignin, starches, and lipids. These compounds go through multiple steps of decomposition when subject to the biofractionation "pyrolysis" process. For example, hemicelluloses comprise C5 sugars such as fructose and xylose, which yield furfural and hydroxymethylfurfurals upon thermolysis. The latter compounds can be further converted to fuel intermediates furan and tetrahydrofuran. The narrow temperature windows experienced within a processing station allows for the collection of these useful intermediates.

The processing stations are connected to a series of catalyst components comprising a dehydration catalyst, an aromatization catalyst, and a gas-upgrading catalyst. The main catalysts are proprietary modified ZSM-5 catalysts. The output from each catalytic array when cooled is comprised of volatile gases, renewable fuel and water. The volatile gases are then programmed to pass through subsequent catalytic columns. Depending on the temperature (300C to 600C) of the processing stations and the biomass composition, the volatile components from any given processing station can be channeled through one or more catalysts using program selector switches.

Biochar

The last step as heat energy transfer is completed to the biomass is the production of biochar (BMF char). The BMF char remains after the previous processing steps, and the physical characteristics of the char will differ depending on the starting biomass material, which can include any of the above-identified cellulosic materials such as wood, grasses, energy crops, and agricultural waste. Different biomass feedstocks are expected to produce different types of BMF chars, varying in porosity and other physical characteristics. The biomass feedstocks can be fed individually or as mixtures of different feedstocks to produce chars containing different physical characteristics. After the BMF char is formed, a last processing step is to transfer the BMF char out of the reaction chamber for a subsequent storage or filling into a bio-char reactor for subsequent optional activation. The transfer may be accomplished via any number of mechanical means, including a press bar outfitted with a scraping knife. The Company has developed a novel process for rendering the BMF char suitable as a soil growth agent. This process comprises several steps including creating the BMF char, expelling detrimental agents within the BMF char, rendering the internal surface area of the BMF char hydrophilic, and adding suitable nutrients and microorganisms to the BMF char to promote plant growth. In addition to acting as a soil enhancing agent, the BMF char can sequester carbon for long periods of time.

Biochar Co-product – Return Carbon to the Soil

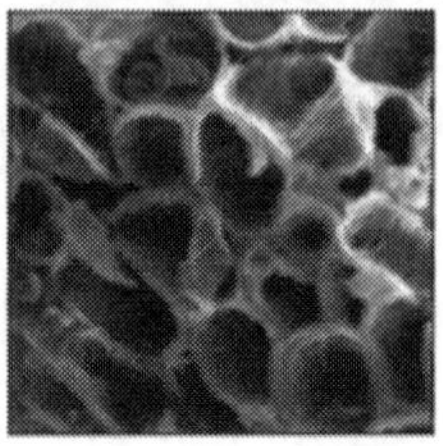

Biochar is a highly porous form of carbon produced by the selective heating of biomass, in the absence of oxygen.

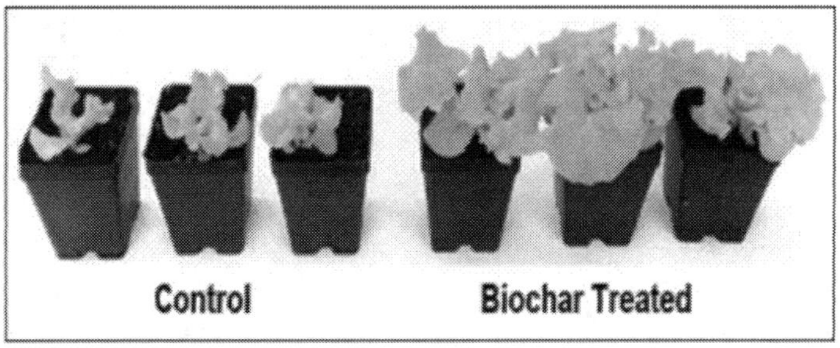

Biochar is very promising as a soil amendment that can retain water and inputs in the soil. Biochar can help maintain a nutrient rich environment and increase productivity of a wide variety of crops.

coolplanet ENERGY SYSTEMS

Drop in Hydrocarbon Fuel Results

Upon completion of catalytic conversion, the devolatilized organic components convert to various liquid fuels such as biodiesel, hydrocarbons, BTX aromatics, jet fuel, light hydrocarbons, gasoline, diesel, methanol, and DME. The organic chemicals can also be useful on their own as co-produced chemicals. By contrast, typical pyrolysis processes do not exhibit a clear volatilization profile such as this. 3rd party testing by independent laboratories shows the compounds consistent with a high octane gasoline as shown in the next graph.

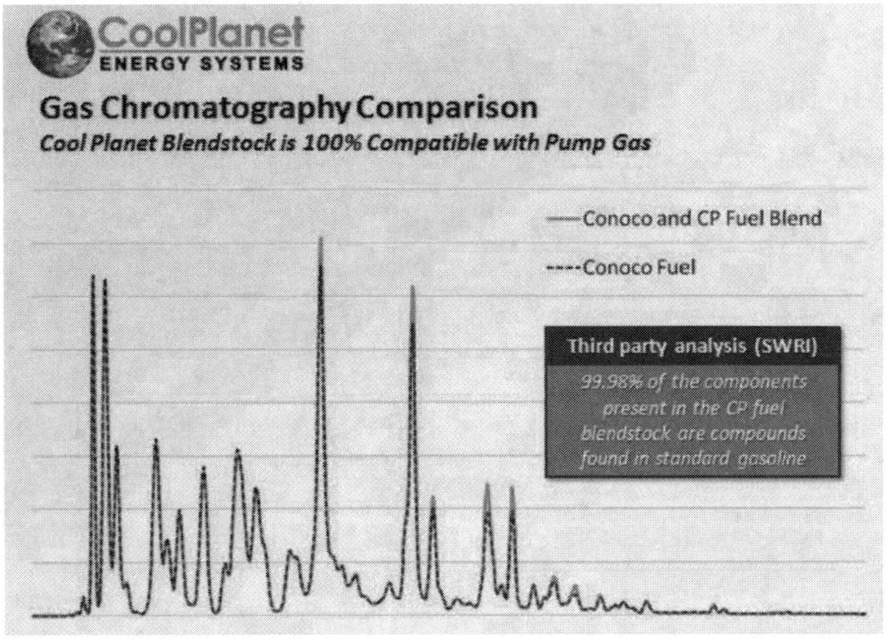

Mass Production of Distributed Units

The Cool Planet refinery equipment is modular, with fractionator and catalyst equipment connected as pairs of processing equipment. This is transported to various biomass locations where the modules are placed within a 30 mile radius of the biomass to be processed. Capex is lower by a factor of 10 compared to existing biomass refineries.

The modular refining units will be manufactured and assembled at a central plant manufacturing facility. Coordinated team will then roll out the modular facilities across the US.

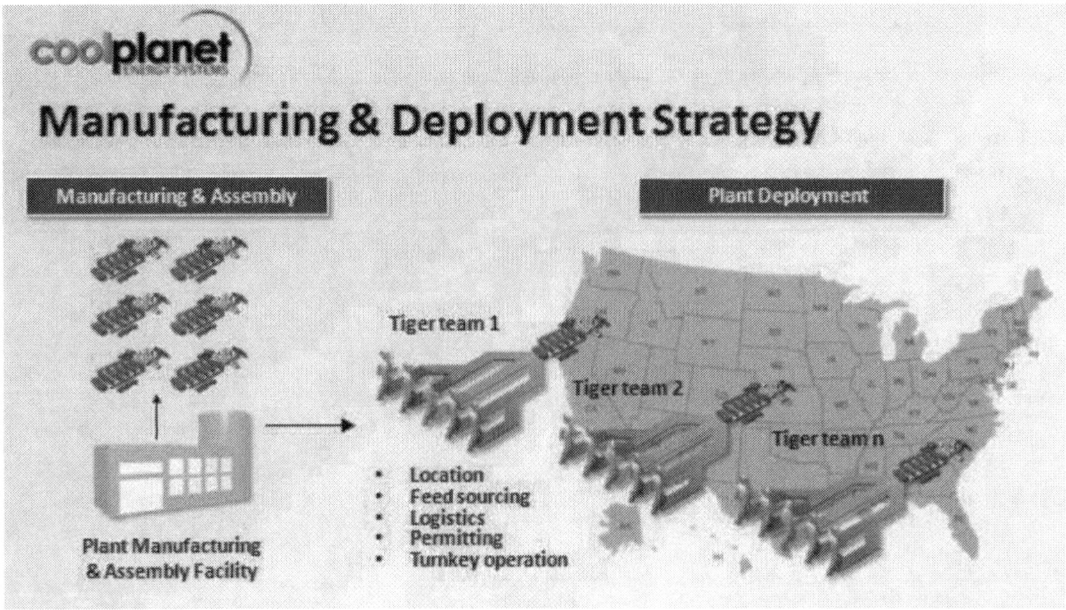

US and Global Roll-out Plan

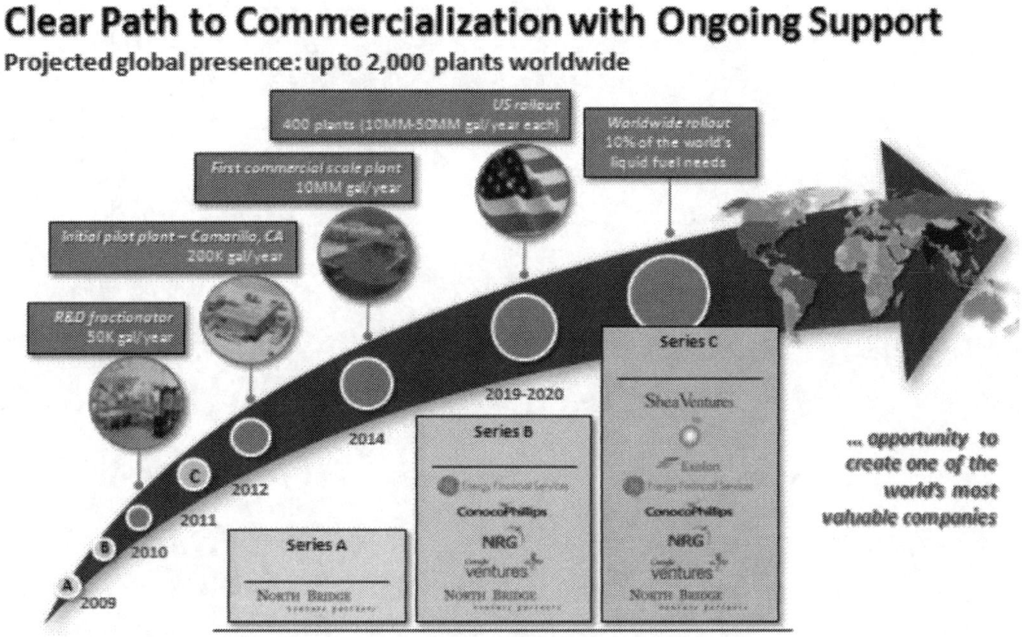

AUTHOR INDEX

AUTHOR INDEX